P. J. Lea · J.-F. Morot-Gaudry (Eds.)

Plant Nitrogen

Springer

Berlin
Heidelberg
New York
Barcelona
Hong Kong
London
Milan
Paris
Singapore
Tokyo

Peter J. Lea · Jean-F. Morot-Gaudry   (Eds.)

Plant Nitrogen

With 74 Figures

 Springer

Professor Dr. PETER J. LEA
Lancaster University
Department of Biological Science
Lancaster LA1 4YQ
UK

JEAN-FRANÇOIS MOROT-GAUDRY
Directeur Recherche
INRA
Unité de Nutrition Azotée des Plantes
Route de Saint-Cyr
78026 Versailles
France

ISBN 3-540-67799-2 Springer-Verlag Berlin Heidelberg New York

Library of Congress Cataloging-in-Publication Data

Plant nitrogen / Peter J. Lea ; Jean-F. Morot-Gaudry (eds.).
   p. cm.
Includes bibliographical references.
ISBN 3450677992
   1. Nitrogen-Metabolism. 2. Plants-Assimilation. I. Lea, Peter J. II. Morot-Gaudry,
Jean-Francois.
QK898.N6 P63 2001

Springer-Verlag Berlin Heidelberg New York
a member of BertelsmannSpringer Science+Business Media GmbH

© Springer-Verlag Berlin Heidelberg 2001, INRA Paris
Printed in Germany

Typesetting: Camera ready by the editors
Cover design: Design & Production, Heidelberg
Cover photograph: © J. Weber
SPIN 10714407   31/3130 - 5 4 3 2 1 0 - Printed on acid-free paper

Preface

The use of fertilisers to control crop growth and productivity has been one of the key contributing factors for the incremental improvement in agricultural production. However, the success of the application of the nitrogen fertilisers is associated with the environmental hazard of nitrate. This problem has stimulated a considerable global research interest in plant physiology, biochemistry, molecular biology, genetics and agronomy, in both academia and industry.

During the past 10 years, major advances have been made in our understanding of the transport and metabolism of nitrogen in plants. However, the information gathered is often scattered in the proceedings of different congresses or as reviews in different publications. The present book on the assimilation of nitrogen by plants provides an exhaustive, comprehensive and authoritative review concerning this field of research.

We are thankful to the authors for their time and their effort. We believe that this book, written by experts in different areas of research, will be useful to undergraduate and graduate students, professors and scientists in biology and agronomy.

Peter J. LEA and Jean-François MOROT-GAUDRY

Introduction

Jean-François MOROT-GAUDRY[1] and Peter J. LEA[2]

Nitrogen fertilisation is known to make a major contribution to the yield of many crops. Grain yield increases linearly with the dose of fertilisers applied to cereals, which can account for the main part of the variation in yield. Consequently, nitrogen is considered, after water deficiency, to be the main controlling factor of plant growth, and an essential element of plant productivity. To obtain the maximum yield permitted by climatic conditions, and to be sure of avoiding crop N deficiency in any situation, farmers apply large amounts of N fertiliser. Although adding fertilisers generally results in enhanced yield, the efficiency of nitrate uptake decreases with the level of fertilisation. The result is an accumulation of nitrate in plant tissues and large amounts of unused fertilisers, which leads to the environmental hazard of nitrate (NO_3^-) and nitrite (NO_2^-) contamination of ground water. Nitrate can seriously compromise drinking water quality and cause toxic algal blooms in river and oceanic waters, disrupting fisheries and tourism. Nitrate can also be lost to the atmosphere by the process of denitrification. Certain bacteria, which, under normal aerobic conditions use oxygen (O_2) as their electron acceptor in respiration, switch to using nitrate when oxygen is in short supply; for example, when the soil becomes waterlogged. The nitrate becomes reduced to either nitrogen gas (N_2), or the gaseous oxides of nitrogen, nitrous oxide (N_2O), nitric oxide (NO) or nitrogen dioxide (NO_2), which are volatilised and lost from the soil. These gaseous compounds are considered major air pollutants in industrial countries. In addition, plants grown with high nitrate fertiliser can contain high tissue concentrations of nitrate and nitrite which are nutritionally undesirable. Elevated levels of nitrate and nitrite in foodstuffs are implicated in several health-care problems, e.g. the formation of nitrosamine, a carcinogen implicated in the etiology of human cancers, methaemoglobinaemia etc. A decrease in undesirable environmental effects and reduced dietary nitrate, could reduce several human and animal health problems.

1. Unité de Nutrition Azotée des Plantes, INRA, route de Saint-Cyr, 78026 Versailles. France. *E-mail :* morot@versailles.inra.fr
2. Department of Biological Sciences, Lancaster University, Lancaster LA1 4YQ, United Kingdom. *E-mail :* p.lea@lancaster.ac.uk

Nowadays, farmers must increasingly take into account the environmental consequences of the application of fertiliser, and hence adjust nitrogen inputs according to the nitrogen requirement of the crops corresponding to the target yield. Under new economic and environmental constraints, the target yield does not correspond exactly to the maximum potential yield allowed by climatic conditions, but to a lower yield. The farmers have to move to a new nitrogen fertilisation strategy, which implies a determination of both the potential yield of any crop in a given climatic condition and the corresponding minimum crop nitrogen requirement necessary to achieve it. Despite recent improvements in the modelling of crop growth, the prediction of both crop yield and minimum nitrogen fertiliser requirement for maximum growth is not easy at the farm level. This is due to the unpredictable climatic parameters which determine plant growth and soil nitrogen availability, but is also caused by our lack of knowledge about the basic mechanisms which govern nitrogen cycling and metabolism.

The nitrogen supply is characterised by the nitrogen dynamics of the soil-plant system, the mechanisms for absorbing nitrogen by the roots (root architecture and size, activity and concentration of the ion transporters) and the nitrogen demand, which is mainly determined by the area of leaf development. In conditions of high nitrate concentration in the soils of Western Europe, the current rate of nitrate uptake appears to be normally well below the nitrate uptake capacity of the roots. The rate of nitrate uptake by roots of higher plants seems to be strictly regulated by the nitrogen demand of the whole plant and specifically controlled by the current nitrogen status of the plant. This is shown by a positive correlation between nitrate uptake and growth rates, a relationship which ensures that the internal nitrogen concentration is homeostatically maintained. Thus, the nitrogen uptake rate is largely determined by the extent of the repression exerted by the nitrogen status of the plants. However, the manner in which this feedback regulation is achieved is not well understood. Recent advances in plant molecular biology, combined with modern physiological and biochemical studies, have expanded our understanding of the regulatory mechanisms controlling the steps of inorganic nitrogen assimilation and the subsequent biochemical pathways.

Nowadays, nitrate is considered the principal nitrogen source for most non-legume crops. Following its uptake by means of specific transporters located in the root cell membrane (plasmalemma), the assimilation of nitrate is a two-step process. First, the enzyme nitrate reductase catalyses the reduction of nitrate to nitrite. Subsequently, the enzyme nitrite reductase mediates the reduction of nitrite to ammonia. In addition to nitrate reduction, ammonium ions may also be taken up directly from the soil by the root. Ammonia can also be generated inside the plant by a variety of metabolic pathways such as photorespiration, phenylpropanoid metabolism, utilisation of nitrogen transport compounds and amino acid catabolism or from symbiotically fixed nitrogen.

Ammonia is then incorporated into an organic molecule by the enzyme glutamine synthetase (GS). The reaction catalysed by the enzyme GS is now considered to be the major route facilitating the incorporation of inorganic nitrogen into the organic form, in conjunction with glutamate synthase (GOGAT), which recycles glutamate and incorporates carbon skeletons for the transfer of amino groups to other amino acids. Amino acids are transported between different organs through both the xylem and phloem. This distribution of nitrogen requires the activity of amino acid transporters, which regulate the delivery of amino acids to the developing pods and seeds.

The 20 amino acids are of central biological importance as they provide the building blocks of proteins. Higher plants possess the ability to carry out *de novo* synthesis of all amino acids; however, mammals have the capacity to synthesise only ten amino acids, the remainder must be supplemented in their diet. A genetic engineering approach to manipulate the activity of key enzymes involved in amino acid pathways will have a broad impact. This approach allows for modulating the expression to desired levels, in a gene member-specific and cell-specific manner, and for protein engineering. The genetic engineering approach of the regulation of amino acid biosynthesis in plants opens up the possibility of producing crop plants with increased levels of essential amino acids (sulphur, branched and aromatic amino acids) in pods, seeds and potato tubers. Considerable attention has also been paid to engineering crop plants that are tolerant to herbicides (phosphinothricin, glyphosate, imidazolinones and sulphonylureas), which act by inhibition of amino acid biosynthetic pathways. Much of the research effort to impart herbicide tolerance into transgenic plants has concentrated on the introduction of genes encoding enzymes that are no longer inhibited by herbicides.

The recycling of reduced nitrogen in plants during development is also a factor in determining crop yield. Measurements of nitrogen content per unit leaf area of senescent leaves, collected immediately after leaf fall, indicated that 80% of the initial leaf nitrogen is recycled before leaf abscission. In cereals, approximately half of the seed nitrogen is derived from leaf protein degradation during senescence. Thus, the remobilization of nitrogen from the leaves to the seed is an important factor in crop yield. In trees in spring, internal nitrogen cycling is responsible for a large part of the amino acid flux in the xylem sap, thus determining the development of the new leaves and inflorescences. Plant growth and seed and fruit production are a result of the capacity for nitrogen uptake and assimilation and the integrated processes of allocation, accumulation and remobilization of nitrogen at the whole plant and canopy level.

In spite of the significant progress that has been made during the past few years on the biochemistry and molecular biology of inorganic nitrogen metabolism, many aspects of nitrate and ammonia assimilation both at the cellular and whole plant level remain obscure. For example, the manner in which the feedback regulation of nitrogen uptake is achieved is not well understood. In particular, the compounds which are used as sensors of the nitrogen status of plants and which are responsible for the control of the nitrate transport systems of root plasmalemma have not yet been precisely identified. Several nitrogen pools, such as nitrate, ammonia itself and some products of ammonia assimilation, e.g. amino acids, are probably involved in the underlying mechanisms. Changes in the concentrations of these compounds under different environmental conditions act as signals to control plant gene expression at the transcriptional level, or enzyme activities by posttranslational modifications and protein turnover, thus influencing metabolism, development and growth.

The processes involved in remobilization of nitrogen, formed by the hydrolysis of storage proteins during plant development, are still not clear. Until now, it is not known whether amino acids themselves, or ammonia, the ultimate product of protein degradation, are reused for the synthesis of new proteins. Many enzymes such as glutamine synthetase, glutamate synthase and glutamate dehydrogenase could be implicated in these metabolic reactions. Only very careful labelling experiments using ^{15}N will provide solid evidence about this pathway, which is linked to nitrogen reintegration in plants during development.

The complexity in composition of phloem sap, and the multiplicity of the cell types that are involved in long-distance transport systems suggests that there are a number of highly regulated transport systems required for amino acids. The remaining transporters will be identified by the *Arabidopsis thaliana* genome project, with the function of the new genes identified by sequence homology and testing in yeast mutants and *Xenopus laevis* oocytes. However, systematic searches for knockout mutants will be necessary to obtain a full understanding of the role of the individual transporter genes. Important questions waiting to be solved can be approached now, such as whether transporters involved in phloem loading of amino acids might be localised in sieve elements, or which transporters play a role in the uptake of amino acids into the endosperm or transfer cells in developing seeds. Genomics might eventually provide sufficient information for engineering improvements in crop species.

Some plants, belonging especially to the legume family (Fabaceae), are able to attract nitrogen-fixing organisms to their roots, some of which are species of the bacterial genus *Rhizobium*. These bacteria, termed rhizobia, contain the enzyme complex nitrogenase, which can catalyse the reduction of dinitrogen to ammonia. This process, which is termed nitrogen fixation, requires a large amount of energy. Such prokaryotic nitrogen-fixing organisms are localised in nodules on the roots of legumes and are able to draw on the carbohydrates produced by the plant to fuel nitrogen fixation and provide carbon skeletons for the production of nitrogenous compounds. In the field, a good stand of clover or lucerne, which form symbiotic relationships with rhizobia, can fix between 100 and 400 kg N ha^{-1} year^{-1}. It is estimated that the annual production of ammonia by rhizobia can account for almost three times that which is manufactured by the industrial process (Haber-Bosch process). In addition to the rhizobium-legume symbiosis, a number of other nitrogen-fixing symbioses exist. Major examples are Actinomycetes with *Alnus* (Alder) and *Casuarina* and cyanobacteria with gymnosperms, ferns, lichens and liverworts. Legumes also present attractive experimental systems for the study of vesicular-arbuscular mycorrhizal fungi, because mycorrhizal colonisation of roots requires some of the same plant genes necessary for nodulation by rhizobia. Another area of opportunity is the observation of several intriguing parallels between root development and the establishment of rhizosphere symbioses. For example, the initiation of both lateral roots and nodules involves the mitotic reactivation of roots cells in specific regions of root tissues.

The establishment of a nodule is achieved collaboratively by both participants and a flow of signals is transmitted between the plant and bacteria. During the past decade, considerable progress has been made in the molecular dissection of the communication between rhizobia and legume plants. The major achievement has been the characterisation of signal molecules produced by both partners through which the communication is established. The isoflavonoids synthesised and secreted by legume plants induce the expression of rhizobial nodulation genes. The specific lipo-oligosaccharides (Nod factors) secreted by rhizobia provoke responses in the plant at spatially separated sites, the epidermis, the cortical cells and the pericycle. Many plant responses to Nod factors have been characterised, including alteration in the polar growth of root hairs, induction of cell division, and expression of nodulins, but the signal transduction pathway leading to these events remains poorly understood. Although there are still many unanswered questions with respect to the communication between legume plant and bacteria, it probably involves a signal transduction cascade. The Nod factors can be considered as plant growth regulators, affecting development of the plants and insuring the formation of nitrogen-fixing nodules. In the future, cloning the genes affected by Nod factors should help to identify components of the Nod factor signal transduction pathway, as well as regulatory and structural molecules that are involved in symbiotic development.

Symbioses between higher plants and nitrogen-fixing microorganisms provide a niche in which the prokaryote can fix nitrogen in a very efficient manner. However, host specificity provides a serious restraint to the application of symbiotic nitrogen fixation in agriculture, because most major crops are unable to establish such a symbiosis. As molecular genetic research has shown that a relatively high number of specific host functions are involved in forming a nitrogen-fixing organ, it has seemed impossible to obtain new nitrogen-fixing plants with the methodology available. The possibility of reaching this goal has become newly invigorated as a result of research indicating that the mechanisms which control nodule development might be derived from processes common to all higher plants. Studies of these common processes might therefore provide new means to design strategies by which non-legume plants can be given the ability to establish a symbiosis with a nitrogen-fixing bacteria.

In recent years, our understanding of nitrogen assimilation and recycling mechanisms has advanced. However, many key questions still remain to be answered. Unravelling these questions will provide new approaches to modifying plant genomes and consequently the ability of plants to manage nitrogen efficiently and to introduce new fertilisation strategies more adapted to economical and environmental constraints. The goal of this book is to introduce and summarise the major new academic findings that have been published recently, which are directly relevant to agronomic and environmental issues.

Contents

Editors Peter J. LEA and Jean F. MOROT-GAUDRY

Nitrate Uptake and Its Regulation

Bruno TOURAINE[1], Françoise DANIEL-VEDELE[2] and Brian G. FORDE[3]

NO_3^- uptake by the roots of higher plants is the main pathway for entry of N into the global food chain. In quantitative terms, N is the most important element after C, H and O, i.e. the most important of the mineral elements that have to be acquired from the soil. It is thus not surprising that it commonly limits plant growth, especially in agricultural systems, because a large proportion of the N accumulated in crops is removed from the plant-soil system at harvest. Since the mid-19th century, the use of NO_3^- – based fertilisers has continually increased to sustain high crop yields.

Physiological data reveal the existence of NO_3^- transport systems with a very high affinity for their substrate, which should provide sufficient capacity for NO_3^- absorption at external NO_3^- concentrations lower than those usually recorded under field conditions (Robinson 1991). On the other hand, physiological investigations also demonstrate that powerful regulatory mechanisms operate at the whole plant level, so that in the long term, actual NO_3^- uptake depends on internal factors related to N demand of the plant, rather than on NO_3^- availability in the soil. Furthermore, NO_3^- acquisition also depends on the capacity of roots to explore the soil volume, but there are complex interactions between NO_3^- availability and N demand on one hand, and root development and root architecture on the other. The discrepancy between field observations and conclusions drawn from physiological approaches

1. Biochimie et Physiologie Moléculaire des Plantes,
ENSA-M / INRA / CNRS UMR 5004 / UM 2,
Place Viala, 34060 Montpellier Cedex 01, France
tel: +33 (0)499 612604; fax: +33 (0)467 525737
e-mail: touraine@ensam.inra.fr
2. Unité de Nutrition Azotée des Plantes, INRA
Rte de St Cyr
78026 Versailles Cedex, France
e-mail: vedele@versailles.inra.fr
3. Dept. of Biological Sciences
Lancaster University
Lancaster, LAI 4 YQ, UK
Tel: +44 (0)1524 594861
Fax: +44 (0)1524 8
E-mail: b.g.Forde@lancaster.ac.uk

probably results from our lack of knowledge on the processes involved in these inter-actions and in the regulation of NO_3^- uptake.

Physiological and electrophysiological approaches have provided precise data and hypotheses on NO_3^- transport systems and their regulation. However, in order to decipher the mechanisms and the key compounds and pathways responsible for the control of NO_3^- absorption, the identification of NO_3^- transporters became a prerequisite. This step is currently ongoing, and should provide tools (specific nucleic acid probes, antibodies, transgenic plants and mutants with altered expres-sion of NO_3^- transporters) to characterise the specific role of each separate trans-porter and identify molecules involved in the sensing of plant N demand and the regulation of NO_3^- transport. In this chapter we present both aspects: the physiolog-ical platform that provides the basis of our present vision of root NO_3^- transport and its regulation at the whole plant level, and the molecular data derived from the clon-ing and sequencing of genes coding for NO_3^- transporters in higher plants. A number of other recent reviews have dealt with various aspects of NO_3^- transport, sometimes in more detail than is possible here (Glass and Siddiqi 1995; Trueman et al. 1996a; von Wiren et al. 1997; Crawford and Glass 1998; Daniel-Vedele et al. 1998; Forde 2000; Forde and Clarkson 1999).

Physiological Analysis of Nitrate Uptake: Evidence for the Occurrence of Multiple Transport Systems

The rate of NO_3^- absorption is the balance of two opposite fluxes: influx from apo-plasm to cytoplasm, and efflux in the reverse direction. However, due to the lack of an appropriate tracer, individual measurements of these two unidirectional fluxes were very rare until the mid-1980s. The first attempts to determine NO_3^- influx and efflux circumvented the problem of lack of suitable N isotope by using $^{36}ClO_3^-$ as an analogue of NO_3^- (Deane-Drummond and Glass 1982; Deane-Drummond and Glass 1983a and b). Although conflicting results have been published about the similarities or differences between NO_3^- and ClO_3^- uptakes depending on specific conditions (essentially external concentration range), overall the literature con-cludes that ClO_3^- is a rather poor analogue for NO_3^- (Guy et al. 1988; Siddiqi et al. 1992; Kosola and Bloom 1996; Touraine and Glass 1997). The solution for influx and efflux measurements for some plant physiology laboratories came from the possibility of using the short-lived radioisotope ^{13}N, produced in local cyclotrons, and from technical improvements in ^{15}N analysis that allowed application to short-term labelling periods (s to min). Since then, a long series of kinetic studies have been performed, using either $^{15}NO_3^-$ (e.g. Teyker et al. 1988; Devienne et al. 1994; Delhon et al. 1995; Muller et al. 1995) or $^{13}NO_3^-$ (e.g. Lee and Clarkson 1986; Oscarson et al. 1987; Siddiqi et al. 1989, 1990; Wieneke 1994; Kronzucker et al. 1995a, b, c; Touraine and Glass 1997). In most cases, NO_3^- influx is determined by tracer incorporation from labelled medium for a period of a few minutes (typically 5 min), followed by washing periods to remove the tracer accumulated in the liquid film at the root surface and in the apoplasm. Also, more refined studies (compart-mental analysis) have been performed in order to measure both influx and efflux across the plasma membrane of root cells, as well as the different NO_3^- pools and fluxes in root cells (see p. 6).

Numerous reports have been published of the occurrence of a NO_3^- efflux that recycles out of the cell a significant proportion of the NO_3^- absorbed (Minotti et al. 1969; Morgan et al. 1973; Jackson et al. 1976; Deane-Drummond and Glass 1982; Lee and Clarkson 1986; Oscarson et al. 1987; Teyker et al. 1988; Lee 1993; Devienne et al. 1994; Muller et al. 1995). Because of the small size of the cytosolic pool of NO_3^-, this recycling is especially rapid (see p. 6). However, most of the literature on NO_3^- uptake deals with NO_3^- influx specifically, while the efflux mechanism is poorly characterised. Thus any suggestions concerning its role would just be a matter of speculation. Furthermore, in most instances, the changes in net NO_3^- uptake (e.g. during the day/night cycle or in response to N availability) are reflected by changes in NO_3^- influx, while efflux responses are relatively minor. Overall, the patterns observed in earlier net uptake studies have been confirmed by NO_3^- influx measurements in the last two decades, thus supporting the idea that it is the regulation of influx that plays the major role in the control of NO_3^- uptake *in planta*. This led to efflux being ignored in most of the studies on NO_3^- uptake. Another consequence of this general consensus is the dismissal of the pump-and-leak model (Scaife 1989). A cause for the lack of knowledge on NO_3^- efflux is that its measurement is far more difficult than the measurement of the influx component, for two reasons. Firstly, an even labelling of the internal NO_3^- pool that is the source of effluxed NO_3^- cannot be easily achieved due to cell and tissue compartmentation. Secondly, in contrast to influx, efflux is affected by the physical manipulation of the plant (Delhon et al. 1995). Since the labelling protocol involves transferring plants from medium to medium, efflux is therefore most likely overestimated, and its responses to treatments applied may be difficult to interpret. In this chapter, as in others, we will discuss both efflux and influx, but focus largely on NO_3^- influx.

Although the physiological function of NO_3^- efflux is not understood, efflux seems to be far more constant. Using RNA and protein synthesis inhibitors, Aslam and co-workers have shown that the efflux is inducible by NO_3^- (Aslam et al. 1996). Furthermore, indications that NO_3^- efflux increases with tissue NO_3^- have been obtained in corn (Teyker et al. 1988) and barley (Van der Leij et al., 1998). In barley, NO_3^- efflux was blocked when NO_3^- was withdrawn from external solution, even though the cytoplasmic NO_3^- concentration did not change, indicating that some regulation of plasma membrane NO_3^- efflux exists (Van der Leij et al. 1998). Studies using plasma membrane vesicles isolated from corn root demonstrated the occurrence of protein-mediated passive efflux of NO_3^- (Grouzis et al. 1997). This transport is saturable (K_m ca. 5 μM), and selective for NO_3^-. Further characterisation of this system showed that the root cell plasma membrane exhibits a large permeability to NO_3^- (permeability coefficient higher than 10^{-9} m s^{-1}, of the same order as that to K^+) at high membrane potential and pH 6.5, and that this permeability is voltage-dependent and decreases with pH increase (Pouliquin et al. 1999). This conclusion fits with thermodynamic calculations which predict that NO_3^- efflux is downhill (see p. 8). Comparing plasma membrane vesicles isolated from tobacco cultured cells grown with or without NO_3^- demonstrated the existence of both a constitutive and an inducible NO_3^- efflux systems. The constitutive system is specifically recovered in a HPLC-purified fraction (Mériam Espuna, Brigitte Touraine, Jean-Pierre Grouzis and Rémy Gibrat, pers. comm.). Thus far, little is known of the localisation or the role in roots of the systems characterised *in vitro* (efflux systems are likely to be involved not only at the outer end of the symplasm, but also at its inner end, in the stele, for NO_3^- secretion in the xylem).

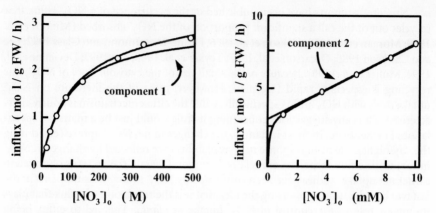

Fig. 1 : Relationship between the rate of NO_3^- influx and the external NO_3^- concentration. Two separate components to the NO_3^- uptake system are identifiable from the kinetics: one in the low (< 1 mM) concentration range, which is saturable, and another in the high concentration (> 1 mM) range, which is linear. These are referred to respectively as the high-affinity transport system (HATS) and the low-affinity transport system (LATS). Plots represent typical influx values measured in *A. thaliana* (Bruno Touraine, unpublished data).

Kinetic experiments on net NO_3^- uptake or NO_3^- influx have shown that there are distinct high-affinity and low-affinity components to the NO_3^- uptake system in the plasma membrane of root cells (see Fig. 1). At low external NO_3^- concentrations (< 0.5 mM), NO_3^- influx is essentially mediated by a saturable system, referred to as the high affinity transport system (HATS). Plants that have not been exposed to NO_3^- for several days prior to the uptake experiment absorb NO_3^- at a low rate within the first hour of exposure to NO_3^-. This rate subsequently increases to peak at 10- to 30-fold the initial level after a period that varies from 3 h to a few days depending on plant species (Lee and Drew 1986; Siddiqi et al. 1990; Aslam et al. 1992, 1993; Kronzucker et al. 1995a and c).

Both the constitutive and inducible uptake systems follow saturable kinetic patterns and present low K_m values, which could be explained by a unique HATS, constitutively present at a low level and expressed at a high level upon exposure to NO_3^- (Siddiqi et al. 1990). Nevertheless, inhibitors of transcription and translation blocked induction (Tompkins et al. 1978; Lainé et al. 1995), suggesting *de novo* synthesis of transporter protein. Moreover, precise determination of the kinetic parameters in roots of uninduced and induced plants have shown that not only the V_{max} value, but also the K_m, increased after exposure to NO_3^-. For instance, in three different studies, supplying NO_3^- to barley seedlings resulted in increasing the K_m from 7 to 13 µM (Lee and Drew 1986), 7 to 35 µM (Aslam et al. 1992) or 20 to 79 µM (Siddiqi et al. 1990). Similar patterns have been obtained in other species, e.g. corn (MacKown and McClure 1988; Hole et al. 1990) and spruce (Kronzucker et al. 1995c).

Based on these observations, two different transport systems, namely a low capacity constitutive high-affinity transport system (cHATS) and a high(er) capacity inducible high-affinity transport system (iHATS), are distinguished. Considering the impermeability of the plasma membrane to ions and the energy required for

NO_3^- absorption into a root cell even at very low cytosolic NO_3^- concentration (see calculations in p. 8), the role of the cHATS may be to enable the cytoplasmic concentration of NO_3^- to rise to a level sufficient for induction of the higher capacity iHATS (Behl et al. 1988).

Induction of influx by NO_3^- has been the subject of numerous reports. It has thus been established that NO_3^- itself, but not NH_4^+ or any product of its assimilation pathway, is the inducer (Aslam et al. 1993; King et al. 1993). NO_2^-, which is not found in significant amounts under natural conditions, is able to induce NO_3^- uptake (Aslam et al. 1993). Induction of the NO_3^- transport system by its substrate, which is unique among the ion absorption systems recognised in plant roots, has been an attractive model for the search of ion uptake proteins. Two-dimensional electrophoretic analyses of microsomal or plasma membrane fractions revealed polypeptides which specifically appear in induced corn roots (Dhugga et al. 1988; Ageorges et al. 1996). To our knowledge, no transport protein involved in NO_3^- influx across the plasma membrane has been isolated using this strategy. This approach has been abandoned because intrinsic membrane proteins like transporters are unlikely to be found due to the exclusion of the majority of hydrophobic proteins from the 2-D gels. Another biochemical approach, using electrophysiological characterisation of plasma membrane vesicles and HPLC fractionation is more promising, as shown by recent success obtained in the search for NO_3^- efflux transporters (see above). Furthermore, most of the laboratories now use genetic or molecular approaches, which turn out to be easier and consequently more efficient, to clone transporters (see p. 9).

At $[NO_3^-]_o$ above 200 µM (the HATS plateau), another uptake system, referred to as the low affinity transport system (LATS), becomes apparent. According to most reports, this transport system is linearly correlated to external NO_3^- concentration, with no saturation up to 50 mM (Pace and McClure 1986; Siddiqi et al. 1990; Aslam et al. 1992; Kronzucker et al. 1995c). In contrast, a study performed in *Arabidopsis* (Doddema and Telkamp 1979) reported that the LATS follows a saturable kinetics. However, it seems that the data could have been misinterpreted due to the unusual way in which they were plotted (log-log scale), and a best fit might be found with a linear regression model if the data are replotted in a standard form (Touraine and Glass 1997). Whatever the species considered, the significance of K_m values cited for NO_3^- LATS therefore seems doubtful. The LATS is constitutive in the sense that it does not require induction by NO_3^-, as shown by both kinetic (Siddiqi et al. 1990) and electrophysiological (Glass et al. 1992) studies. This is not to say that the LATS is not subject to some form of regulation, such as responsiveness to N demand (see p. 15).

At this point of the chapter, it should be noticed that the term transport system refers to a specific process that contributes to NO_3^- transport across the plasmalemma, as characterised by its kinetic parameters and its relationship to NO_3^- availability (e.g. affinity, inducibility). However, *a priori*, different transport systems are not necessarily genetically separate. Indeed, there is no definite evidence inferring that a given transport system corresponds to a single specific transport protein or vice-versa. On the contrary, investigations performed on the *chl1-5* mutant of *Arabidopsis thaliana* where the *AtNRT1.1* gene is deleted have demonstrated that the product of this gene participates in the LATS, but that other transport proteins must also participate to this transport system (Touraine and Glass 1997). Conversely, evidence has been recently obtained in favour to the involvement of AtNRT1.1 (CHL1) in both the LATS and HATS in *A. thaliana* (Wang et al. 1998; Liu et al. 1999; see p. 11). This does not preclude the relevance and importance of the transport system concept (kinetically defined component of transmembrane flux), but stresses that transport systems and transporters (transport proteins) must not be confused, because each transporter might have different properties under different conditions (corresponding to different transport systems) whilst each transport system is probably composed of more than one transporter.

Compartmentation of Nitrate in Root Cells

NO_3^- ions entering the root symplasm may follow any of the four following routes: (1) efflux back to the apoplasm; (2) reduction to NO_2^- and then to NH_4^+ by the enzymes nitrate reductase (NR) and nitrite reductase (NiR) in root cells and further assimilation leading to production of amino acids; (3) accumulation into vacuole of root cells, involving transport across the tonoplast; (4) secretion into the xylem vessels and long-distance transport to shoots (see Fig. 2). Since this chapter is focusing on NO_3^- uptake, fates (2), (3) and (4) are not tackled. Efflux is discussed, but we primarily focus on NO_3^- influx for reasons discussed p. 3.

Determination of the sizes of the different NO_3^- pools in root cells is crucial to address the energetic aspects of NO_3^- uptake by plants. Three different methods, discussed below, have been used for this purpose.

Since ^{13}N and ^{15}N became available for short-term labelling experiments, they have been widely used for so-called compartmental analysis. This approach consists of loading the tissues with an isotopic tracer, and measuring the time course of tracer release after transfer of the labelled tissues in an unlabelled solution. The curves obtained by plotting the loss of tracer from the tissues versus time are mathematically described as the sum of exponential components (Cram 1988). To account for this pattern, it is assumed that the root behaves as compartments arranged in series with each other and exchanging ions in both directions. The root is compared to a single cell, and the successive compartments are likened to the surface liquid film, cell wall, cytoplasm and vacuole. Despite some limitations due to the complexity of the method and the requirement that roots are at a steady state during the experiment, accurate estimates of the ion content and kinetic constant of ion exchange are provided for each compartment. In the case of NO_3^-, both $^{13}NO_3^-$ and $^{15}NO_3^-$ have been used. The radioactive $^{13}NO_3^-$ has the advantage of a very high level of sensitivity, but the duration of the experiments is reduced to ca. 30 min because of the ^{13}N short half-life (less than 10 min). This prevents the determination of vacuole NO_3^- content and kinetic constant using $^{13}NO_3^-$ (the time of half-time of decay for this compartment is of the order of 10 h; Devienne et al. 1994). By contrast, the half-life of the cytoplasmic pool is as short as 2 to 7 min (Presland and MacNaughton 1984; Lee and Clarkson 1986; Devienne et al. 1994; Muller et al. 1995), which explains why measuring the influx component requires very short-term labelling experiments. The estimates for cytoplasmic NO_3^- concentration are in the millimolar range, though exhibiting a rather high level of variability, due to the difficulties of the approach and some slight differences in the procedures used. Whether these differences can be associated to differences in growth conditions, and therefore have physiological meaning is unclear. For instance, in the roots of barley seedlings, the cytoplasmic NO_3^- concentration was found to be 25 mM when fed with 1.5 mM $[NO_3^-]_o$ (Lee and Clarkson 1986) and 10 to 40 mM when fed with 0.01 to 1 mM (Siddiqi et al. 1991). Recently, estimates of 40, 50 and 75 mM have been calculated for barley seedlings grown on 0.1, 1 or 10 mM NO_3^- (Kronzucker et al. 1999). In one study, the cytoplasmic NO_3^- concentration in root cells of young corn plants was estimated at 5 to 500 mM depending on the external NO_3^- concentration, which varied from 14 µM to 70 mM (Presland and MacNaughton 1984). In older corn plants grown on a nutrient solution containing 0.4 mM, this concentration has been estimated at 10-20 mM (Devienne et al. 1994). In soybean seedlings grown on a 0.5 mM NO_3^--containing solution, $[NO_3^-]_{cyt}$ has been calculated to be 4 to 8 mM (Muller et al. 1995).

The concentration of the cytoplasmic pool of NO_3^- has also been estimated indirectly, from measurements of NR activity *in situ*, assuming that this activity is related to the concentration of NO_3^- at the enzyme site in root cells through the K_m and V_{max} values derived from *in vitro* NR assay (Robin et al. 1983; King et al. 1992). The values obtained are in the micromolar to low millimolar range. For the same barley cultivar grown under similar conditions, the figures are lower than those derived from compartmental analysis. To explain this discrepancy, it has been proposed that, due to the uneven distribution of NR and the relative sizes of epidermis and cortex, the two methods determine two different cytoplasmic NO_3^- pools (Siddiqi et al. 1991). The technique based on NR assay would measure a metabolic (NR-containing) pool, pos-

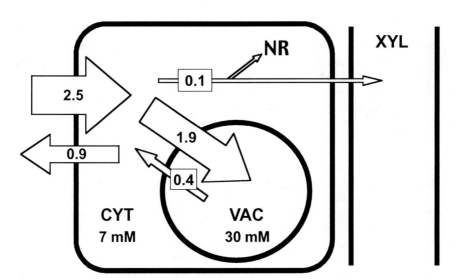

Fig. 2 : Typical fluxes and compartmentation of NO_3^- ions in a root cell of a young seedling. The flux values, $\mu mol\ g^{-1}$ FW h^{-1} and the cytosolic and vacuolar concentrations are those estimated in soybean seedlings by Muller et al. (1995). Note that in older plants, secretion into the xylem would be the major fate of the cytosolic NO_3^- pool, with accumulation in the vacuole being less important than in young seedlings. CYT, cytoplasm; VAC, vacuole; XYL, xylem vessels; NR, nitrate reductase (after Muller et al. 1995).

sibly located in the epidermal cells (Rufty et al. 1986), while the compartmental analysis approach would better measure the pool, exchanging with the medium on one side and the vacuoles on the other side, including the cortical cells.

Intracellular ion-specific microelectrodes enable direct and accurate measurements of NO_3^- activity (about NO_3^- concentration) in the cytoplasm and the vacuole of specific cells of barley roots (Miller and Zhen 1991; Zhen et al. 1991, 1992). Although this technology is not widely developed and the data are therefore scarce, it seems to be the most accurate. The results obtained for the cytosol of both epidermal and cortical cells all fell in the millimolar range (3 to 5 mM) (Zhen et al. 1991; Miller and Smith 1996; Van der Leij et al. 1998), in agreement with the lower estimates obtained by compartmental analysis studies. A remarkable feature of the microelectrode measurements was the observation that the $[NO_3^-]_{cyt}$ remained stable over a wide range of $[NO_3^-]_o$, despite marked changes in the vacuolar $[NO_3^-]$, which suggested the existence of a homeostatic mechanism to maintain the concentration of the cytosolic NO_3^- pool (Miller and Smith 1996).

Electrophysiological Studies: Energetic Coupling of Nitrate Uptake

Early studies on the effects of anoxia (Neyra and Hageman 1976), low temperature (Clarkson and Warner 1979) or metabolic inhibitors (Jackson et al. 1973; Rao and Rains 1976; Deane-Drummond and Glass 1983a and b; Morgan et al. 1985) indicated that NO_3^- uptake is an energy-dependent process. Analysis of root respiration led to the estimate that the absorption of 1 mol NO_3^- consumes 1 to 3 mol ATP, which represents more than 20% of the global cost of NO_3^- acquisition and assimilation in roots, or 5% of the energy supplied by respiration (Bloom et al. 1992). High values for the energy cost of NO_3^- uptake are likely be due to both a high cost of NO_3^- influx and the occurrence of relatively large efflux. Quantitative differences in this futile cycle across the root cell plasma membrane may be related to ecological differences. Lower respiratory costs associated with NO_3^- uptake in fast-growing grasses than in slow-growing (Scheurwater et al. 1998) would be accounted for by a higher efficiency of NO_3^- uptake, i.e. a lower ratio of NO_3^- efflux to influx, or a lower ratio of NO_3^- influx to net uptake (Scheurwater et al. 1999).

Thermodynamic calculations clearly show that NO_3^- uptake is uphill, indicating that NO_3^- influx is an active process. Based on the estimates of $[NO_3^-]_{cyt}$ obtained by either method, compartmental analysis, NR assay, or ion-specific microelectrodes measurement (see above), the $[NO_3^-]_{cyt}/[NO_3^-]_o$ ratio would vary from 0.2 to 1000. The electrical potential difference of root cells plasma membrane has been measured at -100 to -200 mV (Thibaud and Grignon 1981), and even up to -300 mV (Glass et al. 1992). Using these values for $[NO_3^-]_{cyt}/[NO_3^-]_o$ ratio and electrical potential difference, one can calculate that NO_3^- absorption requires 6 to 46 kJ mol^{-1}. This conclusion would apply to the HATS and LATS as well: for $[NO_3^-]_{cyt}$ in the millimolar range, a passive NO_3^- influx would require both a very high $[NO_3^-]_o$ (of the order of 100 mM) and an unreasonably high membrane potential difference (- 50 mV). Conversely, these calculations indicate that the efflux is downhill. Consistent with this conclusion, it has been shown that the secretion into the xylem, which is a net efflux from the root symplasm, is passive and is driven by the electrical potential difference (Touraine and Grignon 1982), while NO_3^- uptake (predominated by influx) is driven by the pH gradient (Thibaud and Grignon 1981), in accordance with the application of chemiosmotic concepts to the whole root (Hanson 1978).

The uptake of NO_3^- by plant roots typically leads to the alkalinisation of the nutrient solution (Kirkby and Mengel 1967; Marschner and Röhmeld 1983; Keltjens and Nijenstein 1987; Macduff et al. 1987; Touraine et al. 1988), suggesting that the NO_3^- ions transported across the plasma membrane are accompanied by H^+ ions, or exchanged against OH^-. Another possible source for the alkalinisation of nutrient solution could be the extrusion of bicarbonate (HCO_3^-) ions. The question of the nature of the ion transported with NO_3^- cannot be addressed directly. Indeed, pH measurements are unable to distinguish between the decrease in H^+ concentration and increase in OH^- or HCO_3^- concentration. The relationships between the decarboxylation of organic acids, the alkalinisation of the solution, and the absorption of NO_3^- suggests that NO_3^- ions may be exchanged against HCO_3^- (Touraine et al. 1992). Furthermore, the release of ^{14}C from plants loaded with ^{14}C-labelled carbohydrates (grown in a $^{14}CO_2$-containing atmosphere) in the solution appears to be linked to NO_3^- uptake (Touraine et al. 1992). However, since no direct data are available, one cannot favour any of the possible cotransporters: NO_3^-: H^+ symport, NO_3^-: OH^- antiport or NO_3^-: HCO_3^- antiport. Since these systems are equivalent in terms of pH relations, however, they are usually ascribed to NO_3^-: H^+ symport (terminology used below).

The stoichiometry of the cotransport has been a matter of debate. Some authors observed a hyperpolarisation of the membrane potential in roots following NO_3^- supply, and proposed that it is due to the operation of a $2NO_3^-$:1 H^+ cotransport (Thibaud and Grignon 1981). By contrast, other authors observed a depolarisation (Ullrich and Novacky 1981). Further investigations have explained the apparent contradiction between these two observations: the membrane potential pattern in response to NO_3^- exposure presents a transient depolarisation followed by a hyperpolarisation (McClure et al. 1990; Glass et al. 1992). It is thus proposed that NO_3^- transporters be $1NO_3^-$: $2H^+$ (McClure et al. 1990; Ruiz-Cristin and Briskin 1991; Glass et al. 1992). The hyperpolarisation would correspond to repolarisation by enhanced H^+ pumping activity by the plasma membrane ATPase (Glass et al. 1992). Electrophysiological studies made at low and high range of NO_3^- concentrations demonstrated that both the HATS and LATS are mediated by electrogenic $1NO_3^-$: $2H^+$ symport, which exhibit depolarisation patterns similar to NO_3^- uptake kinetics (HATS, saturable and inducible; LATS, linear and constitutive) (Glass et al. 1992).

Identification of Transporters Involved in NO_3^- influx: Genetic and Molecular Approaches

As we have seen (p. 4), physiological studies have led to the conclusion that at least three different NO_3^- transport systems coexist within a root cell (cHATS, iHATS and LATS). During the past decade, the challenge has been to identify the different components of these NO_3^- uptake systems at the molecular and biochemical levels. In this section, we will describe how genetic and molecular approaches have led to the cloning and characterisation of genes for both high-affinity and low-affinity NO_3^- transporters in higher plants (Table 1).

Table 1. Nitrate transporter genes
A) Low-affinity nitrate transporters

Name	Organism	Identity (%)	Regulation	Accession number
AtNRT1.1	A.thaliana	100	Nitrate-inducible	L10357
AtNRT1.2	A.thaliana	36	Constitutive	AF073361
AtNTP2	A.thaliana	51	?	AJ011604
AtNTP3	A.thaliana	42	?	AJ131464
LeNRT1.1	Tomato	65	Constitutive	X92853
LeNRT1.2	Tomato	65	Nitrate-inducible	X92852
BnNRT1.2	B. napus	91	Nitrate-inducible	U17987

B) High-affinity nitrate transporters

Name	Organism	Identity (%)	Regulation	Accession number
HvNRT2.1	Barley	100	Nitrate-inducible	U34198
HvNRT2.2	Barley	92	?	U34290
AtNRT2.1	A.thaliana	73	Nitrate-inducible	AF0937545
AtNRT2.2	A.thaliana	68	Nitrate-inducible	AF019749
AtNRT2.3	A.thaliana	72[a]	?	AB015472
AtNRT2.4	A.thaliana	69[a]	?	AB015472
NpNRT2.1	N. plumbaginifolia	71	Nitrate-inducible	Y08210
GmNRT2.1	Soybean	77	Nitrate-inducible	AF047718
CrNRT2.1	Chlamydomonas	48	Nitrate-/nitrite-inducible	Z25438
CrNRT2.2	Chlamydomonas	45	Nitrate-inducible	Z25439
CrNRT2.3	Chlamydomonas	?	Nitrate-/nitrite-inducible	Ref: Quesada et al. (1998)
CRNA	A. nidulans	34	Nitrate-inducible	M61125

[a] Amino acid sequences were deduced from genomic DNA sequences, putative introns being identified by comparison with the AtNRT2.1 coding sequence and using identifiable donor and acceptor consensus sequences.

Genes Involved in Low-Affinity NO_3^- Transport

Chlorate, a herbicide and defoliant, is an NO_3^- analogue that is absorbed and then reduced by NR to chlorite (ClO_2^-), which is toxic. Most plant mutants resistant to ClO_3^- have been found to be NR-deficient, but an important exception was the *chl1-1* mutant of *Arabidopsis* (Oostindier-Braaksma and Feenstra 1973). This mutant had a wild-type level of NR activity but reduced levels of NO_3^- uptake specifically in the low-affinity range (Doddema and Telkamp 1979). In a screen for chlorate-resistant lines among a T-DNA tagged population, Tsay and collaborators (Tsay et al. 1993) identified a disrupted allele of *CHL1* (*chl1-T*), enabling them to go on and clone the corresponding gene (now redesignated *NRT1.1*, or *AtNRT1.1* when it is necessary to distinguish it from *NRT1* genes in other species). The *AtNRT1.1* cDNA contains an open reading frame encoding a 590-amino acid protein with a predicted molecular mass of 65 kDa (Tsay et al. 1993). Hydropathy plots indicated that the AtNRT1.1 protein has 12 putative membrane-spanning domains consisting of two groups of 6 segments separated by a hydrophilic region with many charged amino acids. This predicted membrane topology is similar to many other co-transporters identified in both animals and plants (Saier 1994).

By expressing its mRNA in *Xenopus* oocytes, evidence was obtained that *AtNRT1.1* encodes a low-affinity, H^+-dependent NO_3^- transporter with a K_m for NO_3^- of 8.5 mM (Tsay et al. 1993; Huang et al. 1996). However, two recent papers have suggested that AtNRT1.1 may actually be a dual-affinity NO_3^- transporter (Wang et al. 1998; Liu et al. 1999). This conclusion was based on the finding that under certain growth conditions, *chl1* mutants can show partial deficiencies in NO_3^- uptake in the high-affinity as well as the low-affinity range (Wang et al. 1998; Liu et al. 1999). Oocyte expression studies provided further evidence to support the dual-affinity hypothesis (Liu et al. 1999). By measuring NO_3^- depletion from the medium bathing oocytes injected with *AtNRT1.1* mRNA, and/or NO_3^- accumulation within the oocytes, rates of NO_3^- uptake were established over a wide range of external NO_3^- concentrations. The resulting plot indicated that there was a saturable high-affinity component ($K_m = 50$ μM) in addition to the low-affinity system ($K_m = 4$ mM). One puzzling aspect to this story is that while AtNRT1.1 is strongly NO_3^- – inducible (see p. 13), and might therefore be expected to contribute to the iHATS, the *in vivo* evidence indicates that it is the constitutive component of the HATS that is affected in *chl1* mutants (Wang et al. 1998).

Arabidopsis is now known to have at least three other genes closely related to *AtNRT1.1* (Table 1). One of these (*NTL1* or *AtNRT1.2*) is quite distantly related to *AtNRT1.1* (36% amino acid identity), but functional analysis in Xenopus oocytes has shown that it too can act as a low-affinity NO_3^- transporter (Liu et al. 1999). Two further members of the *NRT1* family (*AtNTP2* and *AtNTP3*) have been identified from the collection of *A. thaliana* ESTs (expressed sequence tags) (Hatzfeld and Saito 1999). The protein sequences deduced for AtNTP2 and AtNTP3 are, respectively, 51 and 42% identical and 68 and 58% similar to AtNRT1.1 at the amino acid level, and both are predicted to contain the 12 putative transmembrane helices found in AtNRT1.1. However, there is as yet no information on the transport function of these two proteins.

AtNRT1.1 was used as heterologous probe to isolate two NRT1 cDNAs (*LeNRT1.1* and *LeNRT1.2*) from a tomato root hair-specific library (Lauter et al. 1996). Both cDNAs were predicted to encode proteins with the same topology as AtNRT1.1 and with a 65% amino acid sequence identity, but again their role in NO_3^- transport remains unknown.

The *BnNRT1.2* gene, which was one of two *NRT1* cDNAs identified in a *Brassica napus* cDNA library using an *AtNRT1.1* probe, is very closely related to *AtNRT1.1* (91% amino acid identity) (Zhou et al. 1998). Detailed oocyte expression studies with this gene have revealed some unusual transport properties for the BnNRT1.2 protein. Oocytes injected with *BnNRT1.2* mRNA exhibited both NO_3^- – and L-histidine-elicited currents, the amino acid generating even larger currents than NO_3^- (Zhou et al. 1998). The inward cation currents obtained with both NO_3^- and histidine were consistent with an H^+ – coupled system. Other possible substrates tested were D-histidine, NO_2^-, cyanate, ClO_3^- and the dipeptide His-Leu, but none gave significant currents. Surprisingly, the pH optima for NO_3^- and L-histidine were quite different, with L-histidine transport being favoured at alkaline pH and NO_3^- transport being favoured at acidic pH.

The ability of a transporter to mobilise two such different substrates is difficult to understand in mechanistic terms. However, the NRT1 transporters belong to an unusual family of membrane transporters known as the PTR family, most members of which have been shown to transport short peptides (Steiner et al. 1994), and BnNRT1.2 is not unique within this family in being able to transport two very different substrates: the rat PHT1 transporter has been reported to be a high-affinity histidine transporter (K_m = 17 μM) as well as being able to transport peptides (Yamashita et al. 1997). Since histidine is unlikely to be present in soil solutions at the kind of concentrations that would be needed for BnNRT1.2 to contribute significantly to its uptake, any role that it may have in histidine transport is more likely to involve internal trafficking.

Genes Involved in High-Affinity NO_3^- Transport

High-affinity NO_3^- transporters were first cloned in fungi and algae, and this paved the way for the subsequent progress that has been made in identifying high-affinity NO_3^- transporters in higher plants. The *crnA* mutant of *Aspergillus nidulans*, which was isolated in a screen for chlorate-resistant lines, was found to be defective in NO_3^- uptake at the conidiospore and young mycelium stages (Brownlee and Arst 1983). When the CRNA gene was isolated it was found to encode a 507-amino acid protein, containing 12 putative membrane-spanning domains, with two groups of 6 segments separated by a central loop (Unkles et al. 1991, 1995).

In the green alga *Chlamydomonas reinhardtii*, six NO_3^- assimilation genes are clustered and coregulated: *CrNRT2.1* (encoding a high-affinity, NO_3^-/NO_2^- bispecific transporter), *CrNRT2.2* (high-affinity, NO_3^- – specific transporter), *Nar2* (unknown function but necessary in conjunction with *CrNRT2.1* or *CrNRT2.2* for the expression of NO_3^- and NO_2^- transport activities), *Nia1* and *Nii1* (structural loci for NO_3^- and NO_2^- reductases) and *Nar1* (unidentified function) (Quesada et al. 1994, 1998; Galván et al. 1996). The deduced *CrNRT2* gene products were found to be homologous to CRNA, although with some distinctive differences in the distribution of their hydrophilic domains (see below). Recently, a third gene, *CrNRT2.3*, was found to be clustered with another gene, *Nar5*, of unidentified function (Quesada et al. 1998). The expression pattern of *CrNRT2.3* in wild-type or mutant strains suggested that it encodes the high-affinity NO_2^- transport activity.

The molecular cloning of *NRT2* genes from a higher plant was first reported in barley (Trueman et al. 1996b). Two full-length cDNAs, *BCH1* and *BCH2* (now renamed *HvNRT2.1* and *HvNRT2.2*), were found to encode polypeptides which are respectively 41-43% identical to CRNA and 56-57% identical to CrNRT2.1 (Trueman et al. 1996b). Models for the membrane topology of HvNRT2.1 and CRNA polypeptides differ in the size of the predicted central loop (96 amino acids in CRNA versus 32 amino acids in HvNRT2.1) and in the presence in HvNRT2.1, as in CrNRT2.1, of a large C-terminal extension (~70 amino acids) which is absent in CRNA (Trueman et al. 1996a; Forde 2000). This C-terminal extension seems to be a general feature of NRT2 transporters from algae and higher plants since it is also found in NRT2-related sequences from soybean (Amarasinghe et al. 1998), *Nicotiana plumbaginifolia* (Quesada et al. 1997) and *Arabidopsis* (Filleur and

Daniel-Vedele 1999; Zhuo et al. 1999). Estimates for the number of NRT2-related sequences in each species (based on genomic Southern blots) vary from two genes in *N. plumbaginifolia* (Krapp et al. 1998) to a small or large multigene family in, respectively, soybean (which is an ancient tetraploid) (Amarasinghe et al. 1998) or barley (Trueman et al. 1996b). *Arabidopsis* has at least four *NRT2* genes: *AtNRT2.1* and *AtNRT2.2*, which are located near the top of chromosome 1 in a tail-to-tail configuration (Filleur and Daniel-Vedele 1999; Zhuo et al. 1999), and two newly identified *NRT2* genes (*AtNRT2.3* and *AtNRT2.4*), which are located close together on chromosome 5 (accession number AB015472). Whether or not all four *NRT2* genes are expressed, and what their functional role may be, remains to be elucidated.

The functional properties of putative NRT2 sequences have up to now only been demonstrated in algae by complementation analysis using loss-of-function mutants (Galván et al. 1996). Such mutants are still not available in plants. Indications on the possible function of plant NRT2 sequences rely on strict correlations that were observed between NRT2 mRNA steady-state levels and NO_3^- uptake in plants grown under different N regimes (Lejay et al. 1999; Zhuo et al. 1999) or in plants affected in the regulation of NO_3^- uptake process (Krapp et al. 1998). In the near future, two parallel approaches will give further insights into the function of plant *NRT2* genes: the physiological characterisation of transgenic plants specifically overexpressing or underexpressing each of these genes, and of loss-of-function mutants that could be obtained in *A. thaliana* by gene disruption (with T-DNA or transposons).

Regulation of NO_3^- Transport: Molecular Studies

Studies on the regulation of NR and NiR have established that the NO_3^- assimilatory pathway in plants is highly regulated, being inducible by NO_3^- and feedback repressible by products of N assimilation (Daniel-Vedele and Caboche 1996). In this section we consider the recent evidence concerning the regulation of the NRT1 and NRT2 transporters.

NO_3^- Induction

The *NRT1* and *NRT2* genes have been demonstrated to be NO_3^--inducible in many plant species. For example, the steady-state levels of NRT2 mRNA increase rapidly in N-starved roots after treatment with NO_3^- in barley (Trueman et al. 1996b), *N. plumbaginifolia* (Krapp et al. 1998), soybean (Amarasinghe et al. 1998) and *A. thaliana* (Filleur and Daniel-Vedele 1999; Zhuo et al. 1999). Very low concentrations of NO_3^- (10-50 µM) are sufficient to induce NRT2 genes in *N. plumbaginifolia* (Krapp et al. 1998) and *A. thaliana* (Filleur and Daniel-Vedele 1999), which is consistent with their putative role in high-affinity NO_3^- uptake.

Surprisingly, the *AtNRT1.1* and *BnNRT1.2* genes, which supposedly encode low-affinity NO_3^- transporters, are also induced by low as well as high concentrations of NO_3^- (Zhou et al. 1998; Filleur and Daniel-Vedele 1999). This may simply indicate that these genes are coregulated with other NO_3^--inducible genes, or it could be taken as supporting the idea (discussed above) that the *AtNRT1.1* gene product plays an additional role in high-affinity NO_3^- uptake under certain physiological conditions (Wang et al. 1998).

NRT1 genes in tomato seem to be differentially regulated by NO_3^-: LeNRT1.1 mRNA accumulation is restricted to root hairs that had been exposed to NO_3^-, whereas LeNRT1.2 transcripts are detected in root hairs as well as in other root tissues independently of the N supply (Lauter et al. 1996). The latter gene may encode a constitutive component of the LATS system in tomato, but the question of their precise physiological roles remains to be answered.

Feedback Regulation

It is already well known that transcription of NR and NiR genes is feedback-regulated by N-derived metabolites (see Chap. 2). In the same manner, high-affinity NO_3^- uptake activity seems to be susceptible to such regulation (see p. 17). At what level this regulation occurs and which N metabolites are involved are currently being investigated.

The transcription of NRT2 genes has been demonstrated to be a possible target of feedback repression: the abundance of NpNRT2 and GmNRT2.1 mRNAs rapidly declines when plants are supplied with NH_4^+ (Amarasinghe et al. 1998; Krapp et al. 1998) and more dramatically with glutamine (Krapp et al. 1998). In a recent paper, various inhibitors of N assimilation were used to obtain evidence indicating that the internal NH_4^+ pool is responsible for feedback regulation of both the AtNRT2.1 gene and the iHATS activity (Zhuo et al. 1999). This hypothesis is reinforced by the physiological analyses of Fd-GOGAT antisense tobacco plants: there is an accumulation of NH_4^+ pools in roots of these plants and, correlated with this, a reduction in the steady-state level of NRT2 mRNAs when compared to the wild type (S. Ferrario-Méry, pers. comm.). Derepression of transporter gene expression in response to a specific nutrient deficiency is a response that is also seen in the case of other nutrients, such as sulphate (Smith et al. 1997b) and phosphate (Smith et al. 1997a).

Recent results suggest that AtNRT1.1 is less sensitive than NRT2 genes to feedback repression. Firstly, the presence of NH_4^+ in the nutrient solution has little effect on the abundance of AtNRT1.1 mRNA, but leads to a significant decrease in the expression of AtNRT2.1 (Lejay et al. 1999). Furthermore, AtNRT2.1 is transiently derepressed by N-deprivation while the abundance of the AtNRT1.1 mRNA rapidly declines (Filleur and Daniel-Vedele 1999). These contrasting fluctuations in the abundance of AtNRT1.1 and AtNRT2.1 mRNAs have been found to correlate remarkably well with changes in the activities of the LATS and HATS, respectively (Lejay et al. 1999), suggesting that transcriptional or posttranscriptional regulatory mechanisms play a major role in controlling the activity of these transport systems, at least during the response to N withdrawal.

It should be noted that the lack of evidence for feedback regulation of AtNRT1.1 (at least at the mRNA level) should not be taken as necessarily typical of all components of the LATS, or even all NRT1 genes. There is clear kinetic evidence, at least for barley, that the activity of the LATS is derepressed in N-deficient plants (Siddiqi et al. 1990).

The contrasting sensitivities of the AtNRT2.1 and AtNRT1.1 genes to feedback repression could explain the decreased expression of the former and the increased expression of the latter when plants are grown on high NO_3^- concentrations (Filleur and Daniel-Vedele 1999). On the other hand, one could imagine that a high internal concentration of NO_3^- itself might be responsible for the repression of NRT2 genes.

This seems not to be the case: inhibition of NR activity by tungstate had only a small effect on *AtNRT2.1* expression (Zhuo et al. 1999). Moreover, in NR-deficient lines of *A.thaliana* or *N. plumbaginifolia*, the *AtNRT2.1* and *NpNRT2* genes are overexpressed (Krapp et al. 1998; Filleur and Daniel-Vedele 1999; Lejay et al. 1999), probably due to the absence of feedback repression by downstream N metabolites.

Regulation by Light and Sugars

In many plant species, such as soybean (Delhon et al. 1995) or tomato (Cardenas-Navarro et al. 1998), NO_3^- uptake is diurnally regulated. In *Arabidopsis* roots, the transcript levels of both *AtNRT1.1* and *AtNRT2.1* undergo marked diurnal changes and can be rapidly increased in the dark period if sucrose is supplied (Lejay et al. 1999); these variations are correlated with changes in NO_3^- influx. These observations support the hypothesis that the coordination of NO_3^- uptake with photosynthesis is, at least in part, mediated by transcriptional regulation of the root NO_3^- transporters and that there is a direct or indirect regulatory role for C metabolites. This regulation seems to be mechanistically different from regulation by N-status as the expression of *AtNRT1.1* and *AtNRT2.1* is affected in the same way by light/dark transition and sucrose supply, but is affected differently by N deficiency (Lejay et al. 1999; Filleur and Daniel-Vedele 1999).

Regulation at the Whole Plant Level: an Interorgan Signalling Process

Control of NO_3^- Uptake by N Demand

Most studies, whether at the whole plant, cellular or molecular levels, have investigated the effect on NO_3^- uptake of external factors that can be easily changed by the investigator (such as $[NO_3^-]_o$). However, there is overwhelming evidence that NO_3^- uptake by plant roots depends not only on these factors, but also on internal factors which act at the whole plant level to match the rate of N acquisition to the demand for N. In plants grown in the field, NO_3^- uptake appears to be determined by the growth rate, which is determined by the climatic conditions or genotypic differences (Raper et al. 1978; Lemaire and Salette 1984a; Lemaire and Salette 1984b; Swiader et al. 1991).

In laboratory experiments, the independence of NO_3^- uptake to external NO_3^- concentration (consequently, dependence to some other, internal, factors) is supported by two sets of evidence. First, varying the concentration of NO_3^- in the nutrient solution has very little effect on either NO_3^- uptake or growth rates. This is illustrated by a study on perennial ryegrass grown on flowing nutrient solutions containing 1.4 μM to 140 mM $[NO_3^-]_o$ (Clement et al. 1978). Only at the very lowest or highest concentrations tested were differences observed: in the range from 14 μM to 14 mM the rates of plant growth and NO_3^- uptake and the plant N status were unchanged despite the 1000-fold change in $[NO_3^-]_o$. This reveals that, in the long term, NO_3^- uptake (in both the LATS or HATS range) is governed by regulatory processes that maintain N homeostasis, whatever the kinetics seen in short-term experiments (see above).

Further evidence that NO_3^- absorption is tightly regulated is provided by studies where plants are grown at constant $[NO_3^-]_o$ but under differing relative growth rates (RGRs). In tall fescue, changing the photosynthetic photon flux density or the atmospheric concentration of CO_2 resulted in changes in the rates of both net CO_2 fixation and NO_3^- uptake from identical nutrient solutions (Gastal and Saugier 1989).

Further indications that the capacity of NO_3^- uptake is regulated by N demand have been obtained by imposing N deficiency. Plants grown on a N-free nutrient solution for periods of hours to 1 day develop enhanced capacity to absorb NO_3^- when the ion is resupplied. For longer starvation periods, NO_3^- uptake decreases (plotting NO_3^- uptake vs the duration of deficiency forms a parabola), due to the deinduction of NO_3^- transport systems. In order to distinguish between the induction by NO_3^- and the regulation by N status, experiments have been designed where the plants are subjected to various starvation periods, but are all exposed to NO_3^- for a few hours (typically 3 h) prior to absorption measurement. In these experiments, the longer the starvation period (up to several days), the higher the rate of NO_3^- uptake (Doddema and Otten 1979; Lee and Rudge 1986; Chapin et al. 1988; Bowman et al. 1989; Rodgers and Barneix 1989; Lee 1993). Conversely, when the time course of NO_3^- uptake is followed after restoration of NO_3^- to 5-days-starved barley plants, $^{13}NO_3^-$ influx increased rapidly, peaked after a few hours, and declined while the nutritional status is restored by current uptake (Siddiqi et al. 1989). This is indicative of two successive steps: the induction of transport systems by NO_3^-, then the repression of NO_3^- influx by N status. Conversely, the enhanced NO_3^- uptake observed in N-starved plants compared to replete plants is considered to be the relief of this negative feedback by N status (derepression). Whereas the induction of NO_3^- uptake by its substrate appears to be rather unique, the nutritional status-dependent repression/derepression is a common feature for ion acquisition by plant roots. For instance, the starvation in K, P or S similarly leads to enhanced absorption of K^+, $H_2PO_4^-$ and SO_4^{2-}, respectively (Cogliatti and Clarkson 1983; Drew et al. 1984; Lee 1993). Notwithstanding this common pattern, the effect of deficiency on NO_3^- uptake is specific since neither did N starvation stimulate the absorption of other elements, nor did the deficiency in another element lead to enhanced NO_3^- uptake (Lee and Rudge 1986; Lappartient and Touraine 1996).

The signals of N demand do not originate in the roots themselves, as demonstrated by split-root experiments. These consist in dividing the root system of intact plants into two parts, fed independently of each other. In the case of NO_3^- deficiency studies, one part is supplied with an N-free solution, and the other part with an NO_3^- containing solution, where NO_3^- absorption is recorded. Compared to control plants, in which both parts of the root system are fed NO_3^-, the uptake rate of NO_3^- is enhanced (Edwards and Barber 1976; Burns 1991; Lainé et al. 1995). Another indication that leaf activity controls NO_3^- uptake is provided by studies on the effect of light on the uptake process (Gastal and Saugier 1989; Delhon et al. 1995).

Different processes, that can be referred to as either non-specific or specific, are likely to be operating in the interorgan signalling mechanism that regulates NO_3^- uptake. This is illustrated by the effect of light and N starvation: the former affects simultaneously the absorption of several ions (Smith and Cheema 1985; Le Bot and Kirkby 1992), while the latter specifically affects NO_3^- uptake. The two types of regulation could *a priori* be the same (the general effect of light might reflect the concomitant operation of multiple regulatory pathways which each control the uptake of a specific ion). However, the probable involvement of carbohydrates in the regulation of NO_3^- uptake by light seems unlikely to be appropriate for a specific regulatory mechanism, which suggests that at least two types of regulation do operate *in planta*.

Regulatory Signals

Considering the two types of regulation (specific and non-specific), and the occurrence of controls of NO_3^- uptake by both the C and the N status of the plant, several compounds have been envisaged as potential signals involved in the regulation of NO_3^- uptake at the whole-plant level. These include mainly carbohydrates, amino acids, and organic acids which are detailed below. This does not preclude the possibility that other compounds, especially NO_3^- itself and NH_4^+, may play a role in the regulation of NO_3^- transport in root cells. However, because none of these ions is significantly translocated in the phloem sap, their role in interorgan signalling is not considered.

Carbohydrates. The stimulatory effect of light on NO_3^- uptake is attributed to photosynthesis, since it could be prevented by decreasing the atmospheric CO_2 concentration (Delhon et al. 1996). Addition of carbohydrates to the nutrient solution is known to enhance NO_3^- uptake (Hänisch ten Cate and Breteler 1981). Furthermore, blocking phloem translocation by the means of stem girdling led to a decline in NO_3^- uptake, which is partially restored by adding glucose to the nutrient solution (Delhon et al. 1996). All together, these results suggest that the availability of carbohydrates in the root cells is required to maintain the activity of NO_3^- transporters. The underlying mechanism, however, is obscure. An increased energization of the plasma membrane transporters by carbohydrate availability to roots is unlikely to be the explanation. Indeed, no consensus exists that an energetic shortage might occur in roots, even at night (see Delhon et al. 1996). Furthermore, the fact that two NO_3^- transporter genes in *A.thaliana*, *AtNRT1.1* and *AtNRT2.1*, are downregulated at night, and that this downregulation is prevented by sucrose (Lejay et al. 1999) suggests that the underlying mechanism corresponds to a regulation by C status mediated by a carbohydrate signal (see p. 15).

Amino Acids. Concerning the specific mechanisms by which the absorption of different ions is regulated, one hypothesis takes into account both the ion specificity and the interactions between organs. It proposes that the signal integrating the demand of the whole plant for a nutrient is the concentration of this element in the sieve sap. Since the amount of element exported from leaves is determined by the difference between the xylem supply and leaf assimilation requirements, it would be a good estimate of the specific demand for an element. Experimental evidence in favour of this model exist for K^+ and $H_2PO_4^-$ (Drew and Saker 1984). In the case of NO_3^-, however, NO_3^- itself cannot be the inter-organ signal that conveys the message of N demand from shoots to roots because it is not transported in the phloem sap (Allen and Raven 1987). Therefore, it has been suggested that amino acids would be the interorgan signals. They would act as a sensor of integrated N status of the whole plant, inform the absorbing root cells and finally exert a feedback control on NO_3^- transporters activity. This theory is supported by different lines of evidence: (1) an extensive pool of amino-N does circulate continuously between roots and shoots; (2) amino acids are repressors of NO_3^- absorption; (3) increasing the translocation rate of amino acids leads to decline in NO_3^- absorption. The occurrence of reduced N circulation in the xylem and phloem is well established in herbaceous plants, where it has been calculated that the amount of N transported to the roots *via* the phloem represents a large proportion of the N transported to shoots by the xylem (Simpson et al. 1982; Touraine et al. 1988; Larsson 1992). Using $^{15}NO_3^-$ in split root experiments, it has been demonstrated that there is a rapid cycling of N from roots to shoots and back to roots (Cooper and Clarkson 1989). Moreover, this transport pool of amino N appears to be relatively isolated: the exchange rate between this pool and the bulk tissue is slower than the exchange rate between the xylem and phloem saps. This would make its composition sufficiently specific to convey information. In barley, it has been shown that N starvation caused important modifications in the composition of the amino acids accumulated (Barneix et al. 1984). Unfortunately, due to the difficulty in measuring the composition and the translocation rate of phloem sap, the effects of N starvation on the phloem transport of amino acid have not been elucidated. Split root experiments in *Ricinus communis*, a species known for the relative ease with which phloem sap can be collected *via* stem incision, failed to show consistent changes in amino acid composition and translocation rate of phloem sap to NO_3^- – fed roots that would account for the increased NO_3^- influx when the other roots were fed NO_3^- – free solution (Tillard et al. 1998).

Evidence from feeding amino acids to plant roots indicates that a wide range of amino acids is able to downregulate NO_3^- absorption, although some are strong inhibitors, some are weak inhibitors, and some have no effect (Doddema and Otten 1979; Breteler and Arnozis 1985; Lee et al. 1992; Muller and Touraine 1992). However, no conclusion on the actual nature of the amino acid(s) active in the root cells is provided by these data, because of possible rapid conversion of the supplied amino acid into another (Lee et al. 1992; K. Mouline, J. Vidmar and B. Touraine, unpubl.). Another piece of evidence is provided by the observation that inhibitors of glutamine synthetase and glutamate synthase prevent the inhibitory effect of NH_4^+ on NO_3^- absorption (Breteler and Siegerist 1984).

A relationship between the accumulation of an amino acid and a change in the rate of NO_3^- uptake has also been found in *Arabidopsis*, by comparing an NR-deficient mutant to the wild type (Doddema and Otten 1979). In this report, the difference in NO_3^- uptake rate was explained by the difference in internal levels of arginine. Even though these data indicate that amino acids are effectors of NO_3^- uptake, this does not necessarily imply that phloem-translocated amino acids are active. This specific question has been addressed in soybean seedlings, by feeding amino acids to the cotyledons, leading to specific enhancement of the amino acid translocation rate (Muller and Touraine 1992). This resulted in a strong effect, or a weak effect, or no effect, on NO_3^- uptake, depending on the amino acid supplied. The sequence of inhibition was very similar to that obtained when amino acids were directly supplied to the roots.

The inhibition of net NO_3^- uptake by amino acids is due to a reduction in NO_3^- influx (Muller et al. 1995), while efflux is only marginally affected, probably as a consequence of slight diminution of cytoplasmic concentration. The role of amino acids as potential signals regulating NO_3^- influx is further supported by observations in *Arabidopsis* that the *AtNRT2.1* gene, which is derepressed by N starvation (Lejay et al. 1999), is repressed by externally supplied amino acids (Zhuo et al. 1999; K. Mouline, J. Vidmar and B. Touraine, unpubl.), while, on the other hand, the *AtNRT1.1* gene, which is not derepressed by N starvation (Lejay et al. 1999), is unaffected by externally supplied amino acids (Bruno Touraine, unpubl.; see also p. 14).

Organic Acids. A hypothesis that predicts the coordination of NO_3^- uptake to leaf NR activity was developed in the 1970s (Ben Zioni et al. 1971). It derived from studies on the impact of N and S metabolism on ionic and pH balance in plant. The reduction of NO_3^- results in the net release of 1 mol OH^- equivalent per mol NO_3^- reduced (Dijkshoorn 1962), leading to increases in both pH and $[HCO_3^-]$ in the cytoplasm of assimilating cells. In contrast to the root, in the leaf the basic anions (OH^- or HCO_3^-) released in the cytoplasm cannot be expelled from the cells because of the finite size of the extracellular space (Raven and Smith 1976) and the reduced transport capacity of OH^- equivalents in the phloem sap (Raven 1977). Therefore, the alkalinisation of the cytoplasm of shoot cells is avoided by the operation of a biochemical pH-stat, which synthesises strong organic acids (mainly malate) in response to an increase in pH (Davies 1986). As a consequence, NO_3^- assimilation in the shoot results in a stoichiometric synthesis of amino acids and organic acids. Most of these carboxylates are not accumulated in roots, or in any other organ, suggesting that the basic ions released by NO_3^- reduction in leaves are excreted into the external medium. To reconcile these observations, it has been proposed that the carboxylates synthesised in the shoot are transported in the phloem to the roots, where decarboxylation releases HCO_3^- ions that are excreted into the external medium (Ben Zioni et al. 1971). Numerous studies on the fate of the anion charge generated by NO_3^- assimilation have been reported, which demonstrated the validity of this interorgan circulation pattern in the species for which leaves are the major site of NO_3^- reduction (Touraine et al. 1988 and references therein). This model predicts that the higher the reduction rate of NO_3^- in leaves, the higher the synthesis and export rate of carboxylates to roots; the HCO_3^- ions released by their decarboxylation would be exchanged with NO_3^- absorbed, leading to stimulation of NO_3^- influx. Experimental data obtained in soybean using a split root system are in agreement with these assumptions: the supply of malate to part of the root system led to increased transport of malate in the phloem, and enhanced NO_3^- uptake, alkalinisation rate of the medium and release of C by the other roots (Touraine et al. 1992).

Interorgan Signalling Mechanisms

The acquisition of NO_3^- is driven by internal regulatory mechanisms that sense and integrate, at the whole plant level, N requirements for sustaining growth and development of the organism, based on the actual capacities of the plant and the specific environmental conditions. This control appears to involve multiple pathways, some of which are specific for N while others are likely to be affecting the absorption of several different ions. Operation of various regulatory processes in parallel would explain how plants manage to combine N homeostasis with a high degree of plasticity (a key feature for an organism directly in contact with environmental fluctuations and lacking mobility). Although it is likely that totally unknown processes operate in plants to adjust the rate of NO_3^- absorption to accommodate changes in the demand for N, some of these processes have been described in broad terms.

Those that have been recognised thus far depend upon metabolite control and involve phloem-translocated compounds that act as interorgan signals conveying information from shoot to root and ultimately repressing or enhancing NO_3^- uptake. Although the occurrence of other regulatory mechanisms (e.g. based on hormonal balance or transport) is certainly not to be ruled out, above we have concentrated on these feedback and feedforward processes, and discussed the involvement of specific phloem-translocated compounds for which a consensus exists on their likely role in the interorgan signalling regulation of NO_3^- absorption. Several aspects of this interorgan signalling remain to be deciphered. For instance, though we know that some amino acids are phloem-translocated compounds that repress NO_3^- uptake, it is not clear whether several of these compounds act as regulatory signals or whether just one specific amino acid is responsible for this regulation *in planta*. There are other important questions. What are the quantitative changes in phloem translocation flux of these compounds? How do these relate to quantitative changes in N or C status? Which pools are affected in the root tissues? Indeed, one would expect that, due to compartmentation of these compounds, relatively important changes of specific "indicator" pools can occur quite rapidly and play the role of signal before the nutrition of root becomes perturbed.

We would stress two further questions which, if answered, would give important clues for understanding how plants control N acquisition: (1) what are the key indicators of nutritional status that modulate the interorgan signals? (2) what are the targets for these signals?

Concerning the first question, the three signals discussed would address three different aspects of the plant nutritional balance: carbohydrates, amino acids and organic acids would be involved in the regulation of NO_3^- absorption by the C status, N status, and ionic balance respectively (see above). Carbohydrates would stimulate N acquisition when C acquisition increases, amino acids would repress NO_3^- absorption when N feeding exceeds the actual needs of the plant for this specific element, whatever the growth rate, carbohydrates synthesis etc. As for the organic acids, they would ensure a coordination of NO_3^- transport in the root with NO_3^- reduction in shoot. Since the balance between the different N pools in leaves does not depend only on current assimilation of incoming NO_3^-, but also on protein turnover and remobilization of organic N, this system would not be a true indicator of N status. It would rather be an indicator of NO_3^- consumption; it would ensure that every NO_3^- ion reduced in leaves is compensated for by an NO_3^- ion absorbed by roots. In other terms, organic N and NO_3^- homeostasis would depend on amino acid and organic acid signalling, respectively.

Concerning the root targets of the phloem-translocated signals, they have not been identified with certainty. However, pieces of evidence recently obtained suggest that the expression of genes coding for root plasma membrane transporters is affected by treatments altering the C and N status (e.g. light, N starvation), and by provision of carbohydrates or amino acids to roots (see p. 14-15). Furthermore, a recent report provides the first demonstration that genes involved in nutrient acquisition and assimilation in roots can be subject to regulation by a shoot-derived signal (Lappartient et al. 1999). The genes in this case were related to S nutrition, but considering the similarities between demand-driven regulation of SO_4^{2-} and NO_3^- uptake, these observations support the hypothesis that the interorgan signals involved in regulating NO_3^- absorption ultimately trigger changes in genes expression.

NO$_3^-$ Availability and Root Development

One of the most important factors in determining a plant's ability to capture limiting supplies of mineral nutrients is the size and topology of its root system. The development of a root system is partly under direct genetic control but is also highly plastic, being influenced by both a wide range of environmental factors and internal factors related to the plant's physiological status.

Nutrients themselves play a major role in determining both the size and the architecture of a root system. This happens in two main ways. Firstly, nutrients are frequently distributed unevenly within the soil, and many plant species can adapt to this situation by proliferating their roots preferentially within the nutrient-rich patches (Robinson 1994). Secondly, under conditions of nutrient limitation or deficiency, many plant species adapt by promoting root growth at the expense of shoot growth, so that exploration of the soil for more nutrients is given priority, leading to increased root:shoot ratios (Ericsson 1995, Améziane et al. 1997).

Recent studies on tobacco and *A.thaliana*, using mutants and transgenic lines, have been looking at the role of NO$_3^-$ in the regulation of root branching and the allocation of resources between shoots and roots, and have begun to shed light on the mechanisms involved.

The Root Response to Localized Supplies of NO$_3^-$

The developmental response of roots to nutrients can be seen most clearly in the localized proliferation of lateral roots (LRs) which occurs when the root system of a nutrient-limited plant encounters a nutrient-rich patch of soil (Robinson 1994). In fact, this is the second part of a two-stage process of adaptation by which plants attempt to maximise their ability to capture the available nutrients. In the first stage, which begins with 24 h of exposing a root system to a localized nutrient supply, the net absorption rate of the nutrient in the nutrient-rich zone increases markedly in comparison with roots uniformly exposed to the nutrient, a kinetic response which is seen for both phosphate (Drew and Saker 1978; Jackson et al. 1990) and NO$_3^-$ (Drew and Saker 1975; Burns 1991; Lainé et al. 1998).

In the second stage, which takes longer to become apparent, there is an increase in the number of LRs initiated in the nutrient-rich zone and an increase in the growth rate of both new and existing LRs, leading to a greatly increased root density within the nutrient-rich zone. This response has been seen for NO$_3^-$, NH$_4^+$ and phosphate (e.g. Hackett 1972; Drew 1973, 1975; Sattelmacher and Thoms 1995; Lainé et al. 1998), but K$^+$ does not elicit the same effect (Drew 1975). Recent studies have shown that, despite the high mobility of soil NO$_3^-$ (which might be expected to limit the significance of increased root branching for N capture), the proliferation of LRs in N-rich patches can, in fact, confer a significant advantage when plants are in direct competition for limiting N supplies (Robinson et al. 1999).

Insights into the Mechanism by Which a Localized NO$_3^-$ Supply Can Stimulate Lateral Root Proliferation

One of the most popular explanations put forward to explain how LR proliferation might be stimulated specifically within an NO$_3^-$ – rich zone has been the idea that NO$_3^-$ assimilation locally at its site of uptake leads to an increased influx of photosynthate and/or auxin, which then stimulates LR growth in that region (Wiersum 1958; Drew 1973; Granato and Raper 1989; Sattelmacher et al. 1993). An alternative view has been that the NO_3^- ion is somehow acting as a regulatory or signal molecule (Trewavas 1983), implying that the plant has a mechanism for sensing the presence of the external NO_3^- supply and converting that signal into the appropriate developmental response.

This question has recently been addressed using *A.thaliana* as a model system (Zhang and Forde 1998, 2000; Zhang et al. 1999). These studies, in which seedlings were grown under aseptic conditions on vertical agar plates, established that a localized supply of KNO_3 (at concentrations as low as 100 µM) could stimulate a two- to three-fold increase in LR elongation rates in the NO_3^- – rich zone, without affecting LR initiation or primary root growth (Zhang and Forde 1998; Zhang et al. 1999). Since the increased rate of LR elongation was not accompanied by a significant increase in the size of mature root cells, it was concluded that the NO_3^- must be stimulating cell production rates in the LR meristem (Zhang et al. 1999).

Two lines of evidence indicated that the LRs are responding to a signal from the NO_3^- ion itself and not to an N metabolite or to the nutritional effects of NO_3^-: firstly, alternative sources of N (such as NH_4^+ or glutamine) did not stimulate LR growth (Zhang et al. 1999) and secondly, the LR response to NO_3^- was found to be undiminished in an NR-deficient mutant which is defective in NO_3^- assimilation (Zhang and Forde 1998).

The first insight into the signal transduction pathway linking external NO_3^- to increased LR meristematic activity came from the identification of a NO_3^- – inducible root-specific gene (*ANR1*) which encodes a member of the MADS-box family of transcription factors (Zhang and Forde 1998). In plants, MADS-box genes are generally associated with the control of floral organ development, but they also occur in animals and fungi where they are generally associated with regulating metabolism or development in response to environmental signals (Shore and Sharrocks 1995). To investigate the function of ANR1, transgenic *A.thaliana* lines were generated in which *ANR1* was down-regulated (by antisense or cosuppression effects). These lines were found to be markedly altered in their sensitivity to environmental NO_3^-, with LR growth failing to be stimulated in response to a localised NO_3^- supply (Zhang and Forde 1998). Thus, it was proposed that the *ANR1* gene product is involved in transducing the NO_3^- signal and eliciting the developmental response leading to increased LR proliferation.

Evidence for an overlap between the NO_3^- response pathway and the auxin response pathway has come from an analysis of the effect of localized NO_3^- treatments on LR proliferation in three auxin-resistant mutants (Zhang et al. 1999). While two of the mutants (*aux1* and *axr2*) showed wild-type responses to the NO_3^- treatment, indicating that their gene products are not involved in the NO_3^- response, the third (*axr4*) failed to show any increase in LR elongation rate (Zhang et al. 1999). It is well established that auxin plays a key role in regulating both LR initiation and development (Evans et al. 1994; Hobbie 1998), but this seems to be the first indication that the signal transduction pathways for NO_3^- and auxin may share common components. Unfortunately, the *AXR4* gene has not yet been cloned and its precise function is unknown.

Evidence for a Role for Tissue NO_3^- Levels in Regulating Root Development and Shoot: Root Ratios

Scheible and colleagues used partially NR-deficient lines of tobacco to examine the role of tissue NO_3^- levels in regulating gene expression, metabolism and shoot-root allocation (Scheible et al. 1997a, b). By increasing the $[NO_3^-]_0$ it was possible to force NR-deficient lines to assimilate NO_3^- and to grow at a rate similar to wild-type lines grown at a lower $[NO_3^-]_0$. In consequence of the bottleneck in NO_3^- reduction, there was a very marked accumulation of NO_3^- in root and shoot tissues in the NR-deficient lines. Amongst the effects that were attributed to the high levels of tissue NO_3^- was a dramatic increase in the shoot:root ratio, which was at least partly due to a strong inhibition of root growth (Scheible et al. 1997b).

What may be a closely related phenomenon has been observed in *A.thaliana* roots. When *A.thaliana* seedlings were grown on vertical agar plates containing a range of NO_3^- concentrations (from 10 μM to 50 mM), it was noted that root growth began to be inhibited at concentrations above 1 mM (Zhang and Forde 1998). Separate measurements of LR and primary root growth showed that the inhibition was specific to the LRs (Zhang and Forde 1998); at 50 mM NO_3^- growth of the LRs was almost completely blocked, while primary root growth was unaffected. Subsequent studies revealed that the high NO_3^- concentrations did not affect LR initiation or the elongation of mature LRs, but specifically inhibited LR development just after the emergence of the LRs from the primary root (Zhang et al. 1999). It has been established that in *A.thaliana* this period just after emergence coincides with a critical phase of LR development when the newly formed LR meristem becomes activated and elongation of the mature LR begins (Malamy and Benfey 1997). Unlike the stimulatory effect of NO_3^- on LR elongation, which is localised to those LRs directly exposed to the NO_3^- (see above), its inhibitory effect was shown to be systemic (Zhang et al. 1999).

To investigate the role of NO_3^- assimilation in the inhibitory effect of NO_3^-, the sensitivity of a *nia1nia2* (NR-deficient) mutant to high concentrations of $[NO_3^-]_0$ was tested (Zhang et al. 1999). The results showed that LR development in the *nia1nia2* mutant was more sensitive to NO_3^- inhibition than the wild type, supporting the conclusions of Scheible and colleagues that tissue NO_3^- concentrations may play a major role in the regulation of root development and partitioning of resources between shoot and root. Nevertheless, despite the evident importance of tissue NO_3^-, there is evidence in both *A.thaliana* (Zhang et al. 1999) and tobacco (Scheible et al. 1997b) that other, as yet unidentified, N metabolites may also contribute to generating the signals that regulate root development and shoot-root allocation.

Where is the tissue NO_3^- level monitored? In a series of tobacco lines with a range of different NR activities and grown at different $[NO_3^-]_0$ it was found that there was a strong correlation between the leaf NO_3^- content and the shoot:root ratio and that the correlation was much weaker when root NO_3^- levels were considered (Scheible et al. 1997b). The results of a split root experiment provided confirmation that the signal for inhibition of root growth comes from the shoot: when one half of the root system was grown in low $[NO_3^-]_0$ and the other in high $[NO_3^-]_0$, the growth rate of the former root sector was 8-20 times less than was seen when the entire root system of the same line was grown in low $[NO_3^-]_0$ (Scheible et al. 1997b). Thus the effects of a high NO_3^- treatment applied to one part of a root system can be transmitted (*via* the shoot) to the rest of the root system.

As yet there is no clear indication of how NO_3^- levels in the shoot are monitored, nor the nature of the signal that is transmitted to the root to modulate LR development. It is possible that whatever mechanisms are responsible for regulating NO_3^- uptake according to the nutrient status of the shoot might also be involved in regulating root development.

Figure 3 shows a model for how NO_3^- regulates LR development *via* two antagonistic pathways: the localised stimulatory effect, which acts on LR elongation and depends on the *ANR1* and *AXR4* gene products, and the systemic inhibitory effect, which is postulated to act on maturation of the LR meristem through an unidentified shoot-derived signal. This model provides a mechanism by which a plant could modulate the intensity and the pattern of its root branching to take into account both its N status and the external distribution of NO_3^- in the soil (Zhang et al. 1999; Zhang and Forde 2000).

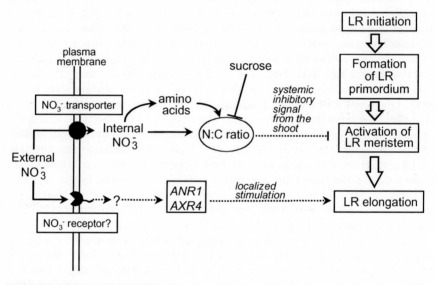

Fig. 3 : Model for regulation of LR growth and development in response to NO_3^- availability in *Arabidopsis*. Broken arrows indicate signalling steps; solid arrows indicate transport or metabolic steps, the open arrows indicate steps in the development of LRs. The NO_3^- sensor is depicted on the plasma membrane, but as yet neither the nature of the NO_3^- sensor(s) nor its cellular location is known. As indicated in the diagram, sucrose has been shown to partially alleviate the inhibitory effect of NO_3^-, indicating that the N:C balance may be important (Zhang et al. 1999). The model, which is modified from that presented by Zhang et al. (1999), is discussed further in the text.

Summary and Conclusions

Our understanding of the mechanism of NO_3^- uptake by plant roots has progressed a very long way in the past 10-15 years. This has come about largely through the application of sophisticated physiological methods for measuring *in vivo* rates of

NO_3^- influx and efflux, and through the application of genetic and molecular approaches which have led to the identification of some of the NO_3^- transporter genes. The physiological studies have established the existence and kinetic properties of at least four components to the NO_3^- uptake machinery: the cHATS, the iHATS, the LATS and the NO_3^- efflux system (which contributes in a negative way to net NO_3^- uptake) (p. 3). They have also revealed much about the differential ways in which the separate systems are regulated by the external NO_3^- supply and by the internal N status of the plant (p. 15).

The cloning of genes for the two different families of NO_3^- transporter (NRT1 and NRT2) has enabled us to begin to address some of the questions that were posed by the results of the physiological analysis. Thus we can now see that, in the main, the different influx systems are specified by different structural genes, with the NRT1 genes being associated with the LATS and the NRT2 genes with the iHATS (table 1). It is also becoming clear, however, that there are additional complexities which were not predictable from the kinetic studies. Firstly, there is the multiplicity of genes of each type: in those species for which data are available, both the NRT1 and NRT2 genes belong to small multigene families (p. 9). This raises the question of the role of each of the different gene products. Do they have different kinetic properties? Are they expressed in different cell types? Are they regulated differently? Secondly, the so-called constitutive LATS in *A.thaliana* contains at least one component that is NO_3^- – inducible. Thirdly, there is the unexpected finding that one of the NRT1 transporters (AtNRT1.1) appears to be a dual-affinity transporter, contributing to both the cHATS and the LATS (p. 11).

The molecular probes have also been used to demonstrate that transcriptional control (or at least control of mRNA abundance) plays a major role in the NO_3^- inducibility and feedback regulation of the transport systems (p. 13-15). The DNA sequences have allowed us to predict the primary structures and membrane topologies of the NRT1 and NRT2 transporters, which has revealed hydrophilic domains in the central loop and the C-terminus that may play a role in posttranslational regulation (p. 10-13 and review by Forde 2000).

There is clearly still much to do before we can say that we have a detailed understanding of the NO_3^- uptake system. The use of knockout mutants of *Arabidopsis* or transgenic lines carrying antisense genes should help to establish the physiological role of individual gene products. Analysis of lines overexpressing different NO_3^- transporter genes should also help to determine the extent to which, and under what conditions, the rate of NO_3^- influx limits the rate of NO_3^- assimilation. This will take us one step further towards the goal of generating transgenic crops with an enhanced ability to capture NO_3^- from the soil, so allowing reduced inputs of fertiliser N without prejudicing yields. However, what we already know about feedback regulation, and the role of N demand in regulating NO_3^- uptake (p. 15-20), gives us reason to expect that this will not be a straightforward task. A major challenge for the future will be to unlock the secrets of the regulatory mechanisms by which the plant monitors its internal N status and transduces these signals to modulate the expression and/or activity of the root NO_3^- transporters.

Acknowledgements. Work in B.T.'s and B.G.F.'s laboratories is funded in part by the European Union (contract no. BIO4-CT97-2231) in the Biotech programme of Framework IV. IACR receives grant-aided support from the Biotechnology and Biological Sciences Research Council.

References

Ageorges A, Morel M-H, Grouzis J-P (1996) High resolution 2-D electrophoresis of polypeptides of maize root plasma membrane: identification of polypeptides whose expression is regulated by nitrate. Plant Physiol Biochem 34: 863-870

Allen S, Raven JA (1987) Intracellular pH regulation in *Ricinus communis* grown with ammonium or nitrate as N source: the role of long-distance transport. J Exp Bot 38: 580-596

Amarasinghe BHRR, de Bruxelles GL, Braddon M, Onyeocha I, Forde BG, Udvardi MK (1998) Regulation of *GmNRT2* expression and nitrate transport activity in roots of soybean (*Glycine max*). Planta 206: 44-52

Améziane R, Deléens E, Noctor G, Morot-Gaudry JF, Limami MA (1997) Stage of development is an important determinant in the effect of nitrate on photoassimilate (^{13}C) partitioning in chicory (*Cichorium intybus*). J Exp Bot 48: 25-33

Aslam M, Travis RL, Huffaker RC (1992) Comparative kinetics and reciprocal inhibition of nitrate and nitrite uptake in roots of uninduced and induced barley (*Hordeum vulgare* L.) seedlings. Plant Physiol 99: 1124-1133

Aslam M, Travis RL, Huffaker RC (1993) Comparative induction of nitrate and nitrite uptake and reduction systems by ambient nitrate and nitrite in intact roots of barley (*Hordeum vulgare* L.) seedlings. Plant Physiol 102: 811-819

Aslam M, Travis RL, Rains DW (1996) Evidence for substrate induction of a nitrate efflux system in barley. Plant Physiol 112: 1167-1175

Barneix AJ, James DM, Watson EF, Hewitt EJ (1984) Some effects of nitrate abundance and starvation on metabolism and accumulation of nitrogen in barley (*Hordeum vulgare* L. cv. Sonja). Planta 162: 469-476

Behl R, Tischner R, Raschke K (1988) Induction of a high capacity nitrate uptake mechanism in barley roots prompted by nitrate uptake through a constitutive low-capacity mechanism. Planta 176: 235-240

Ben Zioni A, Vaadia Y, Lips SH (1971) Nitrate uptake by roots as regulated by nitrate reduction products of the shoot. Physiol Plant 24: 288-290

Bloom AJ, Sukrapanna SS, Warner RL (1992) Root respiration associated with ammonium and nitrate absorption and assimilation by barley. Plant Physiol 99: 1294-1301

Bowman DC, Paul JL, Davis WB (1989) Nitrate and ammonium uptake by nitrogen-deficient perennial ryegrass and Kentucky bluegrass turf. J Am Soc Hortic Sci 114: 421-426

Breteler H, Arnozis PA (1985) Effect of amino compounds on nitrate utilization by roots of dwarf bean. Phytochemistry 24: 653-658

Breteler H, Siegerist M (1984) Effect of ammonium on nitrate utilization by roots of dwarf bean. Plant Physiol 75: 1099-1103

Brownlee AG, Arst HN (1983) Nitrate uptake in *Aspergillus nidulans* and involvement of the third gene of the nitrate assimilation gene cluster. J Bacteriol 155: 1138-1146

Burns IG (1991) Short-term and long-term effects of a change in the spatial distribution of nitrate in the root zone on N uptake, growth and root development of young lettuce plants. Plant Cell Environ 14: 21-33

Cardenas-Navarro R, Adamowicz S, Robin P (1998) Diurnal nitrate uptake in young tomato (*Lycopersicon esculentum* Mill.) plants: test of a feedback-based model. J Exp Bot 49: 721-730

Chapin FS III, Clarkson DT, Lenton LR, Walter CHS (1988) Effect of nitrogen stress and abscisic acid on nitrate absorption and transport in barley and tomato. Planta 173: 340-351

Clarkson DT, Warner A (1979) Relationships between root temperature and transport of ammonium and nitrate ions by Italian and perennial ryegrass *Lolium multiflorum* and *Lolium perenne*. Plant Physiol 64: 557-561

Clement C, Hopper MJ, Jones LHP (1978) The uptake of nitrate by *Lolium* perenne from flowing nutrient solution. I. Effect of NO_3^- concentration. J Exp Bot 29: 453-464

Cogliatti DH, Clarkson DT (1983) Physiological changes in phosphate uptake by potato plants during development of, and recovery from, phosphate deficiency. Physiol Plant 58: 287-294

Cooper HD, Clarkson DT (1989) Cycling of amino-nitrogen and other nutrients between shoots and roots in cereals: a possible mechanism integrating shoot and root in the regulation of nutrient uptake. J Exp Bot 40: 753-762

Cram WJ (1988) Transport of nutrient ions across cell membranes *in vivo*. In: Tinker ED, Läuchli A (eds) Advances in plant nutrition. Praeger, New York, pp 1-53

Crawford NM, Glass ADM (1998) Molecular and physiological aspects of nitrate uptake in plants. Trends Plant Sci 3: 389-395

Daniel-Vedele F, Caboche M (1996) Molecular analysis of nitrate assimilation in higher plants. C R Acad Sci III: Sci Vie 319: 961-968.

Daniel-Vedele F, Filleur S, Caboche M (1998) Nitrate transport: a key step in nitrate assimilation. Curr Opin Plant Biol 1: 235-239

Davies DD (1986) The fine control of cytosolic pH. Physiol Plant 67: 702-706

Deane-Drummond CE, Glass ADM (1982) Nitrate uptake into barley (*Hordeum vulgare*) plants. A new approach using $^{36}ClO_3^-$ as an analog for NO_3^-. Plant Physiol 70: 50-54

Deane-Drummond CE, Glass ADM (1983a) Short-term studies on nitrate uptake into barley plants using ion-specific electrodes and $^{36}ClO_3^-$. I. Control of net uptake by NO_3^- efflux. Plant Physiol 73: 100-104

Deane-Drummond CE, Glass ADM (1983b) Short-term studies on nitrate uptake into barley plants using ion-specific electrodes and $^{36}ClO_3^-$. II. Regulation of NO_3^- efflux by NH_4^+. Plant Physiol 73: 105-110

Delhon P, Gojon A, Tillard P, Passama L (1995) Diurnal regulation of NO_3^- uptake in soybean plants. I. Changes in NO_3^- influx, efflux, and N utilization in the plant during the day/night cycle. J Exp Bot 46: 1585-1594

Delhon P, Gojon A, Tillard P, Passama L (1996) Diurnal regulation of NO_3^- uptake in soybean plants. IV. Dependence on current photosynthesis and sugar availability to the roots. J Exp Bot 47: 893-900

Devienne F, Mary B, Lamaze T (1994) Nitrate transport in intact wheat roots. I. Estimation of cellular fluxes and NO_3^- distribution using compartmental analysis from data of $^{15}NO_3^-$ efflux. J Exp Bot 45: 667-676

Dhugga KS, Waines JG, Leonard RT (1988) Correlated induction of nitrate uptake and membrane polypeptides in corn roots. Plant Physiol 87: 120-125

Dijkshoorn W (1962) Metabolic regulation of the alkaline effect of nitrate utilization in plants. Nature 194: 165-167

Doddema H, Otten H (1979) Uptake of nitrate by mutants of Arabidopsis thaliana, disturbed in uptake or reduction of nitrate. III. Regulation. Physiol Plant 45: 339-346

Doddema H, Telkamp GP (1979) Uptake of nitrate by mutants of Arabidopsis thaliana, disturbed in uptake or reduction of nitrate. II. Kinetics. Physiol Plant 45: 332-338

Drew MC (1973) Nutrient supply and the growth of the seminal root system in barley. I. The effect of nitrate concentration on the growth of axes and laterals. J Exp Bot 24: 1189-1202

Drew MC (1975) Comparison of the effects of a localized supply of phosphate, nitrate, ammonium and potassium on the growth of the seminal root system, and the shoot, in barley. New Phytol 75: 479-490

Drew MC, Saker LR (1975) Nutrient supply and the growth of the seminal root system of barley. II. Localized, compensatory increases in lateral root growth and rates of nitrate uptake when nitrate supply is restricted to only part of the root system. J Exp Bot 26: 79-90

Drew MC, Saker LR (1978) Nutrient supply and the growth of the seminal root system in barley. III. Compensatory increases in growth of lateral roots, and in rates of phosphate uptake in response to a localized supply of phosphate. J Exp Bot 29: 435-451

Drew MC, Saker LR (1984) Uptake and long-distance transport of phosphate, potassium and chloride in relation to internal ion concentrations in barley: evidence for a non-allosteric regulation. Planta 160: 500-507

Drew MC, Saker LR, Barber SA, Jenkins W (1984) Changes in the kinetics of phosphate and potassium absorption in nutrient-deficient barley roots measured by a solution-depletion technique. Planta 160: 490-499

Edwards JH, Barber SA (1976) Nitrogen flux into corn roots as influenced by shoot requirement. Agron J 68: 471-473

Ericsson T (1995) Growth and shoot:root ratio of seedlings in relation to nutrient availability. Plant Soil 168-169: 205-214

Evans ML, Ishikawa H, Estelle MA (1994) Responses of Arabidopsis roots to auxin studied with high temporal resolution: comparison of wild-type and auxin-response mutants. Planta 194: 215-222

Filleur S, Daniel-Vedele F (1999) Expression analysis of a high-affinity nitrate transporter isolated from *Arabidopsis thaliana* by differential display. Planta 207: 461-469

Forde BG (2000) Nitrate transporters in plants: structure, function and regulation. Biochim Biophys Acta 1465: 219-235

Forde BG, Clarkson DT (1999) Nitrate and ammonium nutrition of plants: physiological and molecular perspectives. Adv Bot Res 30: 1-90

Galván A, Quesada A, Fernández E (1996) Nitrate and nitrite are transported by different specific transport systems and by a bispecific transporter in *Chlamydomonas reinhardtii*. J Biol Chem 271: 2088-2092

Gastal F, Saugier B (1989) Relationships between nitrogen uptake and carbon assimilation in whole plants of tall fescue. Plant Cell Environ 12: 407-418

Glass ADM, Siddiqi MY (1995) Nitrogen absorption by plant roots. In: Srivastava HS, Singh RP (eds) Nitrogen nutrition in higher plants. Associated Publishing Co, New Delhi, pp 21-56

Glass ADM, Shaff JE, Kochian LV (1992) Studies of the uptake of nitrate in barley. IV. Electrophysiology. Plant Physiol 99: 456-463

Granato TC, Raper CD Jr (1989) Proliferation of maize (*Zea mays* L.) roots in response to localized supply of nitrate. J Exp Bot 40: 263-275

Grouzis J-P, Pouliquin P, Rigaud J, Grignon C, Gibrat R (1997) In vitro study of passive nitrate transport by native and reconstituted plasma membrane vesicles from corn root cells. Biochim Biophys Acta 1325: 329-342

Guy M, Zabala G, Filner P (1988) The kinetics of chlorate uptake by XD tobacco cells. Plant Physiol 86: 817-821

Hackett C (1972) A method of applying nutrients locally to roots under controlled conditions, and some morphological effects of locally applied nitrate on the branching of wheat roots. Aust J Biol Sci 25: 1169-1180

Hänisch ten Cate CH, Breteler H (1981) Role of sugars in nitrate utilization by roots of dwarf bean. Plant Physiol 52: 129-135

Hanson JB (1978) Application of the chemiosmotic hypothesis to ion transport across the root. Plant Physiol 62: 402-405

Hatzfeld Y, Saito K (1999) Identification of two putative nitrate transporters highly homologous to CHL1 from *Arabidopsis* (PGR 99-018). Plant Physiol 119: 805

Hobbie LJ (1998) Auxin: molecular genetic approaches in *Arabidopsis*. Plant Physiol 36: 91-102

Hole DJ, Emran AM, Fares Y, Drew M (1990) Induction of nitrate transport in maize roots, and kinetics of influx, measured with nitrogen-13. Plant Physiol 93: 642-647

Huang NC, Chiang CS, Crawford NM, Tsay YF (1996) *CHL1* encodes a component of the low-affinity nitrate uptake system in *Arabidopsis* and shows cell type-specific expression in roots. Plant Cell 8: 2183-2191

Jackson RB, Manwaring JH, Caldwell MM (1990) Rapid physiological adjustment of roots to localized soil enrichment. Nature 344: 58-60

Jackson WA, Flesher D, Hageman RH (1973) Nitrate uptake by dark-grown corn seedlings. Some characteristics of apparent induction. Plant Physiol 51: 120-127

Jackson WA, Kwick KD, Volk RJ, Butz RG (1976) Nitrate influx and efflux by intact wheat seedlings: effects of prior nitrate nutrition. Planta 132: 149-156

Keltjens WG, Nijenstein JH (1987) Diurnal variations in uptake, transport and assimilation of NO_3^- and efflux of OH^- in maize plants. J Plant Nutr 10: 887-900

King BJ, Siddiqi MY, Glass ADM (1992) Studies of the uptake of nitrate in barley. V. Estimation of root cytoplasmic nitrate concentration using nitrate reductase activity – implications for nitrate influx. Plant Physiol 99: 1582-1589

King BJ, Siddiqi MY, Ruth TJ, Warner RL, Glass ADM (1993) Feedback regulation of nitrate influx in barley roots by nitrate, nitrite, and ammonium. Plant Physiol 102: 1279-1286

Kirkby EA, Mengel K (1967) Ionic balance in different tissues of the tomato plant in relation to nitrate, urea or ammonium nutrition. Plant Physiol 42: 6-14

Kosola KR, Bloom AJ (1996) Chlorate as a transport analog for nitrate absorption by roots of tomato. Plant Physiol 110: 1293-1299

Krapp A, Fraisier V, Scheible WR, Quesada A, Gojon A, Stitt M, Caboche M, Daniel-Vedele F (1998) Expression studies of Nrt2:1Np, a putative high-affinity nitrate transporter: evidence for its role in nitrate uptake. Plant J 14: 723-731

Kronzucker HJ, Glass ADM, Siddiqi MY (1995a) Nitrate induction in spruce: an approach using compartmental analysis. Planta 196: 683-690

Kronzucker HJ, Siddiqi MY, Glass ADM (1995b) Compartmentation and flux characteristics of nitrate in spruce. Planta 196: 674-682

Kronzucker HJ, Siddiqi MY, Glass ADM (1995c) Kinetics of NO_3^- influx in spruce. Plant Physiol 109: 319-326

Kronzucker HJ, Glass ADM, Siddiqi MY (1999) Inhibition of nitrate uptake by ammonium in barley. Analysis of component fluxes. Plant Physiol 120: 283-291

Lainé P, Ourry A, Boucaud J (1995) Shoot control of nitrate uptake rates by roots of Brassica napus L.: effects of localized nitrate supply. Planta 196: 77-83

Lainé P, Ourry A, Boucaud J, Salette J (1998) Effects of a localized supply of nitrate on NO_3^- uptake rate and growth of roots in Lolium multiflorum Lam. Plant Soil 202: 61-67

Lappartient AG, Touraine B (1996) Demand-driven control of root ATP sulfurylase activity and SO_4^{2-} uptake in intact canola. The role of phloem-translocated glutathione. Plant Physiol 111: 147-157

Lappartient AG, Vidmar JJ, Leustek T, Glass ADM, Touraine B (1999) Inter-organ signaling in plants: regulation of ATP sulfurylase and sulfate transporter genes expression in roots mediated by phloem-translocated compound. Plant J 18: 89-95

Larsson M (1992) Translocation of nitrogen in osmotically stressed wheat seedlings. Plant Cell Environ 15: 447-453

Lauter FR, Ninnemann O, Bucher M, Riesmeier JW, Frommer WB (1996) Preferential expression of an ammonium transporter and of two putative nitrate transporters in root hairs of tomato. Proc Natl Acad Sci USA 93: 8139-8144

Le Bot J, Kirkby EA (1992) Diurnal uptake of nitrate and potassium during the vegetative growth of tomato plants. J Plant Nut 15: 247-264

Lee RB (1993) Control of net uptake of nutrients by regulation of influx in barley plants recovering from nutrient deficiency. Ann Bot 72: 223-230

Lee RB, Clarkson DT (1986) Nitrogen-13 studies of nitrate fluxes in barley roots. I. Compartmental analysis from measurements of ^{13}N efflux. J Exp Bot 37: 1753-1767

Lee RB, Drew MC (1986) Nitrogen-13 studies of nitrate fluxes in barley roots. II. Effect of plant N-status on the kinetic parameters of nitrate influx. J Exp Bot 185: 1768-1779

Lee RB, Rudge K (1986) Effects of nitrogen deficiency on the absorptions of nitrate and ammonium by barley plants. Ann Bot 57: 471-486

Lee RB, Purves JV, Ratcliffe RG, Saker LR (1992) Nitrogen assimilation and the control of ammonium and nitrate absorption by maize roots. J Exp Bot 43: 1385-1396

Lejay L, Tillard P, Lepetit M, Olive FD, Filleur S, Daniel-Vedele F, Gojon A (1999) Molecular and functional regulation of two NO_3^- uptake systems by N- and C-status of Arabidopsis plants. Plant J 18: 509-519

Lemaire G, Salette J (1984a) Relation entre dynamique de croissance et dynamique de prélèvement d'azote pour un peuplement de graminées fourragères. I. Étude de l'effet du milieu. Agronomie 4: 423-430

Lemaire G, Salette J (1984b) Relation entre dynamique de croissance et dynamique de prélèvement d'azote pour un peuplement de graminées fourragères. II. Étude de la variabilité entre génotypes. Agronomie 4: 431-436

Liu K-H, Huang C-Y, Tsay Y-F (1999) CHL1 is a dual-affinity nitrate transporter of Arabidopsis involved in multiple phases of nitrate uptake. Plant Cell 11: 865-874

Macduff JH, Hopper MJ, Wild A, Trim FE (1987) Comparison of the effects of root temperature on nitrate and ammonium nutrition of oilseed rape (Brassica napus L.) in flowing solution culture. II. Cation-anion balance. J Exp Bot 38: 1589-1602

MacKown CT, McClure PR (1988) Development of accelerated net nitrate uptake rate. Plant Physiol 87: 162-166

Malamy JE, Benfey PN (1997) Down and out in Arabidopsis: the formation of lateral roots. Trends Plant Sci 2: 390-396.

Marschner H, Röhmeld V (1983) In vivo measurement of root-induced pH changes at the soil-root interface: effect of plant species and nitrogen source. Z Pflanzenphysiol 111: 241-251

McClure PR, Kochian LV, Spanswick RM, Shaff JE (1990) Evidence for cotransport of nitrate and protons in maize roots. 1. Effects of nitrate on the membrane potential. Plant Physiol 93: 281-289

Miller AJ, Smith SJ (1996) Nitrate transport and compartmentation in cereal root cells. J Exp Bot 47: 843-854

Miller AJ, Zhen RG (1991) Measurement of intracellular nitrate concentrations in *Chara* using nitrate-sensitive microelectrodes. Planta 184: 47-52

Minotti PL, Williams DC, Jackson WA (1969) Nitrate uptake by wheat as influenced by ammonium and other cations. Crop Sci 9: 9-14

Morgan MA, Volk RJ, Jackson WA (1973) Simultaneous influx and efflux of nitrate during uptake by perennial ryegrass. Plant Physiol 51: 267-272

Morgan MA, Volk RJ, Jackson WA (1985) p-Fluorophenylalanine-induced restriction of ion uptake and assimilation by maize roots. Plant Physiol 77: 718-722

Muller B, Touraine B (1992) Inhibition of NO_3^- uptake by various phloem-translocated amino acids in soybean seedlings. J Exp Bot 43: 617-623

Muller B, Tillard P, Touraine B (1995) Nitrate fluxes in soybean seedling roots and their response to amino acids: an approach using ^{15}N. Plant Cell Environ 18: 1267-1279

Neyra CA, Hageman RH (1976) Relationships between carbon dioxide, malate and nitrate accumulation and reduction in corn (*Zea mays* L.) seedlings. Plant Physiol 58: 726-730

Öostindier-Braaksma F, Feenstra W (1973) Isolation and characterisation of chlorate-resistant mutants of *Arabidopsis thaliana*. Mutat Res 19: 175-185

Oscarson P, Ingemarsson B, Af Uglass M, Larsson CM (1987) Short-term studies of NO_3^- uptake in *Pisum* using $^{13}NO_3^-$. Planta 170: 550-555

Pace GM, McClure PR (1986) Comparison of nitrate uptake kinetic parameters across maize inbred lines. J Plant Nutr 9: 1095-1111

Pouliquin P, Grouzis J-P, Gibrat R (1999) Electrophysiological study with oxonol VI of passive NO_3^- transport by isolated plant root plasma membrane. Biophys J 76: 360-373

Presland MR, MacNaughton GS (1984) Whole plant studies using radioactive 13-nitrogen. II. A compartmental model for the uptake and transport of nitrate ions by *Zea mays*. J Exp Bot 35: 1277-1288

Quesada A, Galván A, Fernández E (1994) Identification of nitrate transporter genes in *Chlamydomonas reinhardtii*. Plant J 5: 407-419.

Quesada A, Krapp A, Trueman LJ, Daniel-Vedele F, Fernández E, Forde BG, Caboche M (1997) PCR-identification of a *Nicotiana plumbaginifolia* cDNA homologous to the high-affinity nitrate transporters of the crnA family. Plant Mol Biol 34: 265-274

Quesada A, Hidalgo J, Fernández E (1998) Three Nrt2 genes are differentially regulated in *Chlamydomonas reinhardtii*. Mol Gen Genet 258: 373-377

Rao KP, Rains DW (1976) Nitrate absorption by barley. 1. Kinetics and energetics. Plant Physiol 57: 55-58

Raper CD, Osmond DL, Wann M, Weeks WW (1978) Interdependence of root and shoot activities in determining nitrogen uptake rate of roots. Bot Gaz 139: 289-294

Raven JA (1977) H^+ and Ca^{2+} in phloem and symplast: relation of relative immobility of the ions to the cytoplasmic nature of the transport paths. New Phytol 79: 465-480

Raven JA, Smith FA (1976) Nitrogen assimilation and transport in vascular land plants in relation to intracellular pH regulation. New Phytol 76: 415-431

Robin P, Conejero G, Passama L, Salsac L (1983) Evaluation de la fraction métabolisable du nitrate par la mesure *in situ* de sa réduction. Physiol Vég 21: 115-122

Robinson D (1991) What limits nitrate uptake from soil? Plant Cell Environ 14: 77-85

Robinson D (1994) The responses of plants to non-uniform supplies of nutrients. New Phytol 127: 635-674

Robinson D, Hodge A, Griffiths BS, Fitter AH (1999) Plant root proliferation in nitrogen-rich patches confers competitive advantage. Proc R Soc Lond B 266: 431-435

Rodgers CO, Barneix AJ (1989) The effect of N-deprivation on nitrate uptake and growth rate of two wheat cultivars selected for different fertility levels. Plant Physiol Biochem 27: 387-392

Rufty TW, Thomas JF, Rimmler JL, Campbell WH, Volk RJ (1986) Intercellular localization of nitrate reductase in roots. Plant Physiol 82: 675-680

Ruiz-Cristin J, Briskin DP (1991) Characterization of a H^+/NO_3^- symport associated with plasma membrane vesicles of maize roots using $^{36}ClO_3^-$ as a radiotracer analog. Arch Biochem Biophys 285: 74-82

Saier MH (1994) Computer-aided analyses of transport protein sequences: gleaning evidence concerning function, structure, biogenesis and evolution. Microbiol Rev 58: 71-93.

Sattelmacher B, Thoms K (1995) Morphology and physiology of the seminal root system of young maize (*Zea mays* L.) plants as influenced by a locally restricted nitrate supply. Z Pflan zenemacher Bodenkd 158: 493-497.

Sattelmacher B, Gerendas J, Thoms K, Bruck H, Bagdady NH (1993) Interaction between root growth and mineral nutrition. Environ Exp Biol 33: 63-73.

Scaife A (1989) A pump/leak/buffer model for plant nitrate uptake. Plant Soil 114: 139-141

Scheible WR, González-Fontes A, Lauerer M, Müller-Röber B, Caboche M, Stitt M (1997a) Nitrate acts as a signal to induce organic acid metabolism and repress starch metabolism in tobacco. Plant Cell 9: 783-798

Scheible WR, Lauerer M, Schulze ED, Caboche M, Stitt M (1997b) Accumulation of nitrate in the shoot acts as a signal to regulate shoot-root allocation in tobacco. Plant J 11: 671-691

Scheurwater I, Cornelissen C, Dictus F, Welschen R, Lambers H (1998) Why do fast- and slow-growing grass species differ so little in their rate of root respiration, considering the large differences in rate of growth and ion uptake? Plant Cell Environ 21: 995-1005

Scheurwater I, Clarkson DT, Purves JV, Van Rijt G, Saker LR, Welschen R, Lambers H (1999) Relatively large nitrate efflux can account for the high respiratory costs for nitrate transport in slow-growing grass species. Plant Soil 215: 123-134

Shore P, Sharrocks AD (1995) The MADS-box family of transcription factors. Eur J Biochem 229: 1-13

Siddiqi MY, Glass ADM, Ruth TJ, Fernando M (1989) Studies of the regulation of nitrate influx by barley seedlings using $^{13}NO_3^-$. Plant Physiol 90: 806-813

Siddiqi MY, Glass ADM, Ruth TJ, Rufty T (1990) Studies of the uptake of nitrate in barley. I. Kinetics of $^{13}NO_3^-$ influx. Plant Physiol 93: 1426-1432

Siddiqi MY, Glass ADM, Ruth TJ (1991) Studies of the uptake of nitrate in barley. III. Compartmentation of NO_3^-. J Exp Bot 42: 1455-1463

Siddiqi MY, King BJ, Glass ADM (1992) Effects of nitrite, chlorate and chlorite on nitrate uptake and nitrate reductase activity. Plant Physiol 100: 644-650

Simpson RJ, Lambers H, Dalling MJ (1982) Translocation of nitrogen in a vegetative wheat plant (*Triticum aestivum*). Physiol Plant 56: 11-17

Smith FW, Ealing PM, Dong B, Delhaize E (1997a) The cloning of two Arabidopsis genes belonging to a phosphate transporter family. Plant J 11: 83-92

Smith FW, Hawkesford MJ, Ealing PM, Clarkson DT, Vanden Berg PJ, Belcher AR, Warrilow GS (1997b) Regulation of expression of a cDNA from barley roots encoding a high affinity sulphate transporter. Plant J 12: 875-884

Smith IK, Cheema HK (1985) Sulphate transport into plants and excised roots of soybean (*Glycine max* L.). Ann Bot 56: 219-224

Steiner HY, Song W, Zhang L, Naider F, Becker JM, Stacey G (1994) An *Arabidopsis* peptide transporter is a member of a new class of membrane transport proteins. Plant Cell 6: 1289-1299

Swiader JM, Chyan Y, Splitstoesser WE (1991) Genotypic differences in nitrogen uptake, dry matter production, and nitrogen distribution in pumpkins (*Cucurbita moschata* Poir.). J Plant Nut- 14: 511-524

Teyker RH, Jackson WA, Volk RJ, Moll R (1988) Exogenous $^{15}NO_3^-$ influx and endogenous $^{14}NO_3^-$ efflux by two maize (*Zea mays* L.) inbreds during nitrogen deprivation. Plant Physiol 86: 778-781

Thibaud J-B, Grignon C (1981) Mechanism of nitrate uptake in corn roots. Plant Sci Lett 22: 279-289

Tillard P, Passama L, Gojon A (1998) Are phloem amino acids involved in the shoot to root control of NO_3^- uptake in *Ricinus communis* plants? J Exp Bot 49: 1371-1379

Tompkins GA, Jackson WA, Volk RJ (1978) Accelerated nitrate uptake in wheat seedlings. Effects of ammonium and nitrite pretreatments and of 6-methylpurine and puromycin. Physiol Plant 43: 166-171

Touraine B, Glass ADM (1997) NO_3^- and ClO_3^- fluxes in the *chl1-5* mutant of *Arabidopsis thaliana*. Does the CHL1-5 gene encode a low-affinity NO_3^- transporter? Plant Physiol 114: 137-144

Touraine B, Grignon C (1982) Energetic coupling of nitrate secretion into the xylem of corn roots. Physiol Vég 20: 33-39

Touraine B, Grignon N, Grignon C (1988) Charge balance in NO_3^- – fed soybean. Estimation of K^+ and carboxylate recirculation. Plant Physiol 88: 605-612

Touraine B, Muller B, Grignon C (1992) Effect of phloem-translocated malate on NO_3^- uptake by roots of intact soybean plants. Plant Physiol 93: 1118-1123

Trewavas AJ (1983) Nitrate as a plant hormone. In: Jackson MB (ed) British plant growth regulator group monograph 9. British Plant Growth Regulator Group, Oxford, pp 97-110

Trueman LJ, Onyeocha I, Forde BG (1996a) Recent advances in the molecular biology of a family of eukaryotic high affinity nitrate transporters. Plant Physiol Biochem 34: 621-627

Trueman LJ, Richardson A, Forde BG (1996b) Molecular cloning of higher plant homologues of the high affinity nitrate transporters of *Chlamydomonas reinhardtii* and *Aspergillus nidulans*. Gene 175: 223-231

Tsay YF, Schroeder JI, Feldmann KA, Crawford NM (1993) The herbicide sensitivity gene CHL1 of *Arabidopsis* encodes a nitrate-inducible nitrate transporter. Cell 72: 705-713

Ullrich WR, Novacky A (1981) Nitrate-dependent membrane potential changes and their induction in *Lemna gibba*. Plant Sci Lett 22: 211-217

Unkles SE, Hawker KL, Grieve C, Campbell EI, Montague P, Kinghorn JR (1991) crnA encodes a nitrate transporter in *Aspergillus nidulans*. Proc Natl Acad Sci USA 88: 204-208

Unkles SE, Hawker KL, Grieve C, Campbell EI, Montague P, Kinghorn JR (1995) crnA encodes a nitrate transporter in Aspergillus nidulans (correction). Proc Natl Acad Sci USA 92: 3076-3076

Van der Leij M, Smith SJ, Miller AJ (1998) Remobilization of vacuolar stored nitrate in barley root cells. Planta 205: 64-72

von Wiren N, Gazzarrini S, Frommer WB (1997) Regulation of mineral nitrogen uptake in plants. Plant Soil 196: 191-199

Wang RC, Liu D, Crawford NM (1998) The *Arabidopsis* CHL1 protein plays a major role in high-affinity nitrate uptake. Proc Natl Acad Sci USA 95: 15134-15139

Wieneke J (1994) Nitrate ($^{13}NO_3^-$) flux studies and response to tungstate treatments in wild type barley and nitrate reductase-deficient mutant. J Plant Nutr 17: 127-146

Wiersum LK (1958) Density of root branching as affected by substrate and separate ions. Acta Bot Neerl 7: 174-190

Yamashita T, Shimada S, Guo W, Sato K, Kohmura E, Hayakawa T, Takagi T, Tohyama M (1997) Cloning and functional expression of a brain peptide/histidine transporter. J Biol Chem 272: 10205-10211

Zhang H, Forde BG (1998) An *Arabidopsis* MADS box gene that controls nutrient-induced changes in root architecture. Science 279: 407-409

Zhang H, Jennings AJ, Barlow PW, Forde BG (1999) Dual pathways for regulation of root branching by nitrate. Proc Natl Acad Sci USA 96: 6529-6534

Zhang H, Forde BG (2000) Regulation of lateral root development in *Arabidopsis* by nitrate availability. J Exp Bot 51: 51-59

Zhen RG, Kyoro HW, Leigh RA, Tomos AD, Miller AJ (1991) Compartmental nitrate concentrations in barley root cells measured with nitrate-selective microelectrodes and by single-cell sap sampling. Planta 185: 356-361

Zhen RG, Smith SJ, Miller AJ (1992) A comparison of nitrate-selective microelectrodes made with different nitrate sensors and the measurement of intracellular nitrate activities in cells of excised barley roots. J Exp Bot 43: 131-138

Zhou JJ, Theodoulou FL, Muldin I, Ingemarsson B, Miller AJ (1998) Cloning and functional characterization of a *Brassica napus* transporter that is able to transport nitrate and histidine. J Biol Chem 273: 12017-12023

Zhuo D, Okamoto M, Vidmar JJ, Glass ADM (1999) Regulation of a putative high-affinity nitrate transporter (*NRT2;1At*) in roots of *Arabidopsis thaliana*. Plant J 17: 563-568

Nitrate Reduction and signalling

Christian MEYER[1] and Mark STITT[2]

The ability to use nitrate as sole nitrogen source to sustain growth is a property shared by some bacteria and fungi and by most algae and plants. The biochemical pathway responsible for nitrate assimilation seems to be the same in both prokaryotes and eukaryotes. Soil nitrate is the preferred inorganic nitrogen source for many wild or cultivated plants. It seems, indeed, that most plants, unlike bacteria or fungi, which preferentially use ammonium as a nitrogen source, show better growth when nitrate is present. The generalisation of the use of chemical fertilizers has allowed a tremendous increase in crop yield during the past 50 years. However, there is now a growing concern about the effect of nitrate, on both the environment and human health (Chap. 8). Indeed, nitrate can accumulate in high concentrations in the leaves of edible plants or in drinking water. Once taken up from the soil by an active process (see Chap. 1.1), nitrate is either stored in the plant root system or translocated to aerial parts via the xylem. High concentrations of nitrate can be found in vacuoles and it seems that nitrate, beside its role as a nutrient, participates in the maintenance of the plant osmoticum. The first committed step of the nitrate assimilation pathway is the reduction of nitrate to nitrite, catalysed by assimilatory nitrate reductase (NR, Fig. 1). In some bacteria, dissimilatory nitrate reduction, in which nitrate replaces oxygen as a terminal electron acceptor for respiration, is also found, but the utilisation of nitrate for respiration in anaerobic plant cells is still debated. The nitrite formed by NR activity is translocated to the chloroplast, where it is further reduced to ammonium by nitrite reductase (NiR). Ammonium is subsequently incorporated into the amino acid pool through the action of glutamine synthetase (GS) and glutamate synthase (GOGAT) (see Chap. 2.2 and Fig. 1).

NR is one of the most-studied plant enzymes, since it is considered as a controlling step in nitrate assimilation and is a complex enzyme that is regulated by a number of different processes. Since the early work of Evans and Nason in 1953, who first reported on the isolation of NR from plant tissues, an impressive number of papers have been devoted to studies on NR. NiR has attracted less attention, although it also seems to be finely regulated. Many aspects of nitrate assimilation

1. Laboratoire de Nutrition Azotée des Plantes, INRA, 78026 Versailles, France.
E-mail: meyer@versailles.inra.fr
2. Botanisches Institut, Universität Heidelberg, Im Neuenheimer Feld 360, 69120 Heidelberg, Germany.
E-mail: mstitt@botanikl.bot.uni-heidelberg.de

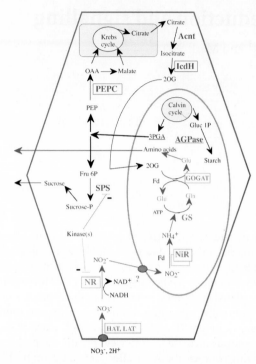

Fig. 1 : Relationships between the nitrate assimilation pathway and carbon metabolism. Enzymes that are inducible by nitrate are boxed, those which are repressed by nitrate are underlined. HAT, LAT respectively high and low affinity nitrate transporters; SPS sucrose phosphate synthase; PEPC phosphoenolpyruvate carboxylase; IcdH isocitrate dehydrogenase; Acnt aconitase; AGPase ADP-glucose pyrophosphorylase; 2OG 2-oxoglutarate; Fd Ferredoxin

have been reviewed in recent years, including biochemistry (Wray 1989; Campbell 1999), regulation (Crawford 1995; Daniel-Vedele and Caboche 1996, MacKintosh 1998; Kaiser et al. 1999) and genetics (Crawford and Arst 1993; Mendel 1997).

This chapter will deal only with the enzymatic steps of the nitrate assimilation pathway, i.e. the reduction of nitrate to nitrite by NR and of nitrite to ammonium by NiR, and will concentrate on the major recent developments in the biochemistry and regulation of NR and NiR.

Nitrate Reduction

Higher plants NR catalyses the following reaction :

$$NO_3^- + NAD(P)H + H^+ \quad \rightarrow \quad NO_2^- + NAD(P)^+ + H_2O.$$

Two forms of NR have been found in plants and the most common form is a NADH-specific NR (EC 1.6.6.1). A NAD(P)H-bispecific NR (EC 1.6.6.2) has also been described in several plant species, either as a second isoform along with NADH-specific NR, as in maize, barley, rice and soybean (Streit et al. 1987; Kleinhofs and Warner 1990), or as the sole isoform, as in *Betula pendula* (Friemann et al. 1991), *Psophocarpus tetragonolobus* and *Erythrina senegalensis* (Kleinhofs and Warner 1990). A third form [NADPH-specific (EC 1.6.6.3)] has not been found in higher plants but is present in fungi and mosses (Padidam et al. 1991). Although the intracellular localisation of NR is still debated, the NR enzyme is thought to be cytosolic (Solomonson and Barber 1990). The NR protein is organised in three domains (Fig. 2), housing the three prosthetic groups of the enzyme, namely: FAD, a haem and a molybdenum cofactor (Fig. 3; Redinbaugh and Campbell 1985). They are used in that order to transfer electrons from NAD(P)H to nitrate.

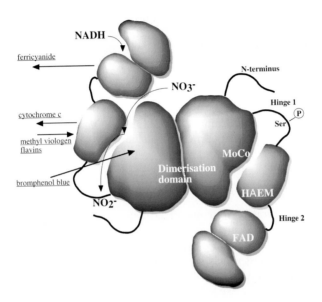

Fig. 2: Structural model of higher plant NR. This model is derived from the structures of the NR FAD domain and from the structure of suphite oxidase. Artificial electron donors and acceptors are *underlined*. The NR regulatory phosphorylation site is shown in *Hinge 1*

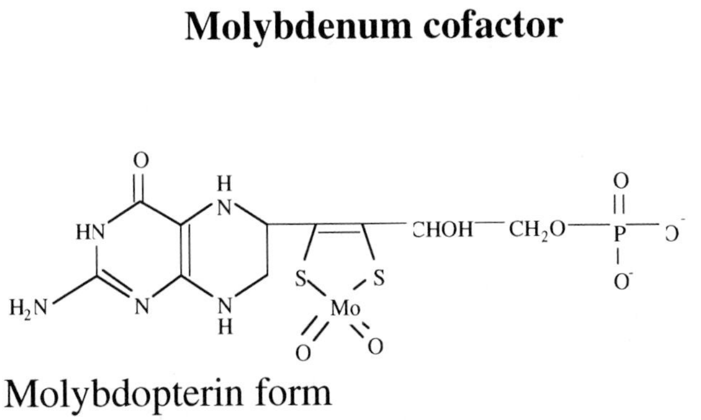

Molybdenum cofactor

Molybdopterin form

Fig. 3: Structure of isolated molybdenum cofactor

Nitrate reductase from higher plants appears to be a dimer of identical subunits of the size 100-120 kDa (Solomonson and Barber 1990). The localisation of each enzyme domain is defined by sequence homology with other enzymes: cytochrome b_5 reductase for the FAD domain, cytochrome b_5 for the haem domain and sulphite oxidase for the molybdenum cofactor (MoCo) domain (Calza et al. 1987; Crawford et al. 1988; Neame and Barber 1989). The N-terminal domain of NR binds MoCo, the central part binds haem, and the C-terminal region binds FAD (Fig. 2). Limited proteolytic analysis confirmed the localisation of the domains and suggested that the domains are linked together by protease sensitive hinges (Fido 1991; Shiraishi et al. 1992).

Besides nitrate, NR is also able to reduce chlorate, a structural analogue of nitrate, to the toxic compound chlorite. In fact, it seems that NR has a rather broad substrate specificity: it has been shown that, for example, NAD(P)H: NR from soybean is able to reduce nitrite to nitrogen oxide(s) (Solomonson and Barber 1990), but the physiological significance of this reducing activity remains to be determined. Very recently, it has also been suggested that NADH:NR from maize is capable of reducing nitrite to NO, which opens up the possibility that NR is somehow involved in NO production in plants (Yamasaki et al. 1999). NO has already been implicated in the regulation of several plant processes, including cell damage and hypersensitive response (Van Camp et al. 1998).

Nitrate Reductase Mutants and Genes

Since the first isolation of mutants affected in nitrate reduction in *Arabidopsis thaliana* (Oostinder-Braaksma and Feenstra 1973), many NR-deficient mutants have been isolated from several plant species (Pelsy and Caboche 1992; Crawford and Arst 1993). The most widely used method to select mutants affected in nitrate assimilation has been on the basis of chlorate resistance, either at the seed level or at the plant cell level (protoplasts). Mutants were selected after mutagenesis of seeds or protoplasts, or appeared spontaneously in protoplast cultures (Pelsy and Caboche 1992). In *Nicotiana tabacum* and *N. plumbaginifolia*, some of the mutations were the result of transposable element insertions and the disrupted NR genes have thus acted as transposable element traps, allowing the identification of mobile elements in the genome of these two species (Grandbastien et al. 1989; Meyer et al. 1994).

NR-defective mutants have been best characterised in *Nicotiana* sp., *Hordeum vulgare* and *A. thaliana* and were shown to belong to two classes: mutations affecting the NR apoenzyme gene(s) (*nia* mutants), or mutations affecting one of the several genes involved in the biosynthesis of MoCo (*cnx* mutants). This biochemical characterisation has been confirmed by a genetic classification into separate complementation groups.

The collections of *nia* mutants turned out to be an invaluable tool, not only for the study of NR structure/function relationships (see Following section), but also for the molecular biology of the *Nia* locus. The number of functional *Nia* loci was found to be different in plants: in *N. plumbaginifolia* all *nia* mutants isolated were shown to belong to one complementation group, indicating a single *Nia* locus in this species (Gabard et al. 1987); in tobacco (amphidiploid) two functional homologous *Nia* loci were found (Müller and Mendei 1989; Vaucheret et al. 1989); in barley and *Arabi-*

dopsis, two *Nia* loci (respectively *Nar1* and *Nar7*, *Nia2* and *Nia1*) have been identified and cloned (Cheng et al. 1988; Schnorr et al. 1991, Wilkinson and Crawford 1993). In barley, the *nar1* mutants were first isolated and shown to lack the NADH-NR activity normally found in roots and leaves of wild-type plants, the *nar7* mutant lacks the NAD(P)H-bispecific NR found in the roots of barley. In *Arabidopsis*, both genes encode NADH:NR and it seems that the *Nia2* gene is responsible for 90% of the total NR activity (Wilkinson and Crawford 1993).

Several NR isoforms are also present in maize and rice, among other species; but the question which remains so far unanswered is what is the physiological significance of these different NR isoforms? For instance the *Arabidopsis* G5 mutant, in which the *Nia2* gene has been deleted, exhibited only about 10% residual NR activity (due to the *Nia1* gene expression) and showed no obvious phenotypic differences when compared to the wild type, apart from a decreased chlorate sensitivity (Wilkinson and Crawford 1993).

It has also been observed that intragenic complementation can occur at the *Nia* locus. This has been demonstrated in *N. tabacum*, *C. reinhardtii* and *N. plumbaginifolia* (Müller and Mendel 1989; Fernandez and Cardenas 1989; Pelsy and Gonneau 1991), and suggests that electrons can flow from one subunit to another, an observation that could be the result of either the formation of heterodimers, or the interaction between different nitrate reductase molecules.

Today, there are more than 30 NR sequences from higher plants, algae or fungi in the nucleotide sequences databases. Comparisons of the deduced protein sequences reveal a high degree of conservation in the putative MoCo-binding domain (Meyer et al. 1995), as well as in the haem and FAD-binding domains. Three regions of the NR protein seem to be less conserved: (1) an N-terminal extension which is poorly conserved, in both length and sequence, between higher-plant NR protein sequences (Pigaglio et al. 1999); (2) two hinge regions separating, respectively, the MoCo and the haem-binding domains and the haem and FAD/NADH-binding modules (Rouzé and Caboche 1992; Campbell 1999). It can be hypothesised that the NR enzyme resulted from the combination of protein units, which have evolved independently. Indeed, it seems that several other soluble redox enzymes, like flavocytochrome b_2 from yeast or sulphite oxidase from animals, have used the same units for their building. Nevertheless, the positions of the intronic sequences which interrupt the higher plants NR genes, do not correspond to boundaries between these functional protein units. Although the positions of the introns in the NR genes are conserved between plant species, only the intron number varies (between one and four).

At least six complementation groups were found, either by somatic hybridisation or by sexual crosses, between MoCo-deficient *cnx* mutants (Gabard et al. 1988; Kleinhofs and Warner 1990; reviewed by Mendel (1997)). The *cnx* mutants often show a pleiotropic phenotype, due to the loss of all MoCo-containing enzymes (including xanthine dehydrogenase, various aldehyde oxidases and NR) and can only be kept alive either *in vitro* with reduced nitrogen sources, or grafted on wild-type stocks. In the *cnxA* mutants of *N. plumbaginifolia*, NR activity is restorable by adding high concentrations of molybdate to the growth medium (Mendel and Müller 1985; Gabard et al. 1988). This led to the hypothesis that the *CnxA* gene product is involved in the insertion of molybdenum into the molybdopterin (Mendel and Müller 1985). This gene corresponds to the *Cnx1* gene of *Arabidopsis* which has been cloned by complementation of the *Escherichia coli mogA* mutant, that is also repairable by high levels of molybdate (Stallmeyer et al. 1995). Two other cDNA clones from *Arabidopsis*, encoding genes involved in early steps of molybdenum cofactor biosynthesis, have also been obtained by functional complementation of two *E. coli* mutants (*moaA* and *moaC*) , deficient in single steps of molybdenum cofactor biosynthesis. These two cDNAs, corresponding to the *Cnx2* and *Cnx3* genes, code for proteins of 43 and 30 kDa, respectively. They have significant identity to the complemented *E. coli* genes, suggesting that the MoCo biosynthesis pathway is conserved between plants and bacteria (Hoff et al. 1995). Since then, other *Arabidopsis* genes involved in MoCo synthesis have been identified (reviewed in Mendel 1997).

NR dimers have been identified in several *cnx* mutants, which suggests that MoCo is probably not involved in the dimerisation of NR. This hypothesis is supported by the fact that expression of NR in yeast, which has no MoCo, leads to synthesis of a dimeric NR protein (Truong et al. 1991). On the other hand, a tobacco NR coding sequence carrying a 56-amino acid deletion in the N-terminal region, was also expressed in yeast and the NADH: NR activity of this truncated protein was very poorly complemented by exogenous MoCo from xanthine oxidase, whereas a significant amount of NADH:NR activity could be restored in wild-type NR (Nussaume et al. 1995; C. Meyer, unpubl.). This points to a possible role of the N-terminal domain in MoCo insertion or stabilisation.

NR biochemistry

Apart from the physiological reduction of nitrate with NAD(P)H as the electron donor, NR can catalyse several partial activities *in vitro*. The partial activities involve one or more of the prosthetic groups and use artificial electron donors or acceptors (Fig. 2). The partial activities are classified as NAD(P)H dehydrogenase (diaphorase), activities with artificial electron acceptors (such as ferricyanide, cytochrome *c* or siderophores), or as terminal nitrate reductase activities with artificial electron donors which include methylviologen, reduced bromphenol blue and flavin nucleotides.

These partial catalytic activities can also be found also in isolated NR domains produced either after mild proteolysis, or resulting from nonsense mutations in *nia* mutants (Meyer et al. 1991, reviewed in Campbell 1999) supporting the view that NR is built from independent units. Fragments of NR (FAD domain or FAD and haem domains) produced in *E. coli* are also active, showing the expected diaphorase activities (for review see Campbell 1999). On the other hand, holo-NR expressed in bacteria did not show NADH:NR activity, whereas sulphite oxidase (another Mo-containing enzyme) was produced in an active form in *E. coli* (Kisker et al. 1997). Tobacco NR was also expressed in the yeast *Saccharomyces cerevisiae*. No NADH NR activity could be detected, but the partial NADH cytochrome *c* reductase activity, which requires both a functional FAD and haem domain, was recovered (Truong et al. 1991). More recently, *Arabidopsis* NR was expressed in an active form in the methylotrophic yeast *Pichia pastoris* (Su et al. 1997).

The analysis of *nia* alleles resulting from point mutations has been extremely fruitful in identifying critical residues in the NR protein sequence (Meyer et al. 1991, 1995; Wilkinson and Crawford 1993; LaBrie and Crawford 1994, reviewed in Rouzé and Caboche 1992; Campbell 1999). Indeed, the selection of NR-deficient null mutants by the chlorate resistance screen has allowed the identification of essential amino acids, mutation of which completely abolishes NADH:NR activity. These results have been presented in the above-mentioned excellent reviews and will not be further discussed here.

The C-terminal FAD-binding domain from maize NR has been crystallised after expression in *E. coli* and its three-dimensional structure was solved by X-ray diffraction analysis (Lu et al. 1994). The structure of this domain shows that it can be subdivided into two smaller domains corresponding to an FAD-binding structural unit (six-stranded β barrel with a α helix involved in FAD binding) and to an NADH-binding domain (six-stranded β barrel similar to the Rossman fold of various dehydrogenases) at the C-terminus of the enzyme (Lu et al. 1994). These two subdomains are separated by a cleft where NADH is thought to transfer electrons to FAD. The overall structure of these two binding domains is similar to that of ferredoxin NADP$^+$ reductase and very similar to that of cytochrome b_5 reductase (Campbell 1999). The structure of the FAD/NADH domain can also be docked to that of the haem domain, which is easily predicted from the solved structure of mammalian cytochrome b_5 (Meyer et al. 1991). Unfortunately, crystallisation of the complete NR protein has not been achieved so far. Nevertheless, the structure of chicken liver sulphite oxidase, which shares a strong homology with the NR MoCo-binding domain, has been recently determined (Kisker et al. 1997). Sulphite oxidase also possesses a cytochrome b_5 domain but at the N terminus of the MoCo domain whereas the cytochrome b_5 domain is located at the C terminus of the MoCo domain in NR. With this restriction, it should now be possible to build an overall structural prediction for holo-NR (Fig. 2), to serve as a paradigm for further studies. Again the MoCo-binding domain of sulphite oxidase can be divided in two subdomains: one binding the MoCo (9 α helices and 13 β strands organised in 3 β sheets) and a C-terminal domain involved in dimerisation, which has the same topology as the immunoglobulin superfamily (Kisker et al. 1997). Interestingly, an *N. plumbaginifolia nia* mutant carrying a point mutation in the NR subdomain corresponding to the one involved in sulphite oxidase dimerisation, contains a monomeric NR (Meyer et al. 1995).

NR is thought to be a cytosolic protein, although the experimental evidence for this assumption is rather scarce. There have also been reports of a plasma-membrane (PM) bound NR, which has some features different from the cytosolic NR (Hoarau et al. 1991; Kunze et al. 1997; Stöhr 1999): the PM-NR is thought to be bound to the PM by a glycosyl-phosphatidylinositol anchor. Recently, it has been proposed that the PM-NR activity is induced by high nitrate concentrations (Stöhr 1999), but molecular data on this PM-specific isoform are still lacking.

Transcriptional Regulation of NR

Regulation by N Metabolites

NR expression is induced by its own substrate, nitrate. Nitrate is the most critical signal leading to the induction of all the components of the nitrate assimilation pathway (for reviews see Crawford 1995; Sivasankar and Oaks 1996; Daniel-Vedele and Caboche 1996). This induction is fast (within minutes) and requires very low concentrations of nitrate (<10 μM), suggesting that nitrate is actually sensed more as a "hormone", than as a nutrient (Crawford 1995). The use of transgenic plants has helped to prove that the regulation by nitrate is at the transcriptional level, thereby confirming the results of run off experiments performed in soybean (Callaci and Smarrelli 1991). For instance, the *N. plumbaginifolia* E23 *nia* mutant was transformed by a construct where the tobacco *Nia2* cDNA was placed under the control of the cauliflower mosaic virus (CaMV) 35S RNA promoter (Vincentz and Caboche 1991). The transgenic plants were restored for growth on nitrate as the sole nitrogen source and exhibited a constitutive expression of the *Nia* mRNA, i.e. even in the absence of nitrate. This suggests that nitrate controls, directly or indirectly, the transcription of the *Nia* genes. Translational and transcriptional fusions were also made between the tobacco *Nia1* promoter and the GUS reporter gene and introduced into the wild-type or *nia* mutant of *N. tabacum* (Vaucheret et al. 1992). In some of the rare cases where the expression of the transgene was still detectable, *GUS* mRNA could be induced by nitrate. These data further support the hypothesis that nitrate regulation is at the transcriptional level.

The detailed analysis of the nitrate-induced NR transcription has been rather complicated, mainly because it was very difficult, at least in tobacco, to obtain expression of reporter genes linked to the NR promoter. This was evident when these constructs were introduced into transformed plants, either after stable transformation, or for transient expression (Vaucheret et al. 1992; Godon et al. 1995). Nevertheless, linker-scanning analysis of the Arabidopsis promoters revealed a putative *cis*-acting element involved in nitrate induction (Hwang et al. 1997). This element, an A/T stretch followed by a A(C/G)TCA motif, was found in NR promoters from other species; but gain-of-function experiments using this motif (i.e. nitrate inducibility of a minimal promoter linked to this sequence) have hitherto been lacking.

In nitrogen-starved tomato plants, NR activity and protein decrease to respectively 3 and 10% of their initial value, while the levels of *Nia* mRNA are maintained at a level close to normal (Galangau et al. 1988). These data suggest that nitrogen availability also affects NR expression at the posttranscriptional level, in addition to transcriptional control.

Mutants deficient in either NR or NiR activities accumulate NR and NiR mRNA (reviewed by Vedele and Caboche 1996). This suggests that N metabolites resulting from nitrate assimilation downregulate NR and NiR expression. Thus in NR- or NiR-deficient plants, this metabolite would be limiting and thus *Nia* and *Nii* expression would be induced. Overaccumulation of NR mRNA in phosphinothricin (a GS inhibitor)-treated tobacco leaves correlated with decreased glutamine accumulation (Deng et al. 1991), suggesting that glutamine may be the effector N metabolite. This has been confirmed by experiments with detached *N. plumbaginifolia* leaves incubated in salt solution or 0.1 M glutamine (Vincentz et al. 1993), where a decrease in NR or NiR activity and mRNA was observed. A barley mutant affected in the leaf-specific NADH-NR expressed root NAD(P)H-NR in leaves (Dailey et al. 1982) and in a *nia* mutant background tobacco root-specific *Nii* mRNA was detected in leaves (Kronenberger et al. 1993). These data may reflect the derepression of root-specific *Nia* and *Nii* genes by the lowered pool of reduced N metabolites. Moreover, it has been shown that glutamine pools seem to fluctuate in an opposite phase to *Nia* mRNA levels during the day (Deng et al. 1991). However, in recent work, it has been shown that in the *gluS* mutant of *Arabidopsis*, which accumulates glutamine due to a defect of the Fd-GOGAT gene, there was no reduction of the NR mRNA pool (Dzuibany et al. 1998), as well as no alteration of the NR circadian rythmicity. These results would suggest that glutamine *per se* , is not the signal regulating NR expression.

Regulation by Hormones

Cytokinins have been shown to increase NR activity in etiolated plants, or in cell suspension cultures in many species (Lu et al. 1992; Suty et al. 1993). Light in many species is often required in addition to cytokinins for the hormonal enhancement to be effective.

In etiolated barley leaves, cytokinin enhancement of NR expression is, in part, transcriptional (Lu et al. 1990). Abscisic acid can suppress the cytokinin enhancement of *Nia* mRNA accumulation, and reduces *Nia* mRNA accumulation in etiolated barley leaves transferred to light (Lu et al. 1992). Thus, the benzyladenine to abscisic acid ratios influence *Nia* mRNA levels. In *Arabidopsis*, it has been shown that the *Nia1* gene expression is specifically induced by exogenous cytokinins (Yu et al. 1998).

Regulation by Light and by Carbohydrates

In addition to nitrate, light is an important signal for NR regulation, at both the transcriptional and posttranscriptional levels. The level of NR mRNA decreased when mature plants were put in darkness (Bowsher et al. 1991) and increased when, for instance, etiolated plants were transferred into the light (Mohr et al. 1992); this induction was mediated by phytochrome (Pilgrim et al. 1993). The *Arabidopsis cop1* mutant, that is affected in light perception and exhibits a deetiolated phenotype in darkness, expressed a number of light-inducible genes in the dark including *Nia2* mRNA (Deng et al. 1991).

In green plants, the role of phytochrome is less clear but transgenic tobacco plants that overexpress the oat *phyA* gene overexpress NR activity in green seedlings, indicating that phytochrome controls NR expression in these light-grown plants (McCormac et al. 1992).

Because nitrate assimilation into glutamine and glutamate is largely dependent on the availability of carbon skeletons, experiments were performed to determine whether the induction by light was mediated by the stimulation of carbohydrate synthesis through photosynthesis. In detached leaves of wild-type dark-adapted *N. plumbaginifolia* plants, incubated in a salt solution for 24 h in the dark, no *Nia* nor *Nii* mRNA was detected. Addition of sucrose, glucose or fructose (0.2 M) to the incubation media induced NR protein activity and mRNA in the dark. Ribose or mannitol could not substitute for these sugars, indicating that this induction was not due to the change in the osmotic strength of the nutrient solution. Growth of plants in the presence of sucrose led to the appearance of NR mRNA in the dark, in both *Arabidopsis* (Cheng et al. 1992) and *N. plumbaginifolia* (Vincentz et al. 1993), which indicates that, at least partly, sugar can substitute for light in mediating NR mRNA induction. Further evidence that light regulation of NR expression is related to sugars is provided by investigations of tobacco growing under short-day conditions (Matt et al. 1998). Under these conditions, sugars fall to very low levels by the middle of the long dark period, and this is followed by a decrease in NR mRNA which, in turn, significantly reduces the extent to which NR protein increases after reillumination of the leaves in the morning.

Transgenic plants in which the tobacco *Nia2* cDNA is under control of the CaMV 35S promoter, express high levels of *Nia* mRNA in darkness, indicating that the light control of *Nia* mRNA is, in part, transcriptional (Vincentz and Caboche 1991). However, light regulation of *Nia* gene expression is not solely transcriptional.

Posttranscriptional Regulation of NR

Regulation of Translation

Several lines of evidence show that NR protein levels are not controlled solely by the level of NR mRNA. Following transfer of wild-type tobacco to light, NR mRNA typically shows a rapid decrease, but NR proteins and NR activity rise during the first hours to reach a peak in the middle of the photoperiod (Galangau et al. 1988; Scheible et al. 1997). Later in the photoperiod there is a decline of NR protein and activity in wild-type tobacco growing in ambient carbon dioxide, which is attenuated or abolished when wild-type plants are grown in elevated carbon dioxide (Geiger et al. 1998) and in mutants with a decreased number of functional *Nia* genes (Scheible et al. 1997), even though there are no obvious differences in the levels of NR mRNA. Even more strikingly, when transgenic plants that express the tobacco *Nia2* cDNA under the control of the constitutive 35S promoter are transferrred to darkness there is a decrease of NR protein and activity, even though NR mRNA remains high (Vincentz and Caboche 1991). Recently, W.M. Kaiser and H. Weiner (pers. comm.) have shown in ^{35}S-Met feeding experiments that light stimulates the rate of *de novo* synthesis and (see below) decreases the rate of degradation of NR protein.

Posttranslational Regulation by Reversible Phosphorylation and Binding of Inhibitory 14-3-3 Proteins

A major breakthrough in the past decade has been the demonstration of light-mediated posttranslational control of NR expression (Kaiser and Förster 1989; Kaiser and Brendle-Behnisch 1991; Kaiser et al. 1992; Huber et al. 1992; MacKintosh 1992). This regulation has been shown to operate by reversible protein phosphorylation: by assaying spinach NR activity in the presence of Mg^{2+}, a reversible inactivation of NR has been demonstrated when plants are submitted to water stress, low CO_2 levels or transferred to the dark (reviewed by Kaiser et al. 1999). Inactivation of NR is linked with phosphorylation both *in vitro* and *in planta*. (Fig. 4) and can only be seen when NR activity is measured in the presence of Mg^{2+} in the millimolar range; indeed, EDTA reactivates the enzyme, which allows an easy measurement of NR activation state (ratio between NR activity measured with Mg^{++} and NR activity measured in the presence of EDTA).

The proteins involved in the inactivation mechanism of spinach NR were identified by different groups, and a two-step regulation model was proposed (Glaab and Kaiser 1995; MacKintosh et al. 1995; Fig. 4). It has thus been shown that spinach NR is first phosphorylated on at least one serine residue (serine 543 in the first hinge separating the MoCo and the haem domains). This serine residue is conserved among higher plant NRs (Douglas et al. 1995; Bachmann et al. 1996a). Phosphorylated NR is still active and becomes inactivated upon the binding of a factor called NIP (nitrate reductase inactivator protein) (Fig. 4). Three peaks of NR kinase activities (named PKI, PKII and PKIII), that both phosphorylated and inactivated NR, were identified in spinach extracts (Bachmann et al. 1996a; Douglas et al. 1997), as well as in tobacco

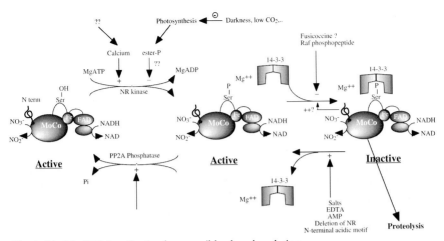

Fig. 4 : Model of NR inactivation by reversible phosphorylation

(E. Pigaglio and C. Meyer, unpubl.). PKI and PKII are calcium-dependent and PKIII appears to be calcium-independent (Bachman et al. 1996a; Douglas et al. 1997). PKI and PKII are members of the plant CDPK (calmodulin-domain protein kinase) family. Spinach PKI has been purified and a partial amino acid sequence was determined showing that this kinase was very similar to *Arabidopsis* CDPK6 (Douglas et al. 1998). PKIII was also identified as a yeast SNF1-related kinase (Douglas et al. 1997).

NIP was identified as a mixture of proteins belonging to the 14-3-3 family, and it was suggested that these 14-3-3 proteins interact with the regulatory phosphorylation site of NR (Bachmann et al. 1996b; Moorhead et al. 1996; Athwal et al. 1998a; Kanamaru et al. 1999). Indeed, the sequence surrounding the NR hinge 1 phosphorylated serine is similar to the consensus 14-3-3-binding site sequence found in many other 14-3-3 targets (Muslin et al. 1996) and a synthetic phospho-peptide corresponding to the Raf-1 14-3-3-binding site is able to disrupt the NR/14-3-3 interaction (Moorhead et al. 1996). It has been shown that divalent cations can bind directly to the 14-3-3 proteins and induce the conformational changes needed for the formation of a NR/14-3-3 complex (Athwal et al. 1998b). *Arabidopsis* contains at least ten different 14-3-3 isoforms and it has been shown that the ω isoform exhibits the strongest binding to the NR phosphorylation site both *in vitro* (Bachman et al. 1996b) and in the yeast two-hybrid system (Kanamaru et al. 1999). Nevertheless, the question of whether there is an isoform specificity *in planta* for inactivation of NR remains open, as yeast 14-3-3s seem as efficient as plant 14-3-3s for inactivating phosphorylated NR. The N-terminal region of NR, the sequence of which is poorly conserved among plants, has been shown to be somehow implicated in the NR inactivation process (Nussaume et al. 1995). Indeed, we have expressed in transgenic *N. plumbaginifolia* plants, a deleted NR lacking 56 amino acids in the N terminus. This truncated protein was found to be still active, apart from a higher thermosensitivity, but its activation state was less affected by light-dark transitions and was also higher in the light than in the wild type, suggesting that the loss of the N-terminal region abolished almost completely the inactivation of NR by phosphorylation (Nussaume et al. 1995). Moreover, we were unable to inactivate the truncated NR *in vitro*, by incubation of a crude plant extract with

MgATP. However, this apparent higher activation state of NR in transgenic plants expressing the truncated protein did not result in a higher nitrate assimilation rate in the dark, probably because other factors, such as substrate or reducing power availibility, limit the rate of nitrate assimilation (Lejay et al. 1997). Still, when these transgenic plants were grown at low CO_2, where reducing power is not limiting, a higher rate of nitrate assimilation was observed, indicating that the truncated NR is indeed less inactivated *in planta*, at least in some circumstances. Surprisingly, the deleted NR protein could be inactivated *in vitro* by exogenous yeast 14-3-3 when purified from dark-grown plants, which also suggests that the deleted NR is normally phosphorylated in the dark (Lillo et al. 1997). Our explanation for this observation is that a putative NR activation factor (NAF) remains bound to the deleted NR protein, thus impeding 14-3-3 binding and inactivation. Upon purification of the truncated phosphorylated NR from dark-exposed leaves, the NAF would be lost, thus allowing 14-3-3 binding and the inactivation of the deleted NR protein (Fig. 5). By expressing in transgenic *N. plumbaginifolia*, modified NRs carrying smaller deletions in the N-terminal domain, we were also able to show that a conserved stretch of acidic amino acids is, in fact, involved in the loss of inactivation in the dark (Pigaglio et al. 1999). Moreover, this acidic motif was also found to be a good substrate for CKII (casein kinase II) phosphorylation.

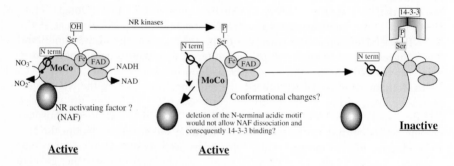

Fig. 5 : Model for NR and deleted NR inactivation

Regulation of NR degradation

The marked two- to three-fold changes in NR protein during the day (see above) imply that NR is degraded at a significant rate.

Correlative evidence indicates that NR degradation is triggered by phosphorylation and binding of 14-3-3 proteins (Huber and Kaiser 1997). Further evidence is supplied by experiments with transgenic tobacco overexpressing the truncated NR lacking the N terminus. This modified form of NR is not susceptible to dark inactivation (see above). When the plants were transferred to prolonged darkness, the decay of NR protein was also abolished (Nussaume et al. 1995). Recently, Weiner and Kaiser (1999) showed that darkening leads to a marked increase in the amount of 14-3-3 protein bound to NR, and that this was accompanied by an increase in the rate of NR degradation as monitored in ^{35}S-Met-labelling experiments.

Nitrite Reduction

Nitrite produced by nitrate reduction is thought to be translocated to the chloroplast, where it is further reduced to ammonium by nitrite reductase (NiR, EC 1.7.7.1) (for review see Wray 1989). This step involves the transfer of six electrons from reduced ferredoxin to nitrite. NiR purified from green leaves has a molecular mass of approximately 63 kDa and is monomeric (Siegel and Wilkerson 1989). The enzyme is a metalloprotein containing as prosthetic groups a sirohaem, to which nitrite binds, and a 4Fe/4S centre, which is probably the initial electron acceptor (Siegel and Wilkerson 1989). In green leaves, nitrite reductase is located within the chloroplast, whereas in roots it is located within plastids (Miflin 1974). However, the presence of an extrachloroplastic form has been proposed in cotyledons of mustard (Schuster and Mohr 1990). The direct electron donor in the chloroplast is reduced ferredoxin, whilst in roots a ferredoxin-like protein has been implicated (Suzuki et al. 1985), which probably obtains reducing power from NAD(P)H generated in the oxidative pentose phosphate pathway.

NiR Genes and Mutants

NiR is a nuclear encoded chloroplast protein. Several *Nii* cDNAs or *Nii* genes (encoding NiR) have been cloned from higher plants such as spinach, maize, birch, rice, *A. thaliana* and tobacco (for reviews see Hoff et al. 1994; Meyer and Caboche 1998). Some higher plants contain only a single *Nii* gene per haploid genome, whereas other plant species contain two copies per haploid genome. Tobacco contains four genes, two from each ancestor (Kronenberger et al. 1993). The two genes coming from each ancestor are expressed differentially in leaves and roots. NiR is made as a precursor protein with an N-terminal transit peptide which directs NiR to the chloroplasts. Sequence comparison of NiR proteins shows high conservation among plant species (75-80% similarity). There is only a small degree of homology between plant NiR and bacterial and fungal NiR, except for cyanobacterial NiR, which seems to be more similar to plant NiR than to enterobacterial NiR (Luque et al. 1993). The C terminal part of the plant and cyanobacterial NiR protein shares homology with a bacterial sulphite reductase, which is also a sirohaem:4Fe/4S redox enzyme. The N-terminal part of plant and cyanobacterial NiR has homology to ferredoxin NADP$^+$ reductase. Based on this homology, the ferredoxin-binding site was assigned to this domain, more precisely to an area which contains a number of positively charged amino acid residues, that could bind the acidic ferredoxin (Friemann et al. 1992).

NiR-deficient mutants (*nii* mutants) have been more difficult to obtain than *nia* mutants because there is no easy selection method, as for NR. Nevertheless, a *nii* mutant has been isolated in barley (Duncanson et al. 1993), by screening an M$_2$ population of mutagenized barley seeds for nitrite accumulation. NiR activity was greatly reduced in this mutant line, which is probably affected in a gene encoding the nitrite reductase apoprotein. On the other hand, a genetically engineered NiR-deficient *N. tabacum* has been constructed (Vaucheret et al. 1992) by introducing an antisense *Nii* coding sequence. The mutant had no detectable NiR activity or mRNA and accumulated five times more nitrite than wild-type plants.

NiR Regulation

In most cases, NiR expression was found to be coregulated with NR in response to endogenous or external factors like nitrate or light (Fig. 6), at least at the transcriptional level. As mentioned earlier, the *Nia* promoter was recalcitrant to studies using conventional approaches, due to the low expression of linked reporter genes. In the case of the *Nii* promoters, however, no problem of gene extinction was encountered. Promoter analysis of the bean *Nii* gene in transgenic tobacco plants showed that the elements involved in nitrate regulation reside in the proximal 0.6-kb region upstream of the translation start (Sander et al. 1995). Deletions of the Nii promoter from spinach fused to the GUS reporter gene and introduced into tobacco indicated that the basic elements required for light- and nitrate-dependent expression of the reporter gene were located within the promoter sequence -200/+131 relative to the transcription-initiation site (Neininger et al. 1994). Moreover, *in vivo* footprinting revealed a nitrate-inducible binding of proteins to GATA elements in the -230 to -181 region of the spinach *Nii* promoter (Rastogi et al. 1997). This suggested that GATA sequences could mediate nitrate regulation of the *Nii* gene. By analysis of the tobacco *Nii* promoter fused to either a *GUS* or luciferase reporter gene, it was shown that the sequences required for nitrate induction of the reporter gene were retained in the proximal 200-bp fragment of the promoter (Dorbe et al. 1998).

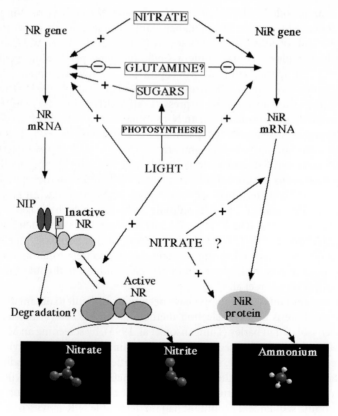

Fig. 6 : Regulation of NR and NiR expression by endogenous and environmental factors

In order to study the posttranscriptional regulation of NiR, *N. plumbaginifolia* and *Arabidopsis* plants were transformed with a 35S-NiR construct, transformed plants overexpressed NiR activity in the leaves (Crété et al. 1997). When these plants were grown *in vitro* on media containing either nitrate or ammonium as sole nitrogen source, NiR mRNA derived from transgene expression was constitutively expressed, whereas NiR activity and protein level were strongly reduced on ammonium containing medium. These results suggest that, together with transcriptional control, posttranscriptional regulation by the nitrogen source is also operating on NiR expression. One explanation for this mechanism could be that a specific enzyme for sirohaem synthesis is induced by nitrate (Sakakibara et al. 1996). This posttranscriptional regulation of NiR expression by the nitrogen source is thus different from the NR posttranscriptional control by light. The reason for this difference is so far unknown, but clearly illustrates the redundancy of the regulation of the nitrate assimilation pathway in plants (Fig. 6).

Concluding remarks

The initial steps in nitrate assimilation are highly regulated, indeed NR is emerging as a paradigm for a plant enzyme that is subject to multifactorial regulation acting at different levels, including transcription, translation, reversible posttranslational modification and protein turnover. However, although it might seem intuitively obvious that NR provides an ideal site to control the rate of nitrate assimilation, the precise physiological function of this sophisticated regulation remains enigmatic. As already noted, mutants with two- to ten-fold less NR activity grow as rapidly as wild-type plants, and transformants that have constitutively high NR activity do not grow faster than wild type plants, and also show no obvious ill effects of possessing NR activity at times and in conditions when NR is repressed and inactivated in wild-type plants. It can be anticipated that NR constitutes one link in a regulatory network connecting the uptake of inorganic nitrogen, its assimilation, and the use of assimilated nitrogen for biosynthesis and growth. In the future, it will be important to understand more about the integration of nitrate assimilation into these processes.

Acknowledgements. This work was partly supported by the EU contract no. BIO4CT97-2231.

References

Athwal GS, Huber JL, Huber SC (1998a) Phosphorylated nitrate reductase and 14-3-3 proteins. Site of interaction, effect of ions, and evidence for an AMP-binding site on 14-3-3 proteins. Plant Physiol 118: 1041-1048

Athwal GS, Huber JL, Huber SC (1998b) Biological significance of divalent metal ion binding to 14-3-3 proteins in relationship to nitrate reductase inactivation. Plant Cell Physiol 39: 1065-1072

Bachmann M, Shiraishi N, Campbell WH, Yoo BC, Harmon AC, Huber SC (1996a) Identification of Ser-543 as the major regulatory phosphorylation site in spinach leaf nitrate reductase. Plant Cell 8: 505-517

Bachmann M, Huber J L, Liao PC, Gage DA, Huber SC (1996b) The inhibitor protein of phosphorylated nitrate reductase from spinach (*Spinacia oleracea*) leaves is a 14-3-3 protein. FEBS Lett 387: 127-131

Bowsher CG, Long DM, Oaks A, Rothstein SJ (1991) Effect of light/dark cycles on expression of nitrate assimilatory genes in maize shoots and roots. Plant Physiol 95: 281-285

Callaci JJ, Smarrelli JJ (1991) Regulation of the inducible nitrate reductase isoform from soybeans. Biochim Biophys Acta 1088: 127-130

Calza R, Huttner E, Vincentz M, Rouzé P, Galangau F, Vaucheret H, Chérel I, Meyer C, Kronenberger J, Caboche M (1987) Cloning of DNA fragments complementary to tobacco nitrate reductase mRNA and encoding epitopes common to the nitrate reductases from higher plants. Mol Gen Genet 209: 552-562

Campbell WH (1999) Nitrate reductase structure, function and regulation: bridging the gap between biochemistry and physiology. Annu Rev Plant Physiol Plant Mol Biol 50: 277-303

Cheng CL, Dewdney J, Nam H-G, den Boer BGW, Goodman HM (1988) A new locus (*Nia1*) in *Arabidopsis thaliana* encoding nitrate reductase. EMBO J 7: 3309-3314

Cheng CL, Acedo GN, Cristinsin M, Conkling MA (1992) Sucrose mimics the light induction of *Arabidopsis* nitrate reductase gene transcription. Proc Natl Acad Sci USA 89: 1861-1864

Crawford N (1995) Nitrate: nutrient and signal for plant growth. Plant Cell 7: 859-868

Crawford N, Arst HN (1993) The molecular genetics of nitrate assimilation in fungi and plants. Annu Rev Genet 27: 115-146

Crawford NM, Smith M, Bellissimo D, Davis RW (1988) Sequence and nitrate regulation of the *Arabidopsis thaliana* mRNA encoding nitrate reductase, a metalloflavoprotein with three functional domains. Proc of the Nat Acad Sci USA 85: 5006-5010

Crete P, Caboche M, Meyer C (1997) Nitrite reductase expression is regulated at the post-transcriptional level by the nitrogen source in *Nicotiana plumbaginifolia* and *Arabidopsis thaliana*. Plant J 11: 625-634

Dailey FA, Warner RL, Somers DA, Kleinhofs A (1982) Characteristics of a nitrate reductase in a barley mutant deficient in NADH nitrate reductase. Plant Physiol 69: 1200-1204

Daniel-Vedele F, Caboche M (1996) Molecular analysis of nitrate assimilation in higher plants. Cr Acad Sci Paris 319: 961-968

Deng M-D, Moureaux T, Leydecker M-T, Caboche M (1990) Nitrate reductase expression is under the control of a circadian rhythm and is light inducible in *Nicotiana tabacum* leaves. Planta 180: 257-261

Deng XW, Caspar T, Quail PH (1991) *cop1*: a regulatory locus involved in light-controlled development and gene expression in *Arabidopsis*. Genes Dev 5: 1172-1182

Dorbe MF, Truong HN, Crété P, Daniel-Vedele F (1998) Deletion analysis of the tobacco *Nii1* promoter in *Arabidopsis thaliana*. Plant Sci 139: 71-82

Douglas P, Morrice N, MacKintosh C (1995) Identification of a regulatory phosphorylation site in the hinge 1 region of nitrate reductase from spinach (*Spinacea oleracea*) leaves. FEBS Lett 377: 113-117

Douglas P, Pigaglio E, Ferrer A, Halford NG, MacKintosh C (1997) Three spinach leaf nitrate reductase-3-hydroxy-3-methylglutaryl-CoA reductase kinases that are regulated by reversible phosphorylation and/or Ca^{2+} ions. Biochem J 325: 101-109

Douglas P, Moorhead G, Hong Y, Morrice N, MacKintosh C (1998) Purification of a nitrate reductase kinase from *Spinacea oleracea* leaves, and its identification as a calmodulin-domain protein kinase. Planta 206: 435-442

Duncanson E, Gilkes AF, Kirk DW, Sherman A, Wray JL (1993) *nir1*, a conditional-lethal mutation in barley causing a defect in nitrite reduction. Mol Gen Genet 236: 275-28

Dzuibany C, Haupt S, Fock H, Biehler K, Migge A, Becker T (1998) Regulation of nitrate reductase transcript levels by glutamine accumulating in the leaves of a ferredoxin-dependent glutamate synthase-deficient *gluS* mutant of *Arabidopsis thaliana* and by glutamine provided by the roots. Planta 206: 515-522

Evans HJ, Nason A (1953) Pyridine nucleotide-nitrate reductase from extracts of higher plants. Plant Physiol 28: 233-254

Fernandez E, Cardenas J (1989) Genetics and regulatory aspects of nitrate assimilation in algae. In: (Wray J L and Kinghorn J R (eds) Molecular and genetic aspects of nitrate assimilation. Oxford Science Publications, Oxford, pp 101-124

Fido RJ (1991) Isolation and partial amino acid sequence of domains of nitrate reductase from spinach. Phytochemistry 30: 3519-3523

Friemann A, Brinkmann K, Hachtel W (1991) Sequence of a cDNA encoding bi-specific NAD(P)H-nitrate reductase from the tree *Betula pendula* and identification of conserved protein regions. Mol Gen Genet 227: 97-105

Friemann A, Brinkmann K, Hachtel W (1992) Sequence of a cDNA encoding nitrite reductase from the tree *Betula pendula* and identification of conserved protein regions. Mol Gen Genet 231: 411-416

Gabard J, Marion-Poll A, Chérel I, Meyer C, Müller A, Caboche M (1987) Isolation and characterization of *Nicotiana plumbaginifolia* nitrate reductase-deficient mutants: genetic and biochemical analysis of the *Nia* complementation group. Mol Gen Genet 209: 596-606

Gabard J, Pelsy F, Marion-Poll A, Caboche M, Sallbach I, Grafe R, Müller AJ (1988) Genetic analysis of nitrate reductase deficient mutants of *Nicotiana plumbaginifolia*: evidence for six complementation groups among 70 classified molybdenum cofactor deficient mutants. Mol Gen Genet 213: 206-213

Galangau F, Daniel-Vedele F, Moureaux T, Dorbe M-F, Leydecker M-T, Caboche M (1988) Expression of leaf nitrate reductase gene from tomato and tobacco in relation to light-dark regimes and nitrate supply. Plant Physiol 88: 383-388

Geiger M, Walch-Piu L, Harnecker J, Schulze E.-D, Ludewig F, Sonnewald U, Stitt M, (1998) Enhanced carbon dioxide leads to a modified diurnal rhythm of nitrate reductase activity in older plants, and a large stimulation of nitrate reductase activity and higher levels of amino acids in higher plants. Plant Cell Environ 21: 253-268

Glaab J, Kaiser WM (1995) Inactivation of nitrate reductase involves NR-protein phosphorylation and subsequent "binding" of an inhibitor protein. Planta 195: 514-518

Godon C, Caboche M, Daniel-Vedele F (1995) Use of biolistic process for the analysis of nitrate-inducible promoters in transient expression assays. Plant Sci 111: 209-218

Grandbastien M-A, Spielmann A, Caboche M (1989) *Tnt1*, a mobile retroviral-like transposable element of tobacco isolated by plant cell genetics. Nature 337: 376-380

Hoarau J, Nato A, Lavergne D, Flipo V, Hirel B (1991) Nitrate reductase activity changes during a culture cycle of tobacco cells: the participation of a membrane-bound form enzyme. Plant Sci 79: 193-204

Hoff T, Truong HN, Caboche M (1994) The use of mutants and transgenic plants to study nitrate assimation. Plant Cell Environ 17: 489-506

Hoff T, Schnorr KM, Meyer C, Caboche M (1995) Isolation of two *Arabidopsis* cDNAs involved in early steps of molybdenum cofactor biosynthesis by functional complementation of *Escherichia coli*. J Biol Chem 270: 6100-6107

Huber JL, Huber SC, Campbell WH, Redinbaugh MG (1992) Reversible light/dark modulation of spinach leaf nitrate reductase activity involves protein phosphorylation. Arch Biochem Biophys 296: 58-65

Huber SC, Kaiser WM (1997) Correlation between apparent activation state of nitrate reductase (NR), NR hysteresis and degradation of NR protein. J Exp Bot 132: 1367-1374

Hwang CF, Lin Y, D'Souza T, Cheng CL (1997) Sequences necessary for nitrate-dependent transcription of *Arabidopsis* nitrate reductase genes. Plant Physiol 113: 853-862

Kaiser WM, Brendle-Behnisch E (1991) Rapid modulation of spinach leaf nitrate reductase activity by photosynthesis. II In vitro modulation by ATP and AMP. Plant Physiol 96: 368-375

Kaiser WM, Förster J (1989) Low CO_2 prevents nitrate reduction in leaves. Plant Physiol 91: 970-974

Kaiser WM, Spill D, Brendle-Behnisch E (1992) Rapid light-dark modulation of assimilatory nitrate reductase in spinach leaves involves adenine nucleotides. Planta 186: 236-240

Kaiser WM, Weiner H, Huber SC (1999) Nitrate reductase in higher plants: a case study for transduction of environmental stimuli into control of catalytic activity. Physiol Plant 105: 385-390

Kanamura K, Wang R, Su W, Crawford NM (1999) Ser-534 in the hinge 1 region of Arabidopsis nitrate reductase is conditionally required for binding of 14-3-3 proteins and *in vitro* inhibition. J Biol Chem 274: 4160-4165

Kisker C, Schindelin H, Pacheco A, Wehbi WA, Garrett RM, Rajagopalan KV, Enemark JH, Rees DC (1997) Molecular basis of sulfite oxidase deficiency from the structure of sulfite oxidase. Cell 91: 973-983

Kleinhofs A, Warner RL (1990) Advances in nitrate assimilation. In: Miflin BJ, Lea PJ (eds) The biochemistry of plants. Academic Press, San Diego, pp 89-120

Kronenberger J, Lepingle A, Caboche M, Vaucheret H (1993) Cloning and expression of distinct nitrite reductases in tobacco leaves and roots. Mol Gen Genet 236: 203-208

Kunze M, Riedel J, Lange U, Hurwitz R, Tischner R (1997) Evidence for the presence of GPI-anchored PM-NR in leaves of *Beta vulgaris* and for PM-NR in barley leaves. Plant Physiol Biochem 35: 507-512

Labrie ST, Crawford NM (1994) A glycine to aspartic acid change in the MoCo domain of nitrate reductase reduces both activity and phosphorylation levels in *Arabidopsis*. J Biol Chem 269: 14497-14501

Lejay L, Quilleré I, Roux Y, Tillard P, Cliquet JB, Meyer C, Morot-Gaudry JF, Gojon A (1997) Abolition of posttranscriptional regulation of nitrate reductase partially prevents the decrease in leaf NO_3^- reduction when photosynthesis is inhibited by CO_2 deprivation, but not in darkness. Plant Physiol 115: 623-630

Lillo C, Kazazaic S, Ruoff P, Meyer C (1997) Characterization of nitrate reductase from light- and dark-exposed leaves. Plant Physiol 114: 1377-1383

Lu J-l, Ertl JR, Chen CM (1990) Cytokinin enhancement of the light induction of nitrate reductase transcript levels in etiolated barley leaves. Plant Mol Biol 12: 585-594

Lu JL, Ertl JR Chen CM (1992) Transcriptional regulation of nitrate reductase messenger RNA levels by cytokinin-abscisic acid interactions in etiolated barley leaves. Plant Physiol 98: 1255-1260

Lu G, Campbell WH, Schneider G, Lindqvist Y (1994) Crystal structure of the FAD-containing fragment of corn nitrate reductase at 2.5 Å resolution: relationships to other flavoprotein reductases. Structure 2: 809-821

Luque I, Flores E, Herrero A (1993) Nitrite reductase gene from *Synechococcus* Sp PCC-7942 - homology between cyanobacterial and higher-plant nitrite reductases. Plant Mol Biol 21: 1201-1205

McCormac A, Whitelam G, Smith H (1992) Light-grown plants of transgenic tobacco expressing an introduced oat phytochrome A gene under the control of a constitutive viral promoter exhibit persistent growth inhibition by far-red light. Planta 188: 173-181

MacKintosh C (1992) Regulation of spinach-leaf nitrate reductase by reversible phosphorylation. Biochim Biophys Acta 1137: 121-126

MacKintosh C (1998) Regulation of plant nitrate assimilation: from ecophysiology to brain proteins. New Phytol 139: 153-159

MacKintosh C, Douglas P, Lillo C (1995) Identification of a protein that inhibits the phosphorylated form of nitrate reductase from spinach (*Spinacia oleracea*) leaves. Plant Physiol 107: 451-457

Matt P, Schurr U, Krapp A, Stitt M (1998) Growth of tobacco in short day conditions leads to high starch, low sugars, altered diurnal changes of the *Nia* transcript and low nitrate reductase activity, and an inhibition of amino acid synthesis. Planta 207, 27-41

Mendel RR (1997) Molybdenum cofactor of higher plants: biosynthesis and molecular biology. Planta 203: 399-405

Mendel RR, Müller AJ (1985) Repair *in vitro* of nitrate reductase-deficient tobacco mutants (*cnx*A) by molybdate and by molybdenum cofactor Planta 163: 370-375

Meyer C, Caboche M (1998) Manipulation of nitrogen metabolism. In: Lindsey K (ed) Transgenic plant research. Harwood Academic Publ, London, pp 125-133

Meyer C, Levin J M, Roussel J-M, Rouzé P (1991) Mutational and structural analysis of the nitrate reductase heme domain of *Nicotiana plumbaginifolia*. J Biol Chem 266: 20561-20566

Meyer C, Pouteau S, Rouzé P, Caboche M (1994) Isolation and molecular characterization of *dTnp1*, a mobile and defective transposable element of *Nicotiana plumbaginifolia*. Mol Gen Genet 242: 194-200

Meyer C, Gonneau M, Caboche M , Rouzé P (1995) Identification by mutational analysis of four critical residues in the molybdenum cofactor domain of eukaryotic nitrate reductase. FEBS Lett 370: 197-202

Miflin BJ (1974) The location of nitrite reductase and other enzymes related to amino acid biosynthesis in the plastids of root and leaves. Plant Physiol 54: 550-555

Mohr H, Neininger A, Seith B (1992) Control of nitrate reductase and nitrite reductase gene expression by light, nitrate and a plastidic factor. Bot Acta 105: 81-89

Moorhead G, Douglas P, Morrice N, Scarabel M, Aitken A, MacKintosh C (1996) Phosphorylated nitrate reductase from spinach leaves is inhibited by 14-3-3 proteins and activated by fusicoccin. Curr Biol 6: 1104-1113

Müller AJ, Mendel RR (1989) Biochemical and somatic cell genetics of nitrate reduction in Nicotiana. In: Wray JL, Kinghorn JR (eds) Molecular and genetic aspects of nitrate assimilation. Oxford Science Publ, Oxford, pp 166-185

Muslin AJ, Tanner JW, Allen PM, Shaw AS (1996) Interaction of 14-3-3 with signaling proteins is mediated by the recognition of phosphoserine. Cell 84:889-897

Neame PJ, Barber MJ (1989) Conserved domains in molybdenum hydroxylases. J Biol Chem 264: 20894-20901

Neininger A, Back E, Bichler J, Schneiderbauer A, Mohr H (1994) Deletion analysis of a nitrite reductase promoter from spinach in transgenic tobacco. Planta 194: 186-192

Nussaume L, Vincentz M, Meyer C, Boutin JP, Caboche M (1995) Post-transcriptional regulation of nitrate reductase by light is abolished by an N-terminal deletion. Plant Cell 7: 611-621

Oostinder-Braaksma FJ, Feenstra WJ (1973) Isolation and characterization of chlorate-resistant mutants of *Arabidopsis thaliana*. Mutat Res 19: 175-185

Padidam M, Venkatesvarlu K, Johri MM (1991) Ammonium represses NADPH-nitratereductase in the moss *Funaria hygrometrica*. Plant Sci 75: 184-194

Pelsy F, Caboche M (1992) Molecular genetics of nitrate reductase in higher plants. Adv Genet 30: 1-40

Pelsy F, Gonneau M (1991) Genetic and biochemical analysis of intragenic complementation events among nitrate reductase apoenzyme-deficient mutants of *Nicotiana plumbaginifolia*. Genetics 127: 199-204

Pigaglio E, Durand N, Meyer C (1999) A conserved acidic motif in the N-terminal domain of nitrate reductase is necessary for the inactivation of the enzyme in the dark by phosphorylation and 14-3-3 binding. Plant Physiol 119: 219-229

Pilgrim ML, Caspar T, Quail PH, McClung CR (1993) Circadian and light-regulated expression of nitrate reductase in *Arabidopsis*. Plant Mol Biol 23: 349-364

Rastogi R, Bate NJ, Sivasankar S, Rothstein S (1997) Footprinting of the spinach nitrite reductase gene promoter reveals the preservation of nitrate regulatory elements between fungi and higher plants. Plant Mol Biol 34: 465-476

Redinbaugh MG, Campbell WH (1985) Quaternary structure and composition of squash NADH:nitrate reductase. J Biol Chem 260: 3380-3385

Rouzé P, Caboche M (1992) Nitrate reduction in higher plants: molecular approaches to function and regulation. In: Wray JL (ed) Inducible plant proteins: their biochemistry and molecular biology. Cambridge University Press, Cambridge, pp 45-77

Sakakibara H, Takei K, Sugiyama T (1996) Isolation and characterization of a cDNA that encodes maize uroporphyrinogen III methyltransferase, an enzyme involved in the synthesis of siroheme, which is a prosthetic group of nitrite reductase. Plant J 10: 883-892

Sander L, Jensen PE, Back LF, Stummann BJ, Henningsen KW (1995) Structure and expression of a nitrite reductase gene from bean (*Phaseolus vulgaris*) and promoter analysis in transgenic tobacco. Plant Mol Biol 27: 165-177

Scheible W-R, Gonzales-Fontes A, Morcuende R, Lauerer M, Geiger M, Glaab J, Schulze E-D, Stitt M (1997) Tobacco mutants with a decreased number of functional *nia*-genes compensate by modifing the diurnal regulation transcription, post-translational modification and turnover of nitrate reductase. Planta 203, 305-319

Schnorr KM, Juricek M, Huang C, Culley D, Kleinhofs A (1991) Analysis of barley nitrate reductase cDNA and genomic clones. Mol Gen Genet 227: 411-416

Schuster C, Mohr H (1990) Appearance of nitrite reductase mRNA in mustard seedling cotyledons is regulated by phytochrome. Planta 181: 327-334

Shiraishi N, Sato T, Ogura N, Nakagawa H (1992) Control by glutamine of the synthesis of nitrate reductase in cultured spinach cells. Plant Cell Physiol 33: 727-731

Siegel LM, Wilkerson JO (1989) Structure and function of spinach ferredoxin-nitrite reductase. In: Wray JL, Kinghorn JR (eds) Molecular and genetic aspects of nitrate assimilation. Oxford Science Publ, Oxford, pp 263-283

Sivasankar S, Oaks A (1996) Nitrate assimilation in higher plants: the effect of metabolites and light. Plant Physiol Biochem 34: 609-620

Solomonson LP Barber MJ (1990) Assimilatory nitrate reductase: functional properties and regulation. Annu Rev Plant Physiol Plant Mol Biol 41: 225-253

Stallmeyer B, Nerlich A, Schiemann J, Brinkman H, Mendel RR (1995) Molybdenum cofactor biosynthesis: the *Arabidopsis* cDNA *cnx1* encodes a multifunctional two-domain protein homologous to a mammalian neuroprotein, the insect protein Cinnamon and three *E. coli* proteins. Plant J 8: 751-762

Stoehr C (1999) Relationships of nitrate supply with grow rate, plasma membrane-bound and cytosolic nitrate reductase, and tissue nitrate content in tobacco plants. Plant Cell Environ 22: 169-177

Streit L, Martin BA, Harper JE (1987) A method for the separation and purification of the three forms of nitrate reductase present in wild-type soybean leaves. Plant Physiol 84: 654-657

Su W, Mertens JA, Kanamaru K, Campbell WH, Crawford NM (1997) Analysis of wild-type and mutant plant nitrate reductase expressed in the methylotrophic yeast *Pichia pastoris*. Pant Physiol 115: 1135-1143

Suty L, Moureaux T, Leydecker MT, Delaserve BT (1993) Cytokinin affects nitrate reductase expression through the modulation of polyadenylation of the nitrate reductase messenger RNA transcript. Plant Sci 90: 11-19

Suzuki A, Oaks A, Jacquot JP, Vidal J, Gadal P (1985) An electron transport system in maize roots for reactions of glutamate synthase and nitrite reductase. Plant Physiol 78: 374-378

Truong H-N, Meyer C, Daniel-Vedele F (1991) Characteristics of *Nicotiana tabacum* nitrate reductase protein produced in *Saccharomyces cerevisiae*. Biochem J 278: 393-397

Van Camp W, Van Montagu M, Inzé D (1998) H_2O_2 and NO: redox signals in disease resistance. Trends Plant Sci 3: 330-334

Vaucheret H, Kronenberger J, Rouzé P, Caboche M (1989) Complete nucleotide sequence of the two homologous tobacco nitrate reductase genes. Plant Mol Biol 12: 597-600

Vaucheret H, Marion-Poll A, Meyer C, Faure JD, Marin E, Caboche M (1992) Interest in and limits to the utilization of reporter genes for the analysis of transcriptional regulation of nitrate reductase. Mol Gen Genet 235: 259-268

Vincentz M, Caboche M (1991) Constitutive expression of nitrate reductase allows normal growth and development of *Nicotiana plumbaginifolia* plants. EMBO J 10: 1027-1035

Vincentz M, Moureaux T, Leydecker M T, Vaucheret H, Caboche M (1993) Regulation of nitrate and nitrite reductase expression of *Nicotiana plumbaginifolia* leaves by nitrogen and carbon metabolites. Plant J 3: 315-324

Weiner H, Kaiser W (1999) 14-3-3 proteins control proteolysis of nitrate reductase in spinach leaves. FEBS Lett 455: 75-78

Wray JL (1989) Molecular and genetic aspects of nitrite reduction in higher plants. In: Wray JL and Kinghorn JR (eds) Molecular and genetic aspects of nitrate assimilation. Oxford Science Publ, Oxford, pp 244-262

Wilkinson JQ, Crawford NM (1993) Identification and characterization of a chlorate-resistant mutant of *Arabidopsis thaliana* with mutations in both nitrate reductase structural genes *NIA1* and *NIA2*. Mol Gen Genet 239: 289-297

Yamasaki H, Sakihama Y, Takahashi S (1999) An alternative pathway for nitric oxide production in plants: new features of an old enzyme. Trends Plant Sci 4: 128-129

Yu X, Sukumaran S, Márton L (1998) Differential expression of the Arabidopsis *Nia1* and *Nia2* genes. Cytokinin-induced nitrate reductase activity is correlated with increased *Nia1* transcription and mRNA levels. Plant Physiol 116: 1091-1096

Mechanisms and Regulation of Ammonium Uptake in Higher Plants

Nicolaus VON WIRÉN[1], Alain GOJON[2], Sylvain CHAILLOU[3] and D. Raper[4]

The significance of ammonium transport for nitrogen nutrition of plants

In soils, ammonium (NH_4^+) mainly results from the mineralisation of organic matter and represents besides nitrate (NO_3^-) the quantitatively most important source of nitrogen (N) for plant nutrition. In well-aerated agricultural soils, however, average annual NH_4^+ concentrations are often 10 to 1000 times lower than those of NO_3^-, rarely exceeding 50 µM (Marschner 1995). Despite these low concentrations in soils, NH_4^+ uptake by plant roots can proceed at very high rates, due to the presence of transport systems in the root plasma membrane with a particularly high substrate affinity. Indeed, NH_4^+ uptake is of major importance for N nutrition under numerous circumstances. On the one hand, NH_4^+ nutrition plays an essential role in waterlogged and acid soils, or in cold climates where nitrification is inhibited (Marschner 1995). On the other hand, under mixed N nutrition (NO_3^- plus NH_4^+), NH_4^+ is often the preferential form of N taken up by the plant (Sasakawa and Yamamoto 1978; Gojon et al. 1986; Glass and Siddiqi 1995; Gazzarrini et al. 1999). NH_4^+ is also probably the main form of N exported from symbiotic N_2-fixing microorganisms to their host plants, thereby making a major contribution to N nutrition in several plant families (Udvardi and Day 1997). In root nodules, NH_4^+ is transported across the symbiosome membrane which segregates the bacteroids from the plant cytosol (see Chap. 3). In this case, NH_4^+ concentrations in the plant cytosol can be about 50-fold lower than in the bacteroids (Streeter 1989), requiring low-affinity and high-capacity transport systems on the plant side, to ensure an efficient import of microbially fixed N (Tyerman et al.

1. Zentrum für Molekularbiologie der Pflanzen, Universität Tübingen, Morgenstelle 1, D – 72076 Tübingen, Germany. *E-mail:* vonwiren@uni-tuebingen.de
2. Biochimie et Physiologie Moléculaire des Plantes, ENSAM/INRA/UM2/CNRS, URA 2133, place Viala, F-34060 Montpellier cedex, France. *E-mail:* gojon@ensam.inra.fr
3. Laboratoire de la Nutrition Azotée des Plantes, INRA, route de St-Cyr, F-78026 Versailles cedex, France. *E-mail:* chaillou@versailles.inra.fr
4. David Raper Jr, Soil Science Department, NCSU, Raleigh, NC, 27695-7619, USA. david_raper@ncsu.edu

1995). Most importantly, NH_4^+ uptake plays a crucial role in compensating losses of root NH_4^+, which occur irrespective of the form of primarily acquired N. Due to intense and continuous generation of NH_4^+ in the root from amino acid catabolism, a significant efflux of NH_4^+ (or NH_3) has been shown to occur permanently, even in NO_3^- – grown plants (Morgan and Jackson 1988; Feng et al. 1994, 1998). Thus, retrieval of NH_4^+ is required as a major house-keeping function of root NH_4^+ transport systems (Fig. 1). The importance of this retrieval function is illustrated by the yeast mutant 31019b, which is deficient in all three endogenous NH_4^+ transporters of the MEP (methylammonium permease) family (Δmep1, Δmep2, Δmep3) and shows reduced growth rates on any N source, as NH_4^+ leaks permanently out of the cell (Marini et al. 1997a).

Several lines of evidence indicate that membrane transport of NH_4^+ in the shoot is also of particular importance. First, plant populations act as sinks for atmospheric NH_3 (Cowling and Lockyer 1981), which is believed to enter the leaves mainly through stomata (Husted and Schjoerring 1996), before subsequent uptake by mes-

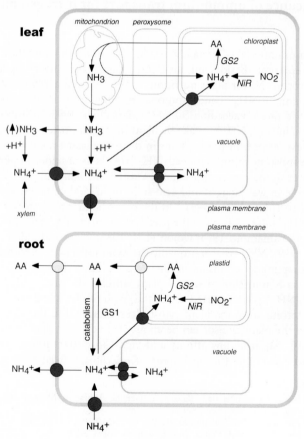

Fig. 1 : Possible localisation of ammonium transporters (*black spots*) in root and leaf cells of plants. Amino acid catabolism also occurs in leaf cells but is not depicted.
GS – Glutamine Synthetase. *NiR* – Nitrate Reductase.

ophyll cells (Yin et al. 1996). Second, in agricultural systems, leaf application of NH_4^+ – based fertilizers is successfully used to alleviate plant N deficiency (Bowman and Paul 1992). Third, NH_4^+ concentrations in the xylem as high as 2 mM, have been reported, even under NO_3^- nutrition (Cramer and Lewis 1993; Husted and Schjoerring 1995). Although such high NH_4^+ concentrations may not be common (Glass and Siddiqi 1995), they indicate that significant amounts of N can be translocated to the shoot in the form of NH_4^+, which is then taken up by mesophyll cells, even against a concentration gradient (Finnemann and Schjoerring 1999). Fourth, leaves release NH_3 originating from photorespiration, degradation of proteins, transamination reactions or the intracellular cleavage of phenylalanine by phenylalanine ammonia lyase (Joy 1998). As a substantial part of this NH_3 is reassimilated (see Chap. 2.2), NH_3 must be reimported, probably via a transport process across the plasma membrane, supposing that part of the NH_3 is reprotonated in the apoplast (Fig. 1).

Molecular Characterisation of Ammonium Transporters

Ammonium Transporters from Higher Plants

Physiological studies in different organisms indicated that, in most cases, NH_4^+ and its toxic analogue methylammonium share the same uptake path (Kosola and Bloom 1994; Venegoni et al. 1997). Thus, screening for mutants resistant to toxic levels of methylammonium appeared to be a promising approach to identify genes involved in NH_4^+ transport, and resulted in the isolation of yeast and *Chlamydomonas reinhardtii* mutants defective in methyl-/ammonium uptake (Dubois and Grenson 1979; Franco et al. 1987). Functional complementation of a yeast double mutant defective in two NH_4^+ uptake systems (*mep1-1, mep2-1*) allowed the isolation of NH_4^+ transporter genes from yeast (*MEP1*, Marini et al. 1994) and in parallel from *Arabidopsis thaliana* (*AtAMT1;1*, Ninnemann et al. 1994). Both transporters encode integral membrane proteins with high amino acid sequence identity, that restore high rates of ^{14}C-methylammonium uptake in the yeast double mutant as well as its growth on a low external NH_4^+ supply. In yeast, MEP1p and AtAMT1;1 are high-affinity transporters for NH_4^+, that exhibit negligible permeability for di- and trimethylamine and for other monovalent cations such as K^+. Their activities strongly depend on the proton motive force and show a pH optimum at 7.0.

As expected from Southern blot analysis, further members of the *AMT1* gene family in *A. thaliana* exist and were recently isolated, namely *AtAMT1;2* and *AtAMT1;3* as full-length clones (Gazzarrini et al. 1999) and *AtAMT1;4* and *AtAMT1;5* as partial sequences (S. Gazzarrini and N. von Wirén, unpubl.). In the yeast triple mutant 31019b (*Δmep1, Δmep2, Δmep3*), AtAMT1;1, 1;2 and 1;3 mediated high-affinity uptake of NH_4^+ and methylammonium. However, different substrate affinities were observed. While AtAMT1;2 and 1;3 showed K_m values between 11 and 40 µM for both substrates, AtAMT1;1 was highly selective with K_m values of 8 µM and lower than 0.5 µM for methylammonium and NH_4^+, respectively (Gazzarrini et al. 1999). Such differences in the sensitivity to methylammonium provide an explanation why, in particular at higher external concentrations, methylammonium can be a competent inhibitor of NH_4^+ uptake (Kosola and Bloom 1994).

Low-affinity NH_4^+ transporters have not yet been isolated from plants. A possible exception are K^+ transporters and K^+ channels, some of which have been shown to possess a considerable conductivity for NH_4^+ (Schachtman and Schroeder 1994; White 1996). However, there are no structural similarities between these transporter classes, and NH_4^+ conductivity in K^+ transporters is likely to represent a transporter side activity rather than an alternative pathway for controlled NH_4^+ entry, at high external concentrations. Furthermore, as NH_4^+ entry by K^+ channels requires NH_4^+ concentrations of 0.1 to 1 mM (White 1996) and average soil concentrations rarely rise beyond 50 µM (Marschner 1995), the ecological significance of this transport path for field-grown plants might be restricted to periods of high N mineralisation, or NH_4^+ fertiliser application.

In yeast, however, a low-affinity NH_4^+ transporter has been identified with the isolation of Mep3p. Mep3p possesses the lowest substrate affinity with a K_m of 1.4-2.1 mM, compared to Mep1p and Mep2p exhibiting K_m values of 5-10 and 1-2 µM, respectively. This shows that the Mep transporter family ensures NH_4^+ uptake over a large concentration range (Marini et al. 1997a). Since the yeast genome sequencing project has now been completed, no further member of the *MEP* gene family is expected.

Recently, a new gene arose from the *A. thaliana* genome sequencing project, preliminarily called *AtAMT2;1* (acc. no. AC003028), which possesses higher sequence similarities to the yeast Mep proteins than to the *A. thaliana* AMT1 transporters. Although functional data are not yet available, it is tempting to speculate whether this gene might represent a member of a new subfamily, with special features.

Further AMT1 homologues have been isolated from other plant species, one from rice (*OsAMT1;1*, von Wirén et al. 1997), and three from tomato (*LeAMT1;1-3*, Lauter et al. 1996; von Wirén et al. 2000). Interestingly, in tomato, *LeAMT1;1* and *LeAMT1;2* are preferentially expressed in root hairs. This confined localisation is considered to be highly favourable for NH_4^+ acquisition at the soil root interface. Since NH_4^+ is strongly adsorbed to soil constituents, especially to clay minerals, and the movement of NH_4^+ in soils is mainly brought about by diffusion (Marschner 1995), NH_4^+ uptake sites in root hairs can cut diffusion distances and thereby potentially raise the NH_4^+ uptake efficiency of the plant (Lauter et al. 1996; von Wirén et al. 2000). In contrast, LeAMT1;3 is a preferentially leaf-expressed NH_4^+ transporter and carries unique characteristics when compared to the other AMTs from tomato, *A. thaliana* or rice (von Wirén et al. 2000). First, with less than 63% sequence homology to the other AMTs, LeAMT1;3 is the most distantly related AMT1 member so far described; second, LeAMT1;3 has an extremely short N terminus of 14 amino acids compared to a range of 35-54 amino acids, as observed in the N termini of other AMT members; and third, the mRNA of LeAMT1;3 possesses two untranslated open reading frames in front of the main open reading frame. Additional untranslated open reading frames on the mRNA are thought to control translational efficiency (Wang and Wessler 1998), a feature which has so far mostly been found in regulatory genes controlling growth and development (Kozak 1987). Although the consequences of these structural differences have to be examined in detail, they may point to a different posttranscriptional regulation of *LeAMT1;3*.

GmSAT1 Does Not Function as an Ammonium Transporter in Yeast

Electrophysiological studies with symbiosome membranes from legumes, identified a monovalent cation channel with preference for NH_4^+ that most likely represents the main route of N export from the symbiosome to the plant (Tyerman et al. 1995). To isolate the corresponding genes, the yeast NH_4^+ transport mutant 26972c (*mep1-1, mep2-1*) was complemented with a cDNA library prepared from soybean nodules allowing the isolation of GmSAT1 (Kaiser et al. 1998). Western blot analysis localised the protein in the symbiosome membrane, and patch-clamp studies with yeast spheroblasts expressing SAT1 showed an inward current of NH_4^+, which was inhibited by Ca^{2+}. However, the unusual structure of SAT1, with only one transmembrane domain and sequence stretches with a high homology to basic helix-loop-helix motifs, raised doubts about its function as an NH_4^+ transporter. These doubts have now been confirmed on an experimental basis, since the complementation of strain 26972c by GmSAT1, is, in fact, dependent on the endogenous Mep3p transporter (Marini et al. 2000), indicating that GmSAT1 is most likely not an NH_4^+ transporter.

Ammonium Transporters as Potential Sensors

In yeast, NH_3 has been shown to act as a signal for long-distance cell-to-cell communication, provoking morphological changes to inhibit growth of competing neighbouring cells (Palkova et al. 1997). Although NH_4^+ is mainly regarded as a nutrient, it is also able to stimulate a morphological adaptation in yeast that causes filamentous outgrowth of pseudohyphae. Recently, the NH_4^+ transporter Mep2p from yeast has been shown to act as an NH_4^+ sensor required for pseudohyphal differentiation, in response to limited NH_4^+ supply. However, only Mep2p, and not Mep1p or Mep3p, is able to generate a signal regulating filamentous growth (Lorenz and Heitman 1998). This is accompanied by a much higher expression of *MEP2* on poor N sources and the fairly high substrate affinity of Mep2p (Marini et al. 1997a). Thus, Mep2p contributes to NH_4^+ uptake in yeast and at the same time elicits a signal transduction cascade, leading to a morphological adaptation to improve nutrient uptake. Whether similar sensing functions of nutrient transporters also exist in plants remains to be shown, but root hair-expressed NH_4^+ transporters, like LeAMT1;1 and LeAMT1;2 in particular, might be candidates for transporters that have physiological functions going beyond NH_4^+ uptake.

The Superfamily of MEP/AMT Ammonium Transporters in Different Organisms

Analysis of sequence databases revealed that genes of the MEP/AMT family can be found in bacteria, fungi, plants, animals and humans. Most of these sequence determinations result from genome sequencing efforts, and functional data are available for a minority only (for example, Marini et al. 1994; Ninnemann et al. 1994; Siewe et al. 1996; Taté et al. 1998; Van Dommelen et al. 1998). Surprisingly, Marini et al. (1997b) reported that the Rhesus blood group polypeptides share significant sequence similarity with the defined MEP/AMT family of NH_4^+ transporters. This raised the possibility that Rhesus proteins might be NH_4^+ transporters or might originate from NH_4^+ transporters and perform some other related function.

Physiology of Ammonium Uptake by Roots

Kinetics

As is the case for many other ions, net uptake of NH_4^+ by plant roots is the difference between concomittant influx and efflux (Morgan and Jackson 1988), and exact assessment usually requires measurements of influx of $^{13}NH_4^+$ or $^{15}NH_4^+$ during short-term uptake experiments (Clarkson et al. 1996). However, compared to other nutrients, the interpretation of tracer incorporation over short periods of time is particularly difficult in the case of NH_4^+. Due to electrical effects in the root-free space and continuous efflux of endogenous NH_4^+ generated by amino acid catabolism in the roots, there is a possibility that short-term incorporation of isoto-pically labelled NH_4^+, significantly underestimates NH_4^+ fluxes across the plasma membrane (Lee and Ayling 1993; Feng et al. 1994). Despite these difficulties, influx of NH_4^+ has been shown to exhibit the classical biphasic uptake pattern that has been described for many ions (Glass and Siddiqi 1995). In *Lemna gibba* and in rice roots, at concentrations below 1 mM NH_4^+, influx via a saturable high-affinity transport system (HATS) could be distinguished from uptake via a nonsaturable low-affinity transport system (LATS), which was active at concentrations above 1 mM external NH_4^+ (Ullrich et al. 1984; Wang et al. 1993). The kinetic parameters (apparent K_m and V_{max}) calculated for the HATS, show a large variability depend-ing on the species and environmental conditions investigated (Table 1). This may indicate that several transporters, with different kinetic properties, contribute to a variable extent to the high-affinity NH_4^+ uptake. This idea is now strongly sup-ported by the fact that the AMT1 families found in higher plants most probably include a multiplicity of transporters displaying significant differences in their affinity for NH_4^+. This may also explain why complex multiphasic patterns have sometimes been observed for NH_4^+ uptake kinetics. Nevertheless, any proposed

Table 1. Diversity of K_m values for root NH_4^+ uptake illustrated by some examples taken from the literature

Species	K_m (μM)	Parameter assayed	Growth conditions	Reference
Arachis hypogaea L.				
Cajanus Cajan L.				
Cicer arietinum L.	In the range of 50-120	Root $^{15}NH_4^+$ influx (2-h labellings)	Diluted Johnson solution	Rao et al. (1993)
Pennisetum glaucum L.				
Sorghum bicolor L.				
Zea mays L.				
Oryza sativa L.	32 90 188	Root $^{13}NH_4^+$ influx (10-min labellings)	1 μM NH_4^+ 100 μM NH_4^+ 1000 μM NH_4^+	Wang et al. (1993)
Hordeum vulgare L.	160	Root net NH_4^+ uptake	N-free solution	Mäck and Tischner (1194)
Oryza sativa L.	360 38	Root $^{13}NH_4^+$ influx (10-min labellings)	100 μM NH_4^+/ 2 μM K^+ 100 μM NH_4^+/ 200 μM K^+	Wang et al. (1996)
Picea glauca (Moench)	20-40	Root $^{13}NH_4^+$ influx (10-min labellings)	N-deprived plants	Kronzucker et al. (1996)

link between molecular and physiological observations on NH_4^+ uptake still remains tentative, because none of the isolated transporters has been functionally characterized *in planta*. Future work using mutants or transgenic lines affected in the expression of the AMTs is required to confirm the interesting similarities found between kinetic properties and regulation (see below) of AMT transporters and NH_4^+ uptake by roots.

Electrophysiology

Electrophysiological studies indicate that HATS and LATS activities are both electrogenic, resulting in a depolarisation of the plasma membrane (Walker et al. 1979; Ullrich et al. 1984; Wang et al. 1994). The kinetics of membrane depolarisation are consistent with those of NH_4^+ influx, but indicate that HATS and LATS have different electrical effects, suggesting that they use different types of energy coupling (Wang et al. 1994). Using protonophores and ATPase inhibitors, a strong dependence of NH_4^+ transport by the HATS on the H^+ electrochemical gradient has been demonstrated in rice (Wang et al. 1994). It is noteworthy that a similar observation has been made for the AtAMT1;1 transport activity in yeast (Ninnemann et al. 1994). However, it is still unclear whether the dependence of NH_4^+ transport on the proton motive force results from a H^+: NH_4^+ symport or an NH_4^+ uniport. Entry of NH_4^+ via LATS might occur through an electrogenic uniport, that could be mediated either by low-affinity NH_4^+ transporters or as a side activity of less specific cation channels, in particular K^+ transporters and channels (Schachtman and Schroeder 1994; Glass and Siddiqi 1995; White 1996).

The energetics of NH_4^+ uptake, which could be helpful to determine the mechanism of NH_4^+ influx (symport or uniport), is still under debate. This is mostly due to the uncertainties related to the determination of the cytoplasmic concentration of NH_4^+. Studies using NMR spectroscopy or capillary electrophoresis reported values ranging from <15 μM to the lower millimolar range (Lee and Ratcliffe 1991; Roberts and Pang 1992; Kawamura et al. 1996). Millimolar concentrations have also been measured electrophysiologically using triple-barrelled microelectrodes, by which NH_4^+ and H^+ could be determined simultaneously (D. Wells and A.J. Miller, unpubl.). In contrast, approaches based on compartmental analysis generally yielded much higher values for cytoplasmic NH_4^+, with values of even up to 76 mM being recorded (Glass and Siddiqi 1995; Kronzucker et al. 1995).

Membrane Transport of Ammonium Versus Ammonia

Recent studies have led to a reconsideration of the question of whether NH_4^+ or NH_3 is the transported form. Using *E. coli* and yeast mutants, Soupene et al. (1998) were able to show that the AmtB and Mep NH_4^+ transporters from *E.coli* and yeast, respectively, are required for growth at low NH_4^+ concentrations only when the external medium pH is lower than 7. This has been taken as evidence for the transport of NH_3 by AmtB and Mep transporters, because NH_3 is the limiting species at low pH. Moreover, a stronger cytosolic alkalinisation in root cells upon addition of NH_3 at pH 7, compared to pH 5, has been interpreted as an enhanced internal protonation of absorbed NH_3 at pH 7, rather than an enhanced proton extrusion resulting from NH_4^+ uptake (Kosegarten et al. 1997). However, these suggestions are at odds with many reports favouring the idea of NH_4^+ rather than NH_3 transport, in plants. Several studies failed to find any increase in NH_4^+/NH_3 uptake at elevated pH (MacFarlane and Smith 1982; Schlee and Komor 1986; Dyr-Jensen and Brix 1996), which should be associated with higher NH_3 concentration in the medium. For example, in rice, transport of NH_4^+ via LATS was reduced by about 30% with a pH shift from 7.5 to 9.0, despite a predicted increase of free NH_3 from less than 1% of total reduced N at pH 7.5, to 36% at pH 9.0 (Wang et al. 1993). Furthermore, the fact that NH_4^+ uptake is electrogenic, argues strongly for a major contribution of NH_4^+ in N transport across the plasmalemma (Ullrich et al. 1984; Wang et al. 1994, Glass and Siddiqi 1995). Experiments with *Chara corallina* clearly showed that a non-electrogenic component of methylamine uptake could be observed only when the pH was raised above 9, which is in accordance with the hypothesis that transport of the charged species is the dominant mechanism at lower pH values (Walker et al. 1979).

Regulation of Ammonium Uptake

Nitrogen nutrition of plants has the almost unique feature that it can be ensured independently by the root uptake of two ions, NH_4^+ and NO_3^-. Since N uptake by the root system is controlled by the N demand of the whole plant, and finely regulated to match the N requirements for growth, the question is raised of a coordination and coregulation of NO_3^- and NH_4^+ uptake (Glass and Siddiqi 1995). The root system has a very high capacity for the uptake of both ions, but this capacity is expressed under only a very limited range of environmental conditions. N starvation of plants for a few days results in an increased capacity for both NO_3^- and NH_4^+ uptake, which corresponds to a specific stimulation at the influx level (Lee and Rudge 1986; Morgan and Jackson 1988; Lee 1993; Glass and Siddiqi 1995; Gazzarrini et al. 1999). This is an adaptative response of the uptake systems, which compensates for the decrease in external availability of either NO_3^- or NH_4^+, and tends to maintain N uptake at a value consistent with the N demand of the plant. Root N demand, however, has a clear priority over shoot N demand, leading to a rapid decrease in N translocation to the shoot in the early stage of N deficiency (Kronzucker et al. 1998). An adaptation to the plant N demand has been described in detail for NH_4^+ influx in rice, where a decrease in external NH_4^+ concentration from 1 mM to 2 μM resulted in both a decrease in K_m (from 188 to 32 μM) and an increase in V_{max} of the HATS (Wang et al. 1993). This demonstrates the involvement of a feedback regulation of high-affinity NH_4^+ uptake at both, the affinity and capacity levels of uptake. Moreover, this notion is well supported by data obtained at the molecular level in *A. thaliana*, where *AMT1* isogene expression was investigated under different physiological conditions and correlated with $^{15}NH_4^+$ influx (Gazzarrini et al. 1999). mRNA transcript levels of *AtAMT1;1*, which was the only transporter exhibiting a substrate affinity in the nanomolar range, steeply increased with $^{15}NH_4^+$ influx in roots when N provision to roots became limiting. Transcript levels of *AtAMT1;2* and *AtAMT1;3* were less or not at all affected by N starvation, suggesting that N deficiency turns on the transporter with the highest substrate affinity. Most probably, high-affinity uptake of NH_4^+ additionally underlies a feedback repression by reduced N, as suggested by physiological data (Wang et al. 1993). In contrast, feedback repression was not observed for low-affinity NH_4^+ uptake, which increased with N supply during preculture. The signals responsible for the strong down-regulation of HATS-mediated influx, in response to elevated NH_4^+ supply during growth, might be complex. Feedback signals may result from accumulated NH_4^+ in root cells, or from other forms of reduced N, since it has been shown that exogenously supplied NH_4^+ and amino acids both decreased net NH_4^+ uptake in wheat (Causin and Barneix 1993; Glass and Siddiqi 1995). Evidence for a regulatory role of glutamine has been obtained by monitoring $^{13}NH_4^+$ influx and *AtAMT1* expression levels in *A. thaliana* roots, which both were negatively correlated with root glutamine concentrations (Rawat et al. 1999). Application of methionine sulphoximine, a potent inhibitor of glutamine synthetase, failed to distinguish between putative repressive effects of NH_4^+ and amino acids. Depending on experimental conditions and plant species, methionine sulphoximine resulted in either an increased or decreased NH_4^+ uptake (Lee et al. 1992; Causin and Barneix 1993; Lee and Ayling 1993; Feng et al. 1994; Rawat et al. 1999).

Another regulatory mechanism modulating the efficiency of NH_4^+ uptake by roots is the stimulation by light and/or photosynthesis. Like NO_3^- uptake, NH_4^+ uptake by root cells is diurnally regulated, with a maximum uptake during the light period, and is stimulated by sugar supply (Hatch et al. 1986; Ourry et al. 1996; Gazzarrini et al. 1999; L. Lejay and A. Gojon, unpubl.). Again, this corresponds to the regulation of *AMT1* gene expression in *A. thaliana*. While all three *AtAMT1* genes showed a diurnal variation in root expression, only *AtAMT1;3* transcript levels peaked simultaneously with $^{15}NH_4^+$ influx at the end of the light period, suggesting that regulation of *AtAMT1;3* may provide a link between NH_4^+ uptake and carbon provision in roots (Gazzarrini et al. 1999). This is confirmed by the observation that *AtAMT1;3* expression is sugar-inducible (L. Lejay and A. Gojon, unpubl.). The fact that N starvation and light do not affect the expression of the same *AtAMT1* genes, suggests that the regulation of NH_4^+ uptake by either N or C status of the plant involves different mechanisms.

Studies in tomato also provided evidence that the AMT1 transporters are strongly regulated at the transcriptional level, and yielded some interesting differences compared to the situation in *A. thaliana*. In tomato roots, *LeAMT1;1* was most abundant under N-limiting growth conditions – similar to *AtAMT1;1* – while *LeAMT1;2* was induced with the supply of NH_4^+ and even more strongly with the supply of NO_3^- (von Wirén et al. 2000). Such a feature has not yet been observed for any of the *AMT1* transporters in *A. thaliana*. The opposite regulation of the two transporter genes in tomato indicates that different NH_4^+ transporters are required to meet the plant N demand under adverse soil N supply. In leaves, expression of *LeAMT1;2* and *LeAMT1;3* was slightly higher under NO_3^- nutrition and low CO_2, suggesting a possible involvement in the retrieval of photorespiratory NH_3/NH_4^+ (Fig. 1). However, expression of *LeAMT1;2* and *LeAMT1;3* showed a reciprocal diurnal regulation with highest transcript levels of *LeAMT1;3* during the dark period and highest levels of *LeAMT1;2* after the onset of light (von Wirén et al. 2000). These results indicated that LeAMT1;2 is most probably involved in the uptake of xylem-derived NH_4^+, or indeed, in the retrieval of photorespiratory NH_3, whereas it is speculated that LeAMT1;3 might retrieve NH_4^+ derived from reactions involving other light-repressed enzymes, such as asparagine synthetase and glutamate dehydrogenase (Lam et al. 1995, 1998).

Physiological Changes in Ammonium-Grown Plants

One of the most prominent physiological reactions of plants supplied with NH_4^+ as the sole N source, is a strong acidification of the root medium (Fig. 2). As indicated by membrane depolarisation, NH_4^+ rather than NH_3 enters the root cells, at least at neutral and acid pHs (Ayling 1993; Glass and Siddiqi 1995). Thus, elevated H^+ concentrations in the rhizosphere are unlikely to result from an NH_4^+ deprotonation prior to transport. Inside the cell, glutamine synthetase assimilates NH_3 into glutamine (see Chap. 2.2) and the released H^+ could be excreted via the H^+– ATPase and compensate for the depolarisation of membrane potential. In contrast, NO_3^- uptake is a H^+ – consuming process, in which $2H^+$ enter the cell via a $2H^+$/ NO_3^- symport (Glass et al. 1992) and one H^+ is consumed to neutralise the hydroxyl ion generated during nitrite reduction. Thus, fewer H^+ can be secreted than taken up, resulting in a pH shift (Fig. 2).

Supply of NH₄+

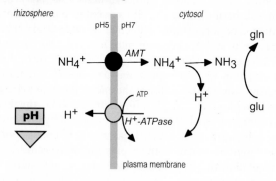

Supply of NO₃-

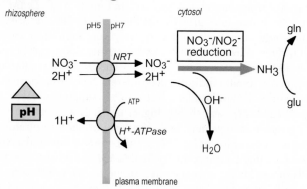

Fig. 2 : Physiological response of plant roots to the supply of ammonium or nitrate as main nitrogen source. The change in rhizosphere pH is a resulting effect from the ion uptake balance and endogenous pH-stat mechanisms affecting proton extrusion. Therefore, the scheme represents only the main physiological reactions leading to a disequilibrium of external protons

The consequences of this acidification may be harmful, particularly in soil-free culture, where accumulating protons cause a decrease in growth rates, mainly through reduced cation uptake to the level where roots become unable to take up further NH_4^+ (Tolley-Henry and Raper 1986). However, in artificial substrates this rhizosphere acidification can be alleviated by frequent renewal of the nutrient solution or by the automatic addition of a base in response to pH shifts. In pH-controlled or buffered media, NH_4^+ – fed plants can grow at rates comparable to NO_3^- – fed plants, although several characteristics of their metabolism are altered, resulting in the so-called ammonium syndrome (Mehrer and Mohr 1989). In comparison to NO_3^- nutrition, NH_4^+ – fed plants show lower contents of mineral cations and organic anions but higher levels of amino acids, particularly in the roots (Chaillou et al. 1991). Since amino acid accumulation is most pronounced in the roots of N-deprived plants that have been resupplied with NH_4^+ as the sole nitrogen source, root NH_4^+ assimilation appears to be rapid (Rideout et al. 1994), while translocation of amino acids from roots to the shoot is limited. These high amino acid concentrations in root cells inhibit NH_4^+ uptake, probably by enhancing NH_4^+ efflux (Rideout et al. 1994) and by downregulating NH_4^+ uptake (Rawat et al. 1999). NH_4^+ alone is thus not an efficient form of N to recover from N starvation.

The low concentration of organic anions in NH_4^+ – supplied plants appears in part to be a consequence of the absence of NO_3^- reduction in the plant, since at least in C_3 plants, the synthesis of organic acids, especially of malic acid, is promoted by NO_3^- reduction, mostly occurring in shoots (Deng et al. 1989). As organic acids have an important role in turgor regulation, NH_4^+ – fed plants may have difficulties in maintaining turgor pressure. Some species, however, circumvent this problem, by either using other osmotically active ions, such as Cl^- in barley, or uncoupling PEP carboxylase activity and nitrate reductase activity, as, for example, in C_4 plants, where organic acid synthesis does not decrease when supplied with NH_4^+ as the sole N source (De Armas et al. 1992). Other types of adaptation to NH_4^+ nutrition can be found in species that grow in areas where nitrification is repressed. For example, the metabolic pathway for synthesis of quinic and shikimic acids allows coniferous trees growing on acid soils to produce organic acids, in the absence of nitrate reduction (Salsac et al. 1987).

Frequently, optimal plant growth occurs with mixed NO_3^- plus NH_4^+ nutrition, when the plant can benefit from the attributes of each N form. A combination of NO_3^- and NH_4^+ uptake usually allows a better pH control within the rhizosphere, because the generation of hydroxyl ions during NO_3^- uptake compensates for the generation of protons during NH_4^+ uptake (Fig. 2). In soil-free substrates a mixture of four parts NO_3^- plus one part NH_4^+ commonly approaches the optimum for growth. When grown on a mixture of NH_4^+ and NO_3^-, plants may take up more N than on either NO_3^- or NH_4^+ alone (Vessey et al. 1990; Kronzucker et al. 1999), and a greater proportion of the NH_4^+ that is taken up is assimilated into the protein fraction than of the NO_3^- (Volk at al. 1992). The uptake and reduction of NO_3^-, on the other hand, contributes to maintenance of turgor pressure both by accumulation of NO_3^- itself within the vacuole and, as shown earlier, by the synthesis and accumulation of organic acids during NO_3^- reduction (Salsac et al. 1987). The organic acids produced by NO_3^- reduction can also serve as carbon skeletons for amino acid synthesis.

Under exclusive supply of NH_4^+, growth of roots and leaves is inhibited and fully expanded leaves show a decreased size and number of cells. This coincides with a strong reduction of cytokinin concentrations in the xylem sap, as observed in tobacco plants, 24 h after transfer to sole NH_4^+ nutrition (Walch-Liu et al. 2000). Since cytokinins are involved in the regulation of both cell division and cell elongation, it seems likely that the presence of NO_3^- is required to maintain biosynthesis and/or root to shoot transfer of cytokinins at a level that is required for full leaf expansion.

Conclusions

The results obtained from the molecular cloning and functional characterisation of NH_4^+ transporter genes and proteins from plants closely match the physiological data obtained from short-term uptake studies. High-affinity transporters for NH_4^+ have been shown to be organised in multigene families, which are highly conserved among the eukaryotes. These gene families consist of several members encoding transport proteins that can differ in localisation, biochemical properties, or regulation such as substrate inducibility. From the expression analysis obtained so far, it can be deduced that the high-affinity systems consist of members that are either constitutive or inducible, while no low-affinity NH_4^+ transporters have been described so far in plants – except K^+ channels, which might trigger NH_4^+ uptake as a side activity. However, in one transporter family, the MEP transporters from yeast, low- and high-affinity transporters for NH_4^+ have been isolated (Marini et al. 1997a). Thus, related members of one family may be responsible for both low- and high-affinity NH_4^+ transport.

Acknowledgements. A large part of the work of the authors was supported by the European Community BIOTECH4 program EURATINE.

References

Ayling SM (1993) The effect of ammonium ions on membrane potential and anion flux in roots of barley and tomato. Plant Cell Environ 16: 297-303

Bowman DC, Paul JL (1992) Foliar absorption of urea, ammonium, and nitrate by perennial ryegrass surf. J Am Soc Hortic Sci 117: 75-79

Causin HF, Barneix AJ (1993) Regulation of NH_4^+ uptake in young wheat plants: effect of root ammonium concentration and amino acids. Plant Soil 151: 211-218

Chaillou S, Vessey JK, Morot-Gaudry JF, Raper CD, Henry LT, Boutin JP (1991) Expression of characteristics of ammonium nutrition as affected by pH of the root medium. J Exp Bot 42: 189-196

Clarkson DT, Gojon A, Saker LR, Wiersema PK, Purves JV, Tillard P, Arnold GM, Paans AJM, Vaalburg W, Stulen I (1996) Nitrate and ammonium influxes in soybean (*Glycine max*) roots: direct comparison of ^{13}N and ^{15}N tracing. Plant Cell Environ 19: 859-868

Cowling DW, Lockyer DW (1981) Increased growth of ryegrass exposed to ammonia. Nature 292: 337-338

Cramer MD, Lewis OAM (1993) The influence of nitrate and ammonium on the growth of wheat (*Triticum aestivum*) and maize (*Zea mays*) plants. Ann Bot 72: 359-365

De Armas R, Valadier MH, Champigny ML, Lamaze T (1992) Influence of ammonium and nitrate on the growth and photosynthesis of sugarcane. J Plant Physiol 140: 531-535

Deng MD, Moureaux T, Lamaze T (1989) Diurnal and circadian fluctuation of malate levels and its close relationship to nitrate reductase activity in tobacco leaves. Plant Sci 65: 191-197

Dubois E, Grenson M (1979) Methylamine/ammonia uptake systems in *Saccharomyces cerevisiae*: multiplicity and regulation. Mol Gen Genet 175: 67-76

Dyr-Jensen K, Brix H (1996) Effects of pH on ammonium uptake by *Typha latifolia* L. Plant Cell Environ 19: 1431-1436

Feng J, Volk RJ Jackson WA (1994) Inward and outward transport of ammonium in roots of maize and sorghum: contrasting effects of methionine sulphoximine. J Exp Bot 45: 429-439

Feng J, Volk RJ, Jackson WA (1998) Source and magnitude of ammonium generation in maize roots. Plant Physiol 118: 835-841

Finnemann J, Schjoerring JK (1999) Translocation of NH_4^+ in oilseed rape plants in relation to glutamine synthetase isogene expression and activity. Physiol Plant 105: 469-477

Franco AR, Cardenas J, Fernandez E (1987) A mutant of *Chlamydomonas reinhardtii* altered in the transport of ammonium and methylammonium. Mol Gen Genet 206: 414-418

Gazzarrini S, Lejay L, Gojon A, Ninnemann O, Frommer WB, von Wirén N (1999) Three functional transporters for constitutive, diurnally regulated and starvation-induced uptake of ammonium into *Arabidopsis* roots. Plant Cell 11: 937-948

Glass ADM, Siddiqi MY (1995) Nitrogen absorption by plant roots. In: Srivastava HS, Singh RP (eds) Nitrogen nutrition of higher plants. Associated Publishing Co, New Dehli, pp 21-56

Glass ADM, Shaff JE, Kochian LV (1992) Studies of uptake of nitrate in barley. IV. Electrophysiology. Plant Physiol 99: 456-463

Gojon A, Soussana JF, Passama L, Robin P (1986) Nitrate reduction in roots and shoots of barley (*Hordeum vulgare* L.) and corn (*Zea mays* L.) seedlings. I. [15]N study. Plant Physiol 82: 254-260

Hatch DJ, Hopper MJ, Dhanoa MS (1986) Measurement of ammonium ions in flowing solution culture and diurnal variation in uptake by *Lolium perenne* L. J Exp Bot 37: 589-596

Husted S, Schjoerring JK (1995) Apoplastic pH and ammonium concentration in leaves of *Brassica napus* L. Plant Physiol 109: 1453-1460

Husted S, Schjoerring JK (1996) Ammonia flux between oilseed rape plants and the atmosphere in response to changes in leaf temperature, light intensity, and air humidity. Plant Physiol 112: 67-74

Joy KW (1998) Ammonia, glutamine and asparagine: a carbon-nitrogen interface. Can J Bot 66: 2103-2109

Kaiser BN, Finnegan PN, Tyerman SD, Whitehead LF, Bergersen FJ, Day DA, Udvardi MK (1998) Characterisation of an ammonium transport protein from the peribacteroid membrane of soybean nodules. Science 281: 1202-1206

Kawamura Y, Takahashi M, Arimura G, Isayama T, Irifune K, Goshima N, Morikawa H (1996) Determination of levels of NO_3^-, NO_2^- and NH_4^+ ions in leaves of various plants by capillary electrophoresis. Plant Cell Physiol 37: 878-880

Kosegarten H, Grolig F, Wieneke J, Wilson G, Hoffmann B (1997) Differential ammonia-elicited changes of cytosolic pH in root hair cells of rice and maize as monitored by 2', 7'-bis-(2-carboxyethyl)-5 (and-6)-carboxyfluorescein-fluorescence ratio. Plant Physiol 113: 451-461

Kosola KR, Bloom AJ (1994) Methylammonium as a transport analog for ammonium in tomato (*Lycopersicon esculentum* L.). Plant Physiol 105: 435-442

Kozak M (1987) An analysis of 5'-noncoding sequences from 699 vertebrate messenger RNAs. Nucleic Acids Res 15: 8125-8148

Kronzucker HJ, Siddiqi MY, Glass ADM (1995) Analysis of $^{13}NH_4^+$ efflux in spruce roots. Plant Physiol 109: 481-490

Kronzucker HJ, Siddiqi MY, Glass ADM (1996) Kinetics of NH_4^+ influx in spruce. Plant Physiol 110: 773-779

Kronzucker HJ, Schjoerring JK, Erner Y, Kirk GJD, Siddiqi MY, Glass ADM (1998) Dynamic interactions between root NH_4^+ influx and long-distance N translocation in rice: insights into feed-back processes. Plant Cell Physiol 39: 1287-1293

Kronzucker HJ, Siddiqi MY, Glass ADM, Kirk JD (1999) Nitrate-ammonium synergism in rice. A subcellular flux analysis. Plant Physiol 119: 1041-1045

Lam HM, Coschigano K, Schultz C, Melo-Olivira R, Tjaden G, Oliveira I, Ngai N, Hsieh MH, Coruzzi G (1995) Use of *Arabidopsis* mutants and genes to study amide amino acid biosynthesis. Plant Cell 7: 887-898

Lam HM, Hsieh MH, Coruzzi G (1998) Reciprocal regulation of distinct asparagine synthetase genes by light and metabolites in *Arabidopsis thaliana*. Plant J 16: 345-353

Lauter FR, Ninnemann O, Bucher M, Riesmeier JW, Frommer WB (1996) Preferential expression of an ammonium transporter and of two putative nitrate transporters in root hairs of tomato. Proc Natl Acad Sci USA 93: 8139-8144

Lee RB (1993) Control of net uptake of nutrients by regulation of influx in barley plants recovering from nutrient deficiency. Ann Bot 72: 223-230

Lee RB, Ayling SM (1993) The effect of methionine sulphoximine on the absorption of ammonium by maize and barley roots over short periods. J Exp Bot 44: 53-63

Lee RB, Ratcliffe RG (1991) Observations on the subcellular distribution of the ammonium ion in maize root tissue using *in vivo* ^{14}N-nuclear resonance spectroscopy. Planta 183: 359-367

Lee RB, Rudge KA (1986) Effects of nitrogen deficiency on the absorption of nitrate and ammonium by barley plants. Ann Bot 57: 471-486

Lee RB, Purves JV, Ratcliffe RG, Saker LR (1992) Nitrogen assimilation and the control of ammonium and nitrate absorption by maize roots. J Exp Bot 43: 1385-1396

Lorenz MC, Heitman J (1998) The MEP2 ammonium permease regulates pseudo-hyphal differentiation in *Saccharomyces cerevisiae*. EMBO J 17: 1236-1247

MacFarlane JJ, Smith FA (1982) Uptake of methylamine by *Ulva rigida*: transport of cations and diffusion of free base. J Exp Bot 33: 195-207

Mäck G, Tischner R (1994) Constitutive and inducible net NH_4^+ uptake of barley (*Hordeum vulgare* L.) seedlings. J Plant Physiol 144: 351-357

Marini AM, Vissers S, Urrestarazu A, André B (1994) Cloning and expression of the *MEP1* gene encoding an ammonium transporter in *Saccharomyces cerevisiae*. EMBO J 13: 3456-3463

Marini AM, Soussi-Boudekou S, Vissers S, André B (1997a) A family of ammonium transporters in *Saccharomyces cerevisiae*. Mol Cell Biol 17: 4282-4293

Marini AM, Urrestarazu A, Beauwens R, André B (1997b) The Rh (Rhesus) blood group polypeptides are related to NH_4^+ transporters. TIBS 22: 460-461

Marini AM, Springael J-Y, Frommer WB, André B (2000) *Cross-talk between ammonium transporters in yeast and interference by the soybean SAT1 protein.* Mol Microbiol 35: 378-385

Marschner H (1995) Mineral nutrition in higher plants. Academic Press, London.

Mehrer I, Mohr H (1989) Ammonium toxicity: description of the syndrome in *Sinapis alba* and the search for its causation. Physiol Plant 77: 545-554

Morgan MA, Jackson WA (1988) Inward and outward movement of ammonium in root systems: transient responses during recovery from nitrogen deprivation in presence of ammonium. J Exp Bot 39: 179-191

Ninnemann O, Jauniaux JC, Frommer WB (1994) Identification of a high affinity ammonium transporter from plants. EMBO J 13: 3464-3471

Ourry A, Macduff JH, Prudhomme MP, Boucaud J (1996) Diurnal variation in the simultaneous uptake and sink allocation of NH_4^+ and NO_3^- by *Lolium perenne* in flowing solution culture. J Exp Bot 47: 1853-1863

Palkova Z, Janderova B, Gabriel J, Zikanova B, Pospisek M, Forstova J (1997) Ammonia mediates communication between yeast colonies. Nature 390: 532-536

Rao TP, Ito O, Matsunga R (1993) Differences in uptake kinetics of ammonium and nitrate in legumes and cereals. Plant Soil 154: 67-72

Rawat SR, Silim SN, Kronzucker HJ, Siddiqi MY, Glass ADM (1999) *AtAMT1* gene expression and NH_4^+ uptake in roots of *Arabidopsis thaliana*: evidence for regulation by root glutamine levels. Plant J 19: 143-152

Rideout JW, Chaillou S, Raper CD, Morot-Gaudry J-F (1994) Ammonium and nitrate uptake by soybean during recovery from nitrogen deprivation. J Exp Bot 45: 23-33

Roberts JKM, Pang MKL (1992) Estimation of ammonium ion distribution between cytoplasm and vacuole using nuclear magnetic resonance spectroscopy. Plant Physiol 100: 1571-1574

Salsac L, Chaillou S, Morot-Gaudry J-F, Lesaint C, Jolivet E (1987) Nitrate and ammonium nutrition in plants. Plant Physiol Biochem 25: 805-812

Sasakawa H, Yamamoto Y (1978) Comparison of the uptake of nitrate and ammonium by rice seedlings – influences of light, temperature, oxygen concentration, exogenous sucrose, and metabolic inhibitors. Plant Physiol 62: 665-669

Schachtman DP, Schroeder JI (1994) Structure and transport mechanism of a high-affinity potassium uptake transporter from higher plants. Nature 370: 655-658

Schlee J, Komor E (1986) Ammonium uptake by *Chlorella*. Planta 168: 232-238

Siewe RM, Weil B, Burkovski A, Eikmanns BJ, Eikmanns M, Krämer R (1996) Functional and genetic characterisation of the (methyl)ammonium uptake carrier of *Corynebacterium glutamicum*. J Biol Chem 271: 5398-5403

Soupene E, He L, Yan D, Kustu S (1998) Ammonia acquisition in enteric bacteria: Physiological role of the ammonium/methylammonium transport B (AmtB) protein. Proc Natl Acad Sci USA 95: 7030-7034

Streeter J (1989) Estimation of ammonium concentration in the cytosol of soybean nodules. Plant Physiol 90: 779-782

Taté R, Riccio A, Merrick M, Patriarca EJ (1998) The *Rhizobium etli amtB* gene coding for an NH_4^+ transporter is down-regulated early during bacteroid differentiation. Mol Plant Microbe Interact 11: 188-98

Tolley-Henry L, Raper CD (1986) Utilisation of ammonium as a nitrogen source. Effects of ambient acidity on growth and nitrogen accumulation by soybean. Plant Physiol 82: 54-60

Tyerman SD, Whitehead LF, Day DA (1995) A channel-like transporter for NH_4^+ on the symbiotic interface of N_2-fixing plants. Nature 378: 629-632

Udvardi MK, Day DA (1997) Metabolite transport across symbiotic membranes of legume nodules. Annu Rev Plant Physiol Plant Mol Biol 48: 493-523

Ullrich WR, Larsson M, Larsson CM, Lesch S, Novacky A (1984) Ammonium uptake in *Lemna gibba* G1, related membrane potential changes, and inhibition of anion uptake. Physiol Plant 61: 369-376

van Dommelen A, Keijers V, Vanderleyden J, de Zamaroczy M (1998) (Methyl)ammonium transport in the nitrogen-fixing bacterium *Azospirillum brasilense*. J Bacteriol 180: 2652-2659

Venegoni A, Moroni A, Gazzarrini S, Marrè MT (1997) Ammonium and methylammonium transport in *Egeria densa* leaves in condition of different H^+ pump activity. Bot Acta 110: 369-377

Vessey JK, Henry LT, Chaillou S, Raper CD (1990) Root-zone acidity affects relative uptake of nitrate and ammonium from mixed nitrogen sources. J Plant Nutr 13: 95-116

Volk RJ, Chaillou S, Mariotti A, Morot-Gaudry JF (1992) Beneficial effects of concurrent ammonium and nitrate nutrition on the growth of *Phaseolus vulgaris*: a ^{15}N study. Plant Physiol Biochem 30: 487-493 von Wirén N, Bergfeld A, Ninnemann O, Frommer WB (1997) *OsAMT1-1*: a high-affinity ammonium transporter from rice (*Oryza sativa* cv. Nipponbare). Plant Mol Biol 3: 681

von Wirén N, Lauter FR, Ninnemann O, Gillisen B, Walch-Liu P, Engels C, Jost W, Frommer WB (2000) Differential regulation of three functional ammonium transporter genes by nitrogen in root hairs and by light in leaves of tomato. Plant J, 21: 167-175

Walch-Liu P, Neumann G, Bangerth F, Engels C. (2000) Rapid effects of nitrogen form on leaf morphogenesis in tobacco. J Exp Bot, 51: 227-237

Walker NA, Smith FA, Beibly MJ (1979) Amine uniport at the plasmalemma of charophyte cells. II. Ratio of matter to charge transported and permeability of free base. J Membr Biol 49: 283-296

Wang L, Wessler SR (1998) Inefficient reinitiation is responsible for upstream open reading frame-mediated translational repression of the maize *R* gene. Plant Cell 10: 1733-1745

Wang MY, Siddiqi MY, Ruth TJ, Glass ADM (1993) Ammonium uptake by rice roots. II. Kinetics of $^{13}NH_4^+$ influx across the plasmalemma. Plant Physiol 103: 1259-1267

Wang MY, Glass ADM, Shaff JE, Kochian LV (1994) Ammonium uptake by rice roots. III. Electrophysiology. Plant Physiol 104: 899-906

Wang MY, Siddiqi YM, Glass ADM (1996) Interactions between K^+ and NH_4^+: effects on ion uptake by rice roots. Plant Cell Env 19: 1037-1046

White PJ (1996) The permeation of ammonium through a voltage-independent K^+ channel in the plasma membrane of rye roots. J Membr Biol 152: 89-99

Yin Z-H, Kaiser WM, Heber U, Raven JA (1996) Acquisition and assimilation of gaseous ammonia as revealed by intracellular pH changes in leaves of higher plants. Planta 200: 380-387

Ammonia Assimilation

Bertrand HIREL[1] and Peter J. LEA[2]

In higher plants, recent advances in plant molecular biotechnology combined with modern physiological and biochemical studies have expanded our understanding of the regulatory mechanisms controlling the primary steps of inorganic nitrogen assimilation and the subsequent biochemical pathways involved (Fig. 1). Nitrate is the principal nitrogen source for most crops. The uptake and reduction of nitrate to ammonia is discussed in detail in Chapter 1.1. In this Chapter, the term ammonia will be used to indicate ammonia and ammonium ions, which are present in equilibrium in solution. In addition to nitrate reduction, ammonia is produced in plant tissues through a variety of processes as well as being taken up directly from the soil (Chap. 2.1). For example, ammonia is generated through the fixation of atmospheric nitrogen by root nodules (Chap. 3.1), by photorespiring leaves and through the phenylpropanoid pathway. Ammonia may also be released for reassimilation by

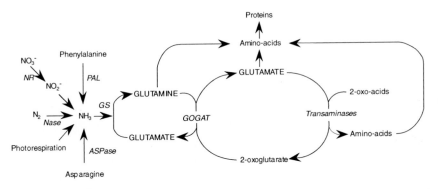

Fig. 1 : Ammonia assimilation in higher plants. *NR* Nitrate reductase; *NiR* nitrite reductase; *Nase* nitrogenase; *PAL* phenylalanine ammonia lyase; *ASPase* asparaginase; *GS*, glutamine synthetase; *GOGAT* glutamate synthase

1. Laboratoire du Metabolisme et de la Nutrition des Plantes, INRA, Route de St Cyr, Versailles Cedex, France. *E-mail:* hirel@versailles.inra.fr
2. Department of Biological Sciences, Lancaster University, Lancaster, LA1 4YQ, UK *E-mail:* p.lea@lancaster.ac.uk

sink tissue from nitrogen transport compounds (e.g. asparagine, arginine and the ureides) and through breakdown of other nitrogenous compounds (Lea et al. 1990; Woodall et al. 1996).

Ammonia is then incorporated into an organic molecule by the enzyme glutamine synthetase (GS). The reaction catalysed by GS is now considered to be the major (if not only) route facilitating the incorporation of inorganic nitrogen into organic molecules. In conjunction with glutamate synthase (GOGAT), which recycles glutamate, amino groups are transferred to other amino acids and utilised for protein synthesis and also for RNA and DNA synthesis (Lea et al. 1990; Lea and Ireland 1999).

Therefore, the regulation of nitrate and ammonia assimilation and incorporation and recycling of nitrogen into organic matter, are of major importance for improving plant nitrogen use efficiency (NUE).

The GS/GOGAT Pathway is the Major Route for Incorporating Reduced Nitrogen into Organic Molecules

The discovery of the major role of the enzyme couple: glutamine synthetase (GS)/ glutamate synthase (GOGAT) in ammonia assimilation in higher plants (Lea and Miflin 1974; Miflin and Lea 1976) has led to a large number of studies on the mechanisms controlling the expression of the genes encoding these proteins. In higher plants, GS and GOGAT are represented by a number of isoenzymes distributed in the cytosol and in the chloroplast. Their relative activity in a particular organ or tissue appears to be tightly linked to a specific role in primary nitrogen assimilation, ammonia recycling during photorespiration or nitrogen remobilisation, and may be subjected to important variations along the plant life cycle, depending on the developmental stage of both vegetative and reproductive organs. In addition to developmental factors, various environmental stimuli including light quantity and quality, or nitrogen feeding conditions have also been found to be important in the regulation of expression of the isoenzymes. Thus, identifying the biochemical and molecular mechanisms controlling this fine tuning is of particular importance if we are to understand and improve nitrogen managment in higher plants, through either genetic manipulation or plant breeding.

Glutamine Synthetase (GS)

GS (EC 6.3.1.2) catalyses the ATP-dependent conversion of glutamine utilizing ammonia as a substrate and is represented by two groups of proteins, a plastidic (GS_1) and cytosolic (GS_2) isoenzyme. The two isoenzymes were originally identified using a combination of ion exchange chromatography and subcellular fractionation experiments on leaf or root extracts (McNally and Hirel 1983). These studies also demonstrated that in the leaf of higher plants, the relative proportions of GS_1 and GS_2 were variable depending on the plant species examined, C_4 plants exhibiting a higher level of GS activity in the cytosol compared to the majority of C_3 plants in which the enzyme activity was substantially lower, or absent (McNally et al. 1983). Later, immunocytochemical experiments using antibodies raised against purified GS protein, confirmed that plastidic GS is located exclusively in chlorophyllous tissues, where it is associated with the stroma matrix (Botella et al. 1988a), although in some species, such as legumes or barley, it has been found to be associated with the plastids in roots (Peat and Tobin 1996) and root nodules (Brangeon et al. 1989). Cytosolic GS is located predominantly in the roots, root nodules (Brangeon et al. 1989; Peat and Tobin 1996) and floral organs (Dubois et al. 1996), whilst in shoots and roots of C_3 plants it is localised in the vascular tissue in which a high proportion of the protein is concentrated in the phloem companion cells (Carvahlo et al. 1992; Dubois et al. 1996; Peat and Tobin 1996; Sakurai et al. 1996). The situation appears to be different in C_4 plants, since a large proportion of GS protein was found in the cytosol of both mesophyll and bundle sheath cells (Becker et al. 1993a).

It is generally considered that the role of plastidic GS is to assimilate ammonia released either through nitrate reduction or during the photorespiratory process (Lea and Ireland 1999). Evidence for the latter has arisen from the studies of photorespiratory mutants which have normal levels of cytosolic GS but lack plastidial GS (Leegood et al. 1995), and more recently from the study of transformed plants with increased enzyme activity (Kozaki and Takeba 1996). Since cytosolic GS is the predominant form in the roots, it is proposed as having a central role in ammonia assimilation from nitrate reduction (Oaks and Hirel 1985). In a number of species such as tomato it was observed that cytosolic GS is induced in leaves during senescence (Pérez-Rodriguez and Valpuesta 1996), water stress (Bauer et al. 1997) or following infection by a pathogen (Pérez-Garcia et al. 1998) and therefore thought to function in the reassimilation of ammonia released from amino acid catabolism in response to either biotic or abiotic stresses. The presence of cytosolic GS in the phloem has also led to suggestions that it may be involved in the generation of glutamine used for nitrogen transport (Lam et al. 1996). However, recent results obtained using transgenic plants with impaired GS activity in the phloem did not confirm this hypothesis. The enzyme was shown to be involved in controlling proline production rather than the synthesis of glutamine for nitrogen export (Brugière et al. 1999). In plant/microorganism symbiotic associations the role of cytosolic GS in assimilating the ammonia resulting from atmospheric dinitrogen reduction is well documented (Hirel et al. 1993; see chap. 3.1).

In the organs of higher plants that have been studied, GS is an octameric protein (Fig. 2). Plastidial type GS (GS_2) generally exhibits a subunit molecular weight of approximately 44-45 kDa (Forde and Cullimore 1989; Hirel et al. 1993). However, in a few species such as tomato, two polypeptides of slightly different sizes were found as components of the plastidic GS holoenzyme. (Migge et al. 1996). Cytosolic GS is generally composed of a single subunit of about 38-40 kDa, which is always smaller when compared to its plastidial counterpart. However, as for plastidic GS, the presence of two polypetides of different sizes has also been found in a number of plant species (Woodall et al. 1996). Several subunits corresponding to electrophoretic variants of a single GS_1 or GS_2 polypeptide have also been identified in the majority of plants examined so far, indicating that the holoenzyme may be composed of a complex assembly of heterooctamers. In addition, it has been observed that the relative proportions of either GS_1 or GS_2 subunits, together with their respective polypeptide composition, may be subjected to important variations depending on the plant species, the organ examined, the developmental stage of a particular organ and/or its physiological status (Mack 1995; Migge et al. 1996). The significance of these variations remains most intriguing and may be interpreted as

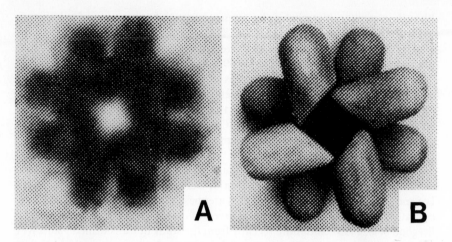

Fig. 2 : Visualisation of pea chloroplast glutamine synthetase. A The purified protein was stained by uranyl acetate and observed by electron microscopy (x 1 200 000). B Schematic representaion of the Quaternary structure of the enzyme. (Tsuprun et al. 1980)

the result of a physiological adaptation for an optimal efficiency of ammonia assimilation suited to a particular environment, or developmental stage of the different plant organs or tissues.

Isolation of the genes encoding plastidic and cytosolic GS followed by expression studies of the corresponding transcripts demonstrated that in a large number of plant species the synthesis of the different GS subunits is mostly controlled at the transcriptional level. However, in a limited number of plants it has been proposed that the protein turnover and stability may be also a means to control the final enzyme activity (Temple et al. 1996; Ortega et al. 1999). A large number of studies on various plant species including both monocots and dicots have shown that the plastidic form of the enzyme is encoded by a single nuclear gene per haploid genome. The corresponding translation product is synthesised as a precursor in the cytosol which is further imported and cleaved in the chloroplast to form an active GS_2 protein (Lightfoot et al. 1988). In addition, chloroplastic GS gene transcription and protein synthesis have been found to be regulated by light in photosynthetic organs involving the action of either a phytochrome (Migge et al. 1996) or a blue/UV-A light receptor (Migge et al. 1998a).

In contrast, cytosolic GS is encoded by a complex multigene family which can be represented by at least two and up to six different members depending on the plant species examined (Ireland and Lea 1999). Each gene encodes a different subunit, which may be assembled either as a homooctamer or a heterooctamer (Forde and Cullimore 1989; Fig. 3). Moreover, the expression of each individual gene may be controlled by the developmental and physiological status of a particular organ or tissue (Hirel et al. 1993). Cytosolic GS gene expression and protein synthesis, in roots and leaves, is generally not greatly influenced by external factors but is developmentally regulated. During the establishment of a symbiotic association between legumes (Hirel et al. 1993) or woody plants (Hirel et al. 1982) and atmospheric nitrogen-fixing microorganisms, GS gene transcription and enzyme activity are

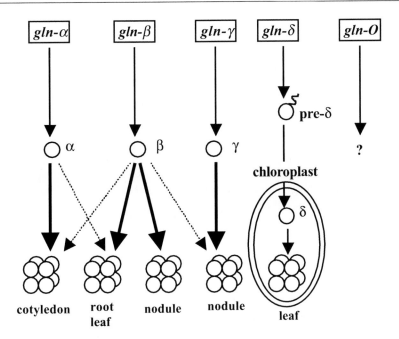

Fig. 3 : Expression of the genes encoding GS in *Phaseolus vulgaris* (Forde and Cullimore 1989)

greatly enhanced to assimilate the extra supply of ammonia produced. Interestingly, it was clearly shown that in the majority of legume species, members of the GS gene family are highly or specifically expressed in the nodules during the establishment of the symbiotic association. In other legumes such as soybean, the transcription of different GS genes expressed in nodules is stimulated in response to the extra supply of ammonia, indicating that in addition to developmental control, metabolic signals may also trigger GS gene expression (Hirel et al. 1993). In more recent studies, the occurrence of a metabolic control of GS gene expression was also demonstrated in non-legume species involving either carbon (Melo-Oliveira et al. 1996) or nitrogen (Marsolier and Hirel 1993) metabolites. The occurrence of an organ-specific expression of the GS multigene family was also shown to occur in a variety of other plant organs, such as flowers (Dubois et al. 1996), or specialised cellular structures such as rubber tree latex vessels (Pujade-Renaud et al. 1997), or maize kernels pedicels (Rastogi et al. 1998)

Earlier investigations have been confirmed by studies monitoring the expression of reporter genes placed under the control of different GS gene promoters in various organs of transgenic plants. These results, combined with sequence comparison analysis, have led to the conclusion that each GS gene promoter is unique and contains the specific regulatory sequences necessary to control the expression of the corresponding members of the GS multigene family (Ireland and Lea 1999). Using this approach, it was shown that the expression of plastidic and cytosolic GS genes is targeted to specific organs or cell types, thus confirming their non-overlapping role in inorganic nitrogen assimilation (Edwards et al. 1990). Studies perfomed on the ammonia-stimulated expression of a soybean gene encoding cytosolic GS in roots and root nodules are also a good example to illustrate the combined metabolic and developmental regulation of a member of the GS multigene family. In this work, it was demonstrated that the transcription of the gene is controlled by the cooperation of multiple *cis* regulatory elements. Moreover, promoter deletion analysis showed that the spatial conformation of the promoter is likely to play an important role allowing both an organ-specific and metabolic regulation of the gene transcription (Tercé-Laforgue et al. 1999). Relatively few studies have been undertaken to isolate *trans*-acting factors interacting with GS promoters and no clearcut evidence has emerged on their specific role in regulating gene transcription (Tjaden and Coruzzi 1994). However, these approaches combined with the use of mutants or transgenic plants modified in their GS activity, will be necessary to identify key elements involved in sensing and channelling the metabolic and developmental signals which control the expression of the different members of the GS multigene family, during plant growth and development.

Glutamate Synthase (GOGAT)

GOGAT catalyses the conversion of glutamine and 2-oxoglutarate to two molecules of glutamate and occurs as two distinct isoforms (ferredoxin-dependent and pyridine nucleotide-dependent), which differ in their molecular mass, subunit composition, reductant specificity and metabolic function (Temple et al. 1998). In C_3 plants, both conventional subcellular fractionation (Lea and Miflin 1974) and immunogold antibody localisation techniques (Botella et al. 1988b) demonstrated that ferredoxin-dependent GOGAT (Fd-GOGAT) (EC 1,4,7,1) is located in chloroplasts of leaves, where it represents the main enzyme activity. Interestingly, in C_4 plants the enzyme is predominantly present in the bundle sheath chloroplasts (Harel et al. 1977; Becker et al. 1993a), the site of photorespiration. Fd-GOGAT is an iron-sulphur flavo-protein composed of a single subunit molecular mass of 130-180 kDa. It has been shown that in leaves the transcription of the Fd-GOGAT gene is light-dependent, leading to an increase in both the corresponding translation product and enzyme activity (Sechley et al. 1992). The light dependent expression of Fd-GOGAT appears to be controlled by the photoreceptor phytochrome (Becker et al. 1993b) and may also be influenced upon illumination by UV-A or UV-B (Migge et al. 1998b). In a number of plant species, the synthesis of both Fd-GOGAT transcripts and the corresponding protein does not strongly depend on the nature or presence of a nitrogen source (Migge et al. 1998b), whereas, in others, nitrogen may be an important factor controlling the final enzyme activity (Sakakibara et al. 1992). It was originally shown that in the tobacco genome two distinct genes encoding Fd-GOGAT were present (Zehnacker et al. 1992). This finding was confirmed later using the model plant *Arabidopsis thaliana*, in which two different cDNA clones (*GLU1* and *GLU2*) encoding Fd-GOGAT were identified. *GLU1* mRNA was highly expressed in photosynthetic tissues and light-responsive, whereas *GLU2* mRNA was preferentially expressed in roots and synthesized at a lower constitutive level in leaves (Coshigano et al. 1998). These findings, together with previous studies of photorespiratory mutants defficient in leaf Fd-GOGAT activity (Lea and Forde 1994; Leegood et al. 1995), strengthen the current consensus that the major role of the enzyme was to reassimilate photorespiratory ammonia and that it may also have an important function in primary nitrogen assimilation. The discovery of the second Fd-GOGAT gene confirmed that the enzyme may also play a role in non-photosynthetic tissue, as some Fd-GOGAT activity has previously been found associated with root plastids (Suzuki et al. 1982) and may also be involved in the assimilation of ammonia derived from soil nitrate (Redinbaugh and Campbell 1993).

In higher plants, the other main GOGAT isoenzyme is represented by NADH-GOGAT (EC 1.4.1.14). Like Fd-GOGAT it is also an iron-sulphur flavoprotein and a monomeric protein but the molecular mass is substantially higher, with values of approximately 220-230 kDa. NADH-GOGAT is found primarily in non-photosynthetic tissue such as nitrogen-fixing legume nodules (Temple et al. 1998), where it is the major form of GOGAT, and where it combines with cytosolic GS to assimilate the ammonia produced by the nitrogen-fixing bacteriods (Chap. 3.1). In *Phaseolus vulgaris* nodules, NADH-GOGAT occurs as two isoforms, one of which is responsible for the increase in GOGAT activity in the root nodules and the assimilation of nitrogen fixed by rhizobium (Chen and Culllimore 1988) . In alfalfa, synthesis of the enzyme and the corresponding mRNA occurred in two phases, one active prior to the synthesis of nitrogenase and the other following the onset of nitrogen fixation (Gregerson et al. 1993). Using immunocytochemical localisation, Trepp et al. (1999a) showed that NADH-GOGAT protein was distributed throughout the infected cell region of both 19- and 33-day-old nodules and was located specifically in the amyloplasts. Promoter analysis indicated that there were at least four regulatory elements involved in the expression of the single NADH-GOGAT gene within the nodules (Trepp et al. 1999b). These results would suggest that in nodules the enzyme plays a dual role, first in nodule development and metabolism, and then in the assimilation of ammonia derived from nitrogen fixation.

NADH-GOGAT is also located in various cell types of the roots (Ishiyama et al. 1998), green leaves, etiolated leaf tissue (Suzuki and Gadal 1984) seeds and the vascular bundles of unexpanded leaf tissues (Hayakawa et al. 1994), where it is localised in the plastids (Hayakawa et al. 1999). In confirmation of this, the NADH-GOGAT cDNA of rice has been shown to have a 99-amino acid presequence at the N-terminal end (Goto et al. 1998). The function of both cytosolic GS and NADH-GOGAT in leaf cells is, as yet, not clearly defined, although their expression patterns are coordinated in non-legumes. A role of these enzymes coacting in a GS/NADH-GOGAT cycle during nitrogen transport via the vascular bundle from roots or senescing tissues has been proposed (Hayakawa et al. 1994).

Asparagine Synthetase

Although asparagine and glutamine are both amides and differ only in chain length, asparagine is more soluble, less reactive and has a higher N:C ration than glutamine, all of which make it a better nitrogen storage and transport compound (Sieciechowicz et al. 1988). Asparagine is often present in very high concentrations in both the xylem and phloem sap (Pate 1989), and is used to carry nitrogen away from source tissues, notable examples being in germinating seedlings, nitrogen-fixing nodules (see Chap. 3.1) and senescing leaves (see Chap. 5.2). Asparagine may also accumulate in plant cells when protein synthesis and photosynthesis are reduced due to either pathogen-induced or abiotic stress (Sieciechowicz et al. 1988; Lea et al . 1990). Following the carbohydrate starvation of cultured sycamore cells, asparagine accumulated at the onset of protein breakdown. This process was, however, reversed by the addition of sucrose and the asparagine concentration then fell to zero (Genix et al. 1994).

Following the incorporation of ammonia into the amide position of glutamine, the nitrogen can be transferred directly to the same position in asparagine by asparagine synthetase (AS; EC 6.3.5.4). The enzyme uses aspartate and glutamine as substrates and ATP is converted to AMP + PPi. AS has proved very difficult to study in higher plants due to an inherent instability of the enzyme and the presence of natural inhibitors (Joy et al. 1983). The most suitable source of the enzyme has proved to be germinating cotyledons (Rognes 1975; Lea and Fowden 1975). The enzyme is also able to use ammonia as a substrate but the K_m value is 40-fold higher than glutamine (Huber and Streeter 1985). AS has also been measured in maize roots, where the activity is lowered in response to light or sucrose (Stulen and Oaks 1977; Brouquisse et al. 1992).

The most important breakthrough in the study of AS was the initial isolation of the two genes AS1 and AS2 , that coded for AS in peas (Tsai and Coruzzi 1990). The proteins derived from the two genes had molecular masses of 65.6 and 66.3 kDa, and were shown to be highly homologous with human AS and to contain specific glutamine-binding sites (Tsai and Coruzzi 1990). Northern analysis indicated that in leaves, the expression of the two genes was repressed by light and stimulated by darkness. In roots, only the expression of AS1 was repressed by light, whilst the expression of AS2 was constitutive (Tsai and Coruzzi 1991). The observation that AS mRNA increases 30-fold in darkened leaves confirms previous observations that the concentration of asparagine increased dramatically in pea leaves in the dark (Joy et al. 1983). The cis-acting elements that conferred light repression of AS1 were identified as a 124 bp promoter sequence. Similarities with the oat phytochrome gene, which is also repressed by light in a similar manner, were noted. Both the AS1 and AS2 pea promoters were able to direct GUS expression in the vascular tissue, cotyledons, leaves and roots of transgenic tobacco. These studies would suggest that at least some of the transported asparagine is synthesised in situ, in the phloem cells (Ngai et al. 1997).

Three AS genes have been identified in A. thaliana (Lam et al. 1996, 1998). The expression of ASN1 in leaves was repressed by light or by the presence of sucrose, which could be reversed by asparagine, glutamine and glutamate. The two additional AS genes (ASN2 and ASN3) were shown to encode proteins that exhibited a greater similarity to the AS enzymes of the monocots, maize and rice. ASN2 was shown to be regulated in a fashion reciprocal to that identified for ASN1, i.e. ASN2 mRNA levels were increased by light or sucrose and repressed by the addition of amino acids. Lam et al. (1998) proposed that the presence of these distinct genes in A. thaliana allowed for the synthesis of asparagine in both a light and dark, particularly under conditions when there is a high rate of ammonia influx or production.

cDNA clones encoding distinct isoenzymes of AS have also been isolated from Lotus japonicus (Waterhouse et al. 1996), Zea mays (Chevalier et al. 1996), Glycine max (Hughes et al. 1997) and P. vulgaris (Osuna et al. 1999). In the majority of the studies, light and/or carbohydrate were shown to repress gene expression; however, in L. japonicus no effect of light was detected (Waterhouse et al. 1996). A gene encoding AS isolated from nitrogen-fixing nodules of alfalfa also contained a promoter sequence that directed expression in dark-treated leaves (Shi et al. 1997).

Asparagine was first isolated from asparagus (*Asparagus sativus*) in 1806. Asparagus tips are an important commercial crop, but are very susceptible to postharvest damage, due to the immediate onset of senescence and the accumulation of asparagine, which is obvious even 12 h after collection from the plant. (King et al. 1990). Although AS activity could not be detected in the asparagus tips, mRNA encoding AS was shown to start to accumulate 2 h after harvesting. The increase in gene expression was accompanied by a decrease in soluble carbohydrates, but was unaffected by light (Davies and King 1993). The stimulation of AS expression by a decrease (and subsequent reversal by an increase) in the concentration of sucrose was confirmed using asparagus cell cultures (Davies et al. 1996).

Glutamate Dehydrogenase

The NADH-dependent glutamate dehydrogenase (GDH; EC 1.4.1.2) catalyses the following reversible reaction:

glutamate + H_2O + NAD^+ ↔ 2-oxoglutarate + NH_3 + NADH + H^+.

Plant biochemists have agonised for over 25 years as to the role of this enzyme in higher plants. As indicated previously, a considerable amount of evidence has built up that demonstrates that over 95% of the ammonia available to higher plants is assimilated via the GS/GOGAT pathway (Lea and Ireland 1999). However, proposals that GDH could operate in the direction of ammonia assimilation have been put forward on a regular basis (Yamaya and Oaks 1987; Oaks 1994, 1995; Melo-Oliveira et al. 1996). Others have argued equally strongly that GDH operates in the direction of glutamate deamination (Robinson et al. 1991, 1992; Fox et al. 1995; Stewart et al. 1995). In this brief section we will describe the properties of plant GDH and discuss possible physiological roles of the enzyme.

Plant GDH has a very high K_m for ammonia, is activated by Ca^{2+} and is localised in the mitochondria of a range of different plant tissues (Srivastava and Singh 1987; Yamaya and Oaks 1987; Turano et al. 1997; Turano 1998). The enzyme protein exists as a hexamer with subunit molecular masses ranging from 41-45 kDa. The presence of seven isoenzymic forms, distinguishable following native polyacrylamide gel electrophoresis, has long been taken as circumstantial evidence that two distinct subunits of GDH are synthesised (Cammaerts and Jacobs 1985; Loulakakis and Roubelakis-Angelakis 1996). Mutants of maize (Magalhaes et al. 1990; Pryor 1990) and *A. thaliana* (Melo-Oliveira et al. 1996) with very low GDH activity have been isolated that contain only one isoenzymic form. Osuji and Madu (1995, 1997) have proposed that in germinating maize and peanut seedlings, three subunits are present that can give rise to 28 possible hexameric structures, that can be separated by Rotofor isoelectric focusing.

The addition of ammonia to plants has frequently stimulated an increase in measurable GDH activity, often combined with changes in the isoenzymic pattern observed following electrophoresis (Cammaerts and Jacobs 1985; Srivastava and Singh 1987; Ireland and Lea 1999). This increase in activity was frequently used as evidence that GDH was operating in the direction of glutamate synthesis, although some of this evidence can now be dimissed, due to difficulties encountered with enzyme assays using crude extracts (Fricke and Pahlich 1992; Loulakakis and Roubelakis-Angelakis 1996). GDH was initially purified to homogeneity from grapevine (*Vitis vinifera*) and antisera raised (Loulakakis and Roubelakis-Angelakis 1990, 1991). Western blot analysis indicated that there were two protein subunits with molecular masses of 43.0 and 42.5 kDa, termed the α- and β-polypetides, respectively. When grapevine callus was transferred from nitrate to ammonia, GDH activity increased threefold and the isoenzyme-banding pattern reversed from high cathodal activity to a high anodal activity. During the same period the concentration of the α-subunit increased fourfold (with the concomitant *de novo* incorporation of ^{35}S-methionine) and the β-subunit decreased (Loulakakis and Roubelakis – Angelakis 1992).

Increases in GDH activity have also been observed following carbohydrate starvation, a process that could be reversed by the addition of a range of soluble sugars (Sahulka and Lisa 1980; Robinson et al. 1991, 1992; Athwal et al. 1997). In carrot cell suspension cultures, a strong correlation was demonstrated between the level of GDH activity and the concentration of glutamate in the cells. Robinson et al. (1992) argued from these data that the primary function of GDH is to carry out the oxidation of glutamate, a hypothesis supported by the later work of Stewart et al. (1995). GDH activity has been shown to increase following the onset of senescence in a range of plant tissues (Srivastava and Singh 1987; Peeters and Van Laere 1994; Bechtold et al. 1998). As senescence is frequently the time when carbohydrate concentrations fall and ammonia increases (Crafts-Brander et al. 1998), a reasonable hypothesis at the present time is that GDH is involved in the metabolism of amino acids liberated during proteolysis.

After some delay, cDNA clones have now been isolated that encode higher-plant GDH; these include maize (Sakakibara et al. 1995), grapevine (Syntichaki et al. 1996), *A. thaliana* (Melo-Oliveria et al. 1996), tomato (Purnell et al. 1997) and tobacco (Ficarelli et al. 1999). Considerable similarities with the amino acid sequence obtained from animals, bacteria and algae GDH were noted and putative binding sites for NADH and glutamate and mitochondrial targeting sequences have been identified. However, by far the most thorough study on the expression of GDH was carried out by Turano et al. (1997). Two distinct cDNA clones isolated from *A. thaliana* that were designated *GDH1* and *GDH2* encoded mitochondrial targeted proteins of molecular mass 43 and 42.5 kDa, respectively. In addition to 2-oxoglutarate and NADH-binding domains, a putative EF hand loop that could be involved in Ca^{2+} binding was also identified in the GDH2 amino acid sequence only. The *GDH2* transcript was readily detectable in all tissues tested, but expression was increased by five fold by ammonia, or 25-fold by dark treatments; much smaller increases were observed in the expression of the *GDH1* gene. However, the expression of the *GDH1* gene was repressed by the presence of sucrose, whilst that of *GDH2* was not. The observations of Turano et al. (1997) taken in conjunction with those of Melo-Oliveria et al. (1996) suggest that in *A. thaliana* the two genes (and presumably the derived enzyme proteins) play distinct roles in carbon nitrogen metabolism, the nature of which still remains to be confirmed.

Modification of Ammonia Assimilation Through Genetic Manipulation

Very recently, significant progress has been made in analysing the physiological impact of altered GS, GOGAT and GDH activities in transgenic legumes and non-legumes. The study of both legumes, where the roots are generally the major sites of primary nitrogen assimilation, and non-legumes, where nitrogen reduction occurs frequently in the shoots, have been particularly useful in elucidating the mechanisms of control of the nitrogen assimilatory pathway. The aim of the studies on the amplification or shift of ammonia assimilation in a particular organ or tissue was first to elucidate some of the aspects of the physiology of ammonia assimilation and then to provide new information on nitrogen management and partitioning between roots and shoots. The ultimate goal of such research work is to examine the impact of modified ammonia assimilation on plant growth, development and plant NUE which could enable crop production under low fertiliser imput, thus reducing cost and environmental hazards.

The original goal of selecting plants modified in their GS gene expression was to obtain plants resistant to herbicides that were structural analogues of glutamate, and which therefore inhibited GS activity. Alfalfa lines resistant to L-phosphinothricin were selected exhibiting a four- to elevenfold increase in one of the GS genes, which resulted in an increase in enzyme activity sufficient to overcome the toxic effect of the herbicide (Donn et al. 1984). Genetically manipulated tobacco plants with additional GS genes have since been used to obtain the same resistance to inhibitors of GS activity (Eckes et al. 1989). These tobacco lines showed no phenotypic effect from the overproduction of GS, and although there was no change in the amino acid composition, both the ammonia and nitrate concentration were reduced.

To further investigate the role of GS in plant nitrogen metabolism with respect to plant productivity, an attempt was made to downregulate GS activity in alfalfa either constitutively or in an organ-specific manner using an RNA antisense strategy (Temple et al. 1994). This experiment resulted either in lethality or no reduction in enzyme activity, suggesting that GS activity below a certain threshold may be detrimental for plant growth and development (Leegood et al. 1995).

Later, a soybean cytosolic GS gene was overexpressed in *Lotus corniculatus* plants. These plants had a 50 to 80% increase in total leaf GS activity. In the transgenic plants cultivated with ammonia as a sole nitrogen source, it was found that both ammonium ion uptake in the roots and the subsequent translocation of amino acids to the shoots were lower in plants overexpressing GS. It was concluded that the buildup of ammonia and the increase in amino acid concentration in the roots were the result of shoot protein degradation. Moreover, early floral development was observed in the transformed plants. As all these properties are characteristic of senescent plants, these findings suggest that expression of cytosolic GS in the shoots may accelerate plant development, leading to early senescence and premature flowering (Vincent et al. 1997) and that GS_1 plays a major role during nitrogen remobilisation (Kamachi et al. 1991).

Another source of reduced inorganic nitrogen is the ammonia produced in the nodules under dinitrogen fixation. Therefore, further studies were performed using nodulated transformed *L. corniculatus* plants. Unexpectedly, GS activity expressed on a fresh weight or protein basis was 40% lower in the nodules of the transformed plants as compared to the nodules of the untransformed plants. The biomass of the shoots and roots was found to be twofold higher. This change in biomass could be explained by more efficient ammonia assimilation in the transgenic nodules, as indicated by a large increase in amino acids with a concomitant decrease in carbohydrate content. Moreover, greater quantities of nitrogen-fixing nodules were produced in the transgenic plants. Therefore the overall levels of nitrogen fixation and assimilation were enhanced, and thus it seems that there are as yet unknown regulatory mechanisms controlling the balance between nodule GS activity and nodule biomass (Hirel et al. 1997).

To further investigate the contribution of cytosolic glutamine synthetase activity to plant biomass production, two other approaches were conducted using L. japonicus as a model plant. In the first series of experiments, it was found that overexpressing GS activity in the roots of transgenic plants caused a decrease in plant biomass production. Using ^{15}N labelling, it was shown that this decrease was likely to be due to a lower rate of nitrate uptake, accompanied by a redistribution of the newly absorbed nitrogen to the shoots, which could not be reduced due to the lack of nitrate reductase activity in this organ. In the second series of experiments, the relationship between plant growth and root GS activity was analysed using a series of recombinant inbred lines obtained by crossing two different L. japonicus ecotypes, Gifu and Funakura. It was confirmed that a negative relationship also exists between root GS expression and plant biomass production in both the two parental lines and their progeny (Limami et al. 1999). Very recently, in poplar, the overexpression of a pine gene encoding cytososolic GS led to an increase in plant biomass production, suggesting that, in perennial species having a long life cycle such as trees, nitrogen is utilised more efficiently (Gallardo et al. 1999)

As decribed above, large amounts of GS activity have been detected in the phloem of most higher-plant species. However, the role of the enzyme in the vascular tissue has never been clearly defined. To determine the effect of decreased levels of GS activity in the phloem on inorganic nitrogen assimilation, an antisense strategy was developed to block the expression of the corresponding gene. In transgenic plants, no major phenotypic effects were observed. However, a decrease in proline content was seen in leaves, stems, roots, phloem and xylem sap of the transformed plants, in comparison to the wild type. In contrast, the relative concentrations of amino acids used for nitrogen transport were similar between both wild type and the transformants, indicating that GS in the phloem did not play a significant role in the export of reduced nitrogen. An ^{15}N-labelling experiment allowed the demonstration that glutamine is the major amino acid used for proline synthesis, and that the lack of GS activity in the phloem is likely to result in a shortage of glutamine, which may be preferentially used for export and/or as a amino group donor rather than for proline synthesis. It was concluded from this study that GS in the phloem plays a major role in regulating proline production, consistent with the function of proline as a nitrogen source and as a protectant in response to water stress (Brugière 1997).

Taken together, these results indicate that in an increasing number of higher-plant species GS activity is an important factor controlling both plant growth and plant development. In the future, modulation of GS expression in an organ-specific and developmental manner will certainly be one of the important targets to improve nitrogen management by plants under adverse environmental conditions.

The generation of plants modified in their GOGAT activity was primarily initiated as a result of its crucial role in reassimilating ammonium produced during photorespiration. Selection of mutant plants was carried out in high CO_2, to avoid the toxic accumulation of ammonia and/or the depletion of amino donors in the photorespiratory nitrogen cycle (Leegood et al. 1995). The leaves of transformed plants were found to contain levels of Fd-GOGAT activities and Fd-GOGAT protein varying between 90 and 10% compared to untransformed control plants. These plants were used to study the metabolism of amino acids, 2-oxoglutarate and ammonia following the transition from CO_2 enrichment (where photorespiration is inhibited) to air (where photorespiration is a major process of ammonia production in leaves). The leaves of the lowest Fd-GOGAT expressors accumulated more foliar glutamine and 2-oxoglutarate than the untransformed controls under both growth conditions. Photorespiration-dependent increases in foliar ammonia, glutamine, 2-oxoglutarate and total amino acids were proportional to the decreases in foliar Fd-GOGAT activity. This finding indicates that several pathways of amino acid biosynthesis are modified when ammonia and glutamine accumulate in leaves and that the reaction catalysed by Fd-GOGAT is a key regulatory element controlling nitrogen partioning and redistribution during plant growth and development (Ferrario-Méry et al. 1999).

Recently, plants overexpressing bacterial GDH have been obtained (Ameziane et al. 1998). In field experiments with transformed tobacco overexpressing GDH, an increase in tolerance both to herbicides that inhibited GS activity and to toxic levels of ammonia were found. Moreover plant biomass and tolerance to water stress were increased. Similar results were seen with transformed corn, where biomass, yield and protein content were all improved. A US patent was recently released by Schmidt and Miller (1999) describing the use of plants transformed with nucleotide sequence encoding the α and β subunits of *Chlorella sorokiniana*. These plants show improved properties such as increased growth and improved stress tolerance. From these studies it appears that altering the GDH activity improves the metabolic efficiency, assimilation of ammonia and stress tolerance. It was proposed that assimilation via GDH may have an advantage over GS assimilation, since the reaction catalysed by the former uses less carbon per ammonia assimilated. However, a comprehensive physiological examination of these plants will be required to determine if the new GDH activity is operating in the direction of glutamate synthesis or breakdown.

Brears et al. (1993) introduced pea asparagine synthetase (*AS1*) into tobacco plants, so that the gene was expressed both in the light and dark, causing a 10-100-fold increase in the asparagine concentration, and an increase in the biomass production of the plants. Brears et al. (1993) also used a second *AS1* construct in which the glutamine-binding domain had been deleted, thus allowing the enzyme to use ammonia directly as a substrate. Although there was evidence of an increase in asparagine concentration, the growth rate of the latter plants was reduced.

These various approaches undertaken to modify ammonia assimilation in transgenic plants have provided new information on nitrogen management and partitioning between roots and shoots. They also demonstrate that the amplification or shift of ammonium assimilation in a particular organ or tissue may have a strong impact on plant growth and development, which opens a bright future for improving NUE in crop plants.

References

Ameziane R, Bernhard K, Bates R, Lightfoot DA (1998) Metabolic engineering of C and N metabolism with NADPH glutamate dehydrogenase. Abstr Annu Mee of the American Society of Plant Physiologists, June 27-July 1, 1988. Madison, Wisconsin, pp 15

Athwal GS, Pearson J, Laurie S (1997) Regulation of glutamate dehydrogenase activity by manipulation of nucleotide supply in *Daucus carota* suspension cultures. Physiol Plant 101: 503-509

Bauer D, Biehler K, Fock H, Carrayol E, Hirel B, Migge A, Becker TW (1997) A role for cytosolic glutamine synthetase in the remobilization of leaf nitrogen during water stress in tomato. Physiol Plant 99: 241-248

Bechtold U, Pahlich E, Lea PJ (1998) Methionine sulphoximine does not inhibit pea and wheat glutamate dehydrogenase. Phytochemistry 49: 347-354

Becker TW, Perrot-Rechenman C, Suzuki A, Hirel B (1993a) Subcellular and immunocytochemical localization of the enzymes involved in ammonia assimilation in mesophyll and bundle sheath strands of maize leaves. Planta 191: 129-136

Becker T, Nef-Campa C, Zehnacker C, Hirel B (1993b) Implication of the phytochrome in light regulation of the tomato gene(s) encoding ferredoxin-dependent glutamate synthase. Plant Physiol Biochem 31: 725-727

Botella R, Verbelen JP, Valpuesta V (1988a) Immunocytolocalization of glutamine synthetase in green leaves and cotyledons of *Lycopersicon esculentum*. Plant Physiol 88: 943-946

Botella R, Verbelen JP, Valpuesta V (1988b) Immunocytolocalization of ferredoxin GOGAT in the cells of green leaves of *Lycopersicon esculentum*. Plant Physiol 87: 255-257

Brangeon J, Hirel B, Forchion A (1989) Immunogold localisation of glutamine synthetase in soybean leaves, roots and nodules. Protoplasma 151: 88-97

Brears T, Lui C, Knight TJ, Coruzzi GM (1993) Ectopic expression of asparagine synthetase in transgenic tobacco. Plant Physiol 103: 1285-1290

Brouquisse R, James F, Pradet A, Raymond P (1992) Asparagine metabolism and nitrogen distribution during protein degradation in sugar-starved maize root tips. Planta 188: 384-395

Brugière N (1997) Etude de l'expression de deux gènes codant pour la glutamine synthetase (GS) cytosolique chez le tabac (*Nicotiana tabacum* L.). Conséquences de l'inhibition de l'activité GS dans le phloème sur la physiologie de la plante. Thèse de Doctorat de l'Université de Paris VI, Paris

Brugière N, Dubois F, Limami AM, Lelandais M, Roux Y, Sangwan RS, Hirel B (1999) Glutamine synthetase in the phloem plays a major role in controlling proline production. Plant Cell 11: 1-19

Cammaerts D, Jacobs M (1985) A study of the role of glutamate dehydrogenase in the nitrogen metabolism of *Arabidopsis thaliana*. Plant Sci Lett 31: 65-73

Carvalho E, Pereira S, Sunkel C, Salema R (1992) Detection of a cytosolic glutamine synthetase in leaves of *Nicotiana tabacum* L. by immunocytochemical methods. Plant Physiol 100: 1591-1594

Chen F, Cullimore JV (1988) Two isoenzymes of NADH-dependent glutamate synthase in root nodules of *Phaseolus vulgaris* L. Plant Physiol 88: 1411-1417

Chevalier C, Bourgeois E, Just D, Raymond P (1996) Metabolic regulation of asparagine synthetase gene expression in maize (*Zea mays* L.) root tips. Plant J 9: 1-11

Coshigano KT, Melo-Oliveira R, Lim J, Coruzzi GM (1998) *Arabidopiss gls* mutants and distinct Fd-GOGAT genes: implication for photorespiration and primary nitrogen assimilation. Plant Cell 10: 741-752

Crafts-Brandner SJ, Holzer R, Feller U (1998) Influence of nitrogen deficiency on senescence and the amounts of RNA and proteins in wheat leaves. Physiol Plant 102: 192-200

Davies KM, King GA (1993) Isolation and characterization of a cDNA clone for a harvest-induced asparagine synthetase from *Asparagus officinalis* L. Plant Physiol 102: 1337-1340

Davies KM, Seelye JF, Irving DE, Borst WM, Hurst PL, King GA (1996) Sugar regulation of harvest-related genes in asparagus. Plant Physiol 111: 877-883

Donn G, Tisher E, Smith JA, Goodman HM (1984) Herbicide resistant alfalfa cells: an example of gene amplification in plants. J Mol Appl Genet 2: 621-635

Dubois F, Brugiere N, Sangwan RS, Hirel B (1996) Localisation of tobacco cytosolic glutamine synthetase enzymes and the corresponding transcripts show organ- and cell-specific pattern of protein synthesis and gene expression. Plant Mol Biol 31: 803-817

Eckes P, Shmidtt P, Daub W, Wegenmayer F (1989) Overproduction of alfalfa glutamine synthetase in transgenic tobacco plants. Mol Gen Genet 217: 263-268

Edwards JW, Walker EL, Coruzzi GM (1990) Cell-specific expression in transgenic plants reveals non-overlapping roles for chloroplast and cytosolic glutamine synthetase. Proc Natl Acad Sci USA 87: 3459-3463

Ferrario-Mery S, Suzuki A, Kunz C, Valadier MH, Roux Y, Hirel B, Foyer C (1999) Modulation of amino acid metabolism in transformed tobacco plants deficient in Fd-GOGAT. Plant Soil (in press)

Ficarreli A, Tassi F, Restivo FM (1999) Isolation and characterisation of two cDNA clones encoding for glutamate dehydrogenase in *Nicotiana plumbaginofolia*. Plant Cell Physioly 40: 339-342

Forde BJ, Cullimore JV (1989) The molecular biology of glutamine synthetase in higher plants. In: Miflin BJ (eds) Oxford surveys of plant molecular biology, vol. 6. Oxford University Press, Oxford, pp 247-296

Fox GG, Ratcliffe RG, Robinson SA, Stewart GR (1995) Evidence for deamination by glutamate dehydrogenase in higher plants: commentary. Can J Bot. 73: 1112-1115

Fricke W, Pahlich E (1992) Malate: a possible source of error in the NAD glutamate dehydrogenase assay. J Exp Bot 43: 1515-1518

Gallardo F, Fu J, Canton FR, Garcia-Gutierez, Canovas F,Kirby E G (1999) Expression of a conifer glutamine sythetase gene in transgenic poplar. Planta 210: 19-26

Genix P, Bligny R, Martin J-B, Douce R (1994) Transient accumulation of asparagine in sycamore cells after a long period of sucrose starvation. Plant Physiol 94: 717-722

Goto S, Akagawa T, Kojima S, Hayakawa T, Yamaya T (1998) Organisation and structure of a NADH-dependent glutamate synthase gene from rice plants. Biochim Biophys Acta 1387: 293-308

Gregerson RG, Miller SS, Twary SN, Grantt JS, Vance CP (1993) Molecular characterisation of NADH-dependent glutamate synthase from alfalfa nodules. Plant Cell 5: 215-226

Harel E, Lea PJ, Miflin BJ (1977) The localisation of enzymes of nitrogen assimilation in maize leaves, and their activity during greening. Planta 134: 195-200

Hayakawa T, Nakamura T, Hattori F, Mae T, Ojima K, Yamaya T (1994) Cellular localisation of NADH-dependent glutamate synthase protein in vascular bundles of unexpanded leaf blades and young grains of rice plants. Planta 193: 455-460

Hayakawa T, Hopkins L, Peat LJ, Yamaya T, Tobin AK (1999) Quantitative intercellular localisation of NADH-dependent glutamate synthase protein in different types of root cells in rice plants. Plant Physiol 119: 409-416

Hirel B, Perrot-Rechenmann C, Maudinas B, Gadal P (1982) Glutamine synthetase in Alder (*Alnus glutinosa*) root nodules. Purification, properties and cytoimmunochemical localization. Physiol Plant 55: 197-203

Hirel B, Miao GH, Verma DPS (1993) Metabolic and developmental control of glutamine synthetase genes in legume and non legume plants. In: Verma DPS (ed) Control of plant gene expression. CRC Press, Boca Raton Florida, pp 443-458

Hirel B, Phillipson B, Murchie E, Suzuki A, Kunz C, Ferrario S, Limami A, Chaillou S, Deléens E, Brugière N, Chaumont-Bonnet M, Foyer CH, Morot-Gaudry J-F (1997) Manipulating the pathway of ammonia assimilation in transgenic nonlegumes and legumes. J Plant Nutr Soil Sci 160: 283-290

Huber TA, Streeter JG (1985) Purification and properties of asparagine synthetase from soybean root nodules. Plant Sci 42: 9-17

Hughes CA, Beard HS, Matthews BF (1997) Molecular cloning and expression of two cDNAs encoding asparagine synthetase in soybean. Plant Mol Biol 33: 301-311

Ireland RJ, Lea PJ (1999) The enzymes of glutamine, glutamate, asparagine and aspartate metabolism. In: Singh K (ed) Plant amino acids. Marcel Dekker, New York, pp 49-109

Ishimaya K, Hayakawa T, Yamaya T (1998) Expression of NADH-dependent glutamate synthase protein in the epidermis and exodermis of rice roots in response to the supply of ammonium ions. Planta 204: 288-294

Joy KW, Ireland RJ, Lea PJ (1983) Asparagine synthesis in pea leaves and the occurrence of an asparaginase inhibitor. Plant Physiol 73: 165-168

Kamachi K, Yamaya T, Maie T, Ojima K (1991) A role for glutamine synthetase in the remobilisation of leaf nitrogen during natural senescence in rice leaves. Plant Physiol 96: 411-417

Keys AF, Bird IF, Cornelius MJ, Lea PJ, Wallsgrove RM, Miflin BJ (1978) Photorespiratory nitrogen cycle. Nature 275: 741-742

King GA, Woollard DC, Irving DE, Borst WM (1990) Physiological changes in asparagus spear tips after harvest. Physiol Plant 80: 393-400

Kozaki A, Takeba G (1996) Photorespiration protects C_3 plants from photooxidation. Nature 384: 557-560

Lam HM, Coschigano KT, Oliveira IC, Melo-Oliveira R, Coruzzi GM (1996) The molecular genetics of nitrogen assimilation into amino acids in higher plants. Annu Rev Plant Physiol Plant Mol Biol 47: 569-593

Lam HM, Hseih MH, Coruzzi G (1998) Reciprocal regulation of distinct asparagine synthetase genes by light and metabolites in *Arabidopsis thaliana*. Plant J 16: 345-353

Lea PJ and Forde B (1994) The use of mutants and transgenic plants to study amino acid metabolism. Plant Cell and Environment 17: 541-556

Lea PJ, Fowden L (1975) The purification and properties of glutamine-dependent asparagine synthetase isolated from Lupinus albus. Proc R Soc Lond B192: 13-26

Lea PJ, Ireland RJ (1999) Nitrogen metabolism in higher plants. In: Singh BK (ed) Plant amino acids. Marcel Dekker, New York, pp 1-47

Lea PJ, Miflin BJ (1974) An alternative route for nitrogen assimilation in plants. Nature 251: 680-685

Lea PJ, Robinson SA, Stewart GR (1990) The enzymology and metabolism of glutamine, glutamate and asparagine. In: Miflin BJ, Lea PJ (eds) The biochemistry of plants, vol 16 Intermediary nitrogen metabolism. Academic Press, New York, pp 147-152

Leegood RC, Lea PJ, Adcock MD, Häusler RE (1995) The regulation and control of photorespiration. J Exp Bot 46: 1397-1414

Lightfoot D, Green NK, Cullimore JV (1988) The chloroplast-located glutamine synthetase of *Phaselus vulgaris* L.: nucleotide sequence, expression in different organs and uptake into isolated chloroplasts. Plant Mol Biol 11: 191-202

Limami A, Phillipson B, Ameziane R, Pernollet N, Jiang Q, Roy R, Deleens E, Chaumont-Bonnet M, Gresshoff PM, Hirel B (1999) Does root glutamine synthetase control plant biomass production in Lotus japonicus L.? Planta 209: 495-502

Loulakakis KA, Roubelakis-Angelakis KA (1990) Intracellular localization and properties of NADH-glutamate dehydrogenase from *Vitis vinifera* L.: purification and characterization of the major leaf isoenzyme. J Exp Bot 41: 1223-1230

Loulakakis KA, Roubelakis-Angelakis KA (1991) Plant NAD(H)-glutamate dehydrogenase consists of two subunit polypeptides and their participation in the seven isoenzymes occurs in an ordered ratio. Plant Physiol 97: 104-111

Loulakakis KA, Roubelakis-Angelakis KA (1992) Ammonium-induced increase in NADH-glutamate dehydrogenase activity is caused by *de novo* synthesis of the α-subunit. Planta 187: 322-327

Loulakakis KA, Roubelakis-Angelakis KA (1996) The seven NAD(H)-glutamate dehydrogenase isoenzymes exhibit similar anabolic activities. Physiol Plant 96: 29-35

Mack G (1995) Organ-specific changes in the activity and subunit composition of glutamine synthetase isoforms of barley (*Hordeum vulgare* L.) after growth on different levels of NH_4^+ Planta 196: 231-238

Magalhaes JR, Ju GC, Rich PJ, Rhodes D (1990) Kinetics of $^{15}NH_4^+$ assimilation in *Zea mays*. Plant Physiol 94: 647-656

Marsolier MC, Hirel B (1993) Metabolic and developmental control cytosolic glutamine synthetase genes in soybean. Physiol Plant 89: 613-617

McNally SF, Hirel B (1983) Glutamine synthetase isoforms in higher plants. Physiol Veg 21: 761-774

McNally SF, Hirel B, Gadal P, Mann AF, Stewart GR (1983) Glutamine synthetase of higher plants: evidence for a specific isoform content related to their possible physiological role and their compartmentation within the leaf. Plant Physiol 72: 22-25

Melo-Oliveria R, Cinha-Oliveria I, Coruzzi GM (1996) Arabidopsis mutant analysis and gene regulation define a non-redundant role for glutamate dehydrogenase in nitrogen assimilation. Proc Natl Acad Sci USA 96: 4718-4723

Miflin BJ, Lea PJ (1976) The pathway of nitrogen assimilation in plants. Phytochemistry 15: 873-885

Migge A, Meya G, Carrayol E, Hirel B, Becker TW (1996) Regulation of subunit composition of tomato plastidic glutamine synthetase by light and the nitrogen source. Planta 200: 213-220

Migge A, Carrayol E, Hirel B, Lohmann M, Meya G, Becker TW (1998a) Regulation of the subunit composition of plastidic glutamine synthetase of the wild-type and of the phytochrome-deficient aurea mutant of tomato by blue/UV-A- or by UV-B-light. Plant Mol Biol 37: 689-700

Migge A, Carrayol E, Hirel B, Lohmann M, Meya G, Becker TW (1998b) Influence of UV-A or UV-B and of the nitrogen source on ferredoxin-dependent glutamate synthase in etiolated tomato cotyledons. Plant Physiol Biochem 36: 789-797

Ngai N, Tsai FY, Coruzzi G (1997) Light-induced transcriptional repression of the pea *AS1* identification of *cis* elements and transfactors. Plant J 12: 1021-1034

Oaks A (1994) Primary nitrogen assimilation in higher plants and its regulation. Can J Bot 72: 739-750

Oaks A (1995) Evidence for deamination by glutamate dehydrogenase in higher plants: Reply. Can J Bot 73: 1116-1117

Oaks A, Hirel B (1985) Nitrogen assimilation in roots. Annu Rev Plant Physiol 36: 345-365

Ortega JL, Roche D, Sengupta-Gopalan C (1999) Oxidative turnover of soybean root glutamine synthetase. *In vitro* and *in vivo* studies. Plant Physiol 119: 1483-1495

Osuji GO, Madu WC (1995) Ammonium-ion-dependent isomerization of glutamate dehydrogenase in relation to glutamate synthesis in maize. Phytochemistry 39: 495-503

Osuji GO, Madu WC (1997) Regulation of peanut glutamate dehydrogenase by methionine sulphoximine.Phytochemistry 46: 817-825

Osuna D, Galvez G, Pineda M, Aguilar M (1999) RT-PCR cloning, characterisation and mRNA expression analysis of a cDNA encoding a type II asparagine synthetase in common bean. Biochim Biophys Acta 1445: 75-85

Pate JS (1989) Synthesis, transport and utilisation of products of nitrogen fixation. In: Poulson JE, Romeo JT, Conn EE (eds) Plant nitrogen metabolism. Plenum Press, New York, pp 65-115

Peat LJ, Tobin A (1996) The effect of nitrogen nutrition on the cellular localization of glutamine synthetase isoforms in barley roots Plant Physiol 111: 1109-1117

Peeters KMU, Van Laere AJ (1994) Amino acid metabolism associated with N-mobilisation from the flag leaf of wheat (*Triticum aestivum* L.) during grain development. Plant Cell Environ 17: 131-141

Pérez-Garcia A, Pereira S, Pissara J, Garcia-Gutierez A, Cazorla FM, Salema R, de Vicente A, Canovas FM (1998) Cytosolic localization in tomato cells of a novel glutamine synthetase induced in response to bacterial infection or phosphinothricin treatment. Planta 206: 426-434

Pérez-Rodriguez J, Valpuesta V (1996) Expression of glutamine synthetase genes during natural senescence of tomato leaves. Physiol Plant 97: 576-582

Pryor AJ (1990) A maize glutamic dehydrogenase null mutant is cold temperature-sensitive. Maydica 35: 367-372

Pujade-Renaud V, Perrot-Rechenmann C, Chrestin H, Lacrotte R, Guern J (1997) Characterization of a full-length cDNA clone encoding glutamine synthetase from rubber tree latex. Plant Physiol Biochem 35: 85-93

Purnell MP, Stewart GR, Botella JR (1997) Cloning and characterization of a glutamate dehydrogenase cDNA from tomato (*Lycopersicon esculentum* L.). Gene 186: 249-254

Rastogi R, Chourey PS, Muhitch MJ (1998) The maize glutamine synthetase GS1-2 gene is preferentially expressed in kernel pedicels and is developmentally regulated. Plant Cell Physiol 39: 443-446

Redinbaugh MG, Campbell WH (1993) Glutamine synthetase and ferredoxin dependant glutamate synthase expression in the maize (*Zea mays*) root primary response to nitrate. Plant Physiol 101: 1249-1255

Robinson SA, Slade AP, Fox GG, Phillips R, Ratcliffe RG, Stewart GR (1991) The role of glutamate dehydrogenase in plant nitrogen metabolism. Plant Physiol 95: 809-816

Robinson SA, Stewart GR, Phillips R (1992) Regulation of glutamate dehydrogenase activity in relation to carbon limitation and protein catabolism in carrot cell suspension cultures. Plant Physiol 98: 1190-1195

Rognes SE (1975) Glutamine-dependent asparagine synthetase from *Lupinus luteus*. Phytochemistry 14: 1975-1982

Sahulka J, Lisa L (1980) Effect of some disaccharides, hexoses and pentoses on nitrate reductase, glutamine synthetase, and glutamate dehydrogenase in excised pea roots. Physiol Plant 50: 32-36

Sakakibara H, Kawabata S, Hase T, Sugiyama T (1992) Differential effect of nitrate and light on the expression of glutamine synthetase and ferredoxin-dependent glutamate synthase in maize. Plant Cell Physiol 33: 1193-1198

Sakakibara H, Fujii K, Sugiyama T (1995) Isolation and characterization of a cDNA that encodes maize glutamate dehydrogenase. Plant Cell Physiol 36: 789-797

Sakurai N, Hayakawa T, Nakamura T, Yamaya T (1996) Changes in the cellular localisation of cytosolic glutamine synthetase protein in vascular bundles of rice leaves at various stages of development. Planta 200: 306-311

Schmidt RR, Miller P (1999) Polypeptides and polynucleotides relating to the α and β subunits of a glutamate dehydrogenase and methods of use. US Patent n° 5, 879,941, Mar 9

Sechley KA, Yamaya T, Oaks A (1992) Compartmentation of nitrogen assimilation in higher plants. Int Rev Cytol 134: 85-163

Shi LF, Twary SN, Yoshioka H, Gregerson RG, Miller SS, Samac DA, Gantt, Unkefer PJ, Vance CP (1997) Nitrogen assimilation in alfalfa: isolation and characterisation of an asparagine synthetase gene showing enhanced expression in root nodules and dark adapted leaves. Plant Cell 9: 1339-1356

Sieciechowicz KA, Joy KW, Ireland RJ (1988) The metabolism of asparagine in plants. Phytochemistry 27: 663-671

Srivastava HS, Singh RP (1987) Role and regulation of L-glutamate dehydrogenase activity in higher plants. Phytochemistry 26: 597-610

Stewart GR, Shatilov VR, Turnbull MH, Robinson SA, Goodall R (1995) Evidence that glutamate dehydrogenase plays a role in oxidative deamination of glutamate in seedlings of Zea mays. Aust J Plant Physiol 22: 805-809

Stulen I, Oaks A (1977) Asparagine synthetase in corn roots. Plant Physiol 60: 680-683

Suzuki A, Gadal P (1984) Glutamate synthase: physiochemical and functional properties of different forms in higher plants and other organisms. Physiol Veg 22: 471-486

Suzuki A, Vidal J, Gadal P (1982) Glutamate synthase isoforms in rice: immunological studies of enzymes of green leaf, etiolated leaf and root tissues. Plant Physiol 70: 827-832

Syntichaki KM, Loulakakis KA, Roubelakis-Angelakis KA (1996) The amino acid sequence similarity of plant glutamate dehydrogenase to the extremophilic archeal enzyme conforms to its stress-related function. Gene 168: 87-92

Temple SJ, Bagga S, Sengupta-Gopalan C (1994) Can glutamine synthetase activity be modulated in transgenic plants by the use of recombinant DNA technology? Biochem Soc Trans 22: 915-920

Temple SJ, Kunjibettu S, Roche D, Sengupta-Gopalan S (1996) Total glutamine synthetase activity during soybean nodule development is controlled at the level of transcription and holoprotein turnover. Plant Physiol 112: 1723-1733

Temple SJ, Vance CP, Gantt JS (1998) Glutamate synthase and nitrogen assimilation. Trends Plant Sci 3: 51-56

Tercé-Laforgue T, Carrayol E, Cren M, Desbrosses G, Hecht V, Hirel B (1999) A strong constitutive positive element is essential for the ammonium-regulated expression of a soybean gene encoding cytosolic glutamine synthetase. Plant Mol Biol 39: 551-564

Tjaden G, Coruzzi G (1994) A novel AT-Rich DNA binding protein that combines an HMG I-like binding domain with putative transcription domain. Plant Cell 6: 1107-1118

Trepp GB, Plank DW, Gantt, Vance CP (1999a) NADH-glutamate synthase in alfalfa root nodules. Immunocytochemical localisation. Plant Physiol 119: 829-837

Trepp GB, van de Mortel M, Yoshioka H, Miller SS, Samac DA, Gantt JS, Vance CP (1999b) NADH-glutamate synthase in alfalfa roots. Genetic regulation and cellular expression. Plant Physiol 119: 817-828

Tsai F-Y, Coruzzi GM (1990) Dark induced and organ-specific expression of two asparagine synthetase genes in *Pisum sativum*. EMBO J 9: 2829-2831

Tsai F-Y, Coruzzi GM (1991) Light represses the transcription of asparagine synthetase genes in photosynthetic and non-photosynthetic organs of plants. Mol Cell Biol 11: 4966-4972

Tsuprun VL, Samsonide TG, Radukina NA, Pushkin AV, Evstigneeva ZG, Kretovich WL (1980) Electron microscopy of glutamine synthetase from pea leaf chloroplast. Biochim Biophys Acta 626: 1-4

Turano FJ (1998) Characterisation of mitochondrial glutamate dehydrogenase from dark-grown soybean seedlings. Physiol Plant 104: 337-344

Turano FJ, Thakkar SS, Fang T, Weisemann JM (1997) Characterisation and expression of NAD(H) dependent glutamate dehydrogenase genes in *Arabidopsis*. Plant Physiol 113: 1329-1341

Vincent R, Fraiser V, Chaillou S, Limami MA, Deleens E, Phillipson B, Couat C, Boutin J, Hirel B (1997) Overexpression of a soybean gene encoding cytosolic glutamine synthetase in shoots of transgenic *Lotus corniculatus* L. plants triggers changes in ammonium assimilation and plant development. Planta 201: 424-433

Waterhouse RN, Smyth AJ, Massonneau A, Prosser IM, Clarkson DT (1996) Molecular cloning and characterization of asparagine synthetase from *Lotus japonicus* – dynamics of asparagine synthesis in N-sufficient conditions. Plant Mol Biol 30: 883-897

Woodall J, Boxall JG, Forde BG, Pearson J (1996) Changing perspectives in plant nitrogen metabolism: the central role of glutamine synthetase. Sci Prog 79: 1-26

Yamaya T, Oaks A (1987) Synthesis of glutamate by mitochondria - an anaplerotic function for glutamate dehydrogenase. Physiol Plant 70: 749-756

Zehnacker C, Becker TW, Suzuki A, Carrayol E, Caboche M, Hirel B (1992) Purification and properties of tobacco ferredoxin-dependent glutamate synthase and isolation of corresponding cDNA clones. Light inducibility and organ specificity in the gene transcription and protein expression. Planta 187: 266-274

Nodule Formation and Function

Anthony J. GORDON[1], Peter J. LEA[2], Charles ROSENBERG[3]
and Jean-Charles TRINCHANT[4]

Introduction

Nitrogen fixation

Nitrogen (N_2) is a very unreactive molecule. The reason for its chemical stability lies in the electronic structure of the molecule, but nitrogen is not totally inert. The industrial reduction of nitrogen to ammonia (principally by the Haber-Bosch process) is very important commercially. The simplest equation for nitrogen fixation is

$$N_2 + 3H_2 \rightarrow 2NH_3$$

This equation describes the Haber-Bosch process. The standard free energy (ΔG^0) for this reaction is -7.7 kcal mol^{-1} at 298 $^{\circ}$C and 1 atm pressure; the negative value of ΔG^0 indicates that energy is released during nitrogen fixation, i.e. the reaction is exergonic. This in turn implies that, in the presence of a suitable catalyst, the reaction should proceed spontaneously at room temperature. Unfortunately, no catalyst exists that allows a reasonable rate of nitrogen reduction at room temperature, so the Haber-Bosch process is carried out at high temperatures (600-800 °C) and pressure as high as 500 atm, with concomitant input of energy. Industrial nitrogen fixation is, therefore, an expensive process. Biological nitrogen fixation, exemplified by the equation

$$N_2 + 10H^+ + 8e^- \rightarrow 2NH_4^+ + H_2$$

is also an exergonic reaction, but in practice requires an input of energy (16 molecules of ATP are hydrolysed per molecule of nitrogen reduced). In addition, since protons replace hydrogen (the reaction takes place in an aqueous environment), a source of electrons (e^-) is required. Note also that hydrogen is a reaction product of biological nitrogen fixation. It has been calculated that biological nitrogen fixation

1. Institute of Grassland & Environmental Research, Aberystwyth, SY23 3EB, U.K. *E-mail:* tony.gordon@bbsrc.ac.uk
2. Biological Sciences, Lancaster University, Lancaster, LA1 4YQ, U.K. *E-mail:* p.lea@lancaster.ac.uk
3. Laboratoire de Biologie Moléculaire des Relations Plantes-Microorganismes, CNRS-INRA, BP 27, 31326 Castanet Tolosan, Cedex, France. *E-mail:* crosen@toulouse.inra.fr
4. Laboratoire de Biologie Végétale et Microbiologie, CNRS URA 1114, Université de Nice-Parc-Valrose, 06108, Nice Cedex 2, France. *E-mail:* trinchan@unice.fr
The authors contributed equally to this chapter and are here listed in alphabetical order.

can account for the annual production of as much as 250×10^9 kg of ammonia, over twice that manufactured by the Haber-Bosch process. The earlier literature on the subject of nitrogen fixation has been covered in a number of excellent books, e.g. Gallon and Chaplin (1987); Postgate (1987); Sprent and Sprent (1990).

Although the total number of organisms so far shown to fix nitrogen is relatively small, they span a wide range of eubacteria and archaebacteria (Sprent and Sprent 1990; Young 1992). Diazotrophs are usually divided into free-living and symbiotic forms, but the position is not always clearcut. Some organisms (notably certain cyanobacteria) appear to be equally able to fix nitrogen independently and in symbiotic association. The free-living diazotrophs include representatives of both archaebacteria and eubacteria. Among the latter, there is an almost random distribution of the ability to fix nitrogen among the various families. Diazotrophy also spans a range of physiological behaviours, covering autotrophy and heterotrophy and also anaerobiosis, facultative anaerobiosis and aerobiosis. As organisms capable of oxygenic photosynthesis, cyanobacteria (blue-green algae) have the distinction of being the only diazotrophic plants, although as prokaryotes they are more properly classified as bacteria. Recent studies on bloom-forming cyanobacteria in the tropical oceans suggest that they can generate as much as 85×10^9 kg of fixed nitrogen per year (Capone et al. 1997; Karl et al. 1997). Symbiotic diazotrophs include a number of genera of the Rhizobiaceae which form the well-documented symbiosis with legumes. For the sake of simplicity, unless a specific organism is mentioned, these genera will be given the general term rhizobium. Infection of legumes by rhizobia results in the formation of nodules, usually on roots. However, some rhizobia, for example *Azorhizobium caulinodans* and the photosynthetic *Photorhizobium thompsonianum*, form stem nodules on certain plants. One non-legume plant, *Parasponia*, is also known to become infected by rhizobium.

A number of other non-legume plants also produce nitrogen-fixing nodules. However, in this case the microsymbiont is not a rhizobium, but an actinomycete, *Frankia*. The macrosymbionts, referred to as actinorhizal plants, are mainly woody species, and belong to a variety of plant families. Finally, cyanobacteria participate in a variety of diazotrophic symbioses (see Gallon and Chaplin 1987; Rai 1990; Sprent and Sprent 1990). The co-symbiont may be an angiosperm, a gymnosperm, a pteridophyte, a bryophyte, a fungus (some lichens can fix nitrogen), a diatom or, in the case of the coral reef sponge *Siphonochalina*, even an animal.

The Rhizobium-Legume Symbiosis

The rhizobium-legume symbiosis plays a major ecological and economical role since, on a global scale, it provides a quantity of fixed nitrogen equivalent to that produced by the entire chemical fertiliser industry. This symbiotic association takes place between most of the plants of the legume family and soil bacteria belonging to the genera *Rhizobium, Sinorhizobium, Mesorhizobium, Allorhizobium, Bradyrhizobium, Azorhizobium* and *Photorhizobium*. These bacteria elicit on the roots or, exceptionally, on the stems of their legume hosts the formation of specialised organs, nodules, which constitute an ecological niche in which the conditions are provided which are necessary for the reduction of atmospheric nitrogen into ammonia. This ammonia is exported into the plant, where it is assimilated. This symbiotic interaction is highly specific, a given rhizobium strain being able to form nodules on a restricted number of plants, which constitute the host range of this strain. This host range can be quite narrow, as in the case of *Sinorhizobium meliloti*, which nodulates only plants of the genera *Medicago*, *Melilotus* or *Trigonella*, or much wider, as for *Rhizobium sp.* strain NGR234, which can nodulate not only plants originating from more than 110 different genera of tropical legumes, but also the non-legume, *Parasponia* (Table 1).

Table 1. Main rhizobium-legume associations

Genera	Species	Host plants
Rhizobium	*R. leguminosarum*	
	bv. viciae	*Vicia, Pisum*
	bv. trifolii	*Trifolium*
	R. etli	*Phaseolus*
	R. tropici	*Phaseolus, Leucaena*
	R. galegae	*Galega*
Sinorhizobium	*R.* sp. NGR234	Broad tropical host range
	S. melitoti	*Medicago*
	S. fredii	*Glycine*
	S. saheli	*Acacia, Sesbania*
	S. teranga	*Acacia, Sesbania*
Mesorhizobium	*M. ciceri*	*Cicer*
	M. huakii	*Astragalus*
	M. loti	*Lotus*
Allorhizobium	*A. undicola*	*Neptunia*
Bradyrhizobium	*B. elkanii*	*Glycine*
	B. japonicum	*Glycine*
	B. lupini	*Lupinus*
	B. sp. cowpea	Broad tropical host range
Azorhizobium	*A. caulinodans*	*Sesbania*

The different steps leading to the establishment of the rhizobium-legume symbiotic interaction are now well documented (for reviews, see Nap and Bisseling 1990; Hadri and Bisseling 1998). Rhizobia, which can survive in the soil as saprophytes, are attracted, and their growth is stimulated by various compounds excreted in the rhizosphere of legumes, such as glycans, carboxylic acids, amino acids or flavonoids (Sadowsky and Graham 1998). When bacteria are in contact with a region of the root susceptible to nodulation, they initiate two processes in parallel: the induction of an organogenesis program leading to the formation of the nodule, and the penetration into root tissues.

Depending on the host plant, the mechanism of root infection varies, ranging from a simple entry through intercellular spaces in the root epidermis (*Arachis*) to the penetration through root hairs (alfalfa, vetch). In this latter case, the attachment of the compatible rhizobia to root hairs results in a characteristic curling, which is followed by an invagination of the cell wall and the formation of an infection thread, a tubular structure containing the bacteria. These infection threads develop within the root hairs, then progress through the root cortex, where they ramify. Good diagrams of proposed infection mechanisms have recently been provided by Pawlowski (1997). In parallel to this infection process, cell divisions are elicited in the root inner cortex, resulting in the formation of a nodule primordium. Development of this primordium leads to the formation of the nodule meristem, which will give rise to a fully differentiated organ, the nodule.

Molecular analysis of the initiation and development of the legume root nodule has concentrated on proteins termed nodulins, that are actively synthesised during this process (Fuller et al. 1983). It was initially suggested that these proteins were located solely in the nodule; however, more recent molecular evidence has confirmed that the relevant genes are also expressed in other parts of the plant (e.g. glutamine synthetase, Cullimore et al. 1992; uricase, Takane et al. 1997). There is also now evidence that related legume nodulin genes are expressed in tobacco (e.g. *ENOD40*, van der Sande et al. 1996), or even in monocots (e.g. haemoglobin, Arredondo-Peter et al. 1998). The genes that are expressed immediately following infection with rhizobia are termed early nodulin (*ENODx*) and encode early nodulin proteins (ENODx), the functions of which are still relatively obscure. However, as will be seen later, prediction of their function has been possible by analysis of the amino acid sequences. Late nodulin genes are first expressed just prior to the onset of nitrogen fixation, and frequently encode the enzymes of carbon and nitrogen metabolism that are involved in ammonia assimilation

The Early Symbiotic Events: a Molecular Dialogue Between the Two Partners

The initiation of this symbiotic interaction involves a specific recognition between the two partners, the differentiation of a new organ, and the invasion of plant roots by the bacteria. The genetic control of this complex biological process involves at least two classes of bacterial genes.

The first class consists of genes involved in the synthesis, the assembly or the transport of cell surface components, such as neutral cyclic β-glucans, acidic exopolysaccharides (EPS), lipopolysaccharides (LPS) and capsular polysaccharides (KPS). Bacterial mutants defective in the synthesis of these components have been shown to form empty nodules, which indicates that these components play a role in the infection process. However, a similar structural defect may or may not impair bacterial invasion, depending on the rhizobial species. The exact functions of these surface components remain to be determined, but it has been proposed that they could act by protecting bacteria from plant defence reactions. This topic has been reviewed by Kannenberg and Brewin (1994); Leigh and Walker (1994); Gonzalez et al. (1996).

The second class of symbiotic bacterial genes are the nodulation genes, which have been shown to play a pivotal role in the signal exchange (Fig. 1) on which the establishment of the symbiotic interaction is based (for reviews, see Mylona et al. 1995; Denarie et al. 1996; Long 1996; Schultze and Kondorosi 1998). Compounds

Host legume   rhizobium

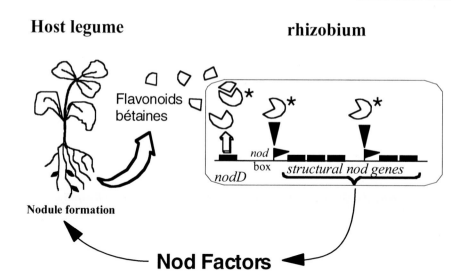

Fig. 1 : Signal exchange in the rhizobium-legume symbiosis. The activated state of the NodD protein is indicated by *

exuded by the roots of the host plant activate the expression of bacterial nodulation genes which, in response, determine the synthesis of lipo-chito-oligosaccharidic signals, the Nod factors.

nod- genes and Nod Factors

Expression of Bacterial Nodulation Genes Is Induced by Plant Compounds Present in Root Exudates

Genetic analysis of the early symbiotic events was based on the identification of the *nod*, *nol* and *noe* (for nodulation) genes, which control host recognition, nodule elicitation and bacterial invasion. The structural nodulation genes, generally located on large plasmids, are clustered in operons, the expression of which is controlled by the regulatory genes of the *nodD* family. The products of these genes are transcriptional activators of the LysR family (Henikoff et al. 1988), able to bind a conserved *cis* regulatory element, called the *nod* box, present upstream of all the *nod* operons (Fisher et al. 1988; for reviews, see Kondorosi 1992; Schlaman et al. 1998). When activated by compounds present in plant root exudates, the NodD proteins activate the transcription of the different structural *nod* genes. These plant compounds are generally flavonoids, which are secondary metabolites of the phenylpropanoid biosynthetic pathway. The NodD proteins of different rhizobium species are able to recognise different flavonoids. For instance, *nod* gene expression is induced by naringenin in *Rhizobium leguminosarum* bv. *viciae*, and by daidzein in *Bradyrhizobium japonicum*. This recognition between a NodD protein and a particular flavonoid determines a first level of specificity in the symbiotic interaction. However, this specificity is not very strict, since a given NodD protein is usually able to recognise several different flavonoids, and a legume plant excretes a mix of different flavonoids. Many rhizobium species contain several copies of *nodD*, with different specificities. For instance, in *S. meliloti*, where three copies of *nodD* are present, the NodD1 protein activates the transcription of the *nod* genes in the presence of luteolin and NodD2 is active in the presence of betains which belong to another category of secondary metabolites. The NodD3 protein regulates *nod* gene expression as a function of the nitrogen status of the bacteria.

The presence of other regulatory elements, such as the two component regulatory system *nodV/nodW* in *B. japonicum*, or the negatively acting elements *nolR* in *S.meliloti*, *nolA* in *B.japonicum*, or *nodD2* in *Rhizobium* sp. NGR 234 suggest that *nod* gene expression could be modulated by additional circuits of regulation, responding to as yet uncharacterised factors (Fisher and Long 1992; Kondorosi 1992; Schlaman et al. 1998).

Rhizobial Nodulation Genes Determine the Production of Extracellular Signals, the Nod Factors

The structural *nod* genes, the number of which varies from 10 to more than 20, depending on the strain, can be divided in two classes:

1. The *nodABC* genes are called common *nod* genes, because they are present in all rhizobia. They play a central role, since the inactivation of these genes results in a complete loss of all symbiotic properties.

2. The host-specific *nod* genes control host specificity. For example, the transfer of the *S. meliloti* host specific *nod* genes confers to *R.leguminosarum* bv.*viciae* the ability to nodulate the *S. meliloti* host alfalfa. Furthermore, mutations in these genes result in modifications of the host range of the strain.

By which mechanism can this limited number of genes control such a complex biological process, including the specific recognition of the host, the induction of plant organogenesis and the infection of the nodule?

It was observed that sterile supernatants of rhizobium cultures are able to induce different types of responses on the roots of their specific host plants, including characteristic deformations and branching of root hairs. Such deformations are not induced by culture supernatants of strains mutated in the common *nodABC* genes, indicating that these genes are involved in the production of extracellular signals responsible for these root hair deformations (van Brussel et al. 1986). Using the ability to induce root hair deformations on alfalfa as a test, the biologically active fractions of the supernatant of a *S. meliloti* culture were purified, and the structure of the active compounds, the Nod factors (NF), was determined for the first time by Lerouge et al. (1990). The *S. meliloti* NFs are lipo-chito-oligosaccharides (LCOs), consisting of a chitin oligomer backbone of four or five β-1,4 linked glucosamine residues, mono-N-acylated by a polyunsaturated C16 fatty acid and O-acetylated at the non-reducing end, and O-sulphated at the reducing end (Fig. 2).

The Common nod Genes Determine the Synthesis of the LCO Backbone, While the Nature of the Substitutions Is Controlled by the Host Specific nod Genes

Genetic and biochemical analysis has shown that the synthesis of the LCO backbone is catalysed by the products of the common *nodABC* genes. NodC, which has homology with chitin synthases, is likely to control the synthesis and the length of the oligo-chitin chain, the non-reducing end of which is deacetylated by NodB, while NodA transfers an acyl chain to this position (for reviews, see Carlson et al. 1994; Spaink 1996; Downie 1998).

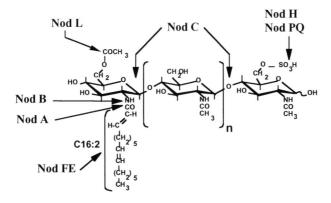

Fig. 2 : Role of *nod* gene products in the synthesis of *Sinorhizobium meliloti* Nod factors (NFs)

Analysis of the structure of NFs produced by different *S. meliloti* mutants showed that the products of the host specific *nod* genes specify the synthesis and the transfer of various substitutions onto this backbone. Thus, an *S. meliloti* strain mutated for one of these host-specific genes produces NFs lacking the corresponding substitution, and exhibits an altered symbiotic phenotype on alfalfa. Mutations in *nodL* result in the production of NFs lacking the O-acetyl group at the non reducing end, while mutations in *nodFE* result in the production of NFs acylated by a fatty acid of general lipid metabolism, instead of the specific poly-unsaturated fatty acid (Dénarié et al. 1996). These two types of mutants show a decreased ability to form infection threads on alfalfa, an *S. meliloti* host plant. A *nodF/nodL* double mutant combining these two structural defects is unable to infect and nodulate alfalfa (Ardourel et al. 1994). A *S. meliloti* strain mutated in *nodH*, which codes for a sulphotransferase, produces NFs lacking the sulphate group. This mutant strain is no longer able to nodulate alfalfa, but has gained the ability to nodulate vetch, which is a non-host plant (Roche et al. 1991). Thus, all three substitutions are important for efficient infection and nodulation of alfalfa. Furthermore, the sulphate group is an essential determinant of the host range, since the presence or absence of the sulphate molecule results in a dramatic switch in the host range of the bacterium. In *R. leguminosarum bv.viciae*, it has been shown that NF structural deficiencies can be compensated by NodO, a secreted bacterial protein not involved in NF production, but which might stimulate uptake of NFs by forming ion channels in the plant plasma membrane (Sutton et al. 1994).

The mechanisms controlling NF secretion are not yet completely understood, but they are likely to involve the products of the *nodIJ* genes. These genes, which are present in all rhizobia, generally as part of the *nodABC* operon, encode proteins which are members of the family of ABC transporters. Mutations in these genes result in a decrease in the amount of NF secreted (Downie 1994). However, depending on the strain, variable levels of NFs are still produced in *nodIJ* mutants, suggesting that redundant or alternative transport mechanisms are present (Downie 1998).

Diversity of Nod Factors

The chemical structures of Nod factors produced by various rhizobia have now been elucidated. Though rhizobia belong to quite distant taxonomic groups, they all produce NFs sharing a basic LCO structure determined by the common *nodABC* genes. The nature of the substitutions present on the oligosaccharidic chain are, as for *S. meliloti*, specified by host-specific *nod* genes, and vary among the strains. The combinations of the various substitutions together with variations in length and in degree of unsaturation of the fatty acyl chain result in a large diversity of structures (Fig. 3). At the genetic level, this diversity can result from either the presence or absence of a host-specific *nod* gene controlling a particular biosynthetic or transfer function, such as *nodL* or *nodH*, or from allelic variations at the same locus, as in the case of the *nodFE* genes, responsible for the synthesis of specific fatty acid chains.

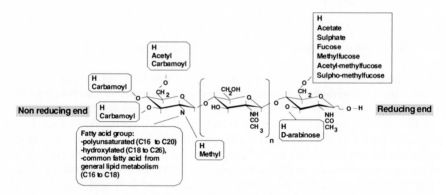

Fig. 3 : Example of Nod factor (NF) substitution diversity. n indicates the number of central glucosamine residues and can vary from one to four. In some rhizobium species, substitutions can be found on non-terminal glucosamine residues [for reviews see Dénarie et al. (1996) and Downie (1998)]

    Allelic variations occurring in the common *nod* genes can also contribute to the definition of the host range by determining the length of the oligochitin chain (*nodC*) or the choice of the fatty acyl chain preferentially transferred (*nodA*) (Roche et al. l996; Peters 1997). A given rhizobium isolate generally does not produce a single Nod factor, but a family of related molecules. In the case of *Rhizobium sp.* NGR234, the production of 18 different LCOs may explain the unique broad host range of this strain (Price et al. 1992).

Nod Factor Structure Is Related to the Taxonomy of the Host Plant, Not to That of the Bacterium.

Determination of NF structures in various rhizobia has revealed that these structures are not related to the taxonomic position of the bacteria from which they originate. For example, it has been shown that *B. japonicum* and *S. fredii*, which are taxonomically quite distant, nodulate the same host, soybean, and produce Nod factors which have very similar structures, characterised by a fucosyl moiety at the non-reducing end (Carlson et al. 1994). A similar example has been reported for two other phylogenetically distant bacteria, *Azorhizobium caulinodans* and *Sinorhizobium saheli* bv. sesbaniae, which both form nodules on roots and stems of the tropical legume *Sesbania rostrata*, and produce similar NFs doubly substituted at the reducing end by fucose and arabinose (Lorquin et al. 1997; Mergaert et al. 1997).

Another illustration of this lack of correlation between NF structure and bacterial phylogeny is provided by a number of rhizobial species which have been shown to produce NFs that are N-acylated with polyunsaturated fatty acids. These species are distributed in two distinct genera, *Rhizobium* and *Sinorhizobium*. However, it has been observed that the legume species nodulated by these rhizobium are members of the *Galegeae*, *Trifolieae* or *Vicieae* tribes, which, according to phylogenetic studies, all derive from a common ancestor (Yang et al. 1999). Thus, a correlation seems to exist between the presence of these peculiar acyl chains and legume phylogeny. This correlation suggests the existence of a coevolution process, in which the plant exerts on the bacteria a selection pressure for a defined type of NF structure, to which the microsymbionts can adapt by means of horizontal genetic transfer of nodulation genes.

Plant Responses to NFs

NFs induce on the roots of host plants a variety of responses similar to those induced by the bacteria from which they originate, such as morphological changes, cell divisions and induction of early nodulin genes. In addition, several electrophysiological responses can be observed in the seconds or minutes following NF application. These responses, which are schematised in Fig. 4, have been reviewed by Mylona et al. (1995); Spaink (1996); Cohn et al. (1998); Hadri and Bisseling (1998); Schultze and Kondorosi (1998).

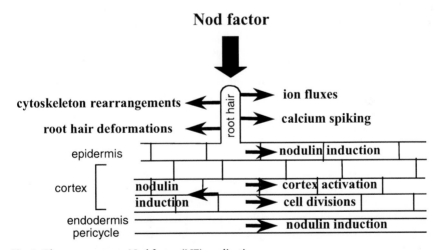

Fig. 4 : Plant responses to Nod factor (NF) application

Electrophysiological Responses

Because of its physical accessibility, the root epidermis has been the focus of numerous studies aimed at characterising early responses to Nod factors. Very fast electrophysiological modifications have been shown to occur in the root epidermis within less than 1 min after NF application. In *Medicago*, these modifications include an increase in cellular calcium concentration in root hairs, rapidly followed by an opposite directed flux of chloride ions, accompanied by a depolarisation of the root hair membrane (Ehrhardt et al. 1992; Felle et al. 1995; Kurkdjian 1995; Felle et al.1998). Subsequently, alkalinisation of the root hair cytoplasm occurs. A few minutes after the application of NFs, periodic oscillations of the cytoplasmic calcium concentration (calcium spiking) can be observed in root hairs (Ehrhardt et al. 1996). The relation between these different physiological responses is not yet known, and it remains to be established whether they are part of a signal pathway leading to other types of responses to NFs.

Developmental Responses

Root hair deformations, such as swelling and branching, are triggered by NF application at concentrations as low as 10^{-12} M (Dénarié et al. 1992). These morphological changes are likely to result from cytoskeletal rearrangements provoked by the reorganisation of microtubules and actin filaments in response to NF application (Timmers et al. 1998; Miller et al. 1999). In the presence of NF at 10^{-7} to 10^{-10} M, cells of the inner cortex, which were in a quiescent state, are mitotically reactivated (Dénarié et al. 1992). These divisions lead to the formation of nodule primordia, which are mainly formed opposite proto-xylem poles. This positioning has been proposed to be negatively regulated by ethylene, and positively regulated by the product of the early nodulin gene *ENOD40* (see following section) and plant compounds such as uridine (Smit et al. 1995; for reviews, see Mylona et al. 1995; Spaink 1996; Cohn et al. 1998; Schultze and Kondorosi 1998).

In some species, the differentiation of these primordia into nodules requires the presence of rhizobia. In contrast, in other plants such as alfalfa, bean, soybean, or *Sesbania*, application of purified NFs is sufficient to induce the differentiation of these primordia into structures having characteristic nodule anatomy. As for bacteria-induced nodules, formation of these structures is repressed in the presence of 20 mM nitrate (Truchet et al. 1991). These structures do not fix nitrogen, since they are devoid of bacteria. These results indicate that the host plant contains the complete developmental genetic program for nodule formation, and that this is switched on by rhizobia through the production of Nod signals. NFs are also involved in the infection process. It has been reported by van Brussel et al. (1992), that in vetch, application of *R. leguminosarum* bv.*viciae* NFs elicits in the outer cortical cells morphological changes, which result in the formation of radially aligned structures, the cytoplasmic bridges. The alignment of these cytoplasmic bridges, called preinfection threads, defines the position of the future infection threads. In alfalfa, an *S.meliloti nodF/nodL* double mutant, which produces NFs lacking the O-acetyl group and acylated by vaccenic acid instead of the specific polyunsaturated fatty acyl chain, is no longer able to initiate the formation of infection threads (Ardourel et al. 1994). Thus, if NFs are not able to elicit the formation of infection threads, which require the presence of the bacteria, they nevertheless play an important role in bacterial invasion.

Early Nodulin Induction

In parallel with these physiological and developmental responses, NFs can trigger the plant to produce a complex tissue-specific response at the transcriptional level, resulting in the induction of plant genes termed early nodulin genes (for reviews, see Long 1996; Hadri and Bisseling 1998; Schultze and Kondorosi 1998).

The ENOD2 protein of soybean is very proline-rich and is composed mainly of two repeating pentapeptides, Pro-Pro-His-Glu-Lys and Pro-Pro-Glu-Tyr-Gln. Similar sequences have been found in ENOD2 proteins from other legumes (Chen et al. 1998). ENOD2 synthesis can be induced by the cooperative action of several LCO molecules (Minami et al. 1996). *In situ* hybridisation and GUS expression studies have shown that the *Enod2* gene is expressed in nodule parenchyma and also in the tissue surrounding the vascular bundle that connects the nodule to the root central cylinder (van der Weil et al. 1990; Chen et al. 1998). Antibodies raised against a synthetic ENOD2 protein have localised the protein in the intercellular spaces of the inner cortical cells. It is has been proposed that ENOD2 protein is targeted to the cell wall and that it could play a role in the formation of a barrier to the diffusion of O_2 to the bacteroids (van der Weil et al. 1990; Wycoff et al. 1998). However, the levels of ENOD2 mRNA or protein in *M. sativa* nodules did not differ significantly when they were grown in 8, 20 or 50% O_2, suggesting that the nodulin is not involved in long-term regulation of nodule permeability (Wycoff et al. 1998).

Two early nodulins genes, *ENOD11* and *ENOD12*, encode proline-rich proteins which are expressed both in epidermal cells and in the root cortex and in the submeristematic zone containing actively growing infection threads (Scheres et al. 1990; Journet et al. 1994; D. Barker, pers. commu.). They are proposed to be associated with infection thread progression. A similar pattern of expression has been described for *MtAnnl*, which shares homology with annexins, a large family of calcium- and phospholipid-binding proteins, involved in a variety of cellular processes, such as membrane fusions, calcium fluxes, cellular signalling or interactions with the cytoskeleton (de Carvalho-Niebel et al. 1998).

Early nodulin genes were initially postulated to play an important role in nodule development. However, it has been shown that a *Medicago* subspecies lacking the *ENOD12* gene is still able to form functional nodules (Csanadi et al. 1994). Thus, the actual importance of the different nodulin genes for the establishment of the symbiotic interaction remains to be determined. Nevertheless, these genes constitute good molecular markers for the study of early symbiotic events. The fusion of the *GUS* reporter gene with the *ENOD12* promoter provides a very sensitive and powerful tool for performing structure/function studies or monitoring the triggering of early symbiotic responses to NFs in different conditions. Using *M.varia* plants transgenic for this gene fusion, it has been shown that *MtENOD12* expression in the epidermis can be induced by *S. meliloti* wild-type NFs at concentrations as low as 10^{-12} to 10^{-13} M (Journet et al. 1994). Structural alterations in NFs, like the absence of the acetate group, or modification of the fatty acyl chain, decrease the inducing properties. However, the presence of the sulphate group seems to be the major determinant of NF recognition by *Medicago*, since non-sulphated factors are 10^3 times less active than sulphated ones for the induction of *MtENOD12* (Journet et al. 1994).

Another interesting early nodulin gene is *ENOD40*, which encodes a small polypeptide (van der Sande et al. 1996). The gene is highly conserved among legumes, and is expressed in the cortex and in pericycle cells opposite the protoxylem poles. Fang and Hirsch (1998) were able to isolate two different clones of *ENOD40* from a *Medicago sativa* library and fused the promoters to *Gus* genes for subsequent transformation of *M. sativa*. GUS histochemical studies indicated that *MsENOD40-1* was not expressed in the absence of rhizobium-stimulated nodulation, whilst

MsENOD40-2 was expressed constitutively in the root. Following inoculation with rhizobium, both promoter-GUS constructs were expressed in the cortical cells/pericycle in the preinfection and infection stages. They were subsequently expressed in the nodule primordium, nodule meristem and vascular bundles of the mature nodule. The expression of both genes could also be induced by purified Nod factors or cytokinin, a process that required the presence of a specific 616-bp region at the distal 5 end of the promoters (Fang and Hirsch 1998).

Overexpression of *ENOD40* or ballistic introduction of an *ENOD40*-expression construct induces cell divisions in the inner root cortex, indicating that this gene can act as a growth regulator (Charon et al. 1997). It has been proposed that this phytohormone-like effect, which has also been observed when *ENOD40* was introduced in the non-legume tobacco (Mylona et al. 1995) could result from a change in the cytokinin/auxin ratio of cortical cells. Other observations suggest that these two phytohormones can be involved in responses to NFs. Auxin-transport inhibitors induce nodule-like structures that express several early nodulin genes, but which do not have the characteristic nodule anatomy and ontogeny (Hirsch et al. 1989). Cytokinin, like NFs, is able to induce cortical cell divisions and the expression of early nodulin genes, such as *ENOD40* (Bauer et al. 1996). Moreover, *S. meliloti* mutants defective in NF synthesis can be partially rescued by the introduction of a gene encoding cytokinin production (Cooper and Long 1994).

It is noteworthy that several early nodulin genes, such as *ENOD5*, *ENOD12*, *ENOD2 and ENOD40* are also induced upon inoculation with endomycorrhizal fungi (Hirsch and Kapulnik 1998; Albrecht et al. 1999).

Nod Factor Perception and Signal Transduction

How are Nod factors perceived and recognised by host plants? The first question that can be addressed is whether these molecules are modified or degraded before they are in contact with roots. Nod factors are substrates for plant chitinases, but it has been shown that the presence of the specific *S. meliloti* substitutions on the chitin backbone has a protective effect against chitinases of alfalfa (Schultze et al. 1994). Thus, plant chitinases of a given legume species seem to degrade the NFs produced by heterologous rhizobia more rapidly and efficiently than those produced by the homologous microsymbiont. Such a mechanism could play a role in the control of host specificity. However, the exquisite sensitivity and the speed of response of the plant root epidermis to the application of exogenous NFs with specific structures suggests that their recognition relies on a mechanism involving high affinity receptors. Furthermore, the importance of the specific substitutions present on the oligochitin chain varies, depending on the plant response considered: an *S. meliloti nodF/nodL* double mutant, producing NFs lacking the O-acetyl group and acylated by vaccenic acid, is no longer able to initiate the formation of infection threads or to enter into the plant, but still elicits very striking root hair deformations and cortical cell activation (Ardourel et al. 1994). Thus, certain plant responses, whose induction requires a very specific NF structure, can be uncoupled from other responses, the induction of which has less stringent NF structural requirements. This observation led Ardourel et al. (1994) to postulate that more than one type of receptor is present in the epidermis: a signalling receptor involved in root hair deformation, and a more stringent uptake receptor, that controls bacterial entry into plant roots.

The possibility of obtaining isotope-labelled NF derivatives either by enzymatic modifications or by complete chemical synthesis (Nicolaou et al. 1992) has opened the way to a biochemical approach to identify NF receptors. Two NF-binding sites have been characterised in microsomal preparations of *Medicago* cells (Bono et al. 1995; Niebel et al. 1997), one of low affinity for Nod factors, the other of higher affinity, compatible with concentrations at which NFs are biologically active. However, it remains to be established whether these binding affinities correspond to functional receptors. A root lectin with an apyrase activity and able to bind NFs has been isolated from the tropical legume *Dolichos*, and could be a promising candidate to play a role in NF recognition (Etzler et al. 1999).

NF signal transduction in the root epidermis of *Medicago* plants has been investigated by a pharmacological approach, using transgenic plants carrying a GUS reporter gene under the control of the *MtENOD12* promoter (Pingret et al.1998). Using various antagonist and agonist compounds, the authors proposed a model of signal transduction involving G proteins, phosphoinositide and calcium second messenger pathways, which seems compatible with the ion fluxes observed in response to NF application.

For plant responses occurring in inner root tissues, mechanisms of Nod signal perception and transduction may be even more complex, since these responses could be induced either directly by NFs, or by second messengers resulting from the interaction of NFs with epidermal cells. In parallel with these biochemical and pharmacological studies, genetic approaches have been initiated which should greatly increase our insight into the processes controlling nodule formation and help us dissect the complex mechanisms controlling Nod signal perception and transduction.

In the past decade, genetic studies performed in pea have allowed the identification of plant genotypes which interact differently with *R. leguminosarum* Nod factors, depending on whether or not these factors have a 6-O-acetyl group on the reducing terminal residue of the oligochitin chain (Firmin et al. 1993; Kozik et al. 1995). The plant gene *SYM2*, responsible for this phenotype, is likely to be involved in NF recognition, and thus is a good candidate for encoding an NF receptor, or part of one.

Plant mutational analysis, aimed at isolating plants defective in the nodulation and infection process, has also been performed in pea. Interestingly, a set of Nod⁻ mutants blocked at early stages of the symbiotic process were also shown to be impaired in the establishment of another symbiotic interaction of a very ancient nature, the arbuscular vesicular endomycorrhizal symbiosis (Duc and Messager 1989). The existence of such Nod⁻ Myc⁻ mutants implies that the establishment of these two endosymbioses shares common steps. Furthermore, the induction of the pea early nodulins *PsENOD5* and *PsENOD12A*, which are normally expressed during the early stages of the interaction with either rhizobium or the endomycorrhizal fungus, is blocked by a mutation in the *SYM8* locus. This result suggests that the signal transduction cascades activated by rhizobium and by the mycorrhizal fungus share at least one common step (Albrecht et al. 1998).

Nod⁻ Myc⁻ mutants have since been isolated in other legume species, like vetch and bean. However, all these species, like other major crop legumes such as alfalfa or soybean, are not easily amenable to genetic analysis and positional cloning, due to features such as ploidy level, allogamous mode of reproduction, genome size and/or inefficient methods for transformation and regeneration. Two legume species, *Lotus japonicus* (Jiang and Gresshoff 1997), which forms determinate nodules, and the alfalfa relative *Medicago truncatula* (Barker et al. 1990), which forms indeterminate nodules, both of which are genetically amenable, have been proposed as model systems. Key attributes of these species include their diploid and autogamous nature, small genomes, short life cycle and the availability of efficient transformation procedures. Furthermore, the bacterial symbiont of *M.truncatula* is the well-characterised species *S. meliloti*. Mutational analysis has been performed in these systems, and a large collection of plant mutants affected in different steps of nodule development are now available (Sagan et al. 1995; Penmetsa and Cook 1997; Szczyglowski et al. 1998b). As for pea, a significant proportion of mutants blocked at an early stage of the nodulation process are also blocked at an early stage of the establishment of the endomycorrhizal symbiosis. In *M. truncatula*, key technologies such as efficient protocols for transgenesis, EST libraries, BAC library, and a genetic map have been developed, which should allow in the coming years the cloning of genes involved in Nod signal perception and transduction. In *Lotus japonicus*, a T-DNA insertion mutagenesis has allowed the isolation of two putatively tagged mutants deficient in nodule establishment (Schauser et al. 1998).

NFs and evolution

Although bacteria had not previously been reported to synthesise chitin polymers or oligomers, the discovery of NFs shows that rhizobia produce lipochito-oligosaccharidic signals, which are able to induce on their host plants, at very low concentrations, processes including cell divisions and organogenesis. NFs thus act as plant growth regulators. Several hypotheses can be proposed to explain the evolutionary origin of this rhizobium-legume signalling system.

NFs could mimic a class of native plant hormones, as yet undiscovered, which would be active in legumes and in other plants. Although there is no experimental evidence for the presence of NF-like compounds in plants, the fact that some alfalfa lines form spontaneous nodules without inoculation by rhizobium or application of NFs could suggest the presence of such an endogenous signal. Furthermore, NFs have been proposed to have a developmental effect on non-legume plants. Application of NFs can rescue somatic embryogenesis in a mutant line of carrot (de Jong et al. 1993), and the development of transgenic tobacco lines expressing the *S. meliloti nodA* and *nodB* genes is affected (Schmidt et al. 1993). Developmental effects of lipochito-oligosaccharides have also been proposed in *Xenopus* (Semino and Robbins 1995) and zebrafish (Bakkers et al. 1997; Kamst et al. 1999), suggesting that this type of molecule could constitute a novel class of developmental signal active in both plants and animals. Alternatively, the rhizobium-legume symbiosis could have recruited genes involved in the more ancient and widespread endomycorrhizal symbiosis. In favour of this second hypothesis is the ability of fungi to synthesise chitin, and the fact that the establishment of the rhizobial and endomycorrhizal symbioses seem to share common early signalling steps. Further studies are necessary to choose between these hypotheses.

The Role played by Rhizobium

Nitrogenase

There are three distinct nitrogenase enzymes, (1) contains molybdenum and iron (Mo-nitrogenase); (2) contains vanadium and iron (V-nitrogenase) and (3) which contains iron alone (Fe-nitrogenase). All three, are cold-labile and sensitive to O_2 and share many common features. The structure of these enzymes has been studied in detail using crystallographic techniques. The more serious investigator is encouraged to read the excellent series of review articles that appeared in *Chemical Reviews* (Burgess and Lowe 1996; Eady 1996; Howard and Rees 1996). Very recently, a molybdenum-containing nitrogenase has been isolated that is not only tolerant to O_2, but actually uses superoxide as a source of electrons (Ribbe et al. 1997). As the Mo-nitrogenase is present in both the rhizobium-legume symbiosis and in the cyanobacteria, it will be described in detail here.

The Mo-nitrogenase consists of two proteins a molybdenum-iron (MoFe) protein and an iron (Fe) protein. The MoFe protein is an $\alpha_2\beta_2$ tetramer with a molecular mass of approximately 220 kDa, with subunits of 50 kDa (α) and 60 kDa (β). The MoFe protein has associated with it two groups of metal clusters or cofactors. The FeMoco cofactor comprises [4Fe-3S] and [1Mo-3Fe-3S] clusters bridged by sulphide with the Mo atom attached to (R)-homocitrate by hydroxyl and carboxyl oxygen atoms. FeMoco is bound to the protein by a cysteine residue ligated to an Fe atom and a histidine residue to the Mo atom. The FeMo cofactor is buried 10 Å beneath the surface of an α-subunit, surrounded by hydrophilic amino acid residues. The P cluster is an [8Fe-7S] cluster that could be generated from two bridged [4Fe-4S] clusters. The two clusters are attached to thiol side chains of cysteine residues in both the α and β subunits. The Fe protein of nitrogenase is made up of two subunits (γ) of molecular mass 57-72 kDa, depending on the source. It contains a single [4Fe-4S] cluster that is attached to cysteine residues on both of the subunits. From a computerised comparison of protein sequences, it has been established that the Fe protein subunits share many common sequences with other nucleotide binding proteins.

Many heterologous nitrogenases, consisting of the MoFe protein from one diazotroph incubated with the Fe protein from another, can catalyse nitrogen fixation. This demonstrates the high degree of structural homology among the nitrogenases of different organisms. In addition to reducing nitrogen to ammonia, nitrogenase can also reduce a variety of other substrates. These substrates have in common their small size and, in many cases, the possession of a triple bond, e.g. azide and cyanide. Notable among these substrates is acetylene (ethyne), because this forms the basis of a widely used and convenient assay for nitrogenase. Mo-nitrogenases reduce acetylene to ethylene (ethene) but no further; ethylene formed in this way can readily be measured by gas chromatography. In contrast, measurement of nitrogen fixation itself, by monitoring reduction of $^{15}N_2$ by mass spectrometry, is laborious, time-consuming and expensive. However, the acetylene reduction technique gives only an indirect measurement of nitrogen fixation and, when a quantitative assessment of the absolute amounts of nitrogen fixation is needed, should be used with caution. Unlike the reduction of nitrogen, no hydrogen is generated by nitrogenase as a by-product of the reduction of acetylene or any other substrate. However, in the absence of an alternative substrate, nitrogenase catalyses the reduction of protons (H+) to hydrogen. This is referred to as the ATP-dependent H_2 evolution reaction catalysed by nitrogenase. As mentioned above, nitrogenase is inhibited by oxygen, cold and ADP. The oxygen sensitivity of the enzyme lies in both constituent proteins, though the Fe protein is particularly sensitive to inactivation by oxygen. Carbon monoxide is also a potent inhibitor of all nitrogenase-catalysed reactions except ATP-dependent hydrogen evolution. In addition, hydrogen is an inhibitor of nitrogen reduction, though it does not interfere with the reduction of any other substrate.

Genetic Control of nitrogen fixation by rhizobium

The genes associated with nitrogen fixation are designated *nif*. Studies into the genetics of nitrogen fixation were initiated in *Klebsiella pneumoniae* in which the organization of the *nif* genes is well understood. Fortuitously in *K. pneumoniae*, all the *nif* genes are clustered into a single regulon containing 23 280 base pairs, which consists of 20 genes organised most probably into seven operons , the products of which are shown in Table 2. Each operon is preceded by a promoter region that is recognised by an RNA polymerase that contains the specialised sigma factor (σ^{54}), also known as σ^N. The promoter region contains the consensus sequence tGGc-N8-tGCa, which is located upstream of the initiation of transcription. The σ^{54} promoter sequences are also found preceding other genes involved in nitrogen metabolism in bacteria.

Table 2. Proposed function of *nif* genes of *Klebsiella pneumoniae*

Gene	Proposed function of gene product
J	Electron transport
H	Fe protein of nitrogenase
D	α-subunit of MoFe protein of nitrogenase
K	β-subunit of MoFe protein of nitrogenase
T	Unknown
Y	Processing of MoFe protein
E N	Protein "scaffold" on which FeMoco is assembled
X	Unknown
U S	Formation of FeS clusters
V	Synthesis of FeMoco (homocitrate synthase)
W	Stabilisation of MoFe protein
Z	Unknown
M	Insertion of [4Fe-4S] centre into Fe protein
F	Electron transport (flavodoxin)
L	Negative regulator (inactivator of *nifA* product)
A	Positive regulator (activator of *nif* transcription)
B	Synthesis of NifB-co, a precursor of FeMoCo
Q	Uptake and processing of Mo

The genes *nifH*, *nifD* and *nifK* encode the structural proteins of the Fe- and MoFe-proteins of nitrogenase. These genes are highly conserved among diazotrophs, emphasising the similar nature of the enzyme in various nitrogen-fixing organisms. In addition, this has greatly facilitated the development of *nif* genetics, since the *nif* genes isolated from one diazotroph can be used to probe the DNA of other diazotrophs for homologous regions. Of the other *nif* genes, the majority are required for the processing of the three structural proteins except for *nifJ* and *nifF*, which are involved in the provision of reductant, via pyruvate oxidation, for nitrogen fixation in *K. pneumoniae*, and *nifL* and *nifA*, which are regulatory genes. Despite extensive studies on deletion mutants, no major function for *nifT*, *nifX* and *nifZ* has been established, it may be that there are other proteins in the cell that carry out the same function. It has recently been proposed that the *nifW* gene product may be involved in stabilising the MoFe protein, possibly against oxygen damage. A full discussion of the function of the *nif* genes in anticipation of their introduction into plants by genetic engineering has been provided by Dixon et al. (1997). In the cyanobacteria during heterocyst development, there is an extensive rearrangement of the DNA encoding the *nif* genes that gives rise to a sequence of *nif* operons very similar to that seen in *K. pneumoniae* (Bohme 1998).

As well as carrying the *nif* genes involved in the synthesis and regulation of nitrogenase, *Rhizobium* also contains other groups of genes involved in the nodulation process and the interaction with the plant host. The nodulation (*nod*) genes of the rhizobium are required for infection, nodule formation and the control of host specificity has been discussed in a previous section. The fixation genes (*fix*) of *Rhizobium* are also involved in the symbiotic association and metabolism within the bacteroid. The *fixNOQP* operon encodes the three subunits of a cbb_3 -type cytochrome oxidase that is essential for the maintenance of aerobic respiratory metabolism and ATP synthesis in the microaerobic environment of the nodule (Preisig et al. 1993). In addition the *fixGHIS* genes are required for the correct assembly and insertion of haem and copper into the cytochrome oxidase protein (Preisig et al. 1996). A range of additional *fix* genes may be required for electron transport and membrane bound cation pumps (Fischer 1994;1996).

In *Rhizobium*, *nifA* transcription (and hence all the other *nif* genes) is regulated by oxygen. The product of the *fixL* gene is able to sense the oxygen status of the cell through a bound haem group. At low oxygen concentrations FixL is able to phosphorylate FixJ, which can then bind to the *nifA* DNA sequence and activate transcription. The full details of this regulation are beyond the scope of this chapter, but are described in Fischer (1994;1996) and discussed in Udvardi and Day (1997).

The Further Development of the Nodule

Late Nodulins

Perhaps the most extensively studied late nodulin is glutamine sythetase (GS), the enzyme responsible for the initial step of ammonia assimilation (see chap. 2.2) and examples from two legume plants will be described. Extensive work on GS expression in peas (Brears et al. 1991) and soybean (Temple et al. 1996; Tercé-Laforgue et al. 1999) has also been carried out. In *Phaseolus vulgaris*, four different polypeptide subunits of GS have been identified termed α, β, γ and δ. During nodule development there are major changes in the expression of the four genes encoding GS (Cullimore et al. 1992). Ony low levels of expression of *gln-α* and *gln-δ* genes were detected in *P. vulgaris* nitrogen-fixing nodules; the former decreased and the latter increased during development. At early stages of development the predominant GS mRNA and polypeptide was *gln-β*, and expression of the gene was maintained throughout a 20-day period. Studies with the promoter of the *gln-β* gene fused to GUS have shown that the gene was initially expressed in both the cortical and infected regions of transgenic *Lotus corniculatus* nodules, but as the nodules matured, expression was restricted to the vascular system (Forde et al. 1989). After 8 days, synthesis of the *gln-γ* mRNA commenced, which was closely followed by the appearance of the γ-polypeptide, which established itself as the major GS form 12 days after inoculation with rhizobium. Similar GUS expression studies have shown that the *gln-γ* gene is expressed only in infected cells of the central tissue of the nodule (Forde et al. 1989).

In *M. sativa*, increased GS activity during nodule development was correlated with a rapid increase in the synthesis of GS protein and mRNA for the first 10 days after inoculation with rhizobium and a more gradual increase for the next 30 days, until the commencement of senescence (Vance and Gantt 1992). Two cytosolic GS genes have been shown to be expressed in the nodules of *M. sativa*, *pGS13* was expressed at a level 20-fold higher in the nodules than in the roots or leaves, whereas *pGS100* was expressed at a slightly higher level in the nodules than in the roots. *In situ* hybridisation studies have shown that the *pGS13* gene is induced immediately behind the apical meristem, in the preinfection zone, where cell differentiation begins, just before the release of the rhizobium from the infection thread (Temple et al. 1995).

The remainder of this discussion will concentrate on the nodules of *M. sativa*, as it is in this plant that the most detailed comparative study of enzymes of carbon and nitrogen metabolism has been carried out. NADH-dependent glutamate synthase (GOGAT) catalyses the second step in ammonia assimilation (see Chapt. 2.2), and has been shown by different techniques to be located in the plastids of root nodules (Awonaike et al. 1981; Trepp et al. 1999b). For NADH-GOGAT, the developmental increase in mRNA corresponded well with enzyme activity and occurred in two phases. The initial increase, which started before the onset of nitrogen fixation, coincided with nodule emergence from the roots at days 7 and 8. A second and more dramatic increase in enzyme activity, protein and mRNA occurred at day 9 in effective nodules but not in ineffective nodules. Promoter-GUS expression studies showed that the gene was expressed in the nodules of transgenic plants, but only slightly in the roots and stems and not in the leaves (Vance et al. 1995). Very recent *in situ* hybridisation studies have indicated that the expression of NADH-GOGAT shifted during nodule development. In 9-day-old nodules, all infected cells appeared to contain NADH-GOGAT mRNA, but by day 19, a gradient of expression from high in the early symbiotic zone to low in the late symbiotic zone was observed (Trepp et al. 1999a). In 33-day-old nodules expression was observed only in the early symbiotic zone. However, immunological studies (Trepp et al. 1999a) indicated that the NADH-GOGAT protein was present throughout the whole of the infected region of both 19- and 33-day-old nodules, even in areas that had ceased to carry out nitrogen fixation. Trepp et al. (1999a) have discussed in detail the possible significance of distinct DNA sequences within the promoter region of NADH-GOGAT and compared them to other known nodulin genes.

M. sativa root nodules contain two forms of glutamate: aspartate aminotransferase (AAT), a cytosolic AAT-1 and a plastid localised AAT-2, which are immunologically distinct from one another. In the developing nodules AAT-2 is the major form, but in roots AAT-1 predominates. The increase in AAT enzyme activity followed a biphasic time course similar to that of NADH-GOGAT and correlated well with increases in AAT-2 mRNA and protein. Only the first phase of AAT-2 induction occurred in ineffective nodules (Gantt et al. 1992). *In situ* hybridisation indicated that AAT-2 mRNA transcripts were present in infective cells throughout the infection zone, but that the levels were greatly reduced in ineffective nodules. In contrast AAT-1 mRNA was detected in the uninfected cells, nodule parenchyma and vascular bundles of both effective and ineffective nodules (Yoshioka et al. 1999).

Asparagine synthetase mRNA and enzyme activity increased 20-fold in effective *M. sativa* root nodules; however, there was a delay of 24 h in the increase in mRNA as compared to the other enzymes of nitrogen assimilation discussed previously. Promoter-GUS studies indicated that the gene was expressed in the infected and uninfected cells of the symbiotic zone and the nodule parenchyma, but not in the outer cortex, meristem or invasion zone (Shi et al. 1997)

In addition to investigating the expression and localisation of enzymes involved in ammonia assimilation, the Vance group has also studied enzymes required for the synthesis of the 4-carbon skeleton required for asparagine synthesis. Phosphoe-*nol*pyruvate (PEP) carboxylase activity increases rapidly during the nodulation process, where the enzyme may constitute 2% of the total soluble protein. Immunolocalisation indicated that the PEP carboxylase protein was distributed relatively equally in infected and uninfected cells of the symbiotic zone. A high amount of gold labelling was also detected in the pericycle cells of the vascular system and to a lesser extent in the inner cortical cells (Robinson et al. 1996). PEP-carboxylase mRNA was detected by *in situ* hybridisation in the nodule meristem, the infection and nitrogen fixation zones and both parenchyma and vascular tissue. Analysis of promoter deletions indicated that a region between -634 and -536 was of particular importance in directing transcription to the infected zone of the nodules (Pathirana et al. 1997) . Miller et al. (1998) were able to isolate cDNAs encoding five different forms of malate dehydrogenase (MDH) from *M.sativa*, including one that was a novel nodule enhanced form (*neMDH*) . Utilising antibodies raised to neMDH, it was shown that it was the major form of MDH protein in the nodules and was probably localised in the plastids.

In a very recent systematic study, Trepp et al. (1999c) localised the mRNA transcripts of 13 essential enzymes involved in carbon and nitrogen metabolism in alfalfa nodules by *in situ* hybridisation. A serial section approach allowed the construction of a map that reflected the relative distribution of the transcripts in a 33-day-old nodule (Fig. 5). The data obtained clearly demonstrated that there was only a 5- to 15-cell-wide region in the nodule, in which there was major expression of all the key genes involved in nitrogen fixation and the subsequent assimilation of ammonia. The production of this transcript distribution map will be of considerable value in future studies to enhance the capacity of the nodule to fix nitrogen.

The detailed work of Vance and his colleagues has now provided clear evidence that at least in the *M. sativa* nodule, the synthesis of the enzymes required for carbon and nitrogen metabolism, is regulated in a coordinated (but not identical) manner, in line with the induction of nitrogenase in the rhizobium bacteroids. However, it is

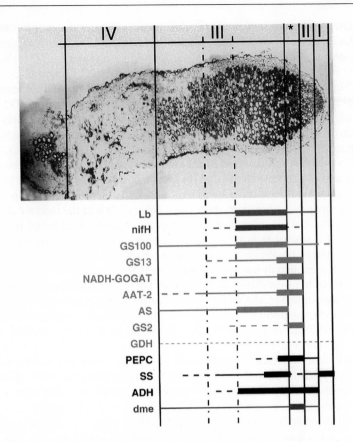

Fig. 5 : Transcript distribution map indicating the qualitative distribution pattern of 13 different mRNAs, localised in different zones of *Medicago sativa* root nodules. *Thick bar* indicates a high abundance of mRNA transcripts; *drawn line* indicates the transcripts were moderately abundant; *dashed line* indicates occasional expression. Nodule zones are indicated: *I.* zone I, meristem; *II.* zone II, invasion zone; * zone II-III, interzone; *III.* zone III, nitrogen-fixing zone; *IV.* zone IV, senescence zone. Reproduced with permission of the American Phytopathological Society and the authors of Trepp et al. (1999c), to whom we are extremely grateful

also clear that the promoter sequences that regulate the genes discussed above are different (Trepp et al. 1999a) and therefore that they presumably respond to different signals from within the nodule. Possible candidates for these signal molecules would need to be able to act as indicators of nitrogen/carbon availability, or of low oxygen partial pressures.

Sucrose synthase (SS) was one of the first nodulins to be identified in soybean (Fuller at al. 1983). In white clover the enzyme activity and protein has been shown to be synthesised before the induction of nitrogenase and leghaemoglobin, but at the same time as GS (Gordon and James 1997). More recently, a genomic clone *MtSuicS1* has been characterised from *M. truncatula* that encodes a nodule-enhanced form of SS, which is distinct from that found in the uninfected root

(Hohnjec et al. 1999). Immunogold labelling has shown that the enzyme protein is more abundant in the uninfected cells than in those involved in nitrogen fixation (Gordon et al. 1992; Zammit and Copeland 1993). In soybean, sucrose synthase (SS) activity, protein and mRNA has been shown to be reduced by drought or dark-induced stress (Gonzalez et al. 1995). Later studies showed that SS mRNA levels were also reduced by salt and nitrate treatments, whilst the majority of other enzymes and leghaemoglobin were unaffected. A key finding was that a stress-induced decrease in nitrogen fixation correlated with the decrease in SS activity, an indication that nitrogen fixation may be regulated indirectly by sucrose metabolism (Gordon et al. 1997). Very recent studies with starch-deficient mutants of pea have shown that sucrose synthase is essential for efficient nitrogen fixation in the nodules (Craig et al. 1999; Gordon et al. 1999).

The expression of leghaemoglobin during the development of the nodule can be monitored by eye due to the characteristic red colour and the high concentration present (20% of soluble protein). Legumes normally contain a number of leghaemoglobin genes which are expressed at high levels in the nodules (Marcker et al. 1984). *In situ* hybridisation indicated that leghaemoglobin mRNA was synthesised in a single cell layer just prior to the start of the nitrogen-fixing zone. Promoters from the leghaemoglobin genes direct high levels of gene expression in the rhizobium-infected cells located in the central region of the nodules. There is now strong evidence that the synthesis of the haem group that is required for the fully functional lehaemoglobin protein is also under the control of plant genes (O'Brian 1996). Five enzymes involved in haem synthesis were shown to increase during the development of soybean nodules (Santana et al. 1998). For coprogen oxidase (Madsen et al. 1993), the increases in enzyme activity also correlated with mRNA and enzyme protein. The time course of induction was virtually identical for each enzyme and paralleled the synthesis of haem and leghaemoglobin.

Nodulin 26 is a major protein component of the symbiotic membrane that encloses the rhizobium bacteroids. The deduced amino acid sequence suggests that it is a member of the major intrinsic protein (MIP) family, which includes aquaporin water transporters. It was originally proposed that Nodulin 26 was involved in dicarboxylate transport. However, recent expression studies in *Xenopus* oocytes and the use of liposomes, have indicated that nodulin 26 is involved in water and glycerol transport to the bacteroids and may play a role in osmoregulation (Dean et al. 1999). Other late nodulin membrane proteins have been shown to have amino sequences that indicate an involvement in organic acid and peptide transport (Szczyglowski et al. 1998a).

The previous discussion of late nodulins has concentrated on the enzymes of carbon and nitrogen metabolism, which have been known from earlier biochemical studies to be important in the nitrogen fixing nodule. In a novel approach, Szczyglowski et al. (1997) used a reverse transcription PCR based differential display procedure, to identify a range of expressed sequence tags (ESTs) associated with the later stages of the development of the *Lotus japonicus* nodule. The kinetics of mRNA accumulation of the ESTs was found to resemble the pattern observed for leghaemoglobin. Of 88 partial cDNA sequences obtained, only 22 shared homology to DNA/amino acid sequences in the available databases. It is clear that there is a lot more work to be done before we fully understand the development of the nitrogen fixing legume root nodule!

Structure and Biochemistry of Nodules in Relation to Function

Nitrogen fixation in legume nodules is a well-documented process and its many features have been extensively and elegantly reviewed over the years. The reader is directed particularly to Brewin (1991), Vance and Heichel (1991), Hunt and Layzell (1993), Batut and Boistard (1994), Fischer (1994), Mylona et al. (1995), Streeter (1995), Udvardi and Day (1997) as well as to the many papers in the volumes arising from the International Congress of Nitrogen Fixation held approximately every 2 years (e.g. Palacios et al. 1993; Tikonovich et al. 1995; Elmerich et al. 1998).

Main Properties of the Nitrogen Fixing System Which Demands a Specialised Organ in the Legume/(brady)rhizobium Symbiosis

Nitrogen fixation in legumes requires the cooperation and coordination of two organisms, a plant and a bacterium, *(brady)*rhizobium. The enzyme nitrogenase is synthesised only in microorganisms (outlined in detail above) and is extremely oxygen-sensitive, yet paradoxically, the bacteria require an aerobic environment in which to generate ATP and reductant for the expensive process of N_2 fixation. The nodule provides this unique environment. A carbon substrate (dicarboxylic acids, such as malate and succinate) and oxygen (at very low concentrations, mediated by a plant haemoglobin and a variable diffusion barrier in the cortex of the nodule) are delivered to the specialised, plant-derived peribacteroid membrane (PBM) for transport into the bacteroids. N_2 is reduced to ammonia in the bacteroids and ammonium ions then enter the plant cytosol. This is assimilated by plant enzymes and used in the synthesis of amino acids, amides and ureides, which are exported from the nodule for use in the growth of the plant. The bacteroids are modified in many respects which fit them for their function in symbiotic nitrogen fixation.

Nodule Structure

Legume nodules are of two types; determinate or indeterminate. The former have no persistent meristem while the latter do. So, whereas determinate nodules (such as soybean) develop to a certain size and cannot grow any more, indeterminate nodules (such as those of pea, clover, etc.) may continue to grow away from their point of attachment to their subtending root. Distinct zones of cells can be distinguished in indeterminate nodules ranging from small rapidly dividing meristem cells at the tip of the nodule, through newly infected cells, mature cells containing active bacteroids, to cells closest to the root attachment which may have senesced. Such nodules are invaluable in the study of the coordinated development of both the host plant and the microsymbiont (e.g. Soupene et al. 1995; Trepp et al. 1999c; Fig. 5). In a cross-section through a zone of the nodule where active N_2 fixation is taking place, characteristic features are evident (Fig. 6). The darkly stained cells in the central region are those containing bacteroids. These are interspersed with fewer uninfected, interstitial cells. The central region is surrounded by a cortex composed of several cell layers which do not contain bacteroids. The cortex also includes the vascular bundles by which sucrose, water and possibly other compounds are delivered to the nodule and nitrogenous products are exported from the nodule. All parts of the nodule are functionally significant.

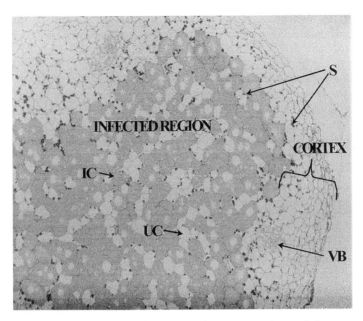

Fig. 6: Light micrograph showing the structure of a *Lotus japonicus* nodule, green cells are infected cells (IC), clear cells in the central region are uninfected cells (UC), vascular bundles (VB) are obvious in the cortex. *Dark spots* are starch grains (S). The section was stained initially with iodine to show starch and then counterstained with Fast Green

Cortex

The cortex usually has a prominent endodermis known as the scleroid layer which separates the outer from the inner cortical cells (Walsh 1995). Vascular bundles lay within the scleroid layer. Although it is not known in detail how materials are imported and exported from the nodule, it is thought that sucrose is delivered down a concentration gradient from the leaves and then unloaded into the cells surrounding the phloem. From there it is thought that sucrose diffuses symplastically through the plasmodesmata of the vascular endodermis, cortical cells, central region interstitial cells and into infected cells (Streeter 1992; Brown et al. 1995). It has been postulated that water delivered with sucrose via the phloem would remain within the vascular endodermis and may be essential to drive export of nitrogen products via the xylem (Walsh et al. 1989; Walsh 1995). In some species the vascular pericycle cells contain wall in growths which increase the surface area and may enhance solute loading to the xylem, and from there export to the shoot (Gunning et al. 1974; Walsh 1995).

The cortex is also the site of the gaseous diffusion barrier. There is now extensive evidence describing this variable barrier (Minchin 1997) and the possible means by which it is regulated, although a definitive explanation of the mechanism is still lacking. Resistance to gaseous movement into or out of the nodule may result from a number of contributing factors; the thick-walled scleroid cells, the mid-cortex where intercellular occlusions may be deposited, thus reducing gas movement through air spaces, and the "boundary layer" cells, which appear to have no intercellular air spaces. Since gases diffuse through the gas phase at a rate 10^4 times faster than through liquid, cells with no intercellular air spaces form a very effective barrier. It is thought that the ability to vary gas diffusion may be related to the changes in turgor or to exudation of materials (such as glycoprotein) such that air spaces become restricted. These ideas were extensively reviewed by Minchin (1997) and will not be expanded further here. It is worth noting, however, that the ability to maintain a low internal O_2 concentration also depends on one other major factor – respiration. No barrier exists which can exclude gases completely, so the utilization of O_2 in respiration is a necessary component of the resistance network. Fortunately, O_2 is needed by mitochondria and bacteroids in central region cells to generate ATP in oxidative phosphorylation. Thus metabolism in bacteroids and central region plant cells aides the consumption of O_2, which allows nitrogenase to survive.

Central Region

The central region of nodules of most legume species is mostly composed of cells occupied by bacteroids, with a smaller number of uninfected cells. In some species these latter cells form a continuous network which may be involved with the supply of sucrose to infected cells and/or the export of nitrogenous products towards the cortex and the vascular bundles (cf. Selker and Newcomb 1985; Selker 1988; Gordon 1992). Infected cells are, by definition, characterised by the presence of very large numbers of bacteroids, which appear to occupy most of the cell cytoplasm. Vacuoles are generally small or absent, and organelles such as mitochondria and plastids are confined to the periphery of the cells, concentrated mainly at the points in closest contact with intercellular air spaces (Fig. 7). The bacteroids are surrounded by a plant-derived membrane called the peribacteroid membrane

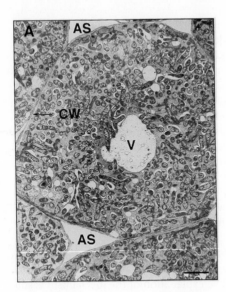 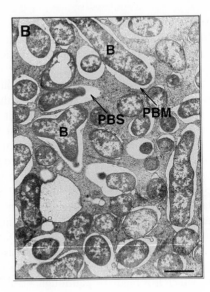

Fig. 7 A, B : Electron micrographs showing a cell in the nitrogen fixing zone of a 5-week-old faba bean nodule (A), and a detail of the structure of the symbiosomes in this zone (B). AS, Air space; CW, cell wall; B bacteroid; PBM peribacteroid membrane; PBS peribacteroid space. *Bars* 5 μm (A) and 1 μm (B)

(PBM) (Verma and Hong 1996) and the space between the bacteroid and the membrane is referred to as the peribacteroid space (PBS) (Fig. 7). Bacteroids plus PBM and PBS form an organelle-like structure in the host cell known as the symbiosome (Roth and Stacey 1988). The PBM located at the interface between the plant cytosol and rhizobia exerts considerable control over the chemical communication between the two partners (see later). The PBM encloses either a single bacteroid (for example, in white clover or faba bean) or many bacteroids (e.g. soybean, french bean). A characteristic of the central region is the presence of large amounts of leghaemoglobin (Lb). The *Lb* genes are plant encoded and are expressed only in nodules. Lb is found predominantly in infected cells, although it has also been observed at lower concentrations in the uninfected interstitial cells (VandenBosch and Newcomb 1988). It is thought that Lb contributes towards the effective delivery of O_2 at low concentrations from the intercellular air spaces to the bacteroids (and possibly to the mitochondria). O_2 levels in infected cells are thought to be in the range 5-40 nM, as determined by spectroscopic observation of the oxygenation state of Lb *in vivo* (Hunt and Layzell 1993). Bacteroids have the remarkable adaptation that they express a terminal oxidase system which has a high affinity for O_2, such that they can function effectively at concentrations of 5-10 nM (Bergersen and Turner 1975b, 1980). There has been some debate about the role of mitochondria in infected cells because of their presence in a low O_2 environment, which could restrict oxidative phosphorylation (Rawsthorne and La Rue 1986). However, reports of soybean nodule mitochondria having a half maximal activity at an O_2 concentration of 50 nM (Millar et al. 1995) suggest that this may not be a problem. However, the data in Millar et al. (1995) also demonstrate that the ATP/O ratio becomes very low at low O_2 concentrations (implying much reduced efficiency in the process of oxidative phosphorylation, or alternatively, ATP hydrolysis within the mitochondrial matrix). The fact that mitochondria are located close to intercellular air spaces, however, may indicate that the oxygen environment is reasonable for effective function. Mitochondria from nodules have been reported to lack malic enzyme and operate a truncated TCA cycle (Day and Mannix 1988; Bryce and Day 1990) and therefore may be limited to oxidising malate to oxaloacetate and the production of ATP, as NADH is oxidised by the electron transport chain. Streeter (1991) argued that one role of mitochondria could be to supply 2-oxoglutarate for use in ammonia assimilation. The recent demonstration of OAA exchange for a number of organic acids in plant mitochondria could be significant (Hanning et al. 1999).

Biochemistry to Support Nodule Nitrogen Fixation

Nodules are primarily dependent on the delivery of sucrose from the shoot to support nitrogen fixation and the maintenance of structural integrity. Approximately 50% of the total daily photosynthetic products may be processed by nodules, about 60% of which is lost as respired CO_2 and the remainder returned to the shoot as nitrogen products (Gordon et al. 1985). Temporary stores of carbohydrate are also accumulated. The location of starch grains is obvious from light or electron micrographs (Fig. 6), although little is known about the cellular or tissue distribution of sucrose. There is evidence that these reserves are drawn upon during the dark periods (e.g. Gordon et al. 1986, 1987; Walsh et al. 1987) and following stress perturbations (Gordon et al. 1986, 1997; Gogorcena et al. 1998).

Prior to the paper by Morrell and Copeland (1984) on SS in legume nodules, it was believed that sucrose was hydrolysed by alkaline invertase (e.g. Robertson and Taylor 1973). Later, Thummler and Verma (1987) established that SS was encoded by a nodule-enhanced gene (nodulin 100). Since then it has remained unclear whether both enzymes are involved with sucrose metabolism. This has now been resolved with the generation of allelic mutants in pea (*Pisum sativum*) in which sucrose synthase is present at only 5-10% of wild-type activity in nodules (Wang and Hedley 1993; Craig et al. 1999; Gordon et al. 1999). When grown from seed in the presence of an effective rhizobium and in the absence of mineral nitrogen, many nodules were formed but these were ineffective in nitrogen fixation. A number of important enzymes of carbohydrate, organic acid and amino acid metabolism were in lower amounts in these mutant nodules than in wild-type nodules, including alkaline invertase. Thus, we concluded that alkaline invertase was not able to substitute or compensate for SS in the metabolism of sucrose to support nodule function (Gordon et al. 1999). In addition, it was apparent that SS was required for normal nodule development and also for the maintenance of nitrogenase activity. However, the precise role of SS in nodule development is not yet known.

The conversion of sucrose to malate through the SS pathway occurs in the cytosol (Gordon et al. 1992) and is summarised in Fig. 8. The products of sucrose

$$Suc + UDP + H^+ \overset{1}{\rightleftharpoons} UDPglucose + Fru$$

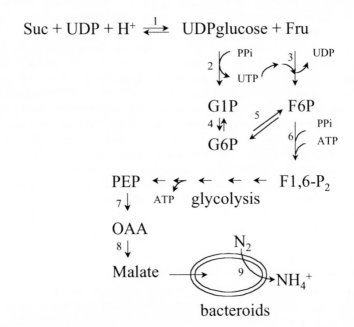

Fig. 8 : Metabolism of sucrose to malate, which may be the main form of carbon metabolised by bacteroids to drive nitrogen fixation. *1.* Sucrose synthase (SS); *2.* UDPglucose pyrophosphorylase; *3.* fructokinase; *4.* phosphoglucomutase; *5.* phosphoglucose isomerase; *6.* pyrophosphate or ATP-dependent phosphofructokinase; *7.* PEP carboxylase; *8.* malate dehydrogenase; *9.* nitrogenase in the bacteroids. The intervening reactions of glycolysis are not shown in detail. If PPi is available for both reactions 2 and 6, 4 mol ATP, in addition to 4 mol malate, are generated from one mol sucrose

hydrolysis by SS (UDP-glucose and fructose) can be converted to phosphorylated sugars without the net expenditure of ATP equivalents in contrast to the products of invertase hydrolysis (Gordon 1992). It is generally considered that metabolism proceeds through the glycolytic or pentose phosphate pathways to PEP (cf. Day and Copeland 1991; Gordon 1992). It has been argued elsewhere that if pyrophosphate is readily available and utilised in both the UDPglucose pyrophosphorylase and the PFK(PPi) reactions, the metabolism of 1 mol sucrose to 4 mol malate could generate 4 mol ATP (Gordon 1992). This may be one reason why the SS route is preferred in the ATP-limited environment of the infected cell cytosol (e.g. Kuzma et al. 1999).

The enzyme PEP carboxylase is significant in the assimilation of some respired CO_2 and the production of oxaloacetate (OAA) (Vance et al. 1994), which can be converted by MDH to malate or by transamination to aspartate via the enzyme AAT. The route to malate is an "anaerobic" process, similar to those catalysed by alcohol dehydrogenase in yeast or lactate dehydrogenase in muscle, all of which are able to reoxidise the NADH formed in the glyceraldehyde dehydrogenase step of glycolysis and so generate limited amounts of ATP. The unique aspect of this nodule "anaerobic" route, however, is that malate is the most likely substrate to be provided by the plant to the bacteroids (Udvardi and Day 1997; see later). However, bacteroids are not limited by O_2 because they have a high-affinity terminal oxidase (Bergersen and Turner 1975b, 1980) and so malate does not accumulate in the same way that alcohol or lactate does in other anaerobic systems.

Properties of Symbiosomes and Bacteroids

Composition of the Mature Peribacteroid Membrane

The phospholipids of PBM from different legumes are predominantly composed of phosphatidylcholine (PC) and phosphatidylethanolamine (PE) (Mellor et al. 1985; Hernandez and Cooke 1996). The PBM has a PC:PE ratio of 2:1, which differs from that of the plasma membrane (1:2). In contrast, these two membranes contained similar fatty acids but the ratio of unsaturated to saturated species is higher in the PBM, explaining perhaps the greater fluidity of this membrane. Another interesting characteristic of the PBM concerns the glycosylation of some lipids, which is similar to that in root plasma membranes. In the PBM, the sugar residues are present on the PBS face, and Perotto et al. (1991) suggested that they could play an important role in the interactions between PBM and bacterial membrane, especially for the coordinated division of PBM and bacteria during nodule formation (Robertson and Lyttleton 1984).

Several non-specific proteins have been identified on the PBM by the fact that they were glycosylated, essentially by arabinose and/or melibiose residues. The PBM exhibits a polypeptide profile (by SDS-PAGE analysis) which is distinct from that of other cell membranes (Mellor and Werner 1986) and specific PBM proteins have been identified either by function or by genetic analysis (but also by both methods in the case of nodulin 26). Different enzyme activities have been recorded on the PBM, the first and most important being ATPase, present in all the legumes studied and resembling a P-type plasma membrane ATPase. Besides ATPase, pyrophosphatase and protein kinase activities have also been detected in the PBM.

Transport Activity of the Peribacteroid Membrane

The main nutrient exchange between the plant and the bacteroid through the PBM is fixed nitrogen to the plant for reduced carbon to the bacteroid. However, many other compounds are also transported by PBM proteins, although only a few of these have been identified so far. As described above, many carbon compounds are plentiful in legume nodules but these vary during nodule development and between species. Symbiosomes isolated from different legumes are able to transport a number of reduced carbon substrates, such as glucose in French bean (Herrada et al. 1989), glutamate in lupin (Radyukina et al. 1992) and oxalate in faba bean (Trinchant et al. 1994). However, all legume symbiosomes are capable of accumulating dicarboxylates at rapid rates, and a PBM dicarboxylate transporter has been characterized (Udvardi et al. 1988; Herrada et al. 1989). In soybean the PBM dicarboxylate carrier exhibits a preference for malate and succinate and acts as a uniport for the monovalent dicarboxylate anion. Thus, malate uptake into symbiosomes is driven by a membrane potential (positive inside) and a pH gradient (more acidic inside the symbiosome), both generated by the proton pumping ATPase located on the PBM (Robertson et al. 1978; Udvardi et al. 1991; Ou Yang and Day 1992; Szafran and Haaker 1995). In relation to substrate specificity, inhibition sensitivity and pH optimum, the dicarboxylate transporter on the PBM is distinct from the DctA protein of the bacteroid. The PBM carrier always exhibits a lower maximum rate of transport that the bacteroid transporter, indicating that the PBM is the limiting step in the transport process.

Ammonia leaving the bacteroid is protonated to the ammonium form due to the acidic pH occurring in the PBS (Fig. 9). In contrast to earlier attempts to demonstrate an ammonium transporter (Udvardi and Day 1990), new methods involving the patch clamp technique have now demonstrated that the ammonium ion is exported across the PBM into the plant cytosol (Tyerman et al. 1995; Mouritzen

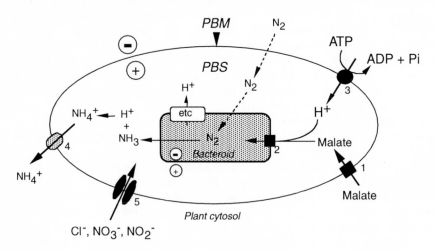

Fig. 9 : Scheme summarising the structure of a typical symbiosome and showing the exchange of nitrogen and carbon through the peribacteroid membrane (PBM). *1*. PBM dicarboxylate carrier; *2*. bacteroid dicarboxylate carrier; *3*. PBM ATPase; *4*. ammonium channel; *5*. anion channel. (etc.) electron transport chain

and Rosendahl 1997; Udvardi and Day 1997). This voltage-gated monovalent cation channel is thought to allow ammonium transport in one direction only (PBS to cytosol) and to be dependent on the generation of a membrane potential probably involving the H^+ ATPase on the PBM. However, another possibility for nitrogen transfer was proposed very recently by Waters et al. (1998). Using soybean bacteroids anaerobically purified by a procedure that removes contaminating plant proteins, these authors demonstrate that the major nitrogen-containing compound excreted by the bacteroids is not ammonia, but alanine generated via an alanine dehydrogenase. This work confirms previous results reported by Rosendahl et al. (1992) demonstrating the efflux of alanine and glutamate from isolated symbiosomes in long-term experiments under nitrogen-fixing conditions. However, from a number of studies using isolated symbiosomes from different legumes (Ou Yang and Day 1992; Pedersen et al. 1996; Trinchant et al. 1998), it appears that transport of amino acids across the PBM is very slow and not mediated by a specific carrier.

Besides the compounds discussed above, probably some other ions can cross the PBM more or less rapidly using channels or carriers. For example, nitrate, nitrite or chloride added to isolated soybean symbiosomes have been shown to rapidly collapse the electrical potential of the PBM (Udvardi and Day 1989). In this way, nodulin 26, an intrinsic PBM protein, is actually considered as an ion channel with a slightly higher permeability to anions than cations (Weaver et al. 1994). It is also possible that nodulin 26 is a stretch-activated channel involved in osmoregulation of the symbiosomes acting as a water channel, as very recently proposed (Dean et al. 1999)

Iron is also an important compound required by bacteroids in nodules for synthesis of different proteins including nitrogenase and cytochromes, but iron would first have to cross the PBM. Two different reports (Moreau et al. 1995; Le Vier et al. 1996) have recently shown that iron is transported across the PBM as ferric citrate. However, bacteroids take up this iron compound less rapidly than symbiosomes, and so it accumulates in the PBS. In this compartment, iron binds further to a low molecular-mass compound excreted by bacteroids (resembling bacterial siderophores) whose precise identity is not known (Wittenberg et al. 1996). These authors proposed that the NADH-ferric chelate reductase identified on the PBM (Le Vier et al. 1996) could reduce the ferric chelate on the PBS side, releasing ferrous ions available for bacteroids.

Despite the considerable work done in this field during the past 10 years (see the summary in Fig. 9), several proteins in the PBM have not yet been characterised at the molecular level. A better understanding of the structure/function relations of these proteins remains of great importance and a target for future research.

If dicarboxylic acids are the sole carbon source supplied by host cells to bacteroids, there is a requirement to convert this to chemical forms which can enter the tricarboxylic acid (TCA) cycle. NAD-dependent malic acid converts malate to acetyl CoA (Driscoll and Finan 1996, 1997), while TCA-cycle enzymes provide oxaloacetic acid (OAA). Both substrates then are completely oxidised in successive turns of the TCA cycle, generating NADH, which is used for ATP synthesis via the bacterial electron transport chain. As indicated earlier, both NADH and ATP are required in large amounts in the fixation of dinitrogen by the enzyme nitrogenase.

Nitrogenase Assays

For several years, symbiosomes have been routinely extracted and purified from different legumes (Price et al. 1987; Herrada et al. 1989; Trinchant et al. 1994; Pedersen et al. 1996). This procedure requires the nodules to be gently crushed and the use of buffers exhibiting strictly controlled osmolarity. The final step of purification is achieved with a discontinuous Percoll gradient in which symbiosomes are located, after centrifugation, at the interface between the 30/60% layers. All the extraction and purification procedures must be carried out anaerobically when nitrogen fixation is measured. To measure nitrogenase activity, symbiosomes or the corresponding naked bacteroids are incubated in stoppered glass cuvettes completely filled with reaction solutions containing an energy-yielding substrate, purified oxyleghaemoglobin, providing O_2 and dissolved C_2H_2. At time intervals, O_2 consumption is monitored spectrophotometrically by leghaemoglobin deoxygenation, and the amount of C_2H_4 in the reaction medium is measured (Bergersen and Turner 1975a; Trinchant et al. 1981). As shown in Fig. 10, bacteroid respiration occurred at very low O_2 concentrations (1 to 10 µM) with the higher rate of C_2H_2 reduction occurring when the free O_2 concentration was lowest (from 3 to 6 min of experiment). C_2H_2 reduction was always higher for free bacteroids than for symbiosomes.

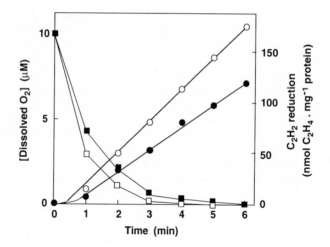

Fig. 10: Time course of the C_2H_2 reduction (*circles*) and O_2 depletion (*squares*) by isolated bacteroids (open symbols) and symbiosomes (*closed symbols*), assayed in a closed cuvette in the presence of succinate and with dissolved oxygen provided by oxyleghaemoglobin

Assimilation of Ammonia

Despite the recent report of alanine export from bacteroids (Waters et al. 1998), the weight of evidence still supports the view that fixed N is most likely transported across the PBM as ammonium and not amino acids (see discussion above). In the host cell cytosol, ammonia is likely to be assimilated rapidly by GS)/GOGAT, as indicated in Fig. 11, in close association with carbohydrate and organic acid metabolism (Vance et al. 1994; Temple et al. 1998). Communication between the various

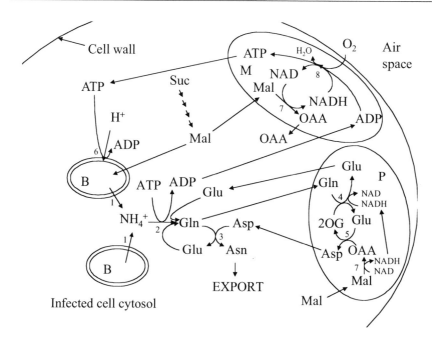

Fig. 11 : Scheme showing some of the interactions and communications between organelles in the infected cells of nodules during nitrogen fixation. Ammonium ions exported (1) from the bacteroids (B) are assimilated by GS (2) and the amide N used for the synthesis of asparagine (Asn) via asparagine synthetase (3) in the cytosol. Glutamine (Gln) is moved to the plastids (P) for conversion to glutamate (Glu) by GOGAT (4). AAT (5) uses Glu and oxaloacetate (OAA) to synthesise aspartate (Asp), which can leave the plastid to be used by asparagine synthetase in the cytosol. The proton ATPase (6) pumps H^+ into the peribacteroid space generating a membrane potential and a pH gradient across the PBM, which allows the transport of malate into and NH_4^+ out of the bacteroids. Malate may also be used as a substrate by the mitochondria (M) to generate NADH (7), for reoxidation by oxidative phosphorylation (8) and the regeneration of ATP. Main points to note are that much ATP is required by the cytosol, at the PBM for the proton ATPase and in ammonia assimilation by GS. Similarly most of the ADP produced will need to be returned to the mitochondria at the periphery of the cell for rephosphorylation. There must also be movement of amino and carboxylic acids to and from the plastids, which are also only located at the periphery of the infected cells. In nodules which export ureides, the plastids are the site of many of the processes involved in ureide synthesis. The last two steps, in fact, take place in uninfected cells, so that movement of metabolites (probably xanthine), also occurs between different cell types. (See Schubert and Boland 1990; Atkins et al. 1997)

reactions is not entirely clear, since the cellular localisation of some of these enzymes is not unequivocally known. The primary assimilation of ammonia by GS probably takes place in the cytosol, close to the symbiosomes from where it is effluxed. Asparagine synthetase and an isoform of AAT may also be cytosolic (Vance et al. 1994). However, the finding that nodule-specific forms of AAT and GOGAT are present in the plastids (Vance et al. 1994; Trepp et al. 1999b) indicates that shuttling of metabolites between the cytosol and peripherally located organelles may be essential. Glutamine would need to move from the cytosol to the plastids to regenerate glutamate. This conversion may be in concert with the formation of Asp from OAA, catalysed by AAT, as shown in Fig. 11. The requirement for NADH (GOGAT reaction) and OAA inside plastids could be accommodated by transport of malate into plastids, and oxidation to OAA by a plastid MDH. The presence of a plastid located MDH in nodules has recently been demonstrated by Miller et al. (1998).

The shuttling of metabolites from the cytosol in the centre of infected cells to an organelle situated at the periphery of the cell seems counter intuitive, but this is no different to the requirement for the continuous regeneration of ATP. ATP is used in the cytosol, for example by GS and the proton ATPase, at the PBM (Fig. 11). The regeneration of ATP from ADP by oxidative phosphorylation must take place predominantly in mitochondria, which also are only observed at the periphery of infected cells. Thus, ADP must move from its point of generation to mitochondria, while ATP moves in the opposite direction. In plastids, malate may be exchanged for aspartate, and glutamine for glutamate. Possible communication between bacteroids, cytosol, mitochondria and plastids is illustrated in Fig. 11. In this scheme it is suggested that malate, generated as indicated in Fig. 8, drives vital processes in mitochondria and plastids, in addition to its central role as the main carbon substrate for bacteroids.

Nitrogenous Compounds for Export

Nodules of temperate legumes mainly export asparagine plus smaller amounts of a number of other amino acids (Pate et al. 1969; Gunning et al. 1974). The synthesis of asparagine and the inter-conversions of some amino acids are shown in Fig. 11. In contrast, tropical legumes export allantoic acid and allantoin, collectively known as ureides. The synthesis of ureides via purine synthesis and breakdown involves remarkable metabolic communication between different organelles and cells within nodules, starting with the cytosol and plastids/mitochondria of infected cells and culminating in the oxidation of uric acid to allantoin in peroxisomes of uninfected cells (Schubert and Boland 1990; Atkins et al. 1997). Conversion of allantoin to allantoic acid also takes place in uninfected cells. From central region uninfected cells, the ureides probably diffuse towards the cortex and to the vascular bundles for export

References

Albrecht C, Geurts R, Lapeyrie F, Bisseling T (1998) Endomycorrhizae and rhizobial Nod factors both require *SYM8* to induce the expression of the early nodulin genes *PsENOD5* and *PsENOD12A*. Plant J 15: 605-614

Albrecht C, Geurts R, Bisseling T (1999) Legume nodulation and mycorrhizae formation; two extremes in host specificity meet. EMBO J 18: 281-288

Ardourel M, Demont N, Debellé F, Maillet F, De Billy F, Promé JC, Dénarié J, Truchet G (1994) *Rhizobium meliloti* lipooligosaccharide nodulation factors: different structural requirements for bacterial entry into target root hair cells and induction of plant symbiotic developmental responses. Plant Cell 6: 1357-1374

Arredondo-Peter R, Hargrove MS, Moran JF, Sarath G, Klucas RV (1998) Plant hemoglobins. Plant Physiol 18: 1121-1125

Atkins CA, Smith PMC, Storer PJ (1997) Re-examination of the intracellular localization of *de novo* purine synthesis in cowpea nodules. Plant Physiol 113: 127-135

Awonaike KO, Lea PJ, Miflin BJ (1981) The localisation of the enzymes of ammonia assimilation in root nodules of *Phaseolus vulgaris*. Plant Sci Lett 23:189-195

Bakkers J, Semino CE, Stroband H, Kijne JW, Robbins PW, Spaink HP (1997) An important developmental role for oligosaccharides during early embryogenesis of cyprinid fish. Proc Natl Acad Sci USA 94: 7982-7986

Barker DG, Bianchi S, Blondon F, Dattee Y, Duc G, Essad S, Flament P, Gallusci P, Génier G, Guy P, Muel X, Tourneur J, Dénarié J, Huguet T (1990) *Medicago truncatula*, a model plant for studying the molecular genetics of the rhizobium-legume symbiosis. Plant Mol Biol Rep 8: 40-49

Batut J, Boistard P (1994) Oxygen control of rhizobium. Antonie van Leeuwenhoek Journal: Antonie varr Leeuwenhoek 66: 129

Bauer P, Ratet P, Crespi MD, Schultze M, Kondorosi A (1996) Nod factors and cytokinins induce similar cortical cell division, amyloplast deposition and *MsEnodl2A* expression patterns in alfalfa roots. Plant J 10: 91-105

Bergersen FJ, Turner GL (1975a) Leghaemoglobin and the supply of O_2 to nitrogen fixing root nodule bacteroids: studies of an experimental system with no gas phase. J Gen Microbiol 89: 31-47

Bergersen FJ, Turner GL (1975b) Leghaemoglobin and the supply of O_2 to nitrogen fixing root nodule bacteroids: presence of two oxidase systems and ATP production at low free O_2 concentration. J Gen Microbiol 91: 345-354

Bergersen FJ, Turner GL (1980) Properties of terminal oxidase systems of bacteroids from root nodules of soybean and cowpea and of N_2-fixing bacteria grown in continuous culture. J Gen Microbiol 118: 235-252

Bohme H (1998) Regulation of nitrogen fixation in heterocyst-forming bacteria. Trends Plant Sci 3: 346-351

Bono J-J, Riond J, Nicolaou KC, Bockovich NJ, Estevez VA, Cullimore JV, Ranjeva R (1995) Characterization of a binding site for chemically synthesized lipo-oligosaccharidic NodRm factors in particulate fractions prepared from roots. Plant J 7: 253-260

Brears T, Walker EL, Coruzzi GM (1991) A promoter sequence involved in cell-specific expression of the pea glutamine synthetase *GS3A* gene in organs of transgenic tobacco and alfalfa. Plant J 1: 235-244

Brewin NJ (1991) Development of the legume root nodule. Annu Rev Cell Biol 7: 191 226

Brown SM, O'Parka KJ, Sprent JI, Walsh KB (1995) Symplastic transport in soybean nodules. Soil Biol Biochem 27: 387-399

Bryce JH, Day DA (1990) Tricarboxylic acid cycle activity in mitochondria from soybean nodules and cotyledons. J Exp Bot 41: 961-967

Burgess BK, Lowe DJ (1996) Mechanism of molybdenum nitrogenase. Chem Rev 96: 2983-3011

Capone DG, Zehr JP. Paerl HW, Bergman B, Carpenter EJ (1997) *Trichodesmium*, a globally significant marine cyanobacterium. Science 276: 1221-1229

Carlson RW, Price NPJ, Stacey G (1994) The biosynthesis of rhizobial lipo-oligosaccharide nodulation signal molecules. Mol Plant-Microbe Interact 7: 684-695

Charon C, Johansson C, Kondorosi E, Kondorosi A, Crespi M (1997) *ENOD40* induces dedifferentiation and division of root cortical cells in legumes. Proc Natl Acad Sci USA 94: 8901-8906

Chen R, Siver DL, de Bruijn FJ (1998) Nodule parenchyma-specific expression of the *Sesbania rostrata* early nodulin gene *SrEnod2* is mediated by its 3 untranslated region. Plant Cell 10: 1585-1602

Cohn J, Day RB, Stacey G (1998) Legume nodule organogenesis. Trends Plant Sci 3: 105- 110

Cooper JB, Long SR (1994) Morphogenetic rescue of *Rhizobium meliloti* nodulation mutants by *trans*-zeatin secretion. Plant Cell 6: 215-225

Craig J, Barratt P, Tatge H, Dejardin A, Handley L, Gardner CD, Barber L, Wang T, Hedley C, Martin C, Smith AM (1999) Mutations at the *rug4* locus alter the carbon and nitrogen metabolism of pea plants through an effect on sucrose synthase. Plant J 17: 353-362

Csanadi G, Szecsi J, Kalo P, Kiss P, Endre G, Kondorosi A, Kondorosi E, Kiss GB (1994) *ENOD 12*, an early nodulin gene, is not required for nodule formation and efficient nitrogen fixation in alfalfa. Plant Cell 6: 201-213

Cullimore JV, Cock JM, Daniell TJ, Swarup R, Bennett MJ (1992) Inducibility of the glutamine synthetase gene family of *Phaseolus vulgaris L.* Inducible plant proteins. J.L. Wray Cambridge University Press, pp 79-95

Day DA, Copeland L (1991) Carbon metabolism and compartmentation in nitrogen-fixing legume nodules. Plant Physiol Biochem 29: 185-201

Day DA, Mannix M (1988) Malate oxidation by soybean nodule mitochondria and the possible consequences for nitrogen fixation. Plant Physiol Biochem 26: 567-273

Dean RM, Rivers, RL, Zeidel, ML, Roberts DM (1999) Purification and functional reconstitution of soybean nodulin 26. An aquaporin with water and glycerol transport properties. Biochemistry 38: 347-353

deBilly F, Barker G, Gallusci P, Truchet G (1991) Leghaemoglobion gene transcrtiption is triggered in a single layer in the indeterminate nitrogen-fixing root nodule of alfalfa. Plant J 1: 27-35

de Carvalho-Niebel F, Lescure N, Cullimore JV, Gamas P (1998) The *Medicago truncatula MtAnnl* gene encoding an annexin is induced by Nod factors and during the symbiotic interaction with *Rhizobium meliloti.* Mol Plant-Microbe Interact 11: 504-513

de Jong AJ, Heidstra R, Spaink HP, Hartog MV, Meijer EA, Hendriks T, Schiavo FL, Terzi M, Bisseling T, Van Kammen A, de Vries SC (1993) *Rhizobium* lipooligosaccharides rescue a carrot somatic embryo mutant. Plant Cell 5: 615-620

Dénarié J, Debellé F, Rosenberg C (1992) Signaling and host range variation in nodulation. Annu Rev Microbiol 46: 497-531

Dénarié J, Debellé F, Promé JC (1996) Rhizobium lipo-chitooligosaccharide nodulation factors: signaling molecules mediating recognition and morphogenesis. Annu Rev Biochem 65: 503-535

Dixon R, Cheng Q, Shen G-F, Day A, Dowson-Day M (1997) *Nif* gene transfer and expression in chloroplasts: prospects and problems. Plant Soil 194: 193-203

Downie JA (1994) Signalling strategies for nodulation of legumes by rhizobia. Trends Microbiol 2: 318-324

Downie JA (1998) Functions of rhizobial nodulation genes. In: Spaink HP, Kondorosi A, Hooykaas PJJ (eds) The Rhizobiaceae, Molecular biology of model Plant-associated bacteria. Kluwer, Dordrecht, pp 387-402

Driscoll BT, Finan TM (1996) NADP-dependent malic enzyme of *Rhizobium meliloti*. J Bacteriol 178: 2224-2231

Driscoll BT, Finan TM (1997) Properties of NAD- and NADP-dependent malic enzymes of *Rhizobium* (*Sinorhizobium* meliloti) and differential expression of their genes in nitrogen-fixing bacteroids. Microbiology 143: 489-498

Dubrovo PN, Krylova W, Livanova GI, Zhiznevskaya GY, Izmailov SF (1992) Properties of ATPases of the peribacteroid membrane in root nodules of yellow lupine. Sov Plant Physiol 39: 318-324

Duc G, Messager A (1989) Mutagenesis of Pea (*Pisum sativum* L.) and the isolation of mutants for nodulation and nitrogen fixation. Plant Sci 60: 207-213

Eady RR (1996) Structure-function relationships of alternative nitrogenases. Chem Rev 96: 3013-3030

Ehrhardt DW, Atkinson EM, Long SR (1992) Depolarization of alfalfa root hair membrane potential by *Rhizobium meliloti* Nod factors. Science 256: 998-1000

Ehrhardt DW, Wais R, Long SR (1996) Calcium spiking in plant root hairs responding to *Rhizobium* nodulation signals. Cell 85 : 673-681

Elmerich C, Kondorosi A, Newton WE (eds) (1998) Biological nitrogen fixation for the 21st century. 11th Int Congr on Nitrogen Fixation. Kluwer, Dordrecht

Etzler ME, Kalsi G, Ewing NN, Roberts NJ, Day RB, Murphy JB (1999) A Nod factor binding lectin with apyrase activity from legume roots. Proc Natl Acad Sci USA 96: 5856-5861

Fang YW, Hirsch AM (1998) Studying early nodulin gene *ENOD40* expression and induction by nodulation factor and cytokinin in transgenic alfalfa. Plant Physiol 116: 53-68

Felle HH, Kondorosi E, Kondorosi A, Schultze M (1995) Nod signal-induced plasma membrane potential changes in alfalfa root hairs are differentially sensitive to structural modifications of the lipochitooligosaccharide. Plant J 7: 939-947

Felle HH, Kondorosi E, Kondorosi A, Schultze M (1998) The role of ion fluxes in Nod factor signalling in *Medicago sativa*. Plant J 13: 455-463

Firmin JL, Wilson KE, Carlson RW, Davies AE, Downie JA (1993) Resistance to nodulation of cv. Afghanistan peas is overcome by *nodX*, which mediates an O-acetylation of the *Rhizobium leguminosarum* lipo-oligosaccharide nodulation factor. Mol Microbiol 10: 351-360

Fischer H-M (1994) Genetic regulation of nitrogen fixation in rhizobia. Microbiol Rev 58: 352-386

Fischer H-M (1996) Environmental regulation of rhizobial symbiotic nitrogen fixing genes.Trends Microbiol 4: 317-319

Fischer H-M, Schneider K, Babst M, Hennecke H (1999) GRoEl chaperon are required for the formation of functional nitrogenase in *Bradyrhizobium japonicum*. Arch Microbiol 171: 279-289

Fisher RF, Long SR (1992) *Rhizobium-plant* signal exchange. Nature 357:655-660

Fisher RF, Egelhoff TT, Mulligan JT, Long SR (1988) Specific binding of proteins from *Rhizobium meliloti* cell-free extracts containing NodD to DNA sequences upstream of inducible nodulation genes. Genes Dev 2: 282-293

Forde BG, Day HM, Turton JF, Shen W-J, Cullimore JV, Oliver JE (1989) Two glutamine synthetase genes from *Phaseolus vulgaris* display contrasting developmental and spatial patterns of expression in transgenic *Lotus corniculatus* plants. Plant Cell 1: 391-401

Fuller F, Künster PW, Nguyen T, Verma DPS (1983) Soybean nodule genes: analysis of cDNA clones reveals several major tissue specific sequences nitrogen-fixing root nodules. Proc Natl Acad Sci USA 80: 2594-2598

Gallon JR, Chaplin AE (1987) An introduction to nitrogen fixation. Cassell, London

Gamas P, Niebel FDC, Lescure N, Cullimore JV (1996) Use of a subtractive hybridization approach to identify new *Medicago truncatula* genes induced during root nodule development. Mol Plant-Microbe Interact 9: 233-242

Gantt JS, Larson RJ, Farnham MW, Parthirana SM, Miller S, Vance C (1992) Aspartate aminotransferase in effective and ineffective root nodules. Plant Physiol 98: 868-878

Gogorcena Y, Gordon AJ, Escuredo PR, Minchin FR, Witty JF, Moran JF, Becana M (1998) N_2 fixation, carbon metabolism and oxidative damage in nodules of dark-stressed common bean plants. Plant Physiol 113: 1193-1201

Gonzalez EM, Gordon AJ, James CL, Arrese-Igor C (1995) The role of sucrose synthase in the response of soybean nodules to drought. J Exp Bot 46: 1515-1523

Gonzalez JE, York GM, Walker GC (1996) *Rhizobium meliloti* exopolysaccharides: synthesis and symbiotic function. Gene 179: 141-146

Gordon AJ (1992) Carbon metabolism in the legume nodule. In: Pollock CJ, Farrar JF, Gordon AJ (eds) Carbon partitioning within and between organisms. Bios, Oxford, pp 33-162

Gordon AJ, James CL (1997) Enzymes of carbohydrate and amino acid metabolism in developing and mature nodules of white clover. J Exp Bot 48: 895-903

Gordon AJ, Ryle GJA, Mitchell DF, Powell CE (1985) The flux of ^{14}C-labelled photosynthate through soyabean root nodules during N_2 fixation. J Exp Bot 36: 756-769

Gordon AJ, Ryle GJA, Mitchell DF, Lowry KH, Powell CE (1986) The effect of defoliation on carbohydrate, protein and leghaemoglobin content of white clover nodules. Ann Bot 58: 141-154

Gordon AJ, Mitchell DF, Ryle, GJA, Powell CE (1987) Diurnal production and utilization of photosynthate in nodulated white clover. J Exp Bot 38: 84-98

Gordon AJ, Thomas BJ, Reynolds PHS(1992) Localisation of sucrose synthase in soybean root nodules. New Phytol 122: 35-44

Gordon AJ, Minchin FR, Skot L, James CL (1997) Stress induced declines in soybean nitrogen fixation are related to nodule sucrose synthase activity. Plant Physiol 114: 937-946

Gordon AJ, Minchin FR, James CL, Komina O (1999) Sucrose synthase in legume nodules is essential for nitrogen fixation. Plant Physiol 120: 867-877

Gunning BES, Pate JS, Minchin FR, Marks I (1974) Quantitative aspects of transfer cell structure in relation to vein loading in leaves and solute transport in legume nodules. In: Transport at the cellular level. SEB Symposium, vol 28. Cambridge University Press, Cambridge, pp 87-126

Hadri AE, Bisseling T (1998) Responses of the plant to Nod factors. In:Spaink HP, Kondorosi A, Hooykaas PJJ (eds) The Rhizobiaceae, molecular biology of model plant-associated bacteria. Kluwer , Dordrecht, pp 403-416

Hanning I, Baumgarten K, Schott K, Heldt HW (1999) Oxaloacetate transport into plant mitochondria. Plant Physiol 119: 1025-1031

Henikoff S, Haughn GW, Calvo JM, Wallace JC (1988) A large family of bacterial activator proteins. Proc Natl Acad Sci USA 85: 6602-6606

Hernandez LE, Cooke DT (1996) Lipid composition of symbiosomes from pea root nodules. Phytochemistry 42: 341-346

Herrada G, Puppo A, Rigaud J (1989) Uptake of metabolites by bacteroid-containing vesicles and by free bacteroids from french bean nodules. J Gen Microbiol 135: 3165-3171

Hirsch AM, Fang YW (1994) Plant hormones and nodulation: what's the connection? Plant Mol Biol 26: 5-9

Hirsch AM, Kapulnik Y (1998) Signal transduction pathways in mycorrhizal associations: comparisons with the rhizobium legume symbiosis. Fungal Genet Biol 23: 205-212

Hirsch AM, Bhuvaneswari TV, Torrey JG, Bisseling T (1989) Early nodulin genes are induced in alfalfa root outgrowths elicited by auxin transport inhibitors. Proc Natl Acad Sci USA 86: 1244-1248

Hohnjec N, Becker JD, Puhler A, Perlick AM, Kuster H (1999) Genomic organisation and expression of the *MtSucS1* gene, which encodes a nodule-enhanced sucrose synthase in the model legume *Medicago truncatula*. Mol Gen Genet 261: 524-522

Howard JB, Rees DC (1996) Structural basis of biological nitrogen fixation. Chem Rev 96: 2965-2982

Hunt S, Layzell DB (1993) Gas exchange of legume nodules and the regulation of nitrogenase activity. Annu Rev Plant Physiol Plant Mol Biol 44: 483-511

Huss-Danell K (1997) Tansley review no 93: actinorhizal symbioses and their N_2 fixation. New Phytol 136: 375-405

Jiang QY, Gresshoff PM (1997) Classical and molecular genetics of the model legume *Lotus japonicus*. Mol Plant-Microbe Interact 10: 59-68

Journet EP, Pichon M, Dedieu A, deBilly F, Truchet G, Barker DG (1994) *Rhizobium meliloti* Nod factors elicit cell specific transcription of the *ENOD12* gene in transgenic alfalfa. Plant J 6: 241-249

Kamst E, Bakkers J, Quaedvlieg NE, Pilling J, Kijne JW, Lugtenberg BJ, Spaink HP (1999) Chitin oligosaccharide synthesis by rhizobia and zebrafish embryos starts by glycosyl transfer to O4 of the reducing-terminal residue. Biochemistry 38: 4045-4052

Kannenberg EL, Brewin NJ (1994) Host-plant invasion by *Rhizobium*: the role of cell-surface components. Trends Microbiol 2: 277-283

Karl D, Letelier R, Tupas L, Dore J, Christian J, Hebel D (1997) The role of nitrogen fixation in biogeochemical cycling in the subtropical North Pacific Ocean. *Nature*, 388: 533-538

Kondorosi A (1992) Regulation of nodulation genes in rhizobia. In: Verma DPS (ed), Molecular signals in plant-microbe communications. CRC Press, Boca Raton, pp 325-340

Kozik A, Heidstra R, Horvath B, Kulikova O, Tikhonovich I, Ellis THN, Vankammen A, Lie TA, Bisseling T (1995) Pea lines carrying *sym1* or *sym2* can be nodulated by *Rhizobium* strains containing *nodX*; *sym1* and *sym2* are allelic. Plant Sci 108: 41-49

Kurkdjian AC (1995) Role of the differentiation of root epidermal cells in Nod factor (from *Rhizobium meliloti*)-induced root hair depolarization of *Medicago sativa*. Plant Physiol 107: 783-790

Kuzma MM, Winter H, Storer P, Oresnik I, Atkins CA, Layzell DB (1999) The site of oxygen limitation in soybean nodules. Plant Physiol 119: 399-407

Leigh JA, Walker GC (1994) Exopolysaccharides of Rhizobium – synthesis, regulation and symbiotic function. Trends Genet 10: 63-67

Lerouge P, Roche P, Faucher C, Maillet F, Truchet G, Promé JC, Dénarié J (1990) Symbiotic host specificity of *Rhizobium meliloti* is determined by a sulphated and acylated glucosamine oligosaccharide signal. Nature 344: 781-784

Le Vier K, Day DA, Guerinot ML (1996) Iron uptake by symbiosomes from soybean root nodules. Plant Physiol 111: 613-618

Long SR (1996) *Rhizobium symbiosis*: Nod factors in perspective. Plant Cell 8: 1885-1898

Lorquin J, Lortet G, Ferro M, Mear N, Dreyfus B, Prome JC, Boivin C (1997) Nod factors from *Sinorhizobium saheli* and S *teranga* bv. *sesbaniae* are both arabinosylated and fucosylated, a structural feature specific to *Sesbania rostrata* symbionts. Mol Plant-Microbe Interact 10: 879-890

Madsen O, Sandal L, Sandal N, Marcker KA (1993) A soybean copropopyrinogen oxidase gene is highly expressed in root nodules. Plant Mol Biol 23: 35-43

Marcker K, Lund M, Jensen EO, Marcker KA (1984) Transcription of soybean leghemoglobin gene during nodule development. EMBO J 3: 1691-1695

Mellor RB, Werner D (1986) The fractionation of Glycine max root nodule cells – a methodological overview. Endocytobios Cell Res 3: 317-336

Mellor RB, Christensen TMIE, Bassarab S, Werner D (1985) Phospholipid transfer from ER to the peribacteroid membrane in soybean nodules. Z Naturforsch 40c: 73-79

Mergaert P, Ferro M, D'Haeze W, van Montagu M, Holsters M, Promé JC (1997) Nod factors of *Azorhizobium caulinodans* strain ORS571 can be glycosylated with an arabinosyl group, a fucosyl group, or both. Mol Plant-Microbe Interact 10: 683-687

Millar AH, Day DA, Bergersen FJ (1995) Microaerobic respiration and oxidative phosphorylation by soybean mitochondria: implications for nitrogen fixation. Plant Cell Environ 18: 715-726

Miller DD, de Ruijter NCA, Bisseling T, Emons AM (I 999) The role of actin in root hair morphogenesis: studies with lipochito-oligosaccharide as a growth stimulator and cytochalasin as an actin-perturbing drug. Plant J 17: 141-154

Miller SS, Driscoll BT, Gregerson RG, Gantt JS, Vance CP (1998) Alfalfa malate dehydrogenase (MDH): molecular cloning and characterisation of five different forms reveals a unique nodule-enhanced MDH. Plant J 15: 173-184

Minami E, Kouchi H, Carlson RW, Cohn JR, Kolli VK, Day RB, Ogawa T, Stacey G (1996) Cooperative action of lipo-chitin nodulation signals on the induction of the early nodulin, ENOD2, in soybean roots. Mol Plant-Microbe Interact 7: 574-583

Minchin FR (1997) Regulation of oxygen diffusion in legume nodules. Soil Biol Biochem 29: 881-888

Moreau S, Meyer JM, Puppo A (1995) Uptake of iron by symbiosomes and bacteroids from soybean nodules. FEBS Lett 361: 225-228

Morrell M, Copeland L (1984) Enzymes of sucrose breakdown in soybean nodules. Plant Physiol 74: 1030-1034

Morrell M, Copeland L (1985) Sucrose synthase of soybean nodules. Plant Physiol 78: 149-154

Mouritzen P, Rosendahl L (1997) Identification of a transport mechanism for NH_4^+ in the symbiosome membrane of pea root nodules. Plant Physiol 115: 519-526

Mylona P, Pawlowski K, Bisseling T (1995) Symbiotic nitrogen fixation. Plant Cell 7: 869-885

Nap JP, Bisseling T (1990) Developmental biology of a plant-prokaryote symbiosis: the legume root nodule. Science 250: 948-954

Nicolaou KC, Bockovich NJ, Carcanague DR, Hummel CW, Even LF (1992) Total synthesis of the NodRm-IV factors, the *Rhizobium* nodulation signals. J Am Chem Soc 114: 8701-8702

Niebel A, Bono JJ, Ranjeva R, Cullimore JV (1997) Identification of a high-affinity binding site for lipo-oligosaccharidic NodRm factors in the microsomal fraction of *Medicago* cell suspension cultures. Mol Plant-Microbe Interact 10: 132-134

O'Brian MR (1996) Heme synthesis in the *Rhizobium-legume* symbiosis: a palette for bacterial and eukaryotic pigments. J Bacteriol 178; 2471-2478

Ou Yang L-J, Day DA (1992) Transport properties of sybiosomes isolated from siratro nodules. Plant Physiol Biochem 30: 613-623

Palacios R, Mora J, Newton WE (1993) (eds) New horizons in nitrogen fixation. 9th International Congress on Nitrogen Fixation. Kluwer, Dordrecht

Pate JS, Gunning BES, Briarty LG (1969) Ultrastructure and functioning of the transport system of the leguminous root nodule. Planta 85: 11-34

Pathirana MS, Samac DA, Roeven R, Yoshioka H, Vance CP (1997) Analysis phosphoenolpyruvate carboxylase gene structure and expression in alfalfa nodules. Plant J 12: 293-304

Pawlowski K (1997) Nodule-specific gene expression. Physiol Plant 99: 617-631

Pedersen AL, Feldner HC, Rosendahl L (1996) Effect of proline on nitrogenase activity in symbiosomes from root nodules of soybean (*Glycine max L.*) subjected to drought stress. J Exp Bot 303: 1533-1539

Penmetsa RV, Cook DR (1997) A legume ethylene-insensitive mutant hyperinfected by its rhizobial symbiont. Science 275: 527-530

Perotto S, Vandenbosch KA, Butcher GW, Brewin NJ (1991) Molecular composition and development of the plant glycocalyx associated with the peribacteroid membrane of pea nodules. Development 112: 763-773

Peters NK (1997) Nodulation: finding the lost common denominator. Curr Biol 7: R223-R226

Pingret JL, Journet EP, Barker DG (1998) Rhizobium Nod factor signaling. Evidence for a G protein-mediated transduction mechanism. Plant Cell 10: 659-672

Postgate J (1987) Nitrogen fixation, 2nd edn. Arnold, London

Preisig O, Anthamatten D, Hennecke H (1993) Genes for a microaerobically induced oxidase complex in *Bradyrhizobium japonicum* are essential for nitrogen-fixing endosymbiosis. Proc Natl Acad Sci USA 90: 3309-3313

Preisig O, Zufferey R, Hennecke H (1996) The *Bradyrhizobium japonicum fixGHIS* genes are required for the formation of the high affinity cbb_3 type cytochrome oxidase. Arch Microbiol 165: 297-305

Price GD, Day DA, Gresshoff PM (1987) Rapid isolation of intact peribacteroid envelopes from soybean nodules and demonstration of selective permeability to metabolites. J Plant Physiol 130: 157-164

Price NPJ, Relic B, Talmont E, Lewin A, Prome D, Pueppke SG, Maillet F, Dénarié J, Promé JC, Broughton WJ (1992) Broad host range *Rhizobium* species strain NGR234 secretes a family of carbamoylated, and fucosylated, nodulation signals that are O-acetylated or sulphated. Mol Microbiol 6: 3575-3584

Radyukina NL, Bruskova RK, Izmailov SF (1992) Transport of ^{14}C substrate through peribacteroidal membrane of yellow-lupin nodules. Dokl Bot Sci 323: 603-606

Rai AN (ed) (1990) A handbook of symbiotic cyanobacteria. CRC Press, Boca Raton, Florida

Rawsthorne S, La Rue TA (1986) Metabolism under microaerobic conditions of mitochondria from cowpea nodules. Plant Physiol 81: 1097-1102

Ribbe M, Gadkari D, Meyer O (1997) N_2 fixation by *Steptomyces thermoautotrophicus* involves a molybdenum-dinitrogenase and a manganese superoxide oxidoreductase that couple N_2 reduction to the oxidation of superoxide produced from O_2 by a molybdenum-Co dehydrogenase. J Biol Chem 272: 26627-26633

Robertson JG, Taylor MP (1973) Acid and alkaline invertases in roots and nodules of *Lupinus augustifolius* infected with *Rhizobium lupini*. Planta 112: 1-6

Robertson JG, Lyttleton P (1984) Division of peribacteroid membranes in root nodules of white clover. J Cell Sci 69: 147-157

Robertson JG, Warburton MP, Lyttleton P, Fordyce AM, Bullivan S (1978) Membranes in lupin root nodules. II. Preparation and properties of peribacteroid membranes and bacteroid envelope inner membranes from developing lupin nodules. J Cell Sci 30: 151-174

Robinson DL, Pathirana SM, Gantt JS, Vance CP (1996) Immunogold localisation of nodule-enhanced phosphoenolpyruvate carboxylase in alfalfa. Plant Cell Environ 19: 602-608

Roche P, Debellé F, Maillet F, Lerouge P, Faucher C, Truchet G, Dénarié J, Promé JC (1991) Molecular basis of symbiotic host specificity in *Rhizobium meliloti*: *nodH* and *nodPQ* genes encode the sulphation of lipo- oligosaccharide signals. Cell 67: 1131-1143

Roche P, Maillet F, Plazanet C, Debellé F, Ferro M, Truchet G, Promé JC, Dénarié J (1996) The common *nodABC* genes of *Rhizobium meliloti* are host-range determinants. Proc Natl Acad Sci USA 93: 15305-15310

Rosendahl L, Dilworth MJ, Glenn AR (1992) Exchange of metabolites across the peribacteroid membrane in pea root nodules. J Plant Physiol 139: 635-638

Roth EJK, Stacey G (1988) Homology in endosymbiotic systems: the term "symbiosome". In: Palacios R, Verma DPS (eds) Molecular genetics of plant-microbe interactions. American Phytopathology Society, St Paul, pp 220-225

Sadowsky MJ, Graham PH (1998) Soil biology of the Rhizobiaceae. In: Spaink HP, Kondorosi A, Hooykaas PJJ (eds) The Rhizobiaceae, molecular biology of model plant-associated bacteria. Kluwer, Dordrecht, pp 155-172

Sagan M, Morandi D, Tarenghi E, Duc G (1995) Selection of nodulation and mycorrhizal mutants in the model plant *Medicago truncatula* (Gaertn) after gamma-ray mutagenesis. Plant Sci 111: 63-71

Santana MA, Pihakaski-Maunsbach K, Sandal N, Marker KA, Smith AG (1998) Evidence that the plant host synthesizes the heme moiety of leghemoglobin in root nodules. Plant Physiol 116: 1259-1269

Schauser L, Handberg K, Sandal N, Stiller J, Thykjaer T, Pajuelo E, Nielsen A, Stougaard J (1998) Symbiotic mutants deficient in nodule establishment identified after T-DNA transformation of *Lotus japonicus*. Mol Gen Genet 259: 414-423

Scheres B, Van De Wiel C, Zalensky A, Horvath B, Spaink HP, van Eck H, Zwartkruis F, Wolters AM, Gloudemans T, Van Kammen A, Bisseling T (1990) The *ENOD12* gene product is involved in the infection process during pea-*Rhizobium* interaction. Cell 60: 281-294

Schlaman HRM, Okker RJH, Lugtenberg BJJ (1992) Regulation of nodulation gene expression by *nodD* in *Rhizobia*. J Bacteriol 174: 5177-5182

Schlaman HRM, Phillips DA, Kondorosi E (1998) Genetic organization and transcriptional regulation of rhizobial nodulation genes. In: Spaink HP, Kondorosi A, Hooykaas PJJ (eds) The Rhizobiaceae, molecular biology of model plant-associated bacteria. Kluwer, Dordrecht, pp 361-386

Schmidt J, Rohrig H, John M, Wieneke U, Stacey G, Koncz C, Schell J (1993) Alteration of plant growth and development by *Rhizobium nodA* and *nodB* genes involved in the synthesis of oligosaccharide signal molecules. Plant J 4: 651-658

Schubert KR, Boland MJ (1990) The ureides. In: Miflin BJ, Lea PJ (eds) The biochemistry of plants, vol 16. Academic Press, San Diego, pp 197-282

Schultze M, Kondorosi A (1998) Regulation of symbiotic root nodule development. Annu Rev Genet 32: 33-57

Schultze M, Kondorosi E, Ratet P, Buire M, Kondorosi A (1994) Cell and molecular biology of Rhizobium-plant interactions. Int Rev Cytol 156: 1-75

Selker JML (1988) Three-dimensional organization of uninfected tissue in soybean root nodules and its relation to cell specialization in the central region. Protoplasma 147: 178-190

Selker JML, Newcomb EH (1985) Spatial relationships between uninfected and infected cells in root nodules of soybean. Planta 156: 446-454

Semino CE, Robbins PW (1995) Synthesis of "Nod"-like chitin oligosaccharides by the *Xenopus* developmental protein DG42. Proc Natl Acad Sci USA 92: 3498-3501

She Q, Lauridsen P, Stougaard J, Marcker KA (1993) Minimal enhancer elements of the leghaemoglobin *lba and lbc3* gene promoters from *Glycine max* have different properties. Plant Mol Biol 22: 945-956

Shi LF, Twary SN, Yoshioka H, Gregerson RG, Miller SS, Samac DA, Gantt, Unkefer PJ, Vance CP (1997) Nitrogen assimilation in alfalfa: isolation and characterisation of an asparagine synthetase gene showing enhanced expression in root nodules and dark adapted leaves. Plant Cell 9: 1339-1356

Smit G, de Koster CC, Schripsema J, Spaink HP, van Brussel AAN, Kijne JW (1995) Uridine, a cell division factor in pea roots. Plant Mol Biol 29: 869-873

Soupene E, Foussard M, Boistard P, Truchet G, Batut J (1995) Oxygen as a key developmental regulator of *Rhizobium meliloti* N_2 fixation gene-expression within the alfalfa root-nodule. Proc Natl Acad Sci USA 92: 3759-3763

Spaink H P (1995) The molecular basis of infection and nodulation by rhizobia – the ins and outs of sympathogenesis. Annu Rev Phytopathol 33: 345-368

Spaink HP (1996) Regulation of plant morphogenesis by lipo-chitin oligosaccharides. Crit Rev Plant Sci 15: 559-582

Sprent JI, Sprent P (1990) Nitrogen-fixing organisms: pure and applied aspects. Chapman and Hall, London

Streeter JG (1991) Transport and metabolism of carbon and nitrogen in legume nodules. Adv Bot Res 18: 129-187

Streeter JG (1992) Analysis of apoplastic solutes in the cortex of soybean nodules. Physiol Plant 84: 584-592

Streeter JG (1995) Recent developments in carbon transport and metabolism in symbiotic systems. Symbiosis 19: 175-196

Sutton JM, Lea EJA, Downie JA (1994) The nodulation-signaling protein NodO from *Rhizobium leguminosarum* biovar viciae forms ion channels in membranes. Proc Natl Acad Sci USA 91: 9990-9994

Szafran MM, Haaker H (1995) Properties of the peribacteroid membrane ATPase of pea root nodules and its effect on the nitrogenase activity. Plant Physiol 108: 1227-1232

Szczyglowski K, Hamburger D, Kapranov P, deBruijn FJ (1997) Construction of a *Lotus japonicus* late nodulin expressed sequence tag library and identification of novel nodule specific genes. Plant Physiol 114: 1335-1346

Szczyglowski K, Kapranov P, Hamburger D, deBruijn FJ (1998a) The *Lotus japonicus NOD70* nodulin gene encodes a protein with similarities to transporters. Plant Mol Biol 37: 651-661

Szczyglowski K, Shaw RS, Wopereis J, Copeland S, Hamburger D, Kasiborski B, Dazzo FB, De Bruijn FJ (1998b) Nodule organogenesis and symbiotic mutants of the model legume *Lotus japonicus*. Mol Plant-Microbe Interact 11: 684-697

Takane K, Tajimas, Kouchi H (1997) Two distinct uricase (nodulin 35) genes are differentially expressed in soybean plants. Mol Plant-Microbe Interactions 10: 735-741

Temple SJ, Heard J, Ganter J, Dunn G, Sengupta-Gopalan C (1995) Characterisation of a nodule-enhanced glutamine synthetase from alfalfa. Mol Plant-Microbe Interact 8: 218-227

Temple SJ, Kunjibettu S, Roche D, Sengupta-Gopalan S (1996) Total glutamine synthetase activity during soybean nodule development is controlled at the level of transcription and holoprotein turnover. Plant Physiol 112: 1723-1733

Temple SJ, Vance CP, Gantt JS (1998) Glutamate synthase and nitrogen assimilation. Trends Plant Sci 3: 51-56

Tercé-Laforgue T, Carrayol E, Cren M, Desbrosses G, Hecht V, Hirel B (1999) A strong constitutive positive element is essential for the ammonium-regulated expression of a soybean gene encoding cytosolic glutamine synthetase. Plant Mol Biol 39: 551-564

Thummler F, Verma DPS (1987) Nodulin-100 of soybean is the subunit of sucrose synthase regulated by the availability of free heme in nodules. J Biol Chem 262: 14730-14736

Tikonovich IA, Provoro NA, Romanov VI, Newton WE (1995) Nitrogen fixation: fundamentals and applications. 10th International Congress of Nitrogen Fixation. Kluwer, Dordrecht

Timmers ACJ, Auriac MC, Debilly F, Truchet G (1998) Nod factor internalization and microtubular cytoskeleton changes occur concomitantly during nodule differentiation in alfalfa. Development 125: 339-349

Trepp GB, van de Mortel M, Yoshioka H, Miller SS, Samac DA, Gantt JS, Vance CP (1999a) NADH-glutamate synthase in alfalfa roots. Genetic regulation and cellular expression. Plant Physiol 119: 817-828

Trepp GB, Plank DW, Gantt, Vance CP (1999b) NADH-glutamate synthase in alfalfa root nodules. Immunocytochemical localisation. Plant Physiol 119: 829-837

Trepp GB, Temple SJ, Bucciarelli B, Shi LF, Vance CP (1999c) Expresssion map for genes involved in nitrogen and carbon metabolism in alfalfa root nodules. Mol Plant-Microbe Interact 12: 526-535

Trinchant J-C, Birot AM, Rigaud J (1981) Oxygen supply and energy-yielding substrates for nitrogen fixation (acetylene reduction) by bacteroid preparations. J Gen Microbiol 125: 159-165

Trinchant J-C, Guérin V, Rigaud J (1994) Acetylene reduction by symbiosomes and free bacteroids from faba-bean (Vicia faba L.) nodules. Plant Physiol 105: 555-561

Trinchant J-C, Yang Y.S., Rigaud J (1998) Proline accumulation inside symbiosomes of faba bean nodules under salt stress. Physiol Plant 104: 38-49

Truchet G, Roche P, Lerouge P, Vasse J, Camut S, De Billy F, Promé JC, Dénarié J (1991) Sulphated lipo-oligosaccharide signals of Rhizobium meliloti elicit root nodule organogenesis in alfalfa. Nature 351: 670-673

Tyerman SD, Whitehead LF, Day DA (1995) A channel-like transporter for NH_4^+ on the symbiotic surface of N_2 fixing plants. Nature 378: 629-632

Udvardi MK, Price GD, Gresshoff PM, Day DA (1988) A dicarboxylate transporter on the peribacteroid membrane of soybean nodules . FEBS Lett 231: 36-40

Udvardi MK, Day DA (1989) Electrogenic ATPase activity on the peribacteroid membrane of soybean (Glycine max L.) root nodules. Plant Physiol 90: 982-987

Udvardi MK, Day DA (1990) Ammonia (^{14}C-methylamine) transport across the bacteroid and peribacteroid membranes of soybean (Glycine max L.) root nodules. Plant Physiol 94: 71-76

Udvardi MK, Day DA (1997) Metabolite transport across symbiotic membranes of legume nodules. Annu Rev Plant Physiol Plant Mol Biol 42: 373-392

Udvardi MK, Lister DL, Day DA (1991) ATPase activity and anion transport across the peribacteroid membrane of isolated soybean symbiosomes. Arch Microbiol 156: 362-366

van Brussel AAN, Zaat SAJ, Canter-Cremers HCJ, Wijffelman CA, Pees E, Tak T, Lugtenberg BJJ (1986) Role of plant root exudate and Sym plasmid-localized nodulation genes in the synthesis by Rhizobium leguminosarum of Tsr factor, which causes thick and short roots on common vetch. J Bacteriol 165: 517-522

van Brussel AAN, Bakhuizen R, Van Spronsen PC, Spaink HP, Tak T, Lugtenberg BJJ, Kijne JW (1992) Induction of preinfection thread structures in the leguminous host plant by mitogenic lipooligosaccharides of Rhizobium. Science 257: 70-72

Vanden Bosch KA, Newcomb EH (1988) The occurrence of leghaemoglobin protein in the uninfected instertitial cells of soybean root nodules. Planta 175: 442-451 van der Sande K, Pawlowski K, Czaja I, Wieneke U, Schmidt J, Walden R, Matvienko M, Wellink J, van Kammen A, Franssen H, Bisseling T (1996) Modification of phytohormone response by a peptide encoded by *ENOD40* of legume and a nonlegume. Science 273: 370-373

van der Weil C, Scheres B, Franssen H, van Lierop MJ, van Lammeren A, van Kammen A, Bisseling T (1990) The early nodulin transcript *ENOD2* is located in the nodule parenchyma (inner cortex) of pea and soybean root nodules. EMBO J 9: 1-7

Vance CP, Gantt JS (1992) Control of nitrogen and carbon metabolism in root nodules. Physiol Plant 85: 266-274

Vance CP, Heichel GH (1991) Carbon in N_2 fixation: limitation or exquisite adaption. Annu Rev Plant Physiol Plant Mol Biol 42: 373-392

Vance CP, Gregerson RG, Robinson SL, Miller SS, Gantt JS (1994) Primary assimilation of nitrogen in alfalfa nodules: molecular features of the enzymes involved. Plant Sci 101: 51-64

Vance CP, Miller SS, Gregerson RG, Samac DA, Robinson DL, Gantt JS (1995) Alfalfa NADH-dependent glutamate synthase: structure of the gene and importance in symbiotic nitrogen fixation. Plant J 8: 345-358

Verma DPS, Hong ZL (1996) Biogenesis of the peribacteroid membrane in root nodules. Trends Microbiol 4: 364-368

Walsh (1995) Physiology of the legume nodule and its response to stress. Soil Biol Biochem 27: 637-655

Walsh KB, Vessey JK, Layzell DB (1987) Carbohydrate supply and N_2 fixation in soybean. The effect of varied daylength and stem girdling. Plant Physiol 85: 137-144

Walsh KB, Canny MJ, Layzell DB (1989) Vascular transport and soybean nodule function: II, a role for phloem supply in product export. Plant Cell Environ 12: 713-723

Wang TL, Hedley CL (1993) Seed mutants in *Pisum*. Pisum Genet 25: 64-70

Waters JK, Hughes BL, Purcell LC, Gerhardt KO, Mawhinney TP, Emerich DW (1998) Alanine, not ammonia, is excreted from nitrogen-fixing soybean nodule bacteroids. Proc Natl Acad Sci USA 95: 12038-12042

Weaver CD, Shomer NH, Louis CF, Roberts DM (1994) Nodulin 26, a nodule-specific symbiosome membrane protein from soybean, is an ion channel. J Biol Chem 269: 1858-1862

Wittenberg JB, Wittenberg BA, Day DA, Udvardi MK, Appelby CA (1996) Siderophore-bound iron in the peribacteroid space of soybean root nodules. Plant Soil 178: 161-169

Wycoff KL, Hunt S, Gonzales MB, VandenBosch KA, Layzell DB, Hirsch AM (1998) Effects of oxygen on nodule physiology and expression of nodulins in alfalfa. Plant Physiol 117: 385-395

Yang GP, Debellé F, Savagnac A, Ferro M, Schiltz O, Maillet F, Promé D, Trilhou M, Vialas C, Lindstrom K, Dénarié J, Promé JC (1999) Structure of the *Mesorhizobium huakuii* and *Rhizobium gaelgae* Nod factors: a cluster of phylogenetically related legumes are nodulated by rhizobia producing Nod factors with α,β-unsaturated N-acyl substitutions. Mol Microbiol 34: 227-237

Yoshioka H, Gregerson RG, Samac DA, Hoevens KCM, Trepp G, Gantt JS, Vance CP (1999) Aspartate aminotransferase in alfalfa nodules. Mol Plant-Microbe Interact 12: 263-274

Young JPW (1992) Phylogenetic classification of nitrogen-fixing organisms. In: Biological nitrogen fixation. Stacey G, Burris RH, Evans HJ (eds) Chapman and Hall, New York pp. 43-86

Zammit A, Copeland L (1993) Immunocytochemical localisation of nodule-specific sucrose synthase in soybean nodules. Aust J Plant Physiol 20: 25-32

Nitrogen Acquisition and Assimilation in Mycorrhizal Symbioses

Francis MARTIN[1], Jean-Bernard CLIQUET[2] and George STEWART[3]

The role of mycorrhizal symbioses as the normal nutrient and water-absorbing organs of the vast majority of species of vascular plants is widely recognised (Smith and Read 1997). The mutualistic symbiosis between roots and soil-borne mycorrhizal fungi, results in an intricate association where the prospecting and absorbing activities of the web of extraradical mycelium are committed to responding to the trophic needs of the plant. On the other hand, the fungal hyphae hosted by the root are protected from competition with other rhizospheric microbes and, consequently, are preferential consumers of plant carbon assimilates. The development of mycorrhiza involves the differentiation of structurally specialised fungal tissues and interfaces between partners (Peterson and Bonfante 1994; Gianinazzi-Pearson 1996) and results from a cascade of polygenic processes (Gianinazzi-Pearson 1996; Martin et al. 1997). Morphological differentiation culminates with the onset of new metabolic organisations in fungal and plant cells, leading to the finished functioning symbiosis. A highly coordinated metabolic interplay involves bidirectional nutrient exchanges, metabolic zonation, and regulation of enzyme activities and metabolite fluxes (Smith and Smith 1990; Martin and Botton 1993; Botton and Chalot 1999).

Several investigations have demonstrated the key ecological role of mycorrhiza in nitrogen (N) cycling in the forests and agroecosystems (Bowen and Smith 1981; Read 1991, 1995). Since ion-absorbing surfaces of the prospecting fungal mycelium penetrate soil more extensively than the host root itself, a larger proportion of N fluxes in the soil is allocated to the plant. In addition, the turnover rate of organic N matter and compounds in the soil is greater for the fungal mycelium than for the roots (Read 1995; Chalot and Brun 1998). The renewed perception of the ecological importance of mycorrhiza in forests and agroecosystems has revived efforts to elucidate the metabolic processes involved in the absorption, assimilation, transport,

1. Équipe de Microbiologie Forestière, Institut National de la Recherche Agronomique, Centre de Recherches Forestières de Nancy, 54280 Champenoux, France. *E-mail:* fmartin@nancy.inra.fr
2. UA INRA Physiologie et Biochimie Végétales, IRBA, Université de Caen, 14032 Caen Cedex, France. *E-mail:* cliquet@ibba.unicaen.fr
3. The Chemistry Building, The University of Western Australia, Nedlands, WA 6907. *E-mail:* g.stewart@botany.uq.edu.au

and partitioning of N in the symbiosis (Cliquet and Stewart 1993; Martin and Botton 1993; Faure et al. 1998; Botton and Chalot 1999). This chapter provides an overview of the N-acquisition processes and N-assimilating pathways operating in ecto- and endomycorrhizal symbioses.

The Ectomycorrhizal Symbiosis

Ectomycorrhizal trees form vast forests, covering much of Eurasia and North America. Low N availability appears to limit the primary production of forests over much of these ecosystems (Read 1991, 1995; Attiwill and Adams 1993). The environmental concentration and forms of N vary widely and the availability of N to plants is tied to the microbial decay cycle. In temperate and boreal forests, the leaf litter produced by ectomycorrhizal tree species is relatively slow to decompose and thus forms a distinctive layer of acidic organically enriched material at the soil interface (Attiwill and Adams 1993). Acidity, high C:N ratio and seasonality of climate with low temperature and surface drying, are all likely to be major constraints upon mineralisation. Indeed, the rates of N mineralisation of many forests is so slow that N becomes the key growth-limiting element in these forests (Read 1991; Attiwill and Adams 1993; Francis and Read 1994). As a consequence, trees have developed ectomycorrhizal associations (Fig. 1) alongside a wide range of alternative trophic adaptations (e.g. N_2-fixing symbioses, cluster roots), when competing for limited resources of specific nutrients (Pate 1994).

About 90-95% of fine roots of hardwood and conifer species are ectomycorrhizal (Dahlberg et al. 1997). Ectomycorrhizal roots dominate in the top 10 cm of the soil profile (*i.e.* the litter layers of the mor and moder humus), where a large proportion of the nutrient pools of the ecosystem is sequestered in organic form (Read 1995). Mycorrhizal fungi are ideally located for acquisition of nutrients by acting at the interface between the host roots and the surrounding soil and by their extensive network of extramatrical hyphae. The constant supply of photoassimilates by the host plant (Högberg et al. 1999) provides a competitive advantage to mycorrhizal fungi against saprophytic soil microbes. As the ectomycorrhiza is a major nutrient-absorbing organ, it is essential to establish the mechanisms and processes whereby N is mobilised, assimilated and transported within the symbiotic tissues and the associated host plant.

Nitrogen Acquisition

A major contributing factor in N acquisition by ectomycorrhizal trees is the existence and continued growth (0.2 to 0.5 m year^{-1}) of the extramatrical mycelium into soil horizons. It provides the contact between the root system and the soil resources which may be discontinuously distributed over large distances from the root (103 to 105 cm cm^{-1} of root; Francis and Read 1994). This extramatrical mycelium thus fulfils the essential role of foraging and exploitation of the nutrient resources of the soil and their translocation to the ectomycorrhizal mantle where storage is taking place. Extramatrical hyphae extend far beyond the depletion zone surrounding the root and thus improve root exploitation of a given soil volume. This hyphal network extends through humified forest soils as a prospecting fan, which provides optimal

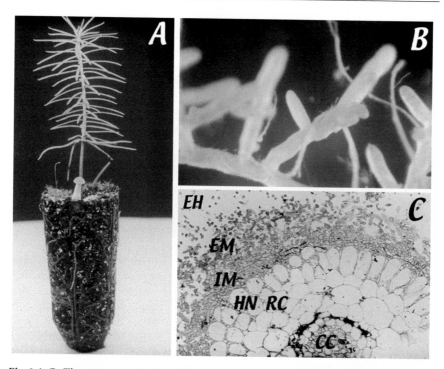

Fig. 1 A-C: The ectomycorrhizal symbiosis. **A.** A seedling of Douglas fir (*Pseudotsuga menziesii*) colonised by the ectomycorrhizal basidiomycete *Laccaria bicolor*. The fungal mycelium has developed ectomycorrhizas on the root system and has produced a basidiocarp above ground. (Photograph courtesy on P. Klett-Frey). **B.** Short roots of spruce (*Picea abies*) ensheathed by the ectomycorrhizal fungus *Amanita muscaria*. (Photograph courtesy J. Garbaye). The white mantle covers the roots and rhizomorphs and extends in to the medium. **C.** Transverse section of a *Eucalyptus grandis-Pisolithus tinctorius* ectomycorrhiza showing the external (*EM*) and internal (*IM*) mantles; the fungal hyphae have begun to penetrate between the epidermal cells of the root cortex (*RC*) to form the Hartig net (*HN*). Epidermal cells are radially enlarged. Extramatrical hyphae (*EH*) are exploring the medium. (Photograph courtesy B. Dell).

efficiency in foraging the patchily distributed resources. Intensive development of mycelium, as fungal mats, is induced by the occurrence of litter organic residues (Bending and Read 1995). Mycorrhizal roots and fungal mats may accumulate in organic residues in order to mobilise N and several investigations are underway to test this hypothesis (Francis and Read 1994; Bending and Read 1995; Read 1995; Colpaert and Van Tichelen 1996). Degradation of organic N residues gives rise to free amino acids and through microbial processes to inorganic N forms, i.e. NH_4^+, NO_3^-. Most fungi are able to break down a large set of compounds (inorganic N, amino acids, peptides, proteins) as N sources by synthesising specific catabolic enzymes (Marzluf 1997). To limit the expression of the genes coding for these enzymes (e.g. proteases) to situations of N limitation, this expression is often controlled by a global N regulatory system that activates the expression of these genes. Assuming that ectomycorrhizal species behave as model fungi (e.g. *Aspergillus nidu-*

lans, Neuropora crassa). When hyphae are progressing through soil regions where readily metabolisable N sources, such as NH_4^+ and glutamine, are available, signals of N sufficiency are released that curtail the metabolism of secondary N compounds (e.g. proteins) (Marzluf 1997). Ectomycorrhizal fungi can contribute to N nutrition of the host plant in two ways: (1) conversion of litter and soil complex N sources into forms which are more readily utilised by the root and (2) absorption, assimilation and translocation of inorganic N compounds from soil to the root. It is likely that different set of enzymes (proteases vs. NH_4^+-assimilating enzymes) are simultaneously expressed in the different mycelial zones of the extensive mycorrhizal web prospecting the various horizons of the soil. Investigations are currently underway to decipher this physiological heterogeneity (Cairney and Burke 1996) within fungal mycelia and some recent data are discussed in the next sections.

Utilisation of Organic Nitrogen

It is generally accepted that 5 to 10% of the soil N is in the form of NH_4^+ and NO_3^-, and the remainder is bound organic forms such as proteins, peptides and free amino acids (Read 1991). In the absence of mycorrhizal associations, trees such as pine, spruce and birch failed to grow either on amino acids or peptides when grown aseptically. In contrast, most ectomycorrhizal fungi readily assimilate amino acids, such as glutamine, glutamate and alanine, which predominate in soil solution, and peptides released from protein degradation (Read 1991, 1995; Chalot and Brun 1998). Arctic strains appeared to be preadapted to the utilisation of seed protein N and glutamic acid N, which is often released in high amounts after soil freezing (Tibbett et al. 1998). This ability to absorb and assimilate amino acids and peptides is retained in the symbiotic state and is likely to be of considerable ecological significance (Fig. 2; Read 1991, 1995; Colpaert and Van Tichelen 1996; Chalot and Brun 1998). However, large inter- and intraspecific variabilities in the ability to use these organic N sources have been noticed within ectomycorrhizal fungi. This may have an impact on the distribution of ectomycorrhizal species through the different horizons of the humified forest soils.

The utilisation of exogenous proteins by ectomycorrhizal fungi requires the enzymatic degradation of proteins to peptides and then amino acids before cellular uptake. A large range of protease activities (*e.g.* serine type) is secreted by ericoid and ectomycorrhizal fungi grown on animal proteins as a substrate (Ramstedt and Söderhäll 1983; Abuzinadah and Read 1986; Zhu et al. 1990; Read 1995). However, protein-enriched fractions purified from forest litter induced a greater secreted proteolytic activity than did commercial preparations of albumin or gelatin (Botton and Chalot 1999). It is evident from these investigations that the extracellular proteinases of ectomycorrhizal and ericoid fungi have an important role in protein N utilisation, as indicated by the close associations among enzyme activities, substrate assimilation and biomass growth. Competition for protein N between saprophytic and ectomycorrhizal fungi is presumably very tight, but symbiotic fungi are favoured by the continuous allocation of carbon from the host plant, which is required to secure the capture of this N form through the synthesis of proteolytic enzymes.

Recent studies by a number of groups have attempted to use the variation in ^{15}N natural abundance to identify the sources of N utilised by species of different natural communities (Högberg 1997). A consistent difference appears in the $\delta^{15}N$ signatures of mycorrhizal and non-mycorrhizal species of different plant communities.

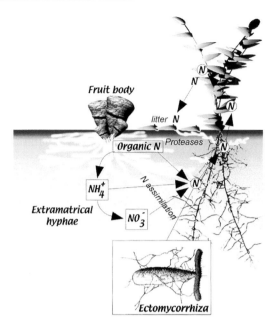

Fig. 2 : The role of the ectomycorrhizal symbiosis in N cycling. Organic and inorganic N produced by the litter decay are taken up by the extramatrical hyphae of the ectomycorrhizal fungi. Symbiotic fungi could play a role in the litter degradation by secreting proteases. Inorganic N is assimilated in extramatrical hyphae and translocated as amino acids to the ectomycorrhizal roots, where they are stored and delivered to the host plant. Amino acids are then translocated to the growing shoot. During fruit body formation, a large amount of amino acids are mobilised and transported to the developing tissues

The leaf tissue of mycorrhizal plants, particularly those in association with ecto-, ericoid or epacrid mycorrhizal fungi have $\delta^{15}N$ values lower than the leaves of non-mycorrhizal species (Högberg 1990; Schmidt and Stewart 1997). The activity of mycorrhizal fungi has the potential to influence plant $\delta^{15}N$ in a number of ways. As stressed above, they can assimilate and transport large amounts of N and enhance N uptake by host plants. In addition, ectomycorrhizal fungi enable host plants to utilise organic forms of N (Read 1991, 1995; Näsholm et al. 1998; Botton and Chalot 1999) and mycelial activity may be concentrated in pockets rich in organic nitrogen (Read 1991). As suggested earlier, mycorrhizal infection may broaden the N source of the host to include a range of inorganic and organic compounds, that may have potentially distinct $\delta^{15}N$ signatures. Michelson et al. (1996) suggest that mycorrhizal species that have low $\delta^{15}N$ values utilise organic N in fresh litter or N in recalcitrant organic matter. The capacity of most ecto- and ericoid mycorrhizal fungi to hydrolyse protein is consistent with this suggestion (see above). However, evidence is lacking to support the view that different soil N sources carry different $\delta^{15}N$ values (Schmidt and Stewart 1997).

Ectomycorrhizal sheaths and fruiting bodies found in European forest systems are ^{15}N-enriched relative to the root and leaf $\delta^{15}N$ values of their higher plant associates (Gebauer and Dietrich 1993; Högberg et al. 1996; Gebauer and Taylor 1999). Similarly, the fine roots of ericoid and ectomycorrhizal species in subtropical heathland were shown to be enriched in ^{15}N (Schmidt and Stewart 1997). Such observations suggest that the depleted ^{15}N signature exhibited by the leaf tissue of mycorrhizal plants may arise as a consequence of fungal N assimilation. If the fungal partner is the predominant site of N assimilation, then discrimination against ^{15}N will occur at the point of N transfer from fungal to plant partner. The result will be enrichment of the fungal tissues and depletion of plant tissues. Differences among species would then reflect the extent to which N assimilation occurs in the fungal components of the association. ^{15}N natural abundance determinations might then provide a means of quantify the contribution of the fungal associates in nitrogen acquisition and assimilation by the association.

In conclusion, most ectomycorrhizal fungi are able to degrade litter proteins and efficiently use the released N. This ability is retained in the symbiotic state and allows the host trees to grow and develop on organic N. This beneficial effect of the ectomycorrhizal associations is most likely of great ecological significance in soils where most of the N is in organic forms, such as in many eucalypt and conifer forests and boreal ecosystems. However, very little investigation has been carried out with intact ectomycorrhizal systems colonising organic substrates that occur in forest ecosystems (Bending and Read 1995; Colpaert and Van Tichelen 1996). There are scant data on either the relative importance of the various organic N sources to mycorrhizal plants or on the contribution of organic N to the total N input of trees (Marschner and Dell 1994). The importance of the saprophytic abilities of ectomycorrhizal fungi for N cycling in forest ecosystems remains to be established (Read 1995).

Nitrate and Ammonium Uptake

Many of the ectomycorrhizal trees are members of closed climax communities, in which low rates of nitrification result in NH_4^+ ions being the predominant available form of inorganic N. Although the NO_3^- level in forest soils is usually low, it is, however, present and its level is increasing as a result of atmospheric deposition (Schulze 1989). The molar NH_4^+ to NO_3^- ratio varies with seasons and precipitation, but it is usually within the range 10 to 1. As a consequence, NH_4^+ ions appear to be the form of N generally utilised by most conifers and hardwood species of temperate ecosystems (Smirnoff et al. 1984; Andrews 1986). The assimilation of NO_3^- and NH_4^+ ions as sole N sources for plant growth carries different potential costs with respect to the reducing power, energy and water requirements, with NH_4^+ being the more cost-effective N source. The lower energy costs incurred in NH_4^+ assimilation may be particularly important with respect to the growth of understorey shrubs and tree species growing in a water- and light-limited environment (Stewart et al. 1989). Non-mycorrhizal trees and free-living ectomycorrhizal fungi generally incorporate NH_4^+ more efficiently than NO_3^- (Plassard et al. 1986); this feature probably reflects an adaptation of these species to NH_4^+ – rich environments. However, this contention has been challenged (Kronzucker et al. 1997). In any case, ectomycorrhizal trees efficiently incorporate NH_4^+ ions (Rygiewicz et al. 1984). The K_m values for NH_4^+ uptake by ectomycorrhizal fungi are within the range of NH_4^+ concentration measured in forest soils (5 to 50 μM) (Jongbloed et al. 1991). Hence, extramatrical mycelia of ectomycorrhizal fungi are likely to be extremely active scavengers of inorganic N in soils. A gene coding for a NH_4^+ transporter, similar to the yeast *MEP1* and *MEP3*, has been cloned from *Paxillus involutus* (M. Chalot, pers. comm.). Investigations on the regulation of this gene will highlight the mechanisms involved in N uptake in the symbiosis.

Most conifer species grow naturally on soils poor in NO_3^- and rich in NH_4^+, as stressed above. However, after forest disturbance, such as fire or clearcutting, novel microbial populations which convert organic N to NO_3^-, appear (Kronzucker et al. 1997). Several groups of woody species, including gymnosperms and Ericaceae, could easily adapt to this new N environment since they have a potential to assimilate NO_3^- in their root systems (Smirnoff et al. 1984; Andrews 1986; Stewart et al. 1989). In contrast to the large body of investigations on herbaceous species, our knowledge of NO_3^- uptake kinetics in woody species is scarce, in spite of the important ecological and economic importance of such species. Only a limited number of studies have been reported (Peuke and Tischner 1991; Plassard et al. 1994; Kronzucker et al. 1995). Ectomycorrhizal fungi (e.g. *Hebeloma cylindrosporum*) have the potential to use NO_3^- efficiently (Scheromm et al. 1990) and the extramatrical hyphae of several types of *Fagus sylvatica* ectomycorrhizas can rapidly take up and assimilate $^{15}NO_3^-$ (Finlay et al. 1989). The NO_3^- uptake system of the extramatrical hyphæ must be versatile and robust because it has to transport sufficient NO_3^- to satisfy the demand of N for the fungus and the host plant in the face of external nitrate concentrations, that can vary by three orders of magnitude between the different soil microsites. A NO_3^- transporter has been cloned and its regulation characterised (Jargeat *et al.*, pers. comm.).

The root system of an adult tree is associated with up to 100 ectomycorrhizal fungi, which presumably have varied abilities to use different forms of N (NH_4^+, NO_3^-, amino acids, proteins) present in the humified soils. It is therefore tempting to speculate that this cortege of ectomycorrhizal associates allows the host tree to exhaustively capture any type of N occurring in the soil (Fig. 3).

The Pathways of N Assimilation in Ectomycorrhiza

The assimilation of NO_3^- into amino acids is best considered in two parts: the NO_3^- reduction to NH_4^+, which is common to plants and fungi, and the two specific pathways that use NH_4^+ as a substrate: the NADP-dependent glutamate dehydrogenase (GDH) and the glutamine synthetase (GS)/glutamate synthase (GOGAT) pathway (see Chap. 2.2).

Nitrate Reduction

The first committed steps in N assimilation are catalysed by nitrate reductase (NADH, EC 1.7.7.1 and NADPH, EC 1.6.6.2; NR) and nitrite reductase (EC 1.7.7.1; NiR). These proteins are involved in the reduction of NO_3^- to NH_4^+. NR and NiR have been measured in a variety of ectomycorrhizal fungi (Sarjala 1990), but only partially characterised from *Hebeloma cylindrosporum* (Plassard et al. 1984a,b). Genes encoding the NO_3^- permease, NR and NiR have been cloned from *H. cylindrosporum* (Jargeat et al., pers. comm.). The sequence of these genes is closely related to those of other fungi and they are associated in a gene cluster. Their expression is induced by high concentration of NO_3^- and on N-depleted medium (Jargeat *et al.*, pers. comm.). To demonstrate the role of the encoded proteins in the symbiosis, gene inactivation studies are currently underway to assess the role of

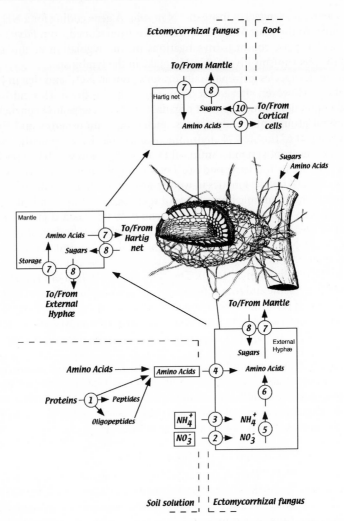

Fig. 3 : The transport of N compounds and sugars at the various interfaces of the ectomycorrhizal symbiosis. Proteins from the litter are degraded by proteases and peptidases (*1*) secreted by ectomycorrhizal fungi and other soil microbes. Amino acids resulting from this protein degradation are taken up from the soil solution by amino acid transporters (*4*) occurring in the plasmamembrane of the external ectomycorrhizal hyphae which permeate the litter. These external hyphae are also able to absorb NO_3^- and NH_4^+ resulting from the litter mineralisation using specific transporters (*2, 3*). Nitrate and nitrite reductases (*5*), NADP-GDH, GS and GOGAT (*6*) assimilate inorganic N and synthesise amino acids. These amino acids (e.g. glutamine, alanine) are transported along the external hyphae to the mantle. Amino acids are then transferred to the mantle (*7*), where they are stored or translocated to the Hartig net. Finally, amino acids are transferred to the plant cells (*9*). At the same time, sugars are transferred from the plant cells (*10*) to the hyphae of the Hartig net, where there are rapidly converted in fungal sugars, such as mannitol and trehalose. These sugars are then translocated to the mantle and the external hyphae to sustain the growth and metabolic activity of the mycobiont

these genes in the NO_3^- acquisition by mycorrhizal pines (Jargeat et al., per. comm.).

Ammonia Assimilation

Although the NADP-GDH is the major pathway of primary N assimilation in several ectomycorrhizal fungi, isotopic experiments are entirely consistent with the exclusive operation of the GS/GOGAT in *Pisolithus tinctorius* and *Paxillus involutus* (Martin and Botton 1993; Turnbull et al. 1995, 1996; Botton and Chalot 1999).

Nitrogen-15 studies showed that NH_4^+ is mostly assimilated by GS in *Cenococcum geophilum*, *Hebeloma* sp. and *Laccaria bicolor* (Martin and Botton, 1993; Chalot and Brun 1998; Botton and Chalot 1999). The prominent role of GS in NH_4^+ assimilation suggested by these isotopic investigations is supported by studies of the purified GS and NADP-GDH of *L. bicolor* mycelium (Brun et al. 1992). GS is a highly abundant protein (3% of the total soluble proteins) with a very high affinity ($K_m = 24~\mu M$) for NH_4^+. The amount and activity of GS are severalfold higher those that of NADP-GDH in the mycelium of *L. bicolor* grown on nitrate and ammonium (Botton and Chalot 1999).

NADP-dependent GDH plays an important position in primary NH_4^+ assimilation in ectomycorrhizal fungi (Martin and Botton 1993). It catalyses either reductive amination of oxoglutarate to yield glutamate, or the reversible oxidative deamination of glutamate to provide NH_4^+. A catabolic role has been assigned to the NAD-specific enzyme (NAD+ oxido-reductase, EC 1.4.1.2.), whereas the NADP-specific enzyme (NADP+ oxidoreductase, EC 1.4.1.4.) has been implicated in glutamate biosynthesis. The NADP-GDH proteins of *Cenococcum geophilum* (Dell et al. 1989) and *Laccaria bicolor* (Brun et al. 1992) have been purified. Considering the K_m values for NH_4^+ and glutamate, and the mycelial NH_4^+ concentration (0.7 to 4 mM; Lorillou et al. 1996), it appears that the NADP-GDH might operate mainly in the direction of glutamate synthesis. NADP-GDH represents about 0.15% of the total soluble proteins (Brun et al. 1992). A cDNA (*gdh-1*) encoding NADP-GDH has been cloned from *L. bicolor* S238 (Lorillou 1995). It shows several regions of high homology with the NADP-GDH of other fungi. The enzyme activity, the protein content and the transcripts of NADP-GDH are increased by 3- to 4.5-fold when the mycelium is transferred from an NH_4^+ – rich medium to either an NO_3^- – rich or a N-depleted medium. (Lorillou 1995; Lorillou et al. 1996). This indicates that the N source regulates the *gdh-1* expression at the transcript level (Lorillou 1995; Lorillou et al. 1996). The expression of the genes coding for the NO_3^- transporter, NR and NiR are similarly regulated in *H. cylindrosporum* mycelium (see above) (Jargeat et al., pers. comm.).

Current data on N assimilation pathways in ectomycorrhizal fungi could be summarised as follows. (1) Absorbed NH_4^+ on entering the mycelium is immediatly incorporated into glutamine. (2) In rapidly growing mycelium, the predominant pathway for NH_4^+ assimilation is the GS/GOGAT pathway; however, the GDH pathway may contribute significantly to synthesis of glutamate and alanine. (3) At low concentrations of NH_4^+, the proportion of N assimilated through the GDH pathway is increased in several ectomycorrhizal species. From these results, we have suggested (Martin et al. 1994) that, as in several non-mycorrhizal fungal species, NADP-GDH and GS/GOGAT are jointly involved in primary assimilation of NH_4^+.

Assimilation Pathways in the Symbiosis

The absorption, assimilation and translocation of N from $^{15}NH_4^+$ and $^{15}NO_3^-$ sources have been followed in observation chambers containing *Pinus sylvestris* (Finlay et al. 1988) or *Fagus sylvatica* (Finlay et al. 1989), presenting an extensive and intact network of extramatrical hyphae. $^{15}NH_4^+$ absorbed by extramatrical hyphae, or resulting from NO_3^- reduction, is rapidly assimilated into glutamate and glutamine. These latter amino acids are then used to synthesize a range of other amino acids (e.g. alanine, arginine and γ-aminobutyrate) within the foraging hyphae. Subsequently, assimilated N is either incorporated into mycelial proteins or translocated to the infected plants, glutamine being considered as the main form of N translocation along the extramatrical hyphæ. These investigations clearly demonstrated that NH_4^+ is assimilated in the fungal hyphæ at considerable distances from the ectomycorrhizal root tips, when an intact hyphal web is present. In natural systems, primary assimilation of NH_4^+ is therefore carried out by the fungal symbiont. N assimilation in root cells is, by inference, mostly involved in the synthesis of glutamine derivatives from the glutamine that is transported from the fungal symbiont. GOGAT and transaminases probably play a major role in these transformations.

Although *in vitro* studies of enzyme activities and ^{15}N tracer studies have pinpointed the major regulatory enzymes of the N-assimilation pathways in a variety of ectomycorrhizas, they have not provided sufficient evidence regarding the exact functional roles of each of these enzymes, nor have these studies shown how the major regulatory enzymes, such as GS, GOGAT and GDH, operate in concert to regulate these complex intermingling pathways. The current picture of the organisation of the enzymes involved in N assimilation is puzzling as a result of the variety of experimental models studied.

Subsequent investigations have revealed a diversity of pathways in which the plant exercises considerable influence over the activities of N-assimilating enzymes of the fungal partner, as well as considerable differences between plant-fungus combinations (Martin and Botton 1993; Botton and Chalot 1999). In ectomycorrhizas involving a conifer (*e.g.* Douglas fir, Norway spruce) as a host plant, the activity of NADPH-GDH in the fungal symbiotic tissues is retained and NH_4^+ is assimilated into amino acids through the cooperative activity of the fungal NADP-GDH and GS. In most hardwood ectomycorrhizas, the activity and amount of the fungal NADP-GDH, if any, are very low. For example, the polypeptide is barely detectable in ectomycorrhiza contracted by *Lactarius subdulcis* or *C. geophilum* with beech (Botton and Chalot 1999). As a consequence, primary assimilation of $^{15}NH_4^+$ is mostly carried out by the GS/GOGAT pathway in these systems (Martin et al. 1986; Botton and Dell 1994; Turnbull et al. 1995, 1996), indicating that the expression of the NADP-GDH activity is not only controlled by the host plant, but may also be depend on the fungal species.

It appears that the development and functioning of the symbiosis alter the biosynthesis and distribution of N-assimilating enzymes and has a profound effect on ability of the fungus to assimilate NH_4^+ from the medium (Martin and Botton 1993; Botton and Chalot 1999). The regulation of these fungal enzymes (*e.g.* NADP-GDH) may be controlled by the host plant and/or the novel environmental cues experienced by the partners in symbiotic tissues (Martin et al. 1992). Studies on the regulation of the *L. bicolor* NADP-GDH and *H. cylindrosporum* NR and NiR indicate that the biosynthesis of these enzymes is influenced by the addition of NO_3^- and NH_4^+ and, more generally by the N status of the cell (Lorillou 1995; Lorillou et al. 1996; P Jargeat et al., pers. comm.).

The synthesis of several amino acids in *Saccharomyces cerevisiea, Neurospora crassa* and *Aspergillus nidulans* is coordinately regulated by nutritional conditions. This regulation is termed the general control (GCN) system. The GCN4 transacting nuclear protein, which binds to specific short consensus sequences in the upstream regions of many genes encoding enzymes of amino acid metabolism, is a major determinant of this regulation (Marzluf 1997). Certain N sources, such as NH_4^+, glutamate and glutamine, are preferentially utilised by fungi, and synthesis of the enzymes needed to utilise NO_3^- or other N source occurs only in the absence of the favoured N sources, i.e. synthesis requires the lifting of N catabolite repression. The cellular level of glutamine (or NH_4^+) most likely controls enzyme repression *via* nuclear transacting factors (*NIT2, AREA*) (Marzluf 1997). A cDNA coding for the WD GTPase Cpc2p, which is required for repression of the GCN4 protein activity in yeast in the absence of amino-acid starvation, has been cloned in *Pisolithus tinctorius* and is a member of the family of symbiosis-regulated genes of the *P. tinctorius-E. globulus* ectomycorrhiza (F Martin, unpubl.). Inactivation of the corresponding gene should provide information on the regulation mechanisms controlling N assimilation in ectomycorrhiza.

The Endomycorrhizal Symbiosis

Nitrogen Transport Through Endomycorrhizal Hyphae

It has long been observed that the arbuscular mycorrhizal (AM) fungi promote not only mineral transport from soil to plants, but also increased nutrient transfer between neighbouring plants (Fig. 4; Ames *et al.* 1983; Bethlenfalvay *et al.* 1991; Frey and Schüepp 1992). Studies with compounds containing isotopic N, C or P provided evidence that elements can move from donor to receiver plants sharing, a common network of mycorrhizal mycelium. While nutrient fluxes of sufficient magnitude to cause important changes in plant growth have yet to be demonstrated, the potential implications of this phenomenon in understanding plant community structure have been recognised (Read 1991) and suggest a concept of resource distribution in plant communities optimised by the movement of nutrients along concentration gradients between the AM donor and receiver plants (Tomm et al. 1994). However, the ecological significance of direct transfer for C and N through the hyphal web is still discussed (Robinson and Fitter 1999). Uptake by the receiver plant of the N excreted by the root system of the donor plant appears to be the major mechanism of short-term N transfer between plants and mycorrhizal hyphae and can thus improve the uptake efficiency of the receiver plant (Ames et al. 1984; Hamel et al. 1991; Ibidjen et al. 1996). For example, AM ryegrass takes up more ^{15}N-labelled amino acids than control plants (Cliquet et al. 1997).

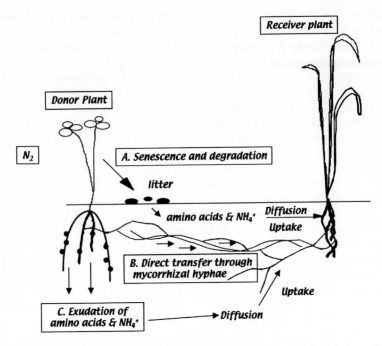

Fig. 4 : Possible pathways of interspecific N transfer. *A.* senescence and degradation of donor plant tissues, followed by reabsorption of resulting N compounds by the receiver plant, *B.* direct transfer through mycorrhizal hyphal bridges, *C.* exudation of N compounds by the donor plant followed by reabsorption by the receiver plant roots colonized by an AM fungus

Absorption of NO_3^- and NH_4^+

Mycorrhizal colonisation enhances plant growth by increasing nutrient uptake via increases in the absorbing surface area, by mobilising sparingly available nutrient sources, or by excretion of chelating compounds or enzymes (Marschner and Dell 1994). In most mycorrhizal types, organic C derived from photosynthesis is transferred in the opposite direction from the plant to the fungus. This bidirectional movement is believed to be the basis for mutualism of the mycorrhizal symbiosis. Arbuscular mycorhizas (AM) are the most common type, occurring in about 80% of plant species (Read 1991, 1995). In this symbiosis the fungal hyphae penetrates the root cell and form vesicles and arbuscules where the bidirectional transfer is thought to occur. Nutrient uptake by the extraradical hyphae from root-distant soil is the most direct contribution of these fungi to plant growth, because the length ratio of AM fungal hyphae to roots in soil can be 100:1 (George *et al.* 1995). The positive effects of the inoculation of AM fungi on plant growth are most generally attributed to the improved uptake by the mycorrhizal roots, of nutrients of low mobility, especially phosphorus (Marschner and Dell 1994). However, there is increasing evidence of a major contribution of AM to plant N uptake (Bago *et al.* 1996; Johansen *et al.* 1996) and assimilation (Cliquet and Stewart 1993).

Bago *et al.* (1996) have obtained evidence of an uptake of NO_3^- by the extraradical mycelium of *Glomus intraradices* grown in monoxenic root cultures, confirming that AM hyphae have the ability to take up NO_3^- and translocate the assimilated N to the host plant (Frey and Schüepp 1992; Johansen $et\ al.$ 1993). Clear quantitative evidence on the uptake of NO_3^-, which is the predominant N form in many agricultural soils, is more difficult to obtain because this ion is much more mobile in soil than PO_4^{3-} or NH_4^+ and is readily transported toward plant roots by mass flow and diffusion. Tobar $et\ al.$ (1994) have, however, provided evidence of higher $^{15}NO_3^-$ uptake by AM $Lactuca\ sativa$ compared to control plants. It appears that this increased uptake was only significant when host plants were submitted to water stress ($i.e.$ diffusion of the ion is reduced).

N Assimilation Pathways in Endomycorrhiza

Although most herbaceous plants are colonised by AM fungi, much current experimental work concerning N metabolism has been carried out with axenic roots and may have little relevance to the performance of root systems in the field. Data concerning the effects of mycorrhizal fungi on N metabolism have been obtained mainly in ectomycorrhizal symbiosis (see above), and a few investigations have been carried out on N assimilation pathways in intact endomycorrhizal systems (Smith $et\ al.$ 1985; Johansen $et\ al.$ 1996). Most of these studies have shown that the colonisation of the plant by an AM fungus does not reduce the N fluxes through the plant NH_4^+ assimilation pathways, but appears to stimulate them. ^{15}N-labelling experiments show that NH_4^+ is exclusively assimilated through the GS/GOGAT cycle. The presence of NR, GS and GOGAT in the AM symbiosis has been demonstrated (Johansen $et\ al.$ 1996). In the associations $Allium\ cepa$ L./$G.\ mosseae.$ and $Zea\ mays$ L./$Glomus\ fasciculatum,$ NH_4^+ is assimilated through the GS/GOGAT cycle (Smith $et\ al.$ 1985; Cliquet and Stewart 1993). These authors have observed higher activities of N assimilatory enzymes, particularly NR and GS, and higher synthesis of ^{15}N amino acids, in AM plants compared to the uninfected controls.

This increase can be the consequence of a direct or indirect contribution of the fungal tissue in the different species and organs studied. Oliver $et\ al.$ (1983) demonstrated an increased NR activity in AM-colonised white clover, attributed to an indirect P-mediated mechanism. In contrast, the higher GS activity found in $Glomus$ $mosseae$ colonised roots of $Allium\ cepa,$ compared to the controls, was attributed to direct contribution of the fungal enzyme (Smith $et\ al.$ 1985). Similarly, the different $\delta\ ^{15}$N values observed in control and AM $Ricinus\ communis$ suggest that the fungus could be the primary site of N assimilation in such a symbiosis (Handley $et\ al.$ 1993; see above). This is consistent with findings by Azcón $et\ al.$ (1992) showing that the positive effect of $Glomus\ fasciculatum$ on NR and GS activity in lettuce leaves could not be attributed to indirect P-mediated mechanisms. The beneficial effects of the mycorrhizal colonisation on NO_3^- assimilation is not mediated via improved phosphate nutrition in AM ryegrass, but is due to an improved N uptake and translocation (Faure et al. 1998).

While evidence has been obtained showing that extraradical hyphae might contribute to the N nutrition of the host, the question of whether or not the intraradical fungal tissue assimilates a significant part of the newly absorbed NH_4^+, and the major form of transfer between the symbionts, is unanswered. It is generally thought that this transfer occurs at the arbuscular interface (Smith and Smith 1990). Determination of ^{15}N enrichment of amino acids into fungal tissue would allow an evaluation of the fungal contribution to the GS/GOGAT cycle. However, the amount of fungal tissue needed for these analyses makes this approach difficult to perform.

Conclusions and Future Prospects

In this chapter we have attempted to highlight the numerous pathways by which organic and inorganic N can be assimilated in endo- and ectomycorrhizas. It is clear that the pathways are many and various, and altered by mycorrhiza development and functioning. Although N metabolism has been most intensively studied in a limited number of mycorrhizal associations, important information is becoming available on a wider set of species. It is also of particular interest to understand how the N concentration of the soil is sensed and how N absorption and assimilation pathways are regulated, both in free-living partners and in symbiotic tissues where source-sink relationships are strikingly modified. Such information will not only contribute to our understanding of basic biochemical and genetic regulatory mechanisms in mycorrhizal systems, but should also provide a better knowledge of N metabolism and cycling in trees and forest ecosystems.

Many questions concerning the metabolic differentiation of plant and fungal symbiotic structures remain unanswered. Regulatory genes controlling the absorption and assimilation of N (e.g. *Gcn4*) have not yet been isolated and the biochemical properties of key N-assimilating enzymes (e.g. GOGAT) are still unknown. Several genes encoding N-assimilating enzymes have, however, been cloned (e.g. NADP-GDH, GS, transaminases) and their experimental manipulation will provide a critical tool to address the question of the co-ordinated regulation of these major enzymes controlling the incorporation of NH_4^+ into amino acids in the free-living mycelium and mycorrhizal tissues. The cloning of the entire set of N-assimilating enzymes in model fungi, such as *H. cylindroporum*, and their potential host plants will be required to address fully the question of the regulation and partitioning of N assimilation pathways in ectomycorrhiza. Using such model systems, it will be possible to thoroughly delineate the mechanisms of genetic regulation of these complex pathways as well as to determine the *in vivo* role of each isoenzyme in the N assimilation pathways distributed in the two interacting symbionts. In addition, the ability to alter the metabolic activities in the fungal partner or its associated plant by using transgenic symbionts will be invaluable in increasing our understanding of the actual regulation of these intermingling pathway in the symbioses.

Acknowledgements. We would like to take this opportunity to thank all those members of our groups who played a part in the development of ideas or gathered experimental data. Of these, Prof. Bernard Botton and Dr. Michel Chalot (University of Henri Poincaré-Nancy I), and Prof. Bernie Dell (Murdoch University, Western Australia) deserve special thanks for their timely suggestions and critical comments.

References

Abuzinadah RA, Read DJ (1986) The role of proteins in the nitrogen nutrition of ectomycorrhizal plants. I. Utilisation of peptides and proteins by ectomycorrhizal fungi. New Phytol 103: 481-493

Ames RN, Reid CP, Porter LK, Cambardella C (1983) Hyphal uptake and transport of nitrogen from two ^{15}N-labelled sources by *Glomus mosseae*, a vesicular-arbuscular mycorrhizal fungus. New Phytol 95: 381-396

Ames RN, Porter LK, St John TV, Reid CPP (1984) Nitrogen sources and "A" values for vesicular-arbuscular and non-mycorrhizal sorghum grown at three rates of ^{15}N -ammonium sulphate. New Phytol 97: 269-276

Andrews M (1986) The partitioning of nitrate assimilation between root and shoot of higher plants. Plant Cell Environ 9: 511-519

Attiwill PM, Adams MA (1993) Nutrient cycling in forests. New Phytol 124: 561-582

Bago B, Vierheilig H, Piché Y, Azcón-Aguilar C (1996) Nitrate depletion and pH changes induced by the extraradical mycelium of the arbuscular mycorrhizal fungus *Glomus intraradices*. New Phytol 133: 273-280

Bending GD, Read DJ (1995) The structure and function of the vegetative mycelium of ectomycorrhizal plants. V. Foraging behaviour and translocation of nutrients from exploited litter. New Phytol 130: 401-409

Bethlenfalvay GJ, Reyes-Solis MG, Camel SB, Ferrera-Cerrota R (1991) Nutrient transfer between the root zones of soybean and maize plant connected by a common mycorrhizal mycelium. Physiol Plant 82: 423-432

Botton B, Chalot M (1999) Nitrogen assimilation: enzymology in ectomycorrhizas. In: Varma AK, Hock B (eds) Mycorrhiza: structure, function, molecular biology and biotechnology. Springer, Berlin Heidelberg New York, pp 325-363

Botton B, Dell B (1994) Glutamate dehydrogenase and aspartate aminotransferase expression in eucalypt ectomycorrhizas. New Phytol 126: 249-257

Bowen GD, Smith SE (1981) The effects of mycorrhizas on nitrogen uptake by plants. Ecol Bull 33: 327-247

Brun A, Chalot M, Botton B, Martin F (1992) Purification and characterization of glutamine synthetase and NADP-glutamate dehydrogenase from the ectomycorrhizal fungus *Laccaria laccata*. Plant Physiol 99: 938-944

Cairney JWG, Burke RM (1996) Physiological heterogeneity within fungal mycelia: an important concept for a functional understanding of the ectomycorrhizal symbiosis. New Phytol 134: 685-695

Chalot M, Brun A (1998) Physiology of organic nitrogen acquisition by ectomycorrhizal fungi and ectomycorrhizas. FEMS Microbiol Rev 22: 21-44

Cliquet JB, Stewart GR (1993) Ammonia assimilation in *Zea mays* L. infected with a vesicular-arbuscular mycorrhizal fungus *Glomus fasciculatum*. Plant Physiol 101: 865-871

Cliquet JB, Murray PJ, Boucaud J (1997) Effect of the arbuscular mycorrhizal fungus *Glomus fasciculatum* on the uptake of amino nitrogen by *Lolium perenne*. New Phytol 137: 345-349

Colpaert JV, Van Tichelen KK (1996) Decomposition, nitrogen and phosphorus mineralisation from beech leaf litter colonized with ectomycorrhizal or litter decomposing basidiomycetes. New Phytol 134: 123-132

Dahlberg A, Jonsson L, Nylund JE (1997) Species diversity and distribution of biomass above and below ground among ectomycorrhizal fungi in an old-growth Norway spruce forest in south Sweden. Can J Bot 75: 1323-1335

Dell B, Botton B, Martin F, Le Tacon F (1989) Glutamate dehydrogenases in ectomycorrhizas of spruce (*Picea excelsa* L.) and beech (*Fagus sylvatica* L.). New Phytol 111: 683-692

Faure S, Cliquet JB, Thephany G, Boucaud J (1998) Nitrogen assimilation in *Lolium perenne* colonized by the arbuscular mycorrhizal fungus *Glomus fasciculatum*. New Phytol 138: 411-417

Finlay RD, Ek H, Odham G, Söderström B (1988) Mycelium uptake, translocation and assimilation of nitrogen from ^{15}N-labelled ammonium by *Pinus sylvestris* plants infected with four different ectomycorrhizal fungi. New Phytol 110: 59-66

Finlay RD, Ek H, Odham G, Söderström B (1989) Uptake, translocation and assimilation of nitrogen from ^{15}N-labelled ammonium and nitrate sources by intact ectomycorrhizal systems of *Fagus sylvatica* infected with *Paxillus involutus*. New Phytol 113: 47-55

Francis R, Read DJ (1994) The contributions of mycorrhizal fungi to the determination of plant community structure. In: Robson, AD, Abbott, LK, Malajczuk, N (eds) Management of mycorrhizas in agriculture, horticulture and forestry. Kluwer, Dordrecht, pp 11-25

Frey B, Schuepp H (1992) Transfer of symbiotically fixed nitrogen from berseem (*Trifolium alexandrinum* L.) to maize via vesicular-arbuscular mycorrhizal hyphae. New Phytol 122: 447-454

Gebauer G, Dietrich P (1993) Nitrogen isotope ratios in different compartments of a mixed stand of spruce, larch and beech trees and of understorey vegetation including fungi. Isotopenpraxis Environ Health Stud 29: 35-44

Gebauer G, Taylor AFS (1999) ^{15}N natural abundance in fruit bodies of different functional groups of fungi in relation to substrate utilisation. New Phytol 142: 93-101.

George E, Marshner H, Jakobsen I (1995) Role of arbuscular mycorrhizal fungi in uptake of phosphorus and nitrogen from soil. Crit Rev Biotechnol 15: 257-270

Gianinazzi-Pearson V (1996) Plant cell responses to arbuscular mycorrhizal fungi: getting to the roots of the symbiosis. Plant Cell 8: 1871-1883

Hamel C, Nesser C, Barantes-Cartin U, Smith DL (1991) Endomycorrhizal fungal species mediate ^{15}N from soybean to maize in non-fumigated soil. Plant Soil 138: 41-47

Handley LL, Daft MJ, Wilson J, Scrimgeour CM, Ingleby K, Sattar MA (1993) Effects of the ecto- and VA-mycorrhizal fungi *Hydnagium carneum* and *Glomus clarum* on the $\partial^{15}N$ and $\partial^{13}C$ values of *Eucalyptus globulus* and *Ricinus communis*. Plant Cell Environ 16: 375-382

Högberg P (1990) ^{15}N natural abundance as a possible marker of the ectomycorrhizal habit of trees in mixed African woodlands. New Phytol 115 : 483-486

Högberg P (1997) ^{15}N natural abundance in soil-plant systems. New Phytol 137: 179-203

Högberg P, Högbom L, Schinkel H, Högberg M, Johannisson C, Wallmark H (1996) ^{15}N natural abundance of surface soils, roots and mycorrhizas in profiles of European forest soils. Oecologia 108: 207-214

Högberg P, Plamboeck AH, Taylor AFS, Fransson PMA (1999) Natural ^{13}C abundance reveals trophic status of fungi and host-origin of carbon in mycorrhizal fungi in mixed forests. Proc Natl Acad Sci USA 96: 8534-8539

Ibibjen J, Urquiaga S, Ismaili M, Alves BJR, Boddey RM (1996) Effect of arbuscular mycorrhizas on uptake of nitrogen by *Brachiria arrecta* and *Sorghum vulgare* from soils labelled for several years with ^{15}N. New Phytol 133: 487-494

Johansen A, Jakobsen I, Jensen ES (1993) Hyphal transport by a vesicular-arbuscular mycorhizal fungus of N applied to the soil as ammonium or nitrate. Biol Fertil Soil 16: 66-70

Johansen A, Finlay RD, Olsson PA (1996) Nitrogen metabolism of external hyphae of arbuscular mycorrhizal fungus *Glomus intraradices*. New Phytol 133: 705-712

Jongbloed RH, Clement JMAM, Borst-Pauwels JK (1991) Kinetics of NH_4^+ and K^+ uptake by ectomycorrhizal fungi. Effect of NH_4^+ on K^+ uptake. Physiol Plant 83: 427-432

Kronzucker HJ, Siddiqi MY, Glass ADM (1995) Compartmentation and flux characteristics of nitrate in spruce. Planta 196: 674-682

Kronzucker HJ, Siddiqi MY, Glass ADM (1997) Conifer root discrimination against soil nitrate and the ecology of forest succession. Nature 385: 59-61

Lorillou S (1995) Contribution à l'étude de la régulation de la glutamate déshydrogénase à NADP chez le basidiomycète ectomycorhizien *Laccaria bicolor* S238N. PhD Thesis, University of Nancy I Nancy

Lorillou S, Botton B, Martin, F (1996) Nitrogen source regulates the biosynthesis of NADP-glutamate dehydrogenase in the ectomycorrhizal basidiomycete *Laccaria bicolor* (Maire) Orton. New Phytol 132: 289-296

Marschner H, Dell B (1994) Nutrient uptake in mycorrhizal symbiosis. Plant Soil 159: 89-102

Martin F, Botton B (1993) Nitrogen metabolism of ectomycorrhizal fungi and ectomycorrhizas. Adv Plant Pathol 9: 83-102

Martin F, Chalot M, Brun A, Lorillou S, Botton B, Dell B (1992) Spatial distribution of nitrogen assimilation pathways in ectomycorrhizas. In: Read D, Lewis D,

Fitter A, Alexander I (eds) Mycorrhizas in ecosystems. CAB International, Wallingford, pp 311-315

Martin F, Côté R, Canet D (1994) NH_4^+ assimilation in the ectomycorrhizal basidiomycete *Laccaria bicolor* (Maire) Orton, a ^{15}N-NMR study. New Phytol 128: 479-485

Martin F, Stewart GR, Genetet I, Le Tacon F (1986) Assimilation of $^{15}NH_4^+$ by beech (*Fagus sylvatica* L.) ectomycorrhizas. New Phytol 102: 85-94

Martin F, Lapeyrie F, Tagu D (1997) Altered gene expression during ectomycorrhiza development. In: Lemke P, Caroll G (eds) The Mycota vol VI. Plant relationships, Springer, Berlin Heidelberg New York pp 223-242

Marzluf GA (1997) Genetic regulation of nitrogen metabolism in the fungi. Microbiol Mol Biol Rev 61: 17-32

Michelson A, Schmidt IK, Jonasson S, Quarmby C, Sleep D (1996) Leaf ^{15}N abundance of subarctic plants provides evidence that ericoid, ectomycorrhizal and non- and arbuscular mycorrhizal species access different sources of soil nitrogen. Oecologia 105: 53-63

Näsholm T, Ekblad A, Nordin A, Giesler R, Högberg M, Högberg P (1998) Boreal forest plants take up organic nitrogen. Nature 392: 914-916

Oliver AJ, Smith SE, Nicholas DJD, Wallace W, Smith FA (1983) Activity of nitrate reductase in *Trifolium subterraneum* L.: effect of mycorrhizal infection and phosphate nutrition. New Phytol 94: 63-79

Pate JS (1994) The mycorrhizal association: just one of many nutrient acquiring specializations in natural ecosystems. In: Robson AD, Abbott LK, Malajczuk N (eds) Management of mycorrhizas in agriculture, horticulture and forestry. Kluwer, Dordrecht, pp 1-10

Pate JS, Stewart GR, Unkovitch M (1993) ^{15}N natural abundance of plant and soil components of a *Banksia* woodland ecosystem in relation to nitrate utilisation, life form, mycorrhizal status and N_2-fixing abilities of component species. Plant Cell Environ 16: 365-373

Peterson RL, Bonfante P (1994) Comparative structure of vesicular-arbuscular mycorrhizas and ectomycorrhizas. Plant Soil 159: 79-88

Peuke AD, Tischner R (1991) Nitrate uptake and reduction of aseptically cultivated spruce seedlings, *Picea abies* (L) Karst. J Exp Bot 42: 723-728

Plassard C, Mousain D, Salsac L (1984a) Mesure *in vitro* de l'activité nitrate réductase dans les thalles de *Hebeloma cylindrosporum*, champignon basidiomycète. Physiol Vég 22: 67-74

Plassard C, Mousain D, Salsac L (1984b) Mesure *in vivo* et *in vitro* de l'activité nitrite réductase dans les thalles de *Hebeloma cylindrosporum*, champignon basidiomycète. Physiol Vég 22: 147-154

Plassard C, Martin F, Mousain D, Salsac L (1986) Physiology of nitrogen assimilation by mycorrhiza. In: Gianinazzi-Pearson V, Gianinazzi S (eds) Physiological and genetical aspects of mycorrhizæ. INRA, Paris, pp 111-120

Plassard C, Barry D, Eltrop L, Mousain D (1994) Nitrate uptake in maritime pine (*Pinus pinaster*) and the ectomycorrhizal fungus *Hebeloma cylindrosporum*: effect of ectomycorrhizal symbiosis. Can J Bot 72: 189-197

Ramstedt M, Söderhäll K (1983) Protease, phenoloxidase and pectinase activities in mycorrhizal fungi. Trans Br Mycol Soc 81: 157-161

Read DJ (1991) Mycorrhizas in ecosystems. Experientia 47: 376-390

Read DJ (1995) Ectomycorrhizas in the ecosystem: structural, functional and community aspects. In: Stocchi V, Bonfante P, Nuti M (eds) Biotechnology of ectomycorrhizae. Molecular approaches. Plenum Press, New York, pp 1-23

Robinson D, Fitter A (1999) The magnitude and control of carbon transfer between plants linked by a common mycorhizal network. J Exp Bot 50: 9-13

Rygiewicz PT, Bledsoe CS, Zasoski RJ (1984) Effects of ectomycorrhizae and solution pH on [^{15}N] ammonium uptake by coniferous seedlings. Can J For Res 14: 885-892

Sarjala T (1990) Effect of nitrate and ammonium concentration on nitrate reductase activity in five species of mycorrhizal fungi. Physiol Plant 79: 65-70

Scheromm P, Plassard C, Salsac L (1990) Regulation of nitrate reductase in the ectomycorrhizal basidiomycete, *Hebeloma cylindrosporum* Romagn., cultured on nitrate and ammonium. New Phytol 114: 441-447

Schmidt S, Stewart GR (1997) Waterlogging and fire impacts on nitrogen availability and utization in a subtropical wet heathland (wallum). Plant Cell Environ 20: 1231-1241

Schulze ED (1989) Air pollution and forest decline in a spruce (*Picea abies*) forest. Science 244: 776-783

Smirnoff N, Todd P, Stewart GR (1984) The occurrence of nitrate reduction in the leaves of woody plants. Ann Bot 54: 363-374

Smith SE, Read DJ (1997) Mycorrhizal symbiosis. Academic Press, San Diego

Smith SE, Smith FA (1990) Structure and function of the interface in biotrophic symbioses as they relate to nutrient transport. New Phytol 114: 1-38

Smith SE, St John BJ, Smith FA, Nicholas DJD (1985) Activity of glutamine synthetase and glutamate dehydrogenase in *Allium cepa* L. and *Trifolium subterraneum* L.: effect of mycorrhizal infection and phosphate nutrition. New Phytol 99: 211-227

Stewart GR, Pearson J, Kershaw JL, Clough ECM (1989) Biochemical aspects of inorganic nitrogen assimilation by woody plants. Ann Sci For 46: 648s-653s

Tibbett M, Sanders FE, Minto SJ, Dowell M, Cairney JWG (1998) Utilisation of organic nitrogen by ectomycorrhizal fungi (*Hebeloma* spp.) of arctic and temperate origin. Mycol Res 102: 1525-1532

Tobar R, Azcón R, Barea JM (1994) Improved nitrogen uptake and transport from N-labelled nitrate by external hyphae of arbuscular mycorrhiza under water-stressed conditions. New Phytol 126: 119-122

Tomm GO, van Kessel C, Slinkard AE (1994) Bi-directional transfer of nitrogen between alfalfa and bromegrass: short-and long-term evidence. Plant Soil 164: 77-86

Turnbull MH, Goodall R, Stewart GR (1995) The impact of mycorrhizal colonization upon nitrogen source utilisation and metabolism in seedlings of *Eucalyptus grandis* Hill ex Maiden and *Eucalyptus maculata* Hook. Plant Cell Environ 18: 1386-1394

Turnbull MH, Goodall R, Stewart GR (1996) Evaluating the contribution of glutamate dehydrogenase and the glutamate synthase cycle to ammonia assimilation by four ectomycorrhizal fungal isolates. Aus J Plant Physiol 23: 151-159

Zhu H, Guo DC, Dancik BP (1990) Purification and characterization of an extracellular acid proteinase from the ectomycorrhizal fungus *Hebeloma crustuliniforme*. Appl Environ Microbiol 56: 837-843

Amino Acid Metabolism

Jean-François MOROT-GAUDRY[1], Dominique JOB[2] and Peter J. LEA[3]

Amino Acids

Amino acids are the building blocks of proteins. They are not only important for cell structure but direct and control the chemical reactions that constitute life processes. Proteins are composed of many copies of 20 amino acids linked in chains by covalent bonds. Thus a 100-unit protein has 20^{100} possible structures. This enormous variability means that cells and organisms can differ greatly in structure and function, even though they are made up of similar macromolecules produced by the same type of reaction.

The monomers that make up proteins are called amino acids because, with one exception, each contains a basic amino group (-NH_2) and an acidic carboxyl group (-COOH). The exception, proline, has an imino group (-NH-) instead of an amino group. All amino acids are constructed according to a general design: a central carbon atom, called the α-carbon (because it is adjacent to the carboxyl group), which is bonded to the amino (or imino) group, the carboxyl group, a hydrogen atom, and to one variable group, called a side chain or R group. This side chain gives the amino acids their individuality. The 20 common amino acids normally occurring in proteins are classified according to whether their side chains are basic, acidic, uncharged polar or non polar. These 20 amino acids are given either three-letter or one-letter abbreviations; thus alanine = Ala = A.

The structures of all amino acids except glycine are asymmetrically arranged around the α-carbon, because it is bonded to four different atoms or groups of atoms (- NH_2, -COOH,-H and-R). Thus all amino acids, except glycine, have at least two stereo-isomeric forms. By convention, these mirror-image structures are called the D and the L forms of amino acids. With very rare exceptions, only the L forms of amino acids are found in proteins.

1. Unité de Nutrition Azotée des Plantes, INRA, route de Saint Cyr, 78026 Versailles France.
   E-mail: morot@versailles.inra.fr
2. UMR 1932 CNRS/AVENTIS CropScience, 14-20 rue Pierre Baizet, 69263 Lyon Cedex 9 France.
   E-mail: dominique.job@rp.fr
3. Department of Biological Sciences, Lancaster University, Lancaster LA1 4YQ, United Kingdom.
   E-mail: p.lea@lancaster.ac.uk

At typical pH values in cells, the amino and carboxyl groups are ionised as NH_3^+ and $-COO^-$. The side chains of some amino acids are also highly ionised and therefore charged at neutral pH. Amino acids are acidic because they have a carboxyl group that tends to dissociate to form the negatively charged $-COO^-$ carboxylate ion. The amino group is a base, because it can take up a hydrogen ion to form the positive $-NH_3^+$ amino ion. The degree to which a dissolved acid releases hydrogen ions or to which a base takes them up, depends partly on the pH of the solution. In neutral solutions, amino acids exist predominantly in the doubly ionised form. Such a dipolar ion is called a zwitterion. In solutions at low pH, carboxylate ions recombine with the abundant hydrogen ions, so that the predominant form of amino acid molecule is $^+H_3N-R-CH-COOH$. At high pH, the scarcity of hydrogen ions decreases the chance that an amino group or a carboxylate ion will pick up a hydrogen ion, so that the predominant form of an amino acid molecule is $H_2N-R-CH-COO^-$.

At neutral pH, arginine and lysine are positively charged, histidine is also positively charged, but only weakly, and aspartic and glutamic acids are negatively charged. Serine and threonine, whose side chains have an $-OH$ group, can interact strongly with water by forming hydrogen bonds. The side chains of asparagine and glutamine have polar amide groups with even more extensive hydrogen-bonding capacities. These nine amino acids constitute hydrophylic, or polar amino acids. The side chains of several other amino acids, alanine, isoleucine, leucine, methionine, phenylalanine, tryptophan, and valine consist only of hydrocarbons, except for the sulphur atom in methionine and the nitrogen atom of tryptophan. These seven non polar amino acids are hydrophobic; their side chains are only slightly soluble in water. Tyrosine is also strongly hydrophobic because of the presence of a benzene ring, but the hydroxyl group allows it to interact with water, making its properties somewhat ambiguous.

Cysteine plays a special role in proteins, because the $-SH$ group allows it to dimerise through an $-S-S-$ bond to a second cysteine to give cystine. When the SH remains free, cysteine is quite hydrophobic. Two other special amino acids are glycine and proline. Glycine has a hydrogen atom as the R group, thus it is the smallest amino acid and has no special hydrophobic or hydrophylic character. Proline, as an imino acid, is very rigid and creates a fixed link in a polypeptide chain, proline is also quite hydrophobic.

The chemical bond that connects two amino acids in a polymer is the peptide bond. It is formed between the α-amino group of the amino acid and the α-carboxyl of another one. This reaction, called condensation, liberates a water molecule. Because the carboxyl carbon and oxygen atoms are connected by a double bond, the peptide bond between C and N exhibits a partial double-bond character. A single linear array of amino acids connected by peptide bonds is called a polypeptide. If the polypeptide is short (i.e. fewer than 30 amino acids long), it may be called an oligopeptide or peptide. Polypeptides in plant cells differ greatly in length. They generally contain between 40 to 1000 amino acids. Each polypeptide has a free amino group at one end (N terminus) and a free carboxyl group at the other end (C terminus). The polypeptides fold up to form fibrous or globular proteins. Proteins give tissues their rigidity or catalyse chemical reactions; these latter proteins are enzymes. The proteins can be modified further by the attachment of additional small molecules (i.e. enzyme cofactors, sugars, phosphate etc.). The amino acids represent the alphabet by which linear proteins are written .

However, a very large number of non-protein nitrogenous compounds are also found in plants. These range from universal nitrogen molecules such as γ-aminobutyric acid, β-alanine, ornithine, hydroxyproline, homocysteine, O-acetylhomoserine,

S-adenosylmethionine, *S*-adenosylhomocysteine and glutathione to some well-identified substances that occur widespread, but not universally (homoserine and pipecolic acid), or to those that occur infrequently in often highly specific but seemingly unpredictable situations (azetidine-2-carboxylic acid and canavanine). Still other substances are present infrequently and apparently at random (γ-methyleneglutamine). There are others that are still regarded as extremely rare in plants (hypoglycine A). Some non-protein nitrogen compounds are considered as intermediates of synthesis. This is the case for homoserine, ornithine, citrulline, diaminopimelate, *O*-acetylserine, cystathionine, β-cyanoalanine and Δ^1-pyrroline-5-carboxylic acid. Some others are products of breakdown of proteins and nucleic acids (γ-aminobutyric acid, β-alanine, β-aminoisobutyric acid). Homoserine is involved, for example, in nitrogen transport in peas; it is also the case in γ-methyleneglutamine, some amides and ureides. Indoleacetic acid, azetidine-2-carboxylic acid and N-formylmethionine are considered as metabolic or growth regulators. Some amino acids are toxic to animals and found in seeds (canavanine in legumes, for example); so they appear to serve both as a storage reserve and as a feeding deterrent to herbivores. Non-protein amino acids generally have a wide range of function in plants. They are involved in the transportation of nitrogen between roots, leaves and harvested organs (fruits, seeds, etc.), and are precursors in the synthesis of chlorophyll and enzyme cofactors (biotin, thiamine pyrophosphate, and coenzyme A). Amino acids are also involved in the production of some secondary natural products such as alkaloids, phenolic acids, etc. (Fowden 1981).

The 20 common amino acids are of central biological importance as they are the constituents of proteins. Considering this importance, it is surprising that mammals can synthesise only ten amino acids by de novo pathways; the remainder, called the essential amino acids, must be supplemented in their diet. Higher plants and many bacteria and fungi possess the ability to synthesise all of the protein amino acids.

By the 1970s, the development of laboratory techniques, particularly in the areas of biochemistry and enzymology, and in the use of labelled precursors (chromatography and all sophisticated and physical techniques of analysis) and instrumentation (automation and miniaturisation), had provided a good understanding of the amino acid biosynthetic pathways in bacteria. However, in plants, the pathways generally accepted for the synthesis of amino acids have been assumed to be the same as those in bacteria and it is only relatively recently that these have been confirmed, or novel pathways demonstrated in plants. However, the synthesis of some amino acids in plants still remains unclear, as is the case with histidine, lysine and many a non-protein amino acids. Moreover, elucidating the regulation of the amino acid biosynthetic mechanisms of plants has come up against some difficulties. While several enzymes involved in the pathways are catalytically similar between plants and bacteria, the mechanisms regulating gene expression and activity of these enzymes are quite distinct. In plants, a number of amino acids can be synthesised by several pathways, depending on the tissue concerned, stage of growth and development and environmental conditions. Consequently, in plants, the branched and the interwoven nature of the pathways for amino acid synthesis require strict and precise control. The carbon and nitrogen destined for one particular amino acid do not all end up in this single amino acid, but must be partitioned between different amino acids, in response to the needs of the plant organs at a particular time of development. Thus, the metabolic pathways of amino acids have to maintain a harmonious balance in the partitioning of C and N fluxes between each amino acid pool. This requires, in a defined manner, a very strict regulation of enzyme activities. Enzymes are regulated to different degrees. For many amino acid biosynthetic pathways, for example, end-product inhibition appears to be the principal mechanism by which amino acids are regulated. There are, however, reported cases where the accumulation of amino acids is regulated at the level of gene expression or by a balance between biosynthesis and degradation.

Research on amino acid metabolism in plants has been complicated by the fact that some of the metabolic reactions are catalysed by multiple isoenzymes located in distinct subcellular compartments. With traditional biochemical approaches, it has been impossible to sort out the function of each isoenzyme. Traditional biochemical methods have involved tissue disruption, which artificially mixes isoenzymes that may not coexist in the same cell types *in vivo*. Moreover, the *in vitro* biochemical methods commonly used to define the controlling enzyme in a pathway in unicellular organisms may lead to erroneous interpretations when employed to study plant metabolic pathways. An alternative way to define the *in vivo* or *in planta* function of a particular isoenzyme is by mutant analysis. Plant mutants, defective in particular isoenzymes in biosynthetic amino acid pathways, have been identified and characterised in *Arabidopsis thaliana*, barley, maize, tobacco, pea, soybean etc., and have greatly added to our understanding of amino acid biosynthetic regulatory mechanisms in plants (Lea and Forde 1994). However, generating mutants is not straightforward for polyploid plants and for specific members of multigene families. The development of molecular genetics and efficient plant transformation techniques has provided transgenic plants containing altered levels of an individual amino acid biosynthetic enzyme in a cell-and tissue-specific manner. Consequently, the reductionist approach that has prevailed recently in molecular biology and has led to the great accumulation of factual minutiae about molecular and cellular mechanisms, now needs to be balanced by the holistic vision of integrated whole-plant biology.

Research in this field is still in its infancy and while we have not made great strides in crop improvement, analysis of the different transformants, with both sense and antisense constructs, has improved our understanding of the regulation of amino acid metabolism in plants. This understanding will also have an impact on genetic engineering strategies in applied research, such as increasing the potential yield, or improving nutritional protein quality.

Because amino acids are derived from a single "head" amino acid, they can be grouped together into different families. Conversely, because more than one amino acid may be involved in the synthesis of another, a single amino acid may be assigned to more than one family. This chapter will be confined to the 20 amino acids normally incorporated into proteins.

Amino acids are synthesised either in roots, leaves, seeds, or in fruits, depending on the sites of nitrate reduction and nitrogen remobilisation. For example, in legumes, much of the nitrate is reduced and converted to organic form (asparagine and glutamine) in the roots prior to transport in the xylem. In developing cereal leaves, nitrate transported in the xylem from the roots is reduced to ammonia and then incorporated into amino acids. Senescing leaves hydrolyse a lot of their proteins and degrade other nitrogenous molecules for transport to the developing fruits and seeds, which receive most of their nitrogen in the form of amino acids translocated in the phloem. Synthesis and metabolism of amino acids may require that carbon skeletons such as α-oxo acids be withdrawn from glycolysis, and that the citric acid and pentose phosphate pathways (reductive and oxidative) replenish the amino acids pools.

Ammonia Assimilation and Transamination

Ammonia assimilation

As described in Chapter 2.2, it appears that nearly all plants assimilate ammonia first into glutamine, owing to the high affinity of glutamine synthetase (EC.6.3.1.2) for ammonia with the concurrent hydrolysis of one molecule of ATP.

$$\text{L-Glutamate} + NH_3 + ATP \quad \rightarrow \quad \text{L-glutamine} + H_2O + ADP + P_i.$$

The ammonia assimilated into the amide group of glutamine is then transferred to the α-oxo position of α-oxoglutarate to form glutamate via the reaction of glutamate synthase NAD(P)H-dependent (EC 1.4.1.14) or ferredoxin dependent (EC 1.4.7.1). This amination reaction requires the oxidation of one molecule of NAD(P)H or equivalent.

$$\text{L-Glutamine} + \alpha\text{-oxoglutarate} + 2H \quad \rightarrow \quad 2 \text{ L-glutamate}.$$

Thus, glutamine synthetase and glutamate synthase are able to act in conjunction with the net production of one molecule of glutamate (the GS/GOGAT cycle) (Ireland and Lea 1999). A similar result is achieved by the reversible enzyme glutamate dehydrogenase (EC 1.4.1.2), which can also produce glutamate from α-oxoglutarate and ammonia, with the oxidation of only one molecule of NADH (Lea and Ireland 1999).

$$\text{L-Glutamate} + H_2O + NAD^+ \quad \leftarrow \quad \alpha\text{-oxoglutarate} + NH_3 + NADH + H^+.$$

Other pathways that have been implicated in ammonia assimilation, include alanine dehydrogenase (EC 1.4.1.1), aspartate dehydrogenase (EC 1.4.3.1), and aspartase (EC 4.3.1.1). Alanine dehydrogenase catalyses the amination of pyruvate to alanine in presence of NAD(P)H. In a similar reaction, aspartate dehydrogenase aminates oxaloacetate to aspartate. Aspartase is another ammonia-assimilating enzyme which catalyses the addition of ammonia to fumarate, yielding aspartate. These latter enzymes have been detected in a few plants, but only the GS/GOGAT is thought to make a significant contribution to ammonia assimilation in plant tissues (Lea and Ireland 1999).

Transamination reactions

These are central to all areas of amino acid metabolism. These reactions result in the redistribution of nitrogen from glutamate to other amino acids. Aminotransferases also known as transaminases, catalyse the transfer of an amino group from the α-carbon of an amino acid to the α-carbon of an α-oxo acid, producing a new amino acid and a new oxo acid (Givan 1980 ; Lea and Ireland 1999).The transamination reactions are considered as an oxidative deamination of the donor (primary amino acid or amine), linked with the reductive amination of the acceptor (α-oxo acid or an aldehyde). Although the aminotransferase reaction might be regarded as oxidoreduction with no net change, the unique and distinctive factor is the transfer of the amino group.The aminotransferases transfer nitrogen from one compound to another in a system described as a donor/acceptor group transferase. The aminotransferases have a tightly bound coenzyme: pyridoxal-5-phosphate. This coenzyme accepts an amino group from the amino acid substrate, becoming aminated to pyridoxamine phosphate. The oxo acid thus produced is released and the aminated form of the coenzyme then undergoes a reversal of the process, giving up the newly acquired amino group to another oxo acid substrate to produce a new amino acid product.

The plant aminotransferases can catalyse the transamination of all of the common protein amino acids, except proline, which is an imino acid and thus has no primary amino group available for transamination. It appears that aminotransferases are not highly specific, and may react with a number of amino acids or oxo acids. In plants, many aminotransferases can use glutamate as amino donor in the synthesis of a wide range of amino acids. The extensive use of glutamate results in the production of large amounts of α-oxoglutarate, which can be reaminated by the GS/GOGAT cycle during ammonia assimilation. Aminotransferases do not exhibit any specific regulatory properties. Their activities are solely dependent on pH and substrate/product concentrations (Ireland and Lea 1999 ; Lea and Ireland 1999).

Aminotransferases serve to redistribute the nitrogen to a range of amino acids as glutamate, aspartate and alanine, which then provide nitrogen for the synthesis of the other amino acids. They contribute to the maintenance of relatively stable amino acid pools. In the plant, for example, aspartate : α-oxoglutarate aminotransferases (EC 2.6.1.1) can catalyse the amination of oxaloacetate to yield aspartate. Alanine : oxoglutarate aminotransferase (EC 2.6.1.2) is the enzyme catalysing the amination of pyruvate to alanine. In the peroxisome, a glutamate : glyoxylate aminotransferase (EC 2.6.1.4) ensures the transfer of nitrogen group from glutamate to glyoxylate to form glycine. Aminotransferases are also implicated in the transfer of many amino acids that are often present in millimolar concentration in the vacuole. These aminotransferases have functions that are unique in leaves, such as those involved in carbon assimilation and other processes requiring the shuttling of metabolites during C_4 photosynthesis and photorespiration (Ireland and Lea 1999 ; Lea and Ireland 1999).

At the molecular level, A.thaliana has been shown to contain five genes encoding the aspartate aminotransferase isoenzymes localised in distinct subcellular compartments, cytosol, chloroplasts, mitochondria, and peroxisomes (Schultz et al. 1998). A study of the evolutionary relationships of tyrosine, histidinol phosphate and aspartate aminotransferases was carried out by Metha et al. (1989). Among the transferases from several species, it was revealed the existence of 12 amino acid residues that were found to be invariant in all the 16 known sequences of α-aminotransferases.

Synthesis of Amino Acids Derived from Glutamate

Glutamate is the precursor of the amino acids glutamine (as previously described), arginine and proline.

Arginine Synthesis

Arginine is a major nitrogen storage compound in plants, where it may form 40% of the nitrogen in seed protein and 50 to 90% of the soluble nitrogen in trees, flower bulbs and tuberised roots of chicory (Ameziane et al. 1997).

The synthesis of arginine proceeds through ornithine, which is derived from glutamate in an acetylated pathway (Fig. 1). In the synthesis of arginine, glutamate

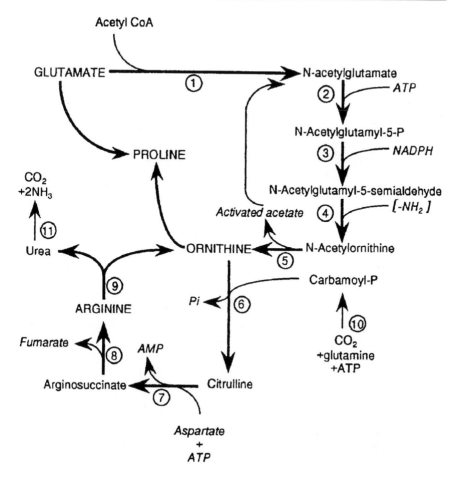

Fig. 1 : The biosynthetic pathway of arginine. Enzymes: *1.* N-acetyl transferase; *2.* N-acetylglutamate kinase; *3.* N-acetylglutamyl-5-phosphate reductase; *4.* N-acetylornithine aminotransferase; *5.* N-acetyl transferase; *6.* ornithine carbamoyl transferase; *7.* arginosuccinate synthetase; *8.* arginosuccinate lyase; *9.* arginase; *10.* carbamoyl phosphate synthetase, *11.* urease.

after acylation is metabolised via phosphorylation, reduction and transamination to the non-protein amino acid, ornithine (Verma and Zhang 1999). The acylation of glutamate to N-acetylglutamyl-5- phosphate, in the presence of acetyl-CoA and ATP, is catalysed by an N-acetyl transferase (acetyl-CoA :glutamate N-acetyl transferase) (EC 2.3.11.1) and N-acetylglutamate kinase (EC 2.7.2.8). Then, two enzymes, N-acetylglutamyl-5-phosphate reductase (EC 1.2.1.38) and N-acetylornithine aminotransferase (EC 2.6.1.11), transform the acetylated intermediate to N-acetylornithine which is converted to ornithine by a N-acetyl transferase (N^2-acetylornithine :glutamate N-acetyl transferase) (EC 2.3.1.35). The acetyl-protecting group can be reused to form N-acetylglutamyl-5-phosphate as described previously (Taylor and Stewart 1981; Micallef and Shelp 1989).

Ornithine reacts with carbamoyl phosphate and yields citrulline. This reaction is catalysed by ornithine carbamoyltransferase (EC 2.1.3.3). Carbamoyl phosphate is formed from glutamine, ATP, and CO_2, via the action of carbamoyl phosphate synthetase (EC 6.3.5.5). Citrulline, in the presence of ATP and aspartate, is converted into arginosuccinate. This reaction is catalysed by arginosuccinate synthetase (EC 4.3.2.1).The arginosuccinate so formed, is subsequently cleaved via the reaction of arginosuccinate lyase (EC 4.3.2.1) to give arginine and fumarate. Then, arginine is broken down by arginase (EC 3.5.3.1) to yield urea and ornithine. Arginase activity was observed to increase in *A. thaliana* seedlings 6 days after germination. The product of hydrolysis, urea, appears to have a function in recycling nitrogen into ammonia via urease (EC 3.5.1.5) (Zonia et al. 1995).

Although there have been a number of studies on the subcellular localisation of arginine-metabolising enzymes in plants, the results have been contradictory. Taylor and Stewart (1981), utilising pea leaf protoplasts, have shown that all the enzymes involved in the metabolism of acetylated intermediates of glutamate, that is glutamate acetyl transferase, carbamoyl phophate synthetase, and ornithine carbamoyl transferase, were localised in the chloroplast, whilst arginosuccinate lyase was present only in the cytoplasm, and arginase was mitochondrial. Arginine is able to regulate the rate of its own synthesis. It inhibits the synthesis of N-acetylglutamate from acetyl-CoA but not from N-acetylornithine. Arginine also appears to control the activity of N-acetylglutamate kinase, another example of an endproduct inhibiting an early reaction in its synthesis.

Glutamate can be decarboxylated into γ-aminobutyrate (GABA).

$$\text{L-Glutamate} + H^+ \rightarrow \text{L-}\gamma\text{-aminobutyrate} + CO_2.$$

This reaction is catalysed by glutamate decarboxylase (EC 4.1.2.15), which can be activated by calcium and calmodulin. GABA is metabolised through a reversible transamination to succinic semialdehyde. This reaction is catalysed by GABA transaminase (EC 2.6.1.19). The product of the reaction is then oxidised to succinate in an irreversible reaction catalysed by succinate semialdehyde dehydrogenase (EC 1.2.1.16). These three reactions constitute a pathway known as the GABA shunt. Evidence indicates that glutamate and GABA are produced during storage protein mobilisation, as a means of recycling arginine-derived N and C. Thus, the GABA shunt could be of considerable importance in the N economy of plants. There is also considerable literature demonstrating the rapid and large accumulation of GABA in response to many stimuli, e.g. cold shock, mechanical stimulation, hypoxia, cytosolic acidification, water stress and phytohormones. GABA accumulation and efflux may play a part in an intercellular signal transduction pathway, leading to the regulation of growth and development (Bown and Shelp 1997; Ireland and Lea 1999).

Proline Synthesis

The accumulation of proline is a striking metabolic response to osmotic stress. Proline has been suggested to function as an osmoticum, as an energy or a reducing power sink, a nitrogen storage compound, a hydroxy-radical scavenger, a compatible solute that protects enzymes, a means of reducing acidity, and a way to regulate cellular redox potentials (Taylor 1996 ; Hua et al. 1997, Hare et al. 1999).

In the first step of proline synthesis, glutamate in the presence of ATP and NAD(P)H is phosphorylated and then reduced to glutamyl-5-semialdehyde via a bifunctional enzyme Δ^1-pyrroline-5-carboxylate synthetase (P5CS) exhibiting both γ-glutamyl kinase (EC 2.7.2.11) and glutamate semialdehyde dehydrogenase activities (EC 1.2.1.41) (Fig. 2). The semialdehyde spontaneously cyclises to give Δ^1-pyrroline-5-carboxylate (P5C) which is then reduced in presence of the NAD(P)H to proline. This reaction is catalysed by the Δ^1-pyrroline-5-carboxylate reductase (P5CR) (EC 1.5.1.2) (Hu et al.1992). Note that the formation of the semialdehyde is similar to the initial stages of ornithine synthesis. However, during ornithine synthesis the presence of the N-acetyl group prevents the non-enzymic cyclisation reaction.

Proline is reported to inhibit its own synthesis though a feedback effect, as the P5CS enzyme is allosterically inhibited by proline (Hu et al. 1992). Proline synthesis probably occurs in the cytosol, and proline degradation in the mitochondria. This pathway of proline synthesis was first established in *Escherichia coli* and *Saccharomyces cerevisiae*, and the corresponding plant genes were identified through the complementation of yeast and bacterial mutants defective in the various steps of the pathway (Delauney and Verma 1993, Verma and Zhang 1999).

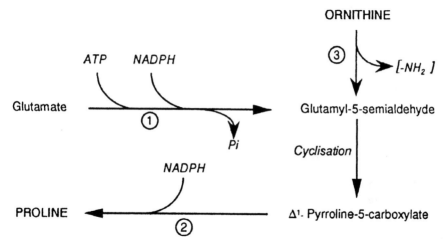

Fig. 2 : The biosynthetic pathway of proline. Enzymes : *1.* Δ^1-pyrroline-5-carboxylate synthetase; *2.* Δ^1-pyrroline-5-carboxylate reductase; *3.* ornithine δ-aminotransferase.

Ornithine can be converted to Δ^1-pyrroline-5-carboxylate via the loss of the δ-amino group by the action of the ornithine δ-aminotransferase (EC 2.6.1.13), as well as to Δ^1-pyrroline-2-carboxylate. There are thus two potential ways of converting ornithine to proline, depending on whether the 2-amino or 5-amino position is transaminated first. Only the route involving the transamination of ornithine into Δ^1-pyrroline-5-carboxylate was strongly indicated by the functional complementation of a defective *E. coli* mutant (Delauney et al.1993).

It was initially suggested that the biosynthesis of proline from glutamate would be enhanced under stress conditions, whereas the ornithine pathway would be inhibited (Delauney et al. 1993). However, Roosens et al. (1998) demonstrated that the ornithine pathway could serve an important role in very young *A. thaliana* plantlets. Under salt-stress conditions, the main task of this pathway is devoted to proline production and contributes, together with the glutamate pathway, to proline accumulation. In later stages of differentiation, the ornithine pathway contributes to a lesser extent to proline biosynthesis. It has been suggested that under normal growth conditions, the role of the ornithine pathway would not be to produce proline, but would be related to the necessity for the plants to synthesise glutamyl semialdehyde via δ-ornithine aminotransferase. Glutamyl semialdehyde is then oxidised to glutamate via Δ^1-pyrroline-5 -dehydrogenase (EC 1.5.1.12). Glutamate can then be used for transamination or to generate new carbon constituents upon entry into the Krebs cycle as α-oxoglutarate.

Less is known about the genes controlling proline catabolism in plants. Proline is oxidised to Δ^1-pyrroline-5-carboxylate by the mitochondrial inner membrane enzyme, proline dehydrogenase (EC 1.5.99.8) (Hare et al. 1999). In plants this enzyme is linked to the respiratory electron transfer system and thus couples proline degradation to ATP formation (Elthon and Stewart 1981). In plants undergoing water deficit, proline degradation in the mitochondria is reduced, whereas proline synthesis is increased in the cytosol. Furthermore, an increase in proline concentration in the phloem of alfalfa under water deficit has been demonstrated, suggesting that increased proline transport may indeed contribute to the ability of the plants to withstand water deficit (Girousse et al. 1996). Recently, Verslues and Sharp (1999) have observed an increase in proline uptake in maize primary roots at low water potentials, which suggests increased proline transport. The molecular mechanisms controlling proline transport during water deficit and salt stress have also been studied in *A. thaliana* (Kiyosue et al. 1996).

Maize mutants, whose growth is limited by the availability of proline, have been isolated by Racchi et al. (1981). The rate of proline synthesis in the maize *Prol 1-1* mutant was not affected, but the rate of proline catabolism seemed to be increased (Dierks-Ventling and Tonelli 1982). The glutamate-synthase-deficient barley mutant which lacks the ability to accumulate proline showed premature symptoms of stress when subjected to a gradual increase in water deficit (Al-Sulaiti et al. 1990).

As proline acts as an osmotic protectant in plants that have been subjected to either drought or high salt-stress, many genetic manipulations have been made to modify the expression of the genes encoding the enzymes involved in proline pathway. Kishor et al. (1995) have shown that the overexpression of P5CS leads to an increase in proline concentration in transgenic tobacco. The higher proline levels correlated with enhanced productivity of the transgenic plants under water deficit conditions. The levels of both P5CS and P5CR are elevated in salt stressed plants (Hu et al. 1992). However, Szoke et al. (1992) have observed that tobacco plants containing a 50-fold enhancement of soybean P5CR, exhibited no significant increase in proline concentration. These results have confirmed the suggestion that P5CR contrary to P5CS, is not a flux-controlling enzyme.

It was observed that high salinity and dehydration resulted in increased accumulation of proline in *A. thaliana* and rice, and have been shown to be accompanied by an increase in the P5CS message level (Verbruggen et al. 1996; Savoure et al. 1997; Strizhov et al. 1997; Igarashi et al. 1997). The expression of the gene encoding P5CS is strongly induced under water deficit and following the addition of NaCl (Hu et al. 1992; Yoshiba et al. 1995). The regulation of expression and structures of two evolutionary divergent genes for Δ^1-pyrroline-5-carboxylate synthetase were studied in tomato by Fujita et al. (1998).

An understanding of the regulation of the proline biosynthesis, transport, and degradation is required for the successful engineering of drought and salinity tolerance in crop plants. It should provide insight into the mechanisms through which proline accumulation contributes to the capacity of plants to survive such stresses.

Synthesis of Amino Acids Derived from Aspartate

Asparagine

Asparagine is almost universally used by higher plants as a storage and transport compound, although glutamine and arginine are also used by some plants (Lea and Ireland 1999). Asparagine synthesis is particularly important in the root nodules of legumes (see Chap. 2.2 and 3.1). Much of the nitrogen fixed by the rhizobium is rapidly transferred to asparagine through the activities of both glutamine synthetase and asparagine synthetase. Thus, much of the nitrogen in legume plants is transported in the xylem, away from the nodule, in the form of asparagine.

The incorporation of ^{15}N into asparagine tends to be much slower than into glutamine, although rapid turnover has been demonstrated (Woo et al. 1982; Ta et al. 1988). Most evidence suggests that ammonia is first incorporated into the amide position of glutamine and is then transferred to aspartate to yield asparagine. Asparagine is synthesised by asparagine synthetase (EC 6.3.5.4) which catalyses the amidation of aspartate by glutamine in an ATP-dependent reaction (Brouquisse et al. 1992; Ireland and Lea 1999).

L-Glutamine + L-aspartate + ATP → L-asparagine + L-glutamate + AMP + PP$_i$.

Asparagine synthetase is present in many plants. The enzyme has been isolated from the cotyledons of germinating seeds, maize roots, and root nodules. Expression of asparagine synthetase is enhanced by low carbohydrate concentration and during senescence. In some cases, there is evidence that the enzyme is able to utilise ammonia. Stulen et al. (1979) have claimed that in maize roots the K_m for ammonia is low enough to allow direct incorporation under certain physiological conditions. In most cases, however, it appears that the *in vivo* substrate for this enzyme is actually glutamine, but not ammonia. Hence, asparagine synthetase does not usually constitute a route for ammonia assimilation in normal conditions (see Hirel and Lea. Chap. 2.2). Asparagine may also be synthesised from the hydrolysis of β-cyanoalanine by β-cyanoalanine hydrolase (EC 4.2.1.65) which is formed from hydrogen cyanide and cysteine, a condensation catalysed by β-cyanoalanine synthase (EC 4.4.1.9). This pathway has been demonstrated in lupin, sorghum, pea and asparagus. This pathway, however, is probably a detoxification mechanism rather than a major pathway of asparagine synthesis. Transamination of α-oxosuccinamate (the β-amide of oxaloacetate) with a suitable amino donor can lead to asparagine synthesis, but this reaction proceeds far more favourably in the reverse direction. Sieciechowicz et al. (1988) have suggested that this pathway is more significant in asparagine catabolism.

Asparagine may be metabolised by two pathways, one involving transamination, and the other immediate deamination by asparaginase (EC 3.5.1.1).

L-Asparagine + H$_2$O → L-aspartate + NH$_3$.

This reaction of deamination by asparaginase is very efficient in seeds. Studies have shown that most asparagine is metabolised in the seed coat, with the embryo receiving the products of asparagine catabolism (Ireland and Lea 1999).

Lysine, Isoleucine, Threonine and Methionine

Considerable attention has been paid to the regulation of the aspartate pathway of amino acid synthesis because lysine, isoleucine, threonine and methionine are essential amino acids which contribute greatly to the nutritive value of plant seed meal. More recently, the conversion of threonine to isoleucine has been the subject of a number of detailed investigations, when it became apparent that it was the target site of a number of extremely potent herbicides (Azevedo et al. 1997).

Lysine, isoleucine, threonine and methionine are synthesised from aspartate (Fig. 3). An interesting point in the pathway is the synthesis of lysine by higher plants from aspartate rather than via the fungal aminoadipic acid pathway. Isotope dilution experiments have indicated that homoserine is an intermediate in threonine and methionine synthesis. There is also evidence that cystathionine and homocysteine are intermediates in methionine formation and that diaminopimelate is on the route to lysine synthesis (Giovanelli et al. 1989).

Both in bacteria and in higher plants, lysine and threonine are synthesised from aspartate by two separate branches of a common pathway, called the aspartate family pathway (for reviews, see Galili 1995; Azevedo et al. 1997; Ravanel et al. 1998). This pathway, which also leads to the synthesis of isoleucine and methionine, is described in Fig. 3. The first two reactions are common to all the aspartate family amino acids. The first reaction is catalysed by aspartate kinase (AK; EC 2.7.2.4), which phosphorylates the β-carboxyl group of aspartate in the presence of ATP to yield 4-aspartyl phosphate and ADP. The second enzyme, aspartate semialdehyde dehydrogenase (EC 1.2.1.11), catalyses the NADPH-dependent reduction of aspartyl phosphate to 4-aspartate semialdehyde (4-ASA). Following this step, there is a branch point that leads on one hand to the synthesis of lysine, and on the other hand to the synthesis of threonine as well as methionine and isoleucine.

Dihydrodipicolinate synthase (DHDPS ; EC 4.2.1.52) is the first specific enzyme for lysine synthesis. Dihydrodipicolinate is formed from the combination of pyruvate and 4-aspartate semialdehyde. The synthesis of lysine from dihydrodipicolinate involves five enzymes, dihydrodipicolinate reductase (EC 1.3.1.26), piperidine dicarboxylate acylase (EC 2.3.1.–), acyldiaminopimelate aminotransferase (EC 2.6.1.17), diaminopimelate epimerase (EC 5.1.1.7), and meso-diaminopimelate decarboxylase (EC 4.1.1.20) (Fig. 3). Although the pathway from dihydrodipicolinate to meso-2,6-diaminopimelate is somewhat obscure in higher plants, circumstantial evidence still indicates that the route shown in Fig. 3 is correct (Azevedo et al. 1997).

Starting from 4-aspartate semialdehyde, the two common enzymes leading to the synthesis of methionine, threonine and isoleucine, are homoserine dehydrogenase (HSDH; EC 1.1.1.3) and homoserine kinase (EC 2.7.1.39). These two consecutive reactions allow the formation of O-phosphohomoserine (OPH) which serves as a second branch point metabolite in the synthesis of aspartate-derived amino acids, thereby leading on one hand to the synthesis of methionine and on the other hand to the synthesis of threonine and isoleucine (Galili 1995 ; Azevedo et al. 1997). In

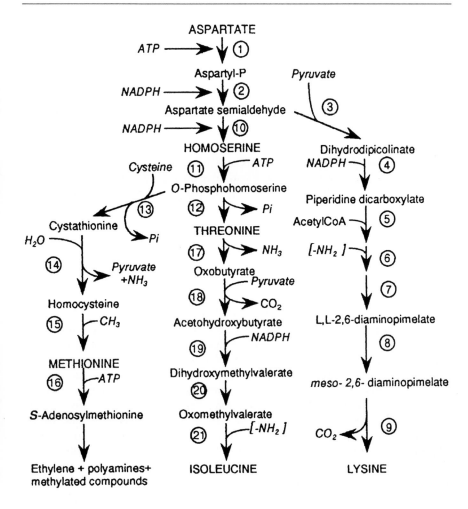

Fig. 3 : The aspartate pathway: biosynthesis of lysine, threonine, methionine, S-adenosylme-thionine and isoleucine. Enzymes : *1.* aspartate kinase; *2.* aspartate semialdehyde dehydroge-nase; *3.* dihydrodipicolinate synthase; *4.* dihydrodipicolinate reductase; *5.* Δ'-piperidine dicarboxylate acylase; *6.* acyldiaminopimelate aminotransferase; *7.* acyldiaminopimelate dea-cylase; *8.* diaminopimelate epimerase; *9. meso-*diaminopimelate decarboxylase; *10.* homose-rine dehydrogenase; *11.* homoserine kinase; *12.* threonine synthase; *13.* cystathionine-γ-synthase; *14.* cystathionine-β-lyase; *15.* methionine synthase; *16.* S-adenosylmethionine syn-thetase; *17.* threonine dehydratase; *18.* acetohydroxyacid synthase; *19.* oxoacid reductoisome-rase; *20.* dihydroxyacid dehydratase; *21.* aminotransferase

certain plants, (e.g. pea) homoserine and OPH may accumulate in high concentra-tions and are used as nitrogen storage and transport compounds (Rochat and Boutin 1991).

The methionine molecule originates from three convergent pathways: the carbon backbone being derived from aspartate, the sulphur atom from cysteine and the methyl group from the β-carbon of serine. First, cystathionine-γ-synthase (EC 4.2.99.9) catalyses the synthesis of cystathionine from cysteine and OPH in a γ-replacement reaction. Cystathionine-β-lyase (EC 4.4.1.8) subsequently catalyses an α,β-elimination of cystathionine to produce homocysteine, pyruvate and ammonia. The terminal step in methionine synthesis involves the transfer of the methyl group from N^5-methyltetrahydrofolate to homocysteine. In plants, this reaction is catalyzed by a cobalamin-independent methionine synthase (EC 2.1.1.14), most presumably localised in the cytosol. Animals have the capacity to convert homocysteine to methionine; however, they are unable to drive the reactions involved in essential amino acids synthesis which, for the most part, take place in the plastids (Ravanel et al. 1998).

Methionine occupies a central position in cellular metabolism, where the processes of protein synthesis, methyl-group transfer through S-adenosylmethionine (AdoMet), polyamine and ethylene syntheses are interlocked. Among these pathways, the synthesis of proteins is the only one consuming the entire methionine molecule. The synthesis of AdoMet, as catalysed by AdoMet synthetase (EC 2.5.1.6) from methionine and ATP, is, however, the major route for methionine metabolism, as 80% of this amino acid is converted to AdoMet. More than 90% of AdoMet is used for transmethylation reactions, in which the methyl group of methionine is transferred to acceptors, the major end products being choline and its derivatives, including phosphatidylcholine (the major polar lipid). These reactions are accompanied by a recycling of the homocysteinyl moiety to regenerate methionine. Briefly, S-adenosyl-L-homocysteine (AdoHcy) produced during the methylation reactions is converted into homocysteine *via* a reaction catalysed by AdoHcy hydrolase (EC 3.3.1.1). Methionine is then regenerated through methylation of homocysteine. Thus, methionine synthase not only catalyses the last reaction in *de novo* methionine synthesis but also serves to regenerate the methyl group of AdoMet. The utilisation of the 4-carbon moiety of AdoMet for the synthesis of polyamines and, in some plant tissues, ethylene is also accompanied by recycling of the methylthio ribose moiety and regeneration of methionine. This route is considerable because it represents approximately 30% of the amount of methionine accumulating in protein (Giovanelli et al. 1985 ; Droux et al. 1995).

OPH can also be rearranged in a single step to threonine by the enzyme threonine synthase (EC 4.2.99.2). A typical feature of the plant enzyme relies upon considerable enhancement of its activity by AdoMet, the end-product of the methionine branch (Giovanelli *et* al. 1984; Curien et al. 1998). Threonine is directly incorporated into proteins, or further deaminated to yield α-oxobutyrate, which serves as a precursor for isoleucine biosynthesis.This reaction is catalysed by threonine deaminase (also called threonine dehydratase, EC 4.2.1.16). Then acetohydroxacid synthase (also known as acetolactate synthase, EC 4.1.3.18) catalyses the condensation of 2-oxobutyrate and pyruvate to yield acetohydroxybutyrate. The latter is converted to isoleucine by a series of reactions involving acetohydroxyacid isomeroreductase (EC 1.1.1.86), dihydroxyacid dehydratase (EC 4.2.1.9), and a branch chain amino acid aminotransferase. The reactions are closely paralleled in the synthesis of valine and leucine, in which two molecule of pyruvate react to form acetolactate (Fig. 3).

The branched nature of the pathway of the amino acids of the aspartate family needs strict regulation, that must occur at several points (Fig. 4). The first enzyme that exhibits regulatory properties is aspartate kinase, existing in at least three isoenzyme forms, which have been isolated from higher plant tissues and their presence confirmed by genetic analyses. One aspartate kinase isoenzyme (AKI), is a bifunctional enzyme containing within a single polypeptide both aspartate kinase and homoserine dehydrogenase activity. This isoenzyme is sensitive to threonine. The two other aspartate kinase isozymes (AKII and AKIII), are monofunctional enzymes exhibiting only aspartate kinase activity and which are lysine-sensitive. Thus, one aspartate kinase is regulated by threonine and two aspartate kinases by lysine. Hence, the presence of high levels of only one of these amino acids is insufficient to shut off the reaction, since the other isoenzymes will still allow the carbon and nitrogen to flow through the pathway. The proportions of activities of the threonine and lysine-sensitive forms appear to vary between plants and more importantly within developmental stages. The AKI isoenzyme is present in only low levels in rapidly growing tissue (Rognes et al. 1983 ; Wallsgrove et al. 1983 ; Arruda et al. 1984 ; Frankard et al. 1997).

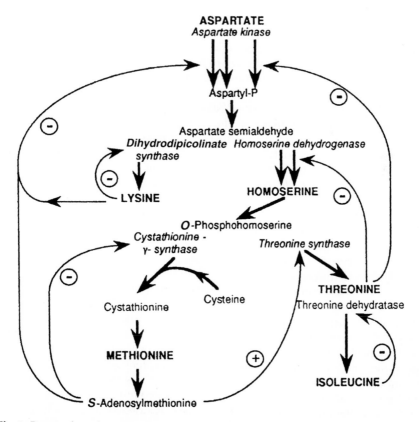

Fig. 4 : Proposed regulatory structure of the biosynthetic pathway of the aspartate family of amino acids. Regulatory interactions are indicated by : (-) inhibition; (+) stimulation; (=) enzyme repression?

Lysine also tightly regulates another key enzyme, dihydrodipicolinate synthase, the first committed enzyme in the lysine biosynthesis pathway. Plant dihydrodipicolinate synthase (DHDPS) is nearly ten-fold more sensitive (IC_{50} of about 10 μM) to lysine inhibition, than are plant lysine-sensitive aspartate kinases. Thus, DHDPS is the major site for lysine regulation in plants, and, therefore, lysine levels should, in theory, never accumulate enough to inhibit aspartate kinase. Genes encoding the 35-38-kDa subunit of the enzyme have been isolated from a range of plants and have been shown to contain a chloroplast transit peptide (Galili 1995; Azevedo et al. 1997).

Threonine inhibits one isoenzyme of aspartate kinase and one isoenzyme of homoserine dehydrogenase (HSDH). The enzyme HSDH exists in two isoenzyme forms, one of which is sensitive to inhibition by threonine and is localised in the chloroplast and the other is insensitive to inhibition by threonine and is localised in the cytosol. A cDNA encoding threonine synthase has been isolated from *A thaliana* and expressed in *E.coli* and the N-terminal domain has been shown be involved in the activation by AdoMet (Azevedo et al. 1997 ; Curien et al. 1998). The first step in isoleucine synthesis, catalysed by threonine dehydratase, is also subject to feedback inhibition by isoleucine in the chloroplast (Singhand Shaner 1995).

It is worth noting that, contrary to the plant enzyme, bacterial threonine synthase, though catalysing the same reaction, is not subject to kinetic modulation by AdoMet, presumably because in bacteria, OPH is not a branch point metabolite. Indeed, in prokaryotes, branching between threonine/isoleucine and methionine pathways occurs at the level of homoserine. In plants, neither cystathionine-γ-synthase nor cystathionine-β-lyase is significantly inhibited by methionine pathway intermediates or end-products such as AdoMet. Also, these enzymes are not sensitive to feedback inhibition by any of the aspartate-derived amino acids. There are strong indications, however, for a control of methionine biosynthesis in plants at the level of gene expression and that the major site for such a regulation is probably cystathionine-γ-synthase (Chiba et al. 1999). For example, when the duckweed *Lemna paucicostata* was grown under culture conditions causing methionine starvation, e.g. in the presence of L-aminoethoxyvinylglycine (AVG; an amino acid first isolated from *Streptomyces* sp. and that inhibits cystathionine-β-lyase), or in the presence of lysine plus threonine, mixtures that inhibit aspartate kinase (EC 2.7.2.4), there was a substantial increase in extractable cystathionine-γ-synthase activity. Furthermore, antisense repression of cystathionine-γ-synthase results in abnormal development of *A. thaliana* (Droux et al. 1995 ; Ravanel et al. 1998).

Isolated intact chloroplasts were observed to carry out light-dependent synthesis of lysine, threonine, homocysteine and isoleucine from ^{14}C-aspartate (Mills et al. 1980). These reactions could not be demonstrated in mitochondria. Initial studies showed that essentially all of the enzymes involved in amino acids formed from aspartate were located in the chloroplast, suggesting that the chloroplast is the sole site of the synthesis of these amino acids, except methionine. Indeed, of the three specific enzymes involved in methionine synthesis in plants, the first two, cystathionine-γ-synthase and cystathionine-β-lyase, are localised to the plastids, whereas the third one, methionine synthase, most presumably is localised in the cytosol (Droux et al. 1995; Ravanel et al. 1998).

Mutants with altered regulation of the aspartate pathway have been isolated. A number of studies have been carried out to obtain tissue culture lines or intact plants, with altered feedback control in the aspartate pathway. Green and Phillips (1974) indicated that the growth of germinating cereal embryos could be inhibited by the presence of lysine and threonine. The inhibition of growth could be alleviated by the presence of low concentrations of methionine or homoserine. Methionine synthesis was thus prevented by the complete inhibition of the aspartate kinase isoenzymes by lysine and threonine.

Three lines of mutant barley plants which are resistant to the toxic action of lysine and threonine, have been isolated at Rothamsted (Bright et al. 1982). An analysis of the properties of the three aspartate kinase isoenzymes following ion exchange chromatography has shown that in the mutant R 3202, the aspartate kinase II peak was not inhibited by lysine even at a high concentration, whilst the aspartate kinase III peak was still inhibited by lysine. Conversely, the mutant R 3004 had an unchanged aspartate kinase II but an aspartate kinase III that was less sensitive to lysine. Genetic analysis of the two mutants suggested that resistance to lysine plus threonine was controlled by two dominant genes, termed *lt1* and *lt2* , which are probably the structural genes for aspartate kinases II and III. A double mutant between R 3202 (*lt1*) and R 3004 (*lt2*) was constructed which contained aspartate kinase II insensitive to lysine inhibition and aspartate kinase III with a decreased sensitivity to lysine (Arruda et al. 1984).

In maize, two separate genes (termed *Ask 1* and *Ask 2*) have also been shown to confer resistance to lysine and threonine (Diedrick et al. 1990). Aspartate kinase purified from the different mutant lines Ask 1 and Ask 2 was less sensitive to inhibition by lysine. Two homologous genes encoding a 52.5-kDa subunit of lysine-sensitive AK with a chloroplast transit peptide have been isolated from *A.thaliana*, confirming earlier biochemical data (Frankard et al. 1997).

Because human food and animal feed derived from many grains are deficient in some of the ten essential amino acids which are required in the animal diet (cereal seeds are deficient in lysine, while legume seeds are deficient in methionine), a number of studies have been conducted to increase the levels of these amino acids in crop seeds. First attempts were based on the characterisation of plant mutants overproducing lysine (and/or methionine, see later). More recent work has been based on molecular genetics and plant transformation (Lea and Forde 1994; Galili 1995).

To obtain mutant plants overproducing lysine, a number of research groups have attempted to select mutant lines that are resistant to the lysine analogue S-(2-aminoethyl)-L-cysteine (AEC). Unfortunately, most of these experiments produced plants with a reduced uptake of AEC. However, one mutant of *Nicotiana sylvestris* (generated from protoplast culture) was shown to be resistant to AEC and able to accumulate lysine. In these mutants, dihydrodipicolinate synthase was insensitive to feedback inhibition by lysine (Negrutiu et al. 1984).

In 1992, Frankard, et al. crossed the homozygous AEC-resistant mutant of tobacco RAEC-1 that overproduced lysine, with the homozygous mutant RLT 70, that was resistant to lysine plus threonine and overproduced threonine. The level of soluble lysine in the leaves of the heterozygous double mutant (RAEC-1 x RLT 70) was 20-fold higher than the wild type and represented 30% of the total soluble amino acid pool. When compared to the lysine-overproducing parent (RAEC-1), the double mutant contained higher levels of soluble lysine (+ 130%). In contrast, the concentrations of threonine and methionine were similar or even reduced when compared to the parent lines. The data confirmed that dihydrodipicolinate synthase exerts a strong control over the pathway, and causes a drain of aspartate semialdehyde to lysine when deregulated.

The higher plant dihydrodipicolinate synthase enzyme is normally very sensitive to feedback inhibition by lysine, as discussed previously. However, the E. coli enzyme is much less sensitive to lysine. Taking into account this information, a construct including the 35S promoter and the *dap A* gene of E. coli that encodes the enzyme, including the pea rbcS-3A chloroplast transit peptide, has been used to transform plants (Shaul and Galili 1992a). An increase in the soluble lysine concentration was detected in tobacco plants that contained E. coli dihydrodipicolinate synthase in the chloroplast. In the homozygous plants, the concentration of soluble lysine was 40-fold higher than the control plants. No increase in soluble methionine or threonine was detected in any transformed plants. Similar series of experiments were conducted by Glassman (1992) using the same constructs. Putative transformed tobacco plants contained an elevated concentration of soluble lysine with a maximum increase of almost 100-fold, as compared to the untransformed plants. Again, no increase in methionine or threonine was observed. Unfortunately, the plants obtained by both groups of scientists, exhibiting a higher level of lysine, were observed to have abnormal leaf morphology that could be attributed to the lysine overproduction (Frankard et al. 1992).

Shaul and Galili (1992b) also introduced a mutant *Lys C* gene from E. coli , that encodes an aspartate kinase insensitive to feedback inhibition by lysine into tobacco. The *Lys C* gene product was attached to a chloroplast transit peptide. The transformed plants showed a tenfold increase in soluble threonine. However, significant increases in soluble threonine concentration were also determined when the *Lys C* gene was expressed in the cytosol (i.e. same construction but without transit peptide). When progeny of the transformed plants were grown in a glasshouse, greater increases of the concentrations of soluble threonine, isoleucine and lysine were noted.

More recently, Karchi, et al. (1993) fused the *Lys C* gene to a chloroplast transit peptide and the seed-specific promoter of the bean phaseolin gene. A 14-17-fold increase in the soluble threonine concentration of the seeds of the homozygous progeny was detected. Karchi et al. (1993) also measured a three fold increase in the concentration of soluble methionine. No alteration in the plant phenotype was detected in the transgenic plants. Similarly, tobacco plants expressing the bacterial feedback-insensitive dihydrodipicolinate synthase in a seed-specific manner synthesised higher than normal levels of free lysine during seed development. However, the level of free lysine was significantly reduced in mature seeds. These results demonstrated that the aspartate family pathway is functional in seeds, and, furthermore, that free lysine levels in mature seeds are controlled not only by biosynthetic enzymes, but more importantly by catabolic activities (Galili et al. 1995). Indeed, increasing the concentration of free lysine in developing tobacco seeds, either by exogeneous administration or by overproducing the bacterial lysine-insensitive dihydropicolinate synthase, causes an approximate ten-fold increase in lysine oxoglutarate reductase (Karchi et al. 1994), the first committed enzyme in lysine: catabolism. Interestingly, this activity is noticeably lower in developing grains of the *opaque-2* maize mutants (this mutation reduces the accumulation of specific zein polypeptides, the most abundant storage proteins of the endosperm that are devoid of lysine residues) than in wild type grains (Kemper et al. 1998). The extent of catabolic degradation of free lysine in seeds seems to depend on the plant species. Thus, by using the same basic approach as that developed by Karchi et al. (1994) for tobacco (i.e. overproduction in transgenic plants of feedback insensitive aspartate kinase and dihydrodipicolinate synthase under the control of a seed specific promoter), Falco et al. (1995) have achieved a considerable degree of success in the production of transgenic canola and soybean seeds with increased lysine. Remarkably, in the best-transformed line analysed, the total (i.e. free plus bound) lysine content increased over fivefold, from 5.9 to 34% of the total amino acids.

Methionine is the first limiting essential amino acid in legume seeds, since the major storage proteins, the globulins, are low in this amino acid. A number of basic strategies are being carried out to increase the methionine content in legumes. They include modifications of the major storage proteins (e.g. introducing methionine residues or methionine-rich peptides into non-conserved, and presumably non-critical, regions of storage proteins); transfer of heterologous genes encoding methionine-rich protein from other species (e.g. Brazil nut); or manipulation of key enzymes in the methionine biosynthetic pathway (De Lumen et al. 1997). As for lysine, another approach is to select mutant lines that are resistant to the methionine analogue, ethionine; *A. thaliana* (Inaba et al. 1994) and soybean (Madison and Thompson 1988) mutant lines overproducing methionine have been obtained in this way. Although such mutants have not yet been characterised at the molecular and biochemical levels, in both cases the results suggested that mutations affect later step(s) in the methionine biosynthesis pathway, occurring after OPH. These genetic attempts open up the possibility of producing crop plants with increased level of essential amino acids in seeds, as described in chapter 6.

Synthesis of Glycine, Serine and Cysteine

L-serine and glycine, as well as being constitutents of proteins, serve as precursors in phospholipid and purine biosynthesis , and are the main sources of one-carbon units in higher plant cells. In green tissues, serine and glycine are also involved in photorespiratory metabolism. Under high irradiance, pool sizes of glycine and serine in leaves vary with conditions affecting the rate of photorespiration. In illuminated leaves serine usually accounts for 8-12% of the free amino acids, whereas glycine accounts for only 1-3%. Glycine and serine are two amino acids that are interconvertible and are distributed in different subcellular compartments, cytosol, chloroplast, mitochondria and peroxisome. The interaction of the different subcellular pools increases the complexity of their metabolism. For example, during photorespiration the mitochondrial pools of glycine and serine are interconnected and, consequently, their synthesis and catabolism are intimately interwoven.

The interconversion of serine and glycine is catalysed by serine hydroxymethyltransferase (SHMT) (EC 2.1.2.1), which is present in the cytosol, chloroplasts and mitochondria. The three-carbon of serine is transferred to tetrahydrofolate to generate methylene tetrahydrofolate and glycine. The role of tetrahydrofolate is to trap the formaldehyde released from serine at the active site of the enzyme. The plant enzyme, which is a homotetramer, similar to bacterial and mammalian SHMT, requires pyridoxal phosphate as a coenzyme. The interconversion of serine to glycine is fully reversible. However, the equilibrium distribution of different substrates shows that the reaction favors serine to glycine conversion. Thus, serine is a major source of glycine and one-carbon units in most living organisms. In green tissues, a high SHMT activity is required in mitochondria to cope with high glycine decarboxylase activity (Bourguignon et al. 1999; Rebeille et al. 1994).

There are two main pathways leading to serine and glycine synthesis. The first route, often referred to as the glycolytic or phosphorylated pathway, is linked to glycolysis and leads to serine formation from 3-phosphoglycerate. The glycolytic pathway involves the conversion of 3-phosphoglycerate to serine through 3-phosphohyroxypyruvate and 3-phosphoserine. These reactions are catalysed by 3-phosphoglycerate dehydrogenase (EC 1.1.1.95); glutamate: phosphohydroxypyruvate aminotransferase (EC 2.6.1.52) and phosphoserine phosphatase (EC 3.1.3.3). Glycine can then be derived from serine by the action of serine hydroxymethyltransferase (Mouillon et al. 1999). The second route linked to photosynthetic metabolism involved transamination of glyoxylate into glycine, which is then converted into serine (Bourguignon et al. 1999). Because of the high flux of carbon through the photorespiratory pathway cycle, the glycolate pathway is assumed to be the major metabolic pathway of glycine and serine synthesis in green tissues. It is still unclear whether the glycolytic pathway is metabolically active one during nonphotosynthetic conditions, or whether it participates in serine synthesis in all physiological situations.

During the course of the photorespiratory cycle, the conversion of glycine to serine involves four subcellular compartments. Glycolate, produced in the chloroplast from phosphoglycolate by the oxygenase activity of Rubisco, is transferred via the cytosol into the peroxisome where it is oxidised to glyoxylate. It is then transaminated to glycine by serine :glyoxylate (EC 2.6.1.45) and glutamate:glyoxylate aminotransferases (EC 2.6.1.4), which may also use alanine as a substrate. Glycine is then transported into the mitochondria, where two glycine molecules react to produced serine. This reaction involves the oxidative decarboxylation and deamination of glycine to yield a C-1 fragment and equal amounts of CO_2, ammonia and NADH (Fig. 5). The available evidence suggests that, in green leaf cells, the complete process occurs within the mitochondrial matrix via a coordinated sequence of reactions (Keys et al. 1978) . Glycine is cleaved in the matrix space by glycine decarboxylase or glycine-cleavage system (EC 2.1.2.10) to CO_2, NH_3 and N^5, N^{10} methylene tetrahydrofolate. The latter compound reacts with a second molecule of glycine to form serine in a reaction catalysed by serine hydroxymethyltransferase (SHMT) (EC 2.1.2.1).

The glycine-cleavage system has been purified from plants, animals and bacteria, and consists of four protein components. They have been named P-protein (a pyridoxal phosphate-containing protein), H-protein (a lipoic acid-containing protein); T-protein (a protein catalysing the tetrahydrofolate-dependent step of the reaction) and L-protein (a lipoamide dehydrogenase). The glycine-cleavage system could be linked to the SHMT system by the soluble pool of reduced folate (Bourguignon et al. 1999). The P, H, T, L proteins are encoded by nuclear genes and synthesised in the cytosol with an N-terminal leader sequence that directs the precursor forms towards the mitochondria. The X-ray structure of these proteins and especially the H protein, the core of this multienzymatic system, has been determined at 2.6 and 2 A resolution by the multiple isomorphous replacement technique (Pares et al.1994 ; Cohen-Addad et al. 1995 ; Macherel et al. 1996). The three-dimensional structure of the oxidised form of the enzyme indicates that the lipoate arm is hanging at the surface of the H protein (Guilhaudis et al. 2000). Indeed, following methylamine transfer, the lipoate cofactor is pivoted about 90° around the lysine linkage (Lys-63), to bind into a cleft at the surface of the protein (Cohen-Addad et al. 1995). This modification of structure allows to the methylamine group to be locked within a hydrophobic pocket, preventing the nonenzymatic release of NH_3 and formaldehyde, owing to nucleophilic attack by water molecules (Cohen-Addad et al. 1997; Bouguignon et al. 1999).

Because high rates of glycine oxidation are required to cope with the flux of photorespiratory glycine entering leaf mitochondria, glycine decarboxylase and SHMT are present at very high concentration in the matrix space of mitochondria (Oliver et al. 1990). These two enzymes systems represent about 40-50% of the soluble proteins in the mitochondria. Vauclare et al. (1996) have also observed that the buildup

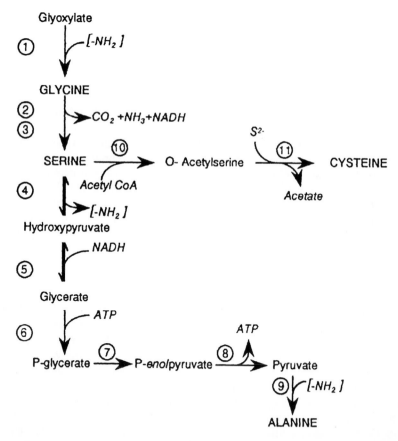

Fig. 5 : The biosynthetic pathway of glycine, serine alanine and cysteine. Enzymes: *1.* serine : glyoxylate and glutamate: glyoxylate transaminases; *2.* glycine decarboxylase; *3.* serine hydroxymethyltranferase; *4.* serine: glyoxylate aminotransferase; *5.* hydroxypyruvate reductase; *6.* glycerate kinase; *7.* enolase; *8.* pyruvate kinase; *9.* aminotransferase.*10.* serine acetyltransferase; *11.* O-acetylserine (thiol)-lyase

of glycine decarboxylase is dependent on leaf development and the genes encoding this enzyme could be under light-dependent transcriptional control.

Ammonia released during glycine deamination is reassimilated by glutamine synthetase in the cytosol or, more likely, in the chloroplast (Leegood et al. 1995). It is important to note that the conversion of glycine to serine produces ammonia at rates tenfold faster than nitrate reduction. Ammonia is recycled through the GS/GOGAT cycle (Keys et al. 1978; Woo et al. 1982), whilst the main part of CO_2 is lost to the atmosphere during photorespiration. The reductant NADH formed is either converted to ATP via the electron transport chain, or is exported from the mitochondria via a malate/oxaloacetate shuttle (Douce 1985).

The serine produced in the mitochondria enters the peroxisomal compartment, where it is converted into the oxo acid analog of serine, hydroxypyruvate by serine:glyoxylate aminotransferase (EC 2.6.1.45). The latter is reduced to glycerate by hydroxypyruvate reductase (EC 1.1.1.81) and converted to phosphoglycerate in the chloroplast by NADH-glycerate kinase (EC 2.7.1.31) (Leegood et al. 1995).

During the course of the photorespiratory cycle, glycine and serine are only intermediary metabolites and not end-products of the pathway, thus photorespiration does not play an important role in the net synthesis of glycine and serine. The glycolate pathway is aimed at the recycling of the two carbons of glycolate into the Calvin-Benson cycle, avoiding the depletion of this cycle when the oxygenase activity of Rubisco is high (Bourguignon et al. 1999).

Barley, tobacco and *A.thaliana* mutants deficient in glycine decarboxylase, serine hydroxymethyltransferase, and serine-glyoxylate aminotransferase have been isolated (Somerville 1986; Blackwell et al. 1990; Lea and Forde 1994). These mutants are viable only in atmospheres of high CO_2. Under these conditions, it was demonstrated that the glycine cleavage step of photorespiration is not necessary for any essential function unrelated to photorespiratory activity. In normal air, the mutants are chlorotic and die prematurely. Blocking the flux of carbon or nitrogen through the photorespiration pathway is an effective way of killing plants, by impairing the recycling of glycolate back to glycerate. Photorespiration mutants have confirmed the view that it will only be possible to decrease photorespiration by genetic manipulation of ribulose biphosphate carboxylase-oxygenase, or the introduction of a CO_2 concentrating mechanism into C_3 plants.

Serine is acetylated to O-acetylserine by serine acetyltransferase (EC 2.3.1.30) prior to sulphydration. Cysteine is then formed by O-acetylserine thiol-lyase (also called cysteine synthase) (EC 4.2.99.8). In plant cells, the existence of several enzymes supporting O-acetylserine (thiol)-lyase activity has been demonstrated (Rolland et al. 1992; Saito et al. 1994; Droux et al. 1998). This subdivision reflects different subcellular localisation of enzyme activity, within the chloroplast, cytosol and mitochondria. Such a differential localisation is also seen for the serine acetyltransferase activity. The reasons for such a compartmentation (which contrasts with that for the aspartate family pathway which is almost entirely localised to the chloroplast) might be that O-acetylserine is a very unstable compound at alkaline pH (the pH of the chloroplast stroma in the light is above 8), being transformed into N-acetylserine which is not a substrate for O-acetylserine thiol-lyase. In addition, the plant cell might be unable to transport cysteine between compartments, so that the cysteine required for protein synthesis must be synthesised *in situ*. Sulphate is reduced to sulphide in presence of ATP and ferredoxin in the chloroplast. The control mechanisms in cysteine synthesis appear to involve feedback inhibition by cysteine at several steps in the reduction of sulphate to sulphide, of sulphate uptake itself, and of the first committed step in cysteine biosynthesis catalysed by serine acetyltransferase (Hell 1997; Droux et al. 1998).

Synthesis of Amino Acids Derived from Pyruvate

Valine and leucine are synthesised from pyruvate through the branched-chain amino acid pathway (Fig. 6). Alanine is synthesised from pyruvate by direct transamination (see above). This reaction is catalysed by alanine aminotransferase (EC 2.6.1.2). Isoleucine and lysine also obtain some of their carbons from pyruvate (Singh and Shaner 1995).

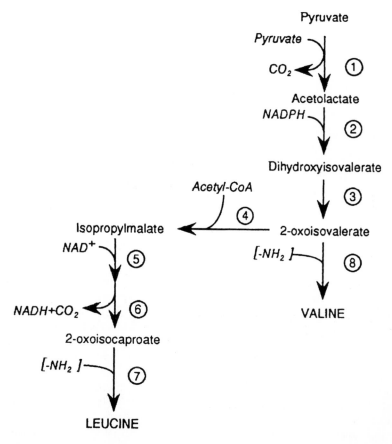

Fig. 6 : The biosynthetic pathway of valine and leucine. Enzymes: *1.* acetohydroxyacid synthase; *2.* acetohydroxyacid reductoisomerase; *3.* dihydroxyacid dehydratase; *4.* isopropylmalate synthase; *5.* isopropylmalate isomerase; *6.* isopropylmalate dehydrogenase; *7.* aminotranferase; *8.* aminotransferase

Pyruvate provides the carbon skeletons for valine and leucine, the branched-chain amino acids. The reaction involves the decarboxylation of pyruvate to form, in the presence of thiamine pyrophosphate, hydroxyethyl thiamine pyrophosphate (activated acetaldehyde). This compound is then accepted by an other molecule of pyruvate or by 2-oxobutyrate (a product derived from the deamination of threonine). The enzyme catalysing the synthesis of acetolactate is acetolactate synthase or acetohydroxyacid synthase (AHAS; EC 4.1.3.18).This enzyme was found in a number of plants to be located in the chloroplast. Acetolactate is transformed, by acetohydroxyacid reductoisomerase (EC 1.1.1.86) and dihydroxyacid dehydratase (EC 4.2.1.9), to oxovalerate and then transaminated to valine. Oxovalerate can be condensed with the methyl group of acetyl-CoA to give isopropylmalate after the action of isopropylmalate synthase (EC.4.1.3.12). This latter compound is then oxidised, isomerised and decarboxylated to α-oxoisocaproate which is transaminated to leucine. The branched-chain α-oxoacid dehydrogenases are associated with lipoic acid acting as a prosthetic group. As with other branched pathways, close regulation is necessary to produce the desired levels of different amino acids; again feedback inhibition is in operation. Valine and leucine inhibit the biosynthetic pathway of their own synthesis at the point of AHAS and leucine inhibits the synthesis of isopropylmalate from oxoisovalerate (Singh and Shaner 1995). All of the enzymes of this pathway are located in chloroplast which has been shown to synthesise valine from $^{14}CO_2$.

A number of mutants have been isolated that lack the enzymes threonine dehydratase and dihydroxyacid dehydratase (Negrutiu et al. 1985; Wallsgrove et al. 1986). These mutants require both valine and isoleucine for growth. Mutants resistant to the toxic action of valine, which acts as a potent feedback inhibitor of the pathway, have been isolated by Bourgin (1978). This class of mutants exhibited a reduced rate of amino acid uptake.

The main interest has centered on the enzyme AHAS, as four active families, the imidazolinones, sulphonylureas, triazolopyrimidines and pyrimidyloxobenzoates, have been observed to inhibit the enzyme (Singh and Shaner 1995). These compounds are among the most advanced herbicides used in agriculture, because they exhibit extremely low mammalian toxicity and high efficacy, resulting in very low application rates (a few grams ha^{-1}!) and thus they have very low environmental impact. The exact mechanism accounting for the lethality of these herbicides is not yet known, that is whether plant death results from a deprivation of branched chain amino acids or from the accumulation of toxic compounds (e.g. oxobutyrate and/or upsteam metabolites and/or by-products) (Epelbaum et al. 1996). Also, by using a luxCDABE reporter gene complex from *Photorhabdus luminescens*, Van Dyk et al. (1998) recently analysed the pattern of gene expression upon treatment of *E. coli* with sulphometuron methyl, a sulphonylurea herbicide. Their results suggested that restricted AHAS activity may cause intracellular acidification and the induction of the σ^S-dependent stress response. Whatever the cause of lethality, the inhibitory action of the herbicides can be reversed by the exogenous application of valine, leucine and isoleucine together. These herbicides are unusual inhibitors in that they bear no obvious similarity to the substrates (pyruvate and α-oxobutyrate), cofactors (thiamine pyrophosphate, FAD and magnesium) or allosteric effectors (valine, leucine and isoleucine) of the enzyme. They behave as noncompetitive slow binding inhibitors, which presumably interact with an evolutionary vestige of the quinone-binding site of pyruvate oxidase (a related enzyme) (Schloss et al. 1988). Owing to high lability and very low abundance, purification of AHAS from plant tissues proved difficult and resulted in very low yields. Recently, the *A. thaliana* enzyme with part of its transit peptide removed (the enzyme is localised to the plastids) has been overexpressed in *E. coli*, allowing purification of the enzyme to homogeneity (Chang and Duggleby 1997). The recombinant enzyme was inhibited by sulphony-lurea and imidazolinone herbicides, but was however insensitive to valine, isoleucine and leucine. This last result indicates strongly that the recombinant plant enzyme contains only the large catalytic subunit and lacks the small regulatory subunit involved in feedback inhibition that has been characterised for the corresponding bacterial enzymes (Vyazmensky et al. 1996).

A large number of mutant plants resistant to the AHAS inhibitors have been isolated and their properties have been discussed in detail (Mazur and Falco 1989). AHAS inhibitor-resistant mutant lines of important crop plants such as soybean, maize, canola and wheat appear to be of uppermost interest. Moreover, populations of weed species resistant to the sulphonylurea herbicides have also occurred in the field (Mallory-Smith et al.1990). In Australia, for example, resistant biotypes of ryegrass have been shown to contain chlorsulphuron-insensitive AHAS and an increased ability to detoxify the herbicide (Christopher et al. 1992).

Many AHAS inhibition resistant mutant lines of *A. thaliana* have been isolated and examined at the molecular level. The chlorsulphuron-resistant mutation gives rise to a proline to serine substitution at amino acid 196 on the AHAS protein (Haughn et al. 1988). The imidazolinone resistant mutation *crs1* causes substitution of an asparagine by a serine at amino acid 653 (Haughn and Somerville 1990). More recently, Mourad and King (1992) have isolated a mutant resistant to triazolopyrimidine and have investigated the cross-resistance properties of all three mutations.

The chlorsulphuron resistant mutant gene, isolated from *A.thaliana*, expressed in tobacco using its own promoter, conferred high levels of herbicide resistance (Haughn *et al.* 1988). This gene was placed under the control of both native and 35S promoters, resulting in a 300-fold increase in herbicide tolerance. The 25-fold increase in AHAS transcript levels correlated with only a twofold increase in AHAS specific activity (Odell et al. 1990). Tourneur et al. (1993) have expressed the *csrl* gene under the control of the 35S promoter with a duplicated enhancer (p70). The transformed plants contained a 12-fold higher level of AHAS activity, and exhibited a 1500-fold increase in resistance to chlorsulphuron and also a resistance to the external application of valine. However, there was no significant increase in the content of free branched-chain amino acids in the leaves, suggesting that subsequent steps in the branched-chain amino acid pathway may also regulate the synthesis of valine and leucine.

Manipulation of branched-chain amino acid biosynthesis has been driven by the search for herbicide-tolerant plants. AHAS appears to be the primary control point of the pathway and its regulation is achieved by feedback inhibition of valine and leucine. AHAS is the site of herbicide action and consequently the interesting enzyme to be genetically manipulated. Other experiments attempting to modulate the gene expression of AHAS are in progress. Potato plants expressing a constitutive antisense AHAS construction have been generated and subjected to a comprehensive analysis (Hofgen et al. 1995). Severe growth retardation and strong phenotypic resemblance to plants treated with the AHAS-inhibiting herbicides were observed.

However, these plants exhibited unexpected elevated levels of valine and leucine. There have been some speculative attempts to elucidate the reasons for the deregulation for this amino acid pathway.

Keto-acid acid isomeroreductase (KARI, also called acetohydroxy acid isomeroreductase, EC 1.1.1.86), which is the second common enzyme of the parallel branched chain amino acid pathway, has been the subject of several studies in both bacteria and plants. This is partly because resistant plants can rapidly emerge under the selective pressure of ALS-inhibiting herbicides, and, therefore, it is important to evaluate the potential of other enzymes in the pathway, to provide new efficient herbicide targets. KARI catalyses a Mg^{2+}-dependent two-step reaction in which the substrate, either 2-acetolactate or 2-aceto-2-hydroxybutyrate is converted via an alkyl migration and an NADPH-dependent reaction to yield 2,3-dihydroxy-3-isovalerate (synthesis of valine and leucine) or 2,3-dihydroxy-3-methylvalerate (synthesis of isoleucine), respectively. The plant enzyme has been purified from spinach chloroplasts and the cDNA has been cloned, allowing investigation of some of its kinetic (Dumas et al. 1994a) and structural (Biou et al. 1997) properties. Site-directed mutagenesis (Dumas et al. 1995) and determination of the crystal structure of the protein (Biou et al. 1997) revealed that the enzyme contains two active site-located magnesium ions, one of which plays a role in the isomerisation reaction and the other in the reduction step. Transition states analogues such as N-hydroxy-N-isopropyloxamate (IpOHA) and 2-(dimethylphosphinoyl)-2-hydroxyacetic acid (Hoe 704) proved to be highly potent inhibitors of the bacterial (Aulabaugh and Schloss 1990;) and plant (Dumas et al.1994b) KARI. However, compared to the ALS-inhibiting herbicides, both IpOHA and Hoe704 showed only poor herbicidal action (Aulabaugh and Schloss 1990). One possible explanation for this behaviour is that these compounds act as extremely slow-binding competitive inhibitors of KARI (Dumas et al.1994b). Furthermore, since inhibition of this enzyme leads *in vivo* to an increase in the concentration of the KARI substrate, these inhibitors bind to the enzyme considerably slower as the substrate concentration increases. Recently, plant KARI has been submitted to high-throughput screening, allowing the discovery of a new class of compounds belonging to the thiadiazole family, which act as either reversible or irreversible noncompetitive inhibitors of the plant enzyme (Halgand et al. 1998).

Synthesis of Histidine

Until recently, very little information was available regarding histidine biosynthesis in plants. It was generally assumed that the route of synthesis is the same as in bacteria. Synthesis starts with ribose-5-phosphate which is transformed into 5-phosphoribosyl-α-1-pyrophosphate, the precursor of N'-[(5'-phosphoribosyl)-formimino]-5-aminoimidazole-4-carboximide ribonucleotide (5'-PRFAR), considered also as the starting point for purine synthesis. In the presence of glutamine, this derivative gives rise to imidazole glycerol phosphate, the precursor of imidazole acetol phosphate. With glutamate, this molecule is transformed to histidinol phosphate, histidinol, histidinal and histidine (Fig. 7).

The enzymes ATP-phosphoribosyl transferase (EC 2.4.2.17), imidazole glycerol phosphate dehydratase (EC 4.2.1.19), and histidinol phosphate phosphatase (EC 3.1.3.15) have been detected in plant extracts. These enzymes catalyse the first,

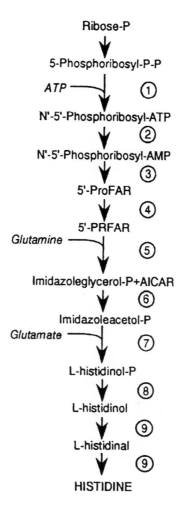

Fig. 7 : The biosynthetic pathway of histidine. Enzymes : *1.* ATP phosphoribosyl transferase; *2.* phosphoribosyl-ATP pyrophosphohydrolase; *3.* phosphoribosyl-AMP cyclohydrolase; *4.* N'-[(5'-phosphoribosyl) formimino]-5-aminoimidazole-4-carboximide ribonucleotide isomerase; *5.* imidazole glycerol phosphate synthase; *6.* imidazole glycerol phosphate dehydratase; *7.* histidinol phosphate aminotransferase; *8.* histidinol phosphate phosphatase; *9.* histidinol dehydrogenase. 5'-ProFAR (cyclic form), N'-[(5'-phosphoribosyl)-formimino]-5-aminoimidazole-4-carboximide ribonucleotide; 5'-PRFAR (linear form), N'-[(5'-phosphoribosyl)-formimino]-5-aminoimidazole-4-carboximide ribonucleotide; AICAR, 5'phosphoribosyl-4-carboximide-5-aminoimidazole

sixth and eighth steps of the pathway, respectively. The imidazole glycerol phosphate dehydratase cDNA has been isolated from *A. thaliana* by Tada et al. (1995). Histidinol dehydrogenase (EC 1.1.1.23) has been purified from wheat germ, exhibiting two forms; the cDNA was isolated from cabbage by Nagai et al. (1991). This enzyme catalyses the last two steps of the pathway. A cDNA encoding histidinol phosphate aminotransferase (EC 2.6.1.9) was also isolated and characterised in *Nicotiana tabacum* (El Malki et al. 1998). This enzyme catalyses the eighth step of the pathway. Aminotriazole inhibits the synthesis of imidazole acetol phosphate, a precursor of histidine. This compound, which exhibits herbicidal activity, affects the activity of the imidazole glycerol phosphate dehydratase, leading to the inhibition of histidine biosynthesis in plants (Mori et al. 1995).

Recently, Fujimori and Ohta (1998) reported the isolation of an *A. thaliana* cDNA that encodes a bifunctional protein (At-IE) that has both phosphoribosyl-ATP pyrophosphohydrolase and phosphoribosyl-AMP cyclohydrolase activities. These enzymes catalyse the second and the third steps, respectively, in histidine biosynthesis. The gene encoding At-IE was expressed ubiquitously throughout development of the plant. Sequence comparison suggested that the At-IE protein has an N-terminal extension with the properties of chloroplast transit peptide.

A number of mutant plants have been isolated which require histidine for growth. The biochemistry of these mutations has not yet been established (Gebhardt et al. 1983). The availability of histidine auxotrophic mutants might represent a valuable tool to elucidate the regulation of the amino acid metabolic pathway in plants. For example, the use of a specific inhibitor to block histidine biosynthesis leads to an increase in the expression of genes involved in the biosynthesis of aromatic amino acids, as well as histidine, lysine and purines ; the addition of histidine reversed the situation. This suggests the existence of a general control system of metabolite biosynthesis in plants, which would represent an efficient way to coordinate the regulation of genes involved in unrelated metabolic pathways, providing evidence for cross-pathway regulation of metabolic gene expression in plants (Guyer et al.1995).

Thus, histidine biosynthesis in plants follows the same pathway as in bacteria and fungi, despite some structural and functional differences between the respective enzymes and genes. Moreover, it is possible that the entire pathway of histidine biosynthesis may be carried out in the chloroplasts, which is an extremely energy-consuming process (41 ATP required for each histidine molecule synthesised).

Synthesis of Aromatic Amino Acids

The group consisting of tryptophan, phenylalanine and tyrosine is synthesised by a common pathway, called the shikimate pathway. Besides leading to the formation of aromatic acids for protein synthesis, the shikimate pathway provides precursors for a number of secondary metabolites in higher plants: such as phenolic acids, flavonoids, isoflavonoids, glucosinolates, phytoalexins, alkaloids, plant growth regulators (indole acetic acid, IAA), suberin and lignin. Several of these compounds are involved in the response of plants to wounding or infection.

The shikimate pathway can be divided into three sections (Fig. 8). The synthesis of chorismate is common to all three aromatic amino acids. Beyond chorismate, there is a branch leading to phenylalanine and tyrosine and one to tryptophan. There is strong evidence that the complete shikimate pathway can operate in the chloroplast, but the existence of a second cytoplasmic pathway is still a matter of debate (Schmid and Amrhein 1995).

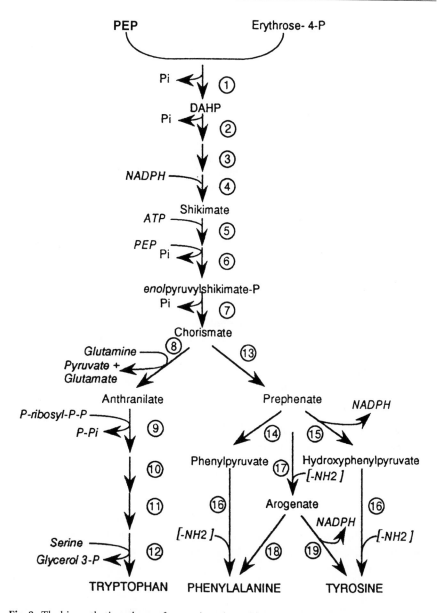

Fig. 8 : The biosynthetic pathway of aromatic amino acids. Enzymes: *1.* 3-deoxy-D-*arabino*heptulosonate-7-P synthase; *2.* 3-dehydroquinate synthase; *3.* 3-dehydroquinate dehydratase; *4.* shikimate dehydrogenase; *5.* shikimate kinase; *6.* 5-*enol*pyruvylshikimitate 3-phosphate synthase; *7.* chorismate synthase; *8.* anthranilate synthase; *9.* phosphoribosylanthranilate transferase; *10.* phosphoribosylanthranilate isomerase; *11.* indole-3-glycerolphosphate synthase; *12.* tryptophan synthase; *13.* chorismate mutase; *14.* prephenate dehydratase; *15.* prephenate dehydrogenase; *16.* phenylalanine and tyrosine aminotranferase; *17.* prephenate aminotransferase; *18.* arogenate dehydratase; *19.* arogenate dehydrogenase. DAHP, 3-deoxy-D-*arabino*-heptulosonate 7-phosphate; PEP, phospho*enol*pyruvate

The first step of this pathway is the condensation of erythrose-4-phosphate, a product of photosynthesis, with phospho*enol*pyruvate (PEP), a product of glycolysis, to give 3-deoxy-D-*arabino*heptulosonate-7-phosphate (DAHP). This reaction is catalysed by 3-deoxy-D-*arabino*heptulosonate-7-phosphate synthase (EC 4.1.2.15). DAHP undergoes a series of subsequent reactions catalysed by 3-dehydroquinate synthase (EC 4.6.1.3), 3-dehydroquinate dehydratase (EC 4.2.1.10) and shikimate dehydrogenase (EC 1.1.1.25) to give shikimate (3,4,5,-trihydroxycyclohexene-1-carboxylic acid). Shikimate is then phosphorylated to 3-phosphoshikimate by shikimate kinase (EC 2.7.1.71), which is condensed with a molecule of phospho*enol*pyruvate (PEP) to give 5-*enol*pyruvylshikimate 3-phosphate. This reaction is catalysed by 5-*enol*pyruvylshikimate 3-phosphate synthase (EC 2.5.1.19) (EPSPS). By the action of chorismate synthase (EC 4.6.1.4), this latter compound is transformed to chorismate. Chorismate is at a branch point in this pathway, one pathway leading to tryptophan, and the other to phenylalanine and tyrosine (Herrmann 1995).

In the tryptophan biosynthetic pathway, chorismate reacts with the amide group of glutamine to produce anthranilate, a reaction catalysed by anthranilate synthase (EC 4.1.3.27). Anthranilate subsequently condenses with phosphoribosylpyrophosphate to give phosphoribosylanthranilate, a reaction catalysed by phosphoribosylanthranilate transferase (EC 2.4.2.18). This product, in turn, is isomerised by phosphoribosylanthranilate isomerase (EC 5.3.1.24) and converted in the presence of pyridoxal phosphate to indole-3-glycerolphosphate (IGP) by indole-3-glycerolphosphate synthase (EC 4.1.1.48). The final reaction is catalysed by tryptophan synthase (EC 4.2.1.20). This enzyme is an $\alpha_2\beta_2$ heterotetramer and each subunit can catalyse a half reaction, first the transformation of indole-3-glycerolphosphate to indole and glyceraldehyde-3-phosphate, and second the addition of serine on indole to produce tryptophan and water (Radwanski and Last 1995).

The synthesis of phenylalanine and tyrosine starts with the rearrangement of chorismate by chorismate mutase (EC 5.4.99.5) to prephenate. In bacteria, the prephenate is either dehydrated to phenylpyruvate by a prephenate dehydratase (EC 4.2.1.51), or oxidatively decarboxylated to hydroxyphenylpyruvate by a prephenate dehydrogenase (EC 1.3.1.13). Both of these oxo acids are subsequently transaminated to phenylalanine and tyrosine, respectively. It has been suggested that there exists in plants another route which involves the transamination of prephenate to arogenate by prephenate aminotransferase. This molecule is then directly converted either to phenylalanine by arogenate dehydratase (EC 4.2.1.91) or tyrosine by arogenate dehydrogenase (EC 1.3.1.43) This pathway is called the arogenate pathway (Siehl 1999).

The enzymes of aromatic amino acid biosynthesic pathways are located in the chloroplast. Labelling with $^{14}CO_2$ has shown that chloroplasts contain the complete pathway for the synthesis of aromatic amino acids. However, it has been demonstrated that at least one enzyme of the prechorismate pathway in the unicellular phytoflagellate *Euglena gracilis*, EPSPS, is expressed in two different molecular forms, one being localised in the chloroplastic compartment, the other in the cytosol (Reinbothe et al. 1994).

The pathway of aromatic amino acid biosynthesis is relatively well regulated. The first enzyme of the shikimate pathway, DAHP synthase, exists as at least two isoenzymes, one requiring Mn^{2+} and the other Co^{2+} for activity. Arogenate and prephenate have been shown to inhibit the Mn^{2+}-dependent form, and in some circumstances tryptophan has been shown to activate the enzyme. This enzyme is known to be induced by environmental factors, including wounding and microbial infection, correlating with an increase in secondary metabolite synthesis. An antisense construct consisting of the 5 end of the wound-inducible DAHP synthase gene from potato under the control of a CaMV 35S promoter has been used to transform potato (Jones et al. 1995). A number of transgenic plants with impaired wound-induced DAHP synthase activity, polypeptide and mRNA were identified. Several plants exhibited altered morphology but no biochemical analyses were reported. It has been suggested that distinct DAHP synthase genes, support different pathways for the synthesis of aromatic nitrogen compounds, destined for proteins and secondary metabolites.

The penultimate step in the shikimate pathway is catalysed by the 5-enolpyruvylshikimate-3-phosphate synthase (EPSPS). This enzyme is considered as the target for the herbicide glyphosate which is marketed by Monsanto as Roundup. Consequently, EPSPS has been the subject of extensive biochemical studies (Steinrucken and Amrhein 1980). Early studies suggested an ordered mechanism for the enzyme-catalysed reaction in which shikimate 3-phosphate binds first, followed by phospho-enolpyruvate (PEP). Furthermore, for all EPSPS enzymes studied to date, glyphosate has been characterised as a reversible competitive inhibitor versus PEP and an uncompetitive inhibitor versus shikimate-3-phosphate. However, more recent results showed unequivocally that the steady-state kinetic mechanism proceeds through random addition of substrates in both the forward and reverse directions (Sammons et al. 1995). Furthermore, recent NMR work indicated that glyphosate is unlikely to bind in the same fashion as PEP, which casts doubts on an earlier proposal that glyphosate acts as a transition-state or intermediate analogue (McDowell et al. 1996). It would appear that glyphosate binding to the binary complex of the enzyme with shikimate-3-phosphate must involve some substantial adventitious interactions with amino acid residues, which are not intimately involved in substrate binding and catalysis (Gruys and Sikorski 1999). As such, part of the glyphosate molecule must bind near to, but outside the EPSPS active site. The activity of the plastid-localised enzyme EPSPS is inhibited by glyphosate and thereby synthesis of the chorismate is prevented. Inhibition of EPSPS also causes a dramatic buildup in shikimate and shikimate-3-phosphate, which may drain carbon from the reductive pentose phosphate cycle (Shaner and Singh 1992).

A number of attempts have been made to select for mutant glyphosate-resistant plant tissue culture cell lines, which have been shown to contain elevated levels of EPSPS (Nafziger et al. 1984). In carrot suspension cultures, glyphosate-resistant cell lines were obtained by a gradual increase in the concentration of the inhibitor (Shyr et al. 1993). The specific activity of EPSPS was observed to increase, along with the mRNA levels and copy number of the genes at each glyphosate selection step. One tissue culture cell line of maize (black mexican sweet) has been selected to have a marked tolerance to glyphosate. After separation of two isoenzymes of EPSPS by anion exchange chromatography, it was observed that one isoenzyme form was resistant to inhibition by glyphosate. Tolerance to the herbicide was accompanied by a reduced affinity of EPSPS for PEP (Forlani et al. 1992).

Genetic engineering of glyphosate tolerance is of great interest to the agriculture and agrochemical industry. A range of mutants of E. coli and Salmonella typhimu-

rium containing glyphosate-resistant EPSPS have been isolated. Transgenic plants containing the mutant bacterial EPSPS construct targeted to the chloroplast have been shown to be tolerant to the application of glyphosate (Fillatti et al. 1987). However, with the mutant bacterial enzyme, a large increase in the K_i for glyphosate was observed that was accompanied by an equivalent increase in the K_m for PEP (Barry et al. 1992). A gene encoding a glyphosate resistant EPSPS (*aro A*) was isolated from an *Agrobacterium* sp. strain CP 4. The encoded enzyme maintained a low K_m for PEP despite an increased K_i for glyphosate. This CP 4 EPSPS gene, fused to the chloroplast targeting sequence and driven by the CAMV 35S promoter, has been introduced into petunia, *A. thaliana*, canola and soybean (Klee et al. 1987). Transgenic canola containing the chloroplast-targeted CP 4 EPSPS enzyme, exhibited high levels of tolerance to the herbicide at both vegetative and reproductive stages. Transgenic soybean transformed with the same construct has been tested extensively under field conditions. For more information on the genetic engineering of glyphosate tolerance, see the recent review of Padgette et al. (1996).

Chorismate mutase occupies a strategic position in catalyzing the conversion of chorismate to prephenate and has been shown to exist in two isoenzymic forms. Chorismate mutase-I is subject to feedback inhibition by tyrosine and phenylalanine and activation by tryptophan, whereas chorismate mutase-II is insensitive to aromatic amino acids (Schmid and Amrhein 1995). In alfalfa, three isoenzymes are present, which can be inhibited by a range of secondary metabolites.

The tryptophan biosynthetic pathway has recently received considerable attention. Initial attempts to obtain tryptophan auxotrophs were carried out using tissue culture cells. However, it is in *A. thaliana* that the first selection of amino acid auxotrophs requiring tryptophan for growth has been carried out, at the whole-plant level (Last 1993). Normal plants convert anthranilate analogue (5-methylanthranilic acid,) into toxic tryptophan analogue (5- methyltryptophan), whereas mutants blocked in this conversion exhibit reduced toxin biosynthesis (Radwanski and Last 1995).

The first tryptophan-requiring mutant line to be characterised was deficient in phosphoribosylanthranilate transferase (PAT). Plants carrying the *trp 1-1* mutation exhibit a blue fluorescence colour under ultraviolet light, because of the accumulation of anthranilate derivatives. These mutant plants appeared smaller with crinkled leaves and an unusual bushiness. They produced larger flower bolts that contained a small number of fertile flowers (Last and Fink 1988). The results suggested that these disorders were a consequence of auxin deficiency, a finding that was confirmed by the addition of IAA conjugates, which caused the mutant to grow larger and form a higher proportion of fertile flowers. The gene encoding the PAT enzyme (*PAT 1*) was introduced into *A. thaliana*. The analysis of the transgenic plants has confirmed that *PAT 1* is a single copy gene and encodes a chloroplast-localised enzyme (Rose et al. 1992).

Anthranilate synthase is subject to feedback regulation by tryptophan, but not by phenylalanine or tyrosine. The enzyme protein contains two different subunits termed α and β, and probably exists in $\alpha_2\beta_2$ formation. In *A. thaliana* two genes encoding the α subunit and three genes encoding the β subunit have been isolated and are differentially expressed (Radwanski and Last 1995). There is evidence that the subunit genes are differentially regulated during plant development and in response to environmental stimuli. Furthermore, accumulation of anthranilate synthase mRNA is induced by pathogenesis and wounding. However, important post-transcriptional control mechanisms in regulating anthranilate synthase expression, have been observed in these elicitor responses (Radwanski and Last 1995).

A. thaliana has two genes encoding the β-subunit of tryptophan synthase, TSB 1 and TSB 2 (Berlyn et al. 1989). Both genes are transcribed, although TSB 1 mRNA levels were determined to be at least tenfold higher in the leaves than TSB 2 mRNA. A trp2-1 tryptophan-requiring mutant has been identified that exhibited about 10% of the wild-type tryptophan synthase β activity. The trp2-1 mutation has been complemented by the TSB 1 transgene and was observed to be linked genetically to a polymorphism in the TSB 1 gene, strongly suggesting that trp2-1 is a mutation in TSB 1. The tpr2-1 mutant required tryptophan for growth under standard illumination but not under very low illumination conditions. Presumably, under low light the poorly expressed gene, TSB2, is capable of supporting growth. The duplication of biosynthetic genes in A. thaliana has important evolutionary, functional, and practical implications for plant molecular genetics. Genetic redundancy has been observed to be common in many aromatic amino acid biosynthetic pathways in plants. The existence of two tryptophan pathways, for example, has important consequences for tissue-specific regulation of amino acid and secondary metabolite biosynthesis (Last et al. 1991).

To date, no mutant has been isolated for phosphoribosylanthanilate isomerase (PAI). This incapacity is probably due to the presence of three PAI genes. Plants constitutively expressing an antisense PAI cDNA had as little as 15% PAI activity, 10% PAI protein and phenotypes consistent with a block early in the tryptophan pathway (blue fluorescence under UV light due to the accumulation of anthranilate compounds) (Li and Last 1996). Studies using tryptophan biosynthetic mutants have challenged the idea that tryptophan is an obligatory intermediate in the biosynthesis of some secondary products. Furthermore, mutants in indole secondary metabolism are being characterised, creating exciting research directions.

Ureide Synthesis

Ureides are nitrogen compounds such as citrulline, allantoin and allantoic acid. These molecules, based on the structure of urea (with a high N/C ratio), are used for the transport of nitrogen in plants, especially in many tropical legumes. The legume plants referred to as the tropical legumes such as soybean, Phaseolus beans and cowpeas, when nodulated, transport organic nitrogen as ureides. Conversely, the temperate legumes such as peas, lupins and alfalfa, synthesise amides, asparagine and glutamine, the more common amides, regardless of whether they are nodulated or not. The rationale for this switch from amide to ureide synthesis in tropical legumes involves the economy of carbon use. It appears that the ureide producers use less organic carbon to transport the same amount of nitrogen than the amide transporters. Moreover, catabolism of ureides appears to be less efficient than catabolism of amides, since some of the carbon is lost as CO_2. The overall cost of using ureides is metabolically more expensive, almost twice as much as using asparagine (Schubert and Boland 1990).

Ureides are synthesised by the oxidation of purines, which are themselves derived from glutamine, glycine, aspartate and ribose-5-phosphate. Purine synthesis from these latter compounds occurs in the plastids of infected root cells of legume plants. Xanthine, one of the purines, is oxidised to uric acid by xanthine dehydrogenase (EC 1.1.1.204) in the cytosol. Uric acid is transformed in the peroxisomes by uricase (EC 1.7.3.3) to allantoin, with the concomitant release of CO_2 and H_2O_2, and then converted to allantoic acid by allantoinase (EC 3.5.3.4), an enzyme of the endoplasmic reticulum. The ureide is cleaved to yield two molecules of urea (or CO_2 and NH_3) and glyoxylate. The cellular and subcellular localisation of these reactions has not yet been completely elucidated and is still the subject of some debate(Schubert and Boland 1990; Stebbins and Polacco 1995; Atkins et al. 1997; Lea and Ireland 1999).

Glutathione Synthesis

Recent years have witnessed an upsurge of interest in the tripetide thiol glutathione (GSH: γ-L-glutamyl-L-cysteine-L-glycine), because it has been shown to have a multiplicity of functions in plant metabolism. Not only is it implicated in protecting the leaf against oxidative stress, particularly in conjunction with ascorbate, but GSH is also the precursor of the phytochelatins, which allow plants to withstand supraoptimal concentrations of heavy metals. Since GSH often constitutes the major pool of non protein reduced sulphur, it may considerably influence sulphur metabolism. Indeed, recent studies provide growing evidence for a coordinating role of this metabolism, owing to the inhibitory effect of GSH upon sulfur uptake at the root level (Lappartient and Touraine 1996).

Glutathione synthesis takes place in two ATP-dependent steps, through reactions catalysed by γ-glutamylcysteine synthetase (EC 6.3.2.2.: γ-ECS) and glutathione synthetase (EC 6.3.2.3). The foliar forms of these enzymes have been partially characterised in chloroplastic and non-chloroplastic compartments. Nevertheless, despite these advances, biochemical data on the two enzymes implicated in glutathione synthesis are still lacking. The extent of glutathione synthesis in leaves is controlled by the availability of cysteine, feedback inhibition of γ-ECS by GSH, and the amount of γ-ECS enzyme. However, assessment of the relative contributions of these factors under given conditions is problematic. An alternative approach is to attempt to influence the synthesis of GSH, through overexpression of the enzymes which catalyse GSH synthesis. Consequently, Foyer et al. (1997) have produced transgenic poplars which constitutively express *E.coli* genes encoding γ-ESC or glutathione synthetase. The lines strongly overexpressing γ-ECS, gg28, ggs11 and ggs5, contained enhanced foliar levels of cysteine (up to twofold), γ-glutamylcysteine (5- to 20-fold) and glutathione (2- to 4-fold) (Arisi et al. 1997). These types of experiments allow an investigation of the contribution of the *in vivo* activities of these enzymes in the regulation of the pathway of GSH biosynthesis and to evaluate the role of GSH in stress resistance (Noctor et al. 1998, 1999).

Polyamines

Spermidine (1,8-diamino-4-azaoctane), spermine (1,12-diamino-4,9-diazododecane) and putrescine (1,4-diaminobutane) are all widely distributed in plants, where they may be involved in many processes, including growth and cell division, embryogenesis, aging and senescence, bud sprouting, response to stress and ethylene synthesis. There are many other diamines and polyamines in plants, including diaminopropane and cadaverine (1,5-diaminopentane), but these tend to have a more restricted distribution. Polyamines are found in vacuoles, cytosol and chloroplasts.

Spermine and spermidine are synthesised from putrescine, after decarboxylation of arginine by arginine decarboxylase, a pyridoxal-phosphate protein, (EC 4.1.1.19) into agmatine which is then converted into putrescine by N-carbamoylputrescine amidohydrolase (EC 3.5.1.53). As in animals and bacteria, S-adenosylmethionine is used as methyl donor. Putrescine is also produced directly by decarboxylation of ornithine by ornithine decarboxylase (EC 4.1.1.17). The biosynthetic mechanism for the production of many amines remains to be established and their metabolism is not completely understood (Bouchereau et al. 1999).

Conclusion

Even though there is a good understanding of the amino acid biosynthetic pathways in bacteria, this is not the case in plants. Although the enzymes involved are catalytically similar in plants and bacteria, the mechanisms regulating gene expression and activities of these enzymes are quite different.

Recently, great strides have been made in increasing our understanding of amino acid metabolism by using mutants of higher plants with alterations in some of the steps of biosynthetic amino acid pathways. More recently, transgenic plants deregulated in gene expression (sense and antisense strategies, for example) have been very useful tools to study the regulation of metabolic pathway of amino acids directly *in planta*. It is now possible to overexpress or downregulate the enzymes of one metabolic pathway. Also, it is possible to redirect the flux of carbon and nitrogen from one metabolic pathway to another, to favour or repress the synthesis of a particular nitrogen compound of economical interest. The genetic manipulation of the pathways of amino acid synthesis in plants is still promising.

Acknowledgements. The authors wish to express their gratitude to M.H. Valadier for supplying figures, and J. Van Camp, F. Maillier and C. Anassalon for assistance in the preparation of the manuscript. They would like to thank R. Douce, J. Bourguignon, G. Noctor and A. Savouré for criticism of this manuscript.

References

Al-Sulaiti A, Lea PJ, Davies WJ (1990) Effects of soil drying on morphogenetic and physiological responses of barley photorespiratory mutants. Br Soc Plant Growth Regul Monogra 21 :347-350

Ameziane R, Richard-Molard C, Deleens E, Morot-Gaudry JF, Limami MA (1997) Nitrate ($^{15}NO_3$) limitation affects nitrogen partitioning between metabolic and storage sinks and nitrogen reserve accumulation in chicory (*Cichorium intybus* L.). Planta 202: 303-312

Arisi ACM, Noctor G, Foyer CH, Jouanin L (1997) Modification of thiol contents in poplars (*Populus Tremula x P. alba*) overexpressing enzymes involved in glutathione synthesis. Planta 203: 362-372

Arruda P, Bright SWJ, Kueh JSH, Lea PJ, Rognes SE (1984) Regulation of aspartate kinase isoenzymes in barley mutants resistant to lysine plus threonine. Plant Physiol 76: 442-446

Atkins CA, Smith PMC, Storer PJ (1997) Reexamination of the intracellular localization of *de novo* purine synthesis in cowpea nodules. Plant Physiol 113: 127-135

Aulabaugh A, Schloss J V (1990) Oxalyl hydroxamates as reaction intermediate analogues for keto-acid reductoisomerase. Biochemistry 29 : 2824-2830

Azevedo RA, Arruda P, Turner WL, Lea PJ (1997) The biosynthesis and metabolism of the aspartate derived amino acids in higher plants. Phytochemistry 46: 395-419

Barry G, Kishore G, Padgette S, Taylor M, Kolacz K, Weldon M, Re D, Eichholtz D, Fincher K, Hallas L (1992) Inhibitors of amino acid biosynthsis : strategies for imparting glyphosate tolerance to crop plants. In: Sing BK, Flores HE, Shannon JC (eds) Biosynthesis and molecular regulation of amino acids in plants. American Society of Plant Physiologists, Rockville, Maryland, pp 139-143

Berlyn MB, Last RL, Fink GR (1989) A gene encoding the tryptophan synthase β subunit of *Arabidopsis thaliana*. Proc Natl Acad Sci USA 86: 4604-4608

Biou V, Dumas R, Cohen-Addad C, Douce R, Job D, Pebay-Peyroula E (1997) The crystal structure of plant acetohydroxy acid isomeroreductase complexed with NADPH, two magnesium ions and a herbicidal transition state analog determined at 1.65 Å resolution. EMBO J 16: 3405-3415

Blackwell RD, Murray AJS, Lea PJ (1990) Photorespiratory mutants of the mitochondrial conversion of glycine to serine. Plant Physiol 94: 1316-1322

Bouchereau A, Aziz A, Larher F, Martin-Tanguy (1999) Polyamines and environmental challenges: recent development. Plant Sci 140: 103-125

Bourgin JP (1978) Valine resistant plants grown *in vitro* selected tobacco cells. Mol Gen Genet 161: 225-230

Bourguignon J, Rebeille F, Douce R (1999) Serine and glycine metabolism in higher plants. In: Singh BK (ed) Plant amino acids. Biochemistry and biotechnology. Marcel Dekker, New York, pp 111-146

Bown AW, Shelp BJ (1997) The metabolism and functions of γ-aminobutyric acid. Plant Physiol 115: 1-5

Bright SWJ, Kueh JSH, Franklin J, Rognes SE, Miflin BJ (1982) Two genes for threonine accumulation in barley seeds. Nature 299: 278-279

Brouquisse R, James F, Pradet A, Raymond P (1992) Asparagine metabolism and nitrogen distribution during protein degradation in sugar-starved maize root tips. Planta 118: 384-395

Chang AK, Duggleby RG (1997) Expression, purification and characterization of *Arabidopsis thaliana* acetohydroxyacid synthase. Biochem J 327, 161-169.

Chiba Y, Ishikawa M, Kijima F, Tyson RH, Kim J, Yamamoto A, Nambara E, Leustek T, Wallsgrove RM, Naito S (1999) Evidence for autoregulation of cystathionine χ-synthase mRNA stability in arabidopsis. Science 286: 1371-1374

Christopher JT, Powles SB, Holtum JAM (1992) Resistance to acetolactate synthase inhibiting herbicides in annual ryegrass (*Lolium rigidum*) involves at least two mechanisms. Plant Physiol 100: 1909-1913

Cohen-Addad C, Pares S, Sieker L, Neuberger M, Douce R (1995) The lipoamide arm in the glycine decarboxylase complex is not freely swinging. Nat Struct Biol 2: 63-68

Cohen-Addad C, Faure M, Neuburger M, Ober R, Sieker L, Bourguignon J, Macherel D, Douce R (1997) Structural studies of the glycine decarboxylase complex from pea leaf mitochondria. Biochimie 79: 637-644

Curien G, Job D, Douce, R Dumas R (1998) Allosteric activation of *Arabidopsis* threonine synthase by S-adenosylmethionine. Biochemistry 37, 13212-13221.

Delaunay AJ, Verma DPS (1993) Proline biosynthesis and osmoregulation in plants. Plant J 4: 215-223

Delauney AJ, Hu C, Kishor K, Verma DPS (1993) Cloning of ornithine delta-aminotransferase cDNA from *Vigna aconitifolia* by *trans*-complementation in *Escherichia coli* and regulation of proline biosynthesis. J Biol Chem 268: 18673-18678.

De Lumen BO, Krenz DC, Revilleza MJ (1997) Molecular strategies to improve the protein quality of legumes. Food Technol 51: 67-70.

Diedrick TJ, Frisch DA, Gengenbach BG (1990) Tissue culture isolation of a second mutant lows for increased threonine accumulation in maize. Theor Appl Genet 79: 209-215

Dierks-Ventling C, Tonelli C (1982) Metabolism of proline, glutamate and ornithine in proline mutant root tips of *Zea mays* L. Plant Physiol 69: 130-134

Douce R (1985) Mitochondria in higher plants: structure, function and Biogenesis. Academic Press, London.

Droux M, Ravanel S, Douce R (1995) Methionine biosynthesis in higher plants. II. Purification and characterization of cystathionine β-lyase from spinach chloroplasts. Arch Biochem Biophys 316: 585-595

Droux M, Ruffet M-L, Douce R, Job D (1998) Interactions between serine acetyltransferase and O-acetylserine (thiol) lyase in higher plants. Structural and kinetic properties of free and bound enzymes. Eur J Biochem 255: 235-245

Dumas R, Job D, Douce R (1994a) Crystallization and preliminary crystallographic data for acethohydroxy acid isomeroreductase from *Spinacea oleracea*. J Mol Biol 242: 578-581

Dumas R, Cornillon-Bertrand C, Guigue-Talet P, Genix P, Douce R, Job D (1994b) Interactions of plant acetohydroxy acid isomeroreductase with reaction intermediate analogues: correlations of the slow competitive inhibition kinetics of enzyme activity and herbicidal effects. Biochem J 301: 813-820

Dumas R, Butikofer MC, Job D, Douce R (1995) Evidence for two catalytically different magnesium binding sites in acetohydroxy acid isomeroreductase by site-directed mutagenesis. Biochemistry 34: 6026-6036

El Malki F, Frankard V, Jacobs M (1998) Molecular cloning and expression of a cDNA sequence encoding histidinol phosphate aminotransferase from *Nicotiana tabacum*. Plant Mol Biol 37: 1013-1022.

Elthon TE, Stewart CR (1981) Submitochondrial location and electron transport characteristics of enzymes involved in proline oxidation. Plant Physiol 67: 492-494

Epelbaum S, Chipman DM, Barak Z (1996) Metabolic effects of two enzymes of the branched-chain amino acid pathway in *Salmonella typhimurium*. J Bacteriol 178: 1187-1196

Falco SC, Guida T, Locke M, Mauvais J, Sanders C, Ward RT, Webber P (1995) Transgenic canola and soybean seeds with increased lysine. Biotechnology 13: 577-582

Fillatti JAJ, Kiser J, Rose R, Comai L (1987) Efficient transfer of a glyphosate tolerance gene into tomato using a binary *Agrobacterium tumefaciens* vector. Biotechnology 5: 726-730

Forlani G, Nielsen E, Racchi ML (1992) A glyphosate resistant 5-*enol*pyruvylshikimate-3-phosphate synthase conferts tolerance to a maize cell line. Plant Sci 85: 9-15

Fowden L (1981) Nonprotein amino acids. In: Conn EE (ed) The biochemistry of plants, vol. 7. Academic Press, New York, pp 215-247

Foyer CH, Lopez-Delgado H, Dat JF, Scott I (1997) Hydrogen peroxide and glutatione associated mechanisms of acclimatory stress tolerance and signalling. Physiol Plant 100: 241-254

Frankard V, Ghislain M, Jacobs M (1992) Two feedback-insensitive enzymes of the aspartate pathway in *Nicotiana sylvestris*. Plant Physiol 99: 1285-1293

Frankard V, Vauterin M, Jacobs M (1997) Molecular characterisation of an *Arabidopsis thaliana* cDNA coding for a monofunctional aspartate kinase. Plant Mol Biol 34: 233-242

Fujita T, Maggio A, Garcia-Rioz M, Bressan RA, Csonka LN (1998) Comparative analysis of the regulation of expression and structures of two evolutionarily divergent genes for delta Δ'-pyrroline-5-carboxylate synthetase from tomato. Plant Physiol 118: 661-674

Fujimori K, Ohta D (1998) Isolation and characterization of a histidine biosynthetic gene in *Arabidopsis* encoding a polypeptide with two separate domains for phophoribosyl-ATP pyrophosphohydrolase and phosphoribosyl-AMP cyclohydrolase. Plant Physiol 118: 275-283

Galili G (1995) Regulation of lysine and threonine synthesis. Plant Cell 7: 899-906

Galili G, Shaul O, Perl A, Karchi H (1995) Synthesis and accumulation of the essential amino acids lysine and threonine in seeds. In: Kigel J, Gazlili G (eds) Seed development and germination. Marcel Dekker , New York, pp 811-831

Gebhardt C, Shimamoto K, Lazar G, Schnebli V, King PJ (1983) Isolation of biochemical mutants using haploid mesophyll protoplasts of *Hyoscyamus muticus* III. Planta 159: 18-24

Giovanelli J, Veluthambi K, Thompson GA, Mudd SH, Datko AH (1984) Threonine synthase of *Lemna paucicostata* Helgem. 6746. Plant Physiol 76: 285-292

Giovanelli J, Mudd SH, Datko AH (1985) Quantitative analysis of pathways of methionine metabolism and their regulation in *Lemna*. Plant Physiol 78: 555-560

Giovanelli J, Mudd Sh, Datko AH (1989) Regulatory structure of the biosynthetic pathway for the aspartate family of amino acids in *Lemna paucicostata* Hegelm. 6746, with special reference to the role of aspartokinase. Plant Physiol 90: 1584-1599

Girousse C, Bournoville R, Bonnemain J-L (1996) Water deficit-induced changes in concentration in proline and some other amino acids in the phloem sap of alfalfa. Plant Physiol 111: 109-113

Givan CV (1980) Aminotransferases in higher plants. In: Miflin BJ (ed) The biochemistry of plants, vol 5. Academic Press, New York, pp 329-357

Glassman KF (1992) A molecular approach to elevating free lysine in plants. In: Singh BK, Flores HE, Shannon JC (eds) Biosynthesis and molecular regulation of amino acids in plants. Americain Society of Plant Physiologists, Rockville Maryland, pp 217-228

Green CE, Phillips RL (1974) Potential selection system for mutants with increased lysine, threonine and methionine in cereal cropss. Crop Sci 14: 827-830

Gruys KJ, Sikorsky JA (1999) Inhibitors of tryptophan, phenylalanine and tyrosine biosynthesis as herbicides. In: Singh BK (ed) Plant amino acids. biochemistry and biotechnology. Marcel Dekker, New York, pp. 357-384

Guilhaudis L, Simorre JP, Blackledge M, Marion D, Gans P, Neuburger M, Douce R (2000) Combined structural and biochemical analysis of the H-T complex in the glycine decarboxylase cycle: Evidence for a destabilization mechanism of the H-protein. Biochemistry 39: 4259-4266

Guyer D, Patton D, Ward E (1995) Evidence for cross-pathway regulation of metabolic gene expression in plants. Proc Nat Acad Sci USA 92: 4997-5000

Halgand F, Vives F, Dumas R, Biou V, Andersen J, Andrieu JP, Cantegril R, Gagnon J, Douce R, Forest E, Job D (1998) Kinetic and mass spectrometric analyses of the interactions between plant acetohydroxy acid isomeroreductase and thiadiazole derivatives. Biochemistry 37: 4773-4781

Hare PD, Cress WA, van Staden J (1999) Proline synthesis and degradation: a model system for elucidating stress-related signal transduction. J Exp Bot 50: 413-434

Haughn GW, Somerville CR (1990) A mutation causing imidazolinone resistance maps to the *csr1* locus of *Arabidopsis thaliana*. Plant Physiol 92: 1081-1085

Haughn GW, Smith J, Mazur B, Somerville CR (1988) Transformation with a mutant *Arabidopsis* acetolactate synthase gene renders tobacco resistant to sulphonylurea herbicides. Mol Gen Genet 211: 266-271

Hell R (1997) Molecular physiology of plant sulphur metabolism. Planta 202: 138-148

Herrmann KM (1995) The shikimate pathway : early steps in the biosynthesis of aromatic compounds. Plant Cell 7: 907-919

Hofgen R, Laber B, Schüttke I, Klonus AK, Streber W, Pohlenz HD (1995) Repression of acetolactate synthase activity through antisense inhibition. Plant Physiol 107: 469-477

Hu CA, Delauney AJ, Verma DPS (1992) A bifunctional enzyme (Δ^1-pyrroline-5-carboxylate synthetase) catalyzes the first two steps in proline biosynthesis in plants. Proc Natl Acad Sci USA 89: 9354-9358

Hua XJ, van de Cotte B, Van Montagu M, Verbruggen N (1997) Developmental regulation of pyrroline-5-carboxylate reductase gene expression in Arabidopsis. Plant Physiol 114: 1215-1224

Igarashi Y, Yoshiba Y, Sanada Y Yamaguchi-Shinozaki K, Wada K, Shinozaki K (1997) Characterization of the gene for Δ^1-pyrroline-5-carboxylate synthetase and correlation between the expression of the gene and salt tolerance in Oriza sativa L. Plant Mol Biol 33: 857-865

Inaba K, Fujiwara T, Hayashi H, Chino M, Komeda Y, Naito S (1994) Isolation of an Arabidopsis thaliana mutant, mto1, that overaccumulates soluble methionine. Temporal and spatial patterns of soluble methionine accumulation. Plant Physiol 104: 881-887

Ireland RJ, Lea PJ (1999) The enzymes of glutamine, glutamate, asparagine, and aspartate metabolism. In: Singh BK (ed) Plant amino acids.Biochemistry and biotechnology. Marcel Dekker , New York, pp 49-109

Jones JD, Henstrand JM, Handa AK, Herrmann KM, Weller SC (1995) Impaired wound induction of 3-deoxy-D-arabino-heptulosonate-7-phosphate (DAHP) synthase and altered stem development in transgenic potato plants expressing a DAHP synthase antisense construct. Plant Physiol 108: 1413-1421

Karchi H, Shaul O, Galili G (1993) Seed-specific expression of the bacterial desensitized aspartate kinase increases the production of seed threonine and methionine in transgenic tobacco. Plant J 3: 721-727

Karchi H, Shaul O, Galili G (1994) Lysine synthesis and catabolism are coordinately regulated during tobacco seed development. Proc Natl Acad Sci USA 91: 2577-2581

Kemper EL, Cord-Neto G, Capella AN, Gonçalves-Butruile, Azevedo RA, Arruda P (1998) Structure and regulation of the bifunctional enzyme lysine-oxoglutarate reductase-saccharopine dehydrogenase in maize. Eur J Biochem 253: 720-729

Keys AJ, Bird IF, Cornelius MJ, Lea PJ, Wallsgrove RM, Miflin BJ (1978) The photorespiratory nitrogen cycle. Nature 275: 741-743

Kishor PB, Hong Z, Miao G-H, Hu C-AA, Verma DPS (1995) Overexpression of Δ^1-pyrroline-5-carboxylate synthetase increases proline production and confers osmotolerance in transgenic plants. Plant Physiol 108: 1387-1394

Kiyosue T, Yoshiba Y, Yamaguchi-Shinozaki K, Shinozaki K (1996) A nuclear gene encoding mitochondrial proline dehydrogenase, an enzyme involved in proline metabolism, is upregulated by proline but downregulated by dehydratation in Arabidopsis. Plant Cell 8: 1323-1335

Klee HJ, Muskopf YM, Gasser CS (1987) Cloning of an *Arabidopsis thaliana* gene encoding 5-enolpyruvylshikimate -3-phosphate synthase: sequence analysis and manipulation to obtain glyphosate tolerant plants. Mol Gen Genet 210: 437-442

Lam HM, Coschigano KT, Oliveira IC, Melo-Oliveira R, Coruzzi GM (1996) The molecular genetics of nitrogen assimilation in to amino acids in higher plants. Annu Rev Plant Physiol Mol Biol 47: 569-593

Lappartient AG, Touraine B (1996) Demand-driven control of root ATP sulfurylase activity and sulfate uptake in intact *Canola*. Plant Physiol 111: 147-157

Last RL (1993) The genetics of nitrogen assimilation and amino acid biosynthesis in flowering plants: progress and prospects. Int Rev Cytol 143: 297-230

Last RL, Fink GR (1988) Tryptophan-requiring mutants of the plant *Arabidopsis thaliana*. Science 240: 305-310

Last RL, Bissinger PH, Mahoney DJ, Radwanski ER, Fink GR (1991) Tryptophan mutants in *Arabidopsis* : the consequences of duplicated tryptophan synthase β genes. Plant Cell 3: 345-358

Lea PJ, Forde BG (1994) The use of mutants and transgenic plants to study anino acid metabolism. Plant Cell Environ 17: 541-556

Lea PJ, Ireland RJ (1999) Nitrogen metabolism in higher plants. In: Singh BK (ed) Plant amino acids. Biochemistry and biotechnology. Marcel Dekker, New York, pp 1-47

Lea PJ, Robinson SA, Stewart GR (1990) The enzymology and metabolism of glutamine, glutamate and asparagine. In: Miflin BJ, Lea PJ (eds) The biochemistry of plants, vol 16. Academic Press, London, pp 121-159

Leegood RC, Lea PJ, Adcock MD, Hausler RE (1995) The regulation and control of photorespiration. J Exp Bot 46: 1363-1376

Li J, Last RL (1996) The *Arabidopsis thaliana* trp5 mutant has a feedback-resistant anthranilate synthase and elevated soluble tryptophan. Plant Physiol 110: 51-59

Macherel D, Bourguignon J, Forest E, Faure M,Cohen-Addad C, Douce R (1996) Expression, lipoylation and structure determination of recombinant pea H-protein in *Escherichia coli*. Eur J Biochem 236: 27-33

Madison JT, Thompson JF (1988) Characterization of soybean tissue culture cell lines resistant to methionine analogs. Plant Cell Rep 7: 473-476

Mallory-Smith C, Thill DC, Dial MJ, Zemetra RS (1990) Inheritance of sulfonylurea herbicide resistance in *Lactuca* spp. Weed Technology 4: 787-790

Mazur BJ, Falco SC (1989) The development of herbicide-resistant crops. Annu Rev Plant Physiol Mol Biol 40: 441-470

McDowell LM, Klug CA, Beusen DD, Schaefer J (1996) Ligand geometry of the ternary complex of 5-enolpyruvylshikimate-3-phosphate synthase from rotational-echo double-resonance NMR. Biochemistry 35: 5395-5403

Metha K, Hale TI, Christen P (1989) Evolutionary relationships among aminotransferases: tyrosine aminotransferase, histidinol-phosphate aminotransferase, and aspartate aminotransferase are homologous proteins. Eur J Biochem 186: 249-253

Micallef BJ, Shelp BJ (1989) Arginine metabolism in developing soybean cotyledons. Plant Physiol 90: 624-630

Mills WR, Lea PJ, Miflin BJ (1980) Photosynthetic formation of the aspartate family of amino acids in isolated chloroplasts. Plant Physiol 65: 1166-1172

Mori I, Fonne-Pfister R, Matsunaga S, Tada S, Kimura Y, Iwasaki G, Mano J, Hatano M, Nakano T, Kolzumi S, Scheidegger A, Hayakawa K, Ohta D (1995) A novel class of herbicides. Plant Physiol 107: 719-723

Mouillon JM, Aubert S, Bourguignon J, Gout E, Douce R, Rebeille F (1999) Glycine and serine catabolism in non-photosynthetic higher plant cells: their role in C1 metabolism. Plant J. 20: 197-205

Mourad G, King J (1992) Effect of four classes of herbicides on growth and acetolactate synthase activity in several varients of *Arabidopsis thaliana*. Planta 188: 491-497

Nagai A, Ward E, Beck J, Tada S, Chang JY, Scheidegger A, Ryals J (1991) Structural and functional conservation of histidinol dehydrogenase between plants and microbes. Proc Nat Acad Sci USA 88: 4133-4137

Nafziger ED, Widholm JM, Steinrucken HC, Kilmer JL (1984) Selection and characterization of a carrot cell line tolerant to glyphosate. Plant Physiol 76: 571-574

Negrutiu I, Cattoir-Reynaerts A, Verbruggen I, Jacobs M (1984) Lysine overproducing mutants with an altered dihydrodipicolinate synthase from protoplast culture of *Nicotiana sylvestris* (Spegg and Comes) Theor Appl Genet 68: 11-20

Negrutiu I, De Bronwer D, Dirks R, Jacobs M (1985) Amino acid auxotrophs from protoplast cultures of *Nicotiana plumbaginifolia*. Mol Gen Genet 199: 330-37

Noctor G, Arisi ACM, Jouanin L, Kunert KJ, Rennenberg H, Foyer CH (1998) Glutathione: biosynthesis, metabolism and relationship to stress tolerance explored in transformed plants. J Exp Bot 49: 623-647

Noctor G, Arisi ACM, Jouanin L, Foyer CH (1999) Photorespiratory glycine enhances glutathione accumulation in both the chloroplastic and cytosolic compartments. J Exp Bot 50: 1157-1167

Odell JT, Caimi PG, Yadav NS, Mauvais CJ (1990) Comparison of increased expression of wild-type and herbicide-resistant acetolactate synthase genes in transgenic plants and indication of posttranscriptional limitation of enzyme activities. Plant Physiol 94: 1647-1654

Oliver DJ, Neuburger M, Bourguignon J, Douce R (1990) Interaction between the component enzymes of the glycine decarboxylase multienzyme complex. Plant Physiol 94: 833-839

Padgette SR, Re DB, Barry GF, Eichholtz DE, Delannay X, Fuchs RL, Kishore GM, Fraley RT (1996) New weed control opportunities : development of soybeans with a roundup Ready gene. In: Duke SO (ed) Herbicide-resistant crops. Lewis, Boca Raton Florida, pp 53-84

Pares S, Cohen-Addad C, Sieker L, Neuburger M, Douce R (1994) X-ray structure determination at 2.6 – A resolution of a lipoate-containing protein: the H-protein of the glycine decarboxylase complex from pea leaves. Proc Natl Acad Sci USA 91: 4850-853

Racchi ML, Gavazzi G, Dierks-Ventling C, King P (1981) Characterization of proline requiring mutants in *Zea mays*. Z Pflanzenphysiol 101: 303-311

Radwanski ER, Last RL (1995) Tryptophan biosynthesis and metabolism : biochemical and molecular genetics. Plant Cell 7: 921-934

Ravanel S, Gakière B, Job D, Douce R (1998) The specific features of methionine biosynthesis and metabolism in plants. Proc Natl Acad Sci USA 95: 7805-7812

Rebeille F, Neuberger M, Douce R (1994) Interaction between glycine decarboxylase, serine hydroxymethyltransferase and tetrahydrofolate polyglutamates in pea leaf mitochondria. Biochem J 302: 223-228

Reinbothe C, Orgel B, Parthier B, Reinbothe S (1994) Cytosolic and plastid forms of 5-enolpyruvylshikimate-3-phosphate synthase in *Euglena gracilis* are differently expressed during light-induced chloroplast development. Mol Gen Genet 245: 616-622

Rochat C, Boutin JP (1991) Metabolism of phloem-borne amino acids in maternal tissues of fruit of nodulated or nitrate-fed pea plants (*Pisum sativum L.*). J Exp Bot 42: 207-214

Rognes SE, Bright SWJ, Miflin BJ (1983) Feedback-insensitive aspartate kinase isoenzymes in barley mutants resistant to lysine plus threonine. Planta 157: 32-38

Rolland N, Droux M, Douce R (1992) Subcellular distribution of O-acetylserine(thiol) lyase in cauliflower (*Brassica oleracea L.*) inflorescence. Plant Physiol 98: 927-935

Roosens NH, Tran TT, Iskandar HM, Jacobs M (1998) Isolation of the ornithine-delta-aminotransferase cDNA and effect of salt stress on its expression in *Arabidopsis thaliana*. Plant Physiol 117: 263-271

Rose AB, Casselman AL, Last RL (1992) A phosphoribosylanthranilate transferase gene is defective in blue fluorescent *Arabidopsis thaliana* tryptophan mutants. Plant Physiol 100: 582-592

Saito K, Kurosawa M, Tatsuguchi K, Takagi Y, Murakoshi I (1994) Modulation of cysteine biosynthesis in chloroplasts of transgenic tobacco overexpressing cysteine synthase [O-Acetylserine (thiol)-lyase].Plant Physiol 106: 887-895

Sammons RD, Gruys KJ, Anderson KS, Johnson KA, Sikorski JA (1995) Reevaluating glyphosate as a transition-state inhibitor of EPSP synthase: identification of an EPSP synthase·EPSP·glyphosate ternary complex. Biochemistry 34: 6433-6440.

Savoure A , Hua XJ, Bertauche N, Van Montagu, Verbruggen N (1997) Abscisic acid-independent and abscisic acid-dependent regulation of proline biosynthesis following cold and osmotic stresses in *Arabidopsis thaliana*. Mol Gen Genet 254: 104-109

Schloss JV, Ciskanik LM, Van Dyk DE (1988) Origin of the herbicide binding site of acetolactate synthase. Nature 331: 360-362

Schmid J, Amrhein N (1995) Molecular organisation of the shikimate pathway in higher plants.Phytochemistry 39: 737-749

Schubert KR, Boland MJ (1990) The ureides. In: Miflin BJ, Lea PJ (eds) The biochemistry of plants, vol 16. Academic Press, New York, pp 197-282

Schultz CJ, Hsu M, Miesak B, Coruzzi G (1998) *Arabidopsis* mutants define an *in vivo* role for isoenzymes of aspartate aminotransferase in plant nitrogen assimilation. Genetics 149: 491-499

Shaner DL, Singh BK (1992) How does inhibition of amino acid biosynthesis kill plants. In: Singh BK,Flores HE, Shannon HC (eds) Biosynthesis and molecular regulation of amino acids in plants. American Society of Plant Physiologists, Rockville, Maryland, pp 174-183

Shaul O, Galili G (1992a) Increased lysine synthesis in tobacco plants that express high levels of bacterial dihydrodipicolinate synthase in their chloroplast. Plant J 2: 203-209

Shaul O, Galili G (1992b) Threonine over production in transgenic tobacco plants expressing a mutant desensitized aspartate kinase of *Escherichia coli*. Plant Physiol 100: 1157-1163

Shyr YYJ, Caretto S, Widholm JM (1993) Characterisation of the glyphosate selection of carrot suspension cultures resulting in gene amplification. Plant Sci 88: 219-228

Sieciechowicz KA, Joy KW, Ireland RJ (1988) The metabolism of asparagine in plants. Phytochemistry 27: 663-671

Siehl DL, (1999) The biosynthesis of tryptophan, tyrosine, and phenylalanine from chorismate. In :Singh BK (ed) Plant amino acids. Biochemistry and biotechnology. Marcel Dekker , New York. pp 171-204

Singh BK, Shaner DL (1995) Biosynthesis of branched chain amino acids: from test tube to field. Plant Cell 7: 935-944

Somerville CR (1986) Analysis of photosynthesis with mutants of higher plants and algae. Annu Rev Plant Physiol 37: 467-507

Stebbins NE, Polacco JC (1995) Urease is not essential for ureide degradation in soybean. Plant Physiol 109: 169-175

Steinrucken HC, Amrhein N (1980) The herbicide glyphosate is a potent inhibitor of 5-enolpyruvyl-shikimic acid-3-phosphate synthase Biochem Biophys Res Commun 94: 1207-1212

Strizhov N, Abraham E, Okresz L, Blicking S, Zilberstein L, Schell J, Koncz C, Szabados L (1997) Differential expression of two P5CS genes controlling proline accumulation during sal-stress requires ABA and is regulated by *ABA1*, *ABI1* and *AXR2* in Arabidopsis. Plant J 12: 557-569

Stulen I, Israelstam GF, Oaks A (1979) Enzymes of asparagine synthesis in maize roots. Planta 146: 237-241

Szoke A, Miao GH, Hong Z, Verma DPS (1992) Subcellular location of Δ'- pyroline-5-carboxylate reductase in root/nodule and leaf of soybean. Plant Physiol 99: 1642-1649

Ta TC, Macdowall FDH, Faris MA (1988) Metabolism of nitrogen fixed by nodules of alfalfa (*Medicago sativa L.*) II. asparagine synthesis. Biochem Cell Biol 66: 1349-1354

Tada SD, Hatano M, Nakayama Y, Volrath S, Guyer O, Ward E, Ohta D (1995) Insect cell expression of recombinant imidazole glycerophosphate dehydratase of *Arabidopsis* and wheat and inhibition by triazole herbicides. Plant Physiol 109: 153-159

Taylor AA, Stewart GR (1981) Tissue and subcellular localization of enzymes of arginine metabolism in *Pisum sativum*. Biochem Biophys Res Commun 101: 1281-1289

Taylor CB (1996) Proline and water deficit: ups, downs, ins, and outs. Plant Cell 8: 1221-1224

Tourneur C, Jouanin L,Vaucheret H (1993) Over-expression of acetolactate synthase confers resistance to valine in transgenic tobacco. Plant Sci 88: 159-168

Van Dyk, TK, Ayers BL, Morgan RW, Larossa RA (1998) Constricted flux through the branched-chain amino acid biosynthetic enzyme acetolactate synthase triggers elevated expression of genes regulated by *rpoS* and internal acidification. J Bacteriol 180: 785-792

Vauclare P, Diallo N, Bourguignon J, Macherel D, Douce R (1996) Regulation of the expression of the glycine decarboxylase complex during pea leaf development. Plant Physiol 112: 1523-1530

Verbruggen N, Hua X-J, May M, Van Montagu M (1996) Environmental and developmental signals modulate proline homeostasis: evidence for a negative transcriptional regulator. Proc Natl Acad Sci USA 93: 8787-8791

Verslues PE, Sharp RE (1999) Proline accumulation in maize (*Zea mays L.*) primary roots at low water potentials. II. Metabolic source of increased proline deposition in the elongation zone. Plant Physiol 119: 1349-1360

Verma DPS, Zhang C-S (1999) Regulation of proline and arginine biosynthesis in plants. In: Singh BK (ed) Plant amino acids. Biochemistry and biotechnolgy. Marcel Dekker, New York, pp. 249-265

Vyazmensky M, Sella C, Barak Z, Chipman DM (1996) Isolation and characterization of subunits of acetohydroxy acid synthase isozyme III and reconstitution of the holoenzyme. Biochemistry 35: 10339-10346

Wallsgrove RM, Lea PJ, Miflin BJ (1983) Intracellular localisation of aspartate kinase and the enzymes of threonine and methionine biosynthesis in green leaves. Plant Physiol 71: 780-784

Wallsgrove RM, Risiott R, King J, Bright SWJ (1986) Biochemical characterization of an auxotroph of *Datura innoxia* requiring isoleucine and valine. Plant Sci 43: 109-114

Woo KC, Morot-Gaudry JF, Summons RE, Osmond B (1982) Evidence for the glutamine synthetase/glutamate synthase pathway during the photorespiratory nitrogen cycle in spinach leaves. Plant Physiol 70: 1514 - 1517

Yoshiba Y, Kiyosue T, Katagiri T, Ueda H, Mizoguchi T, Yamaguchi-Shinozaki K, Wada K, Harada Y, Shinozaki K (1995) Correlation between the induction of a gene for Δ'-pyrroline-5-carboxylate synthetase and the accumulation of proline in *Arabidopsis thaliana* under osmotic stress. Plant J 7: 751-760

Zonia L, Stebbins N, Polacco J (1995) Essential rôle of urease in germination of nitrogen-limited *Arabidopsis thaliana* seeds. Plant Physiol 107: 1097-1103

Amino Acid Transport

Serge DELROT[1], Christine ROCHAT[2], Metchthild TEGEDER[3] and Wolf FROMMER[3]

Depending on the species, nitrogen may be reduced either by symbiosis with bacteria (legumes, see Chap. 3.1), directly after root uptake from the soil (see Chap. 2), or after xylem transport of the root nitrate to the leaf (most herbaceous species, e.g. *Gossypium*, *Xanthium*). Some plants also reduce nitrate partly in the roots and partly in the leaves (*Picea*, barley, maize). Although the different patterns of reduction result in different patterns of nitrogen transport, one feature common to all plants is that the two main conducting systems (xylem and phloem) are involved, and that xylem/phloem exchanges may occur along the path (Pate 1980). This ensures a constant and efficient recycling of nitrogen among the different organs. As will be detailed below, amino acids play a major role in nitrogen transport. In some cases, especially in the remobilisation of nitrogen from the endosperm to the growing embryo, the transport of small peptides (up to four to five amino acids) also has physiological significance. Although the spectra of amino acids found in the xylem and in the phloem are similar, the xylem usually contains low amino acid concentrations (3-20 mM in *Urtica*, Rosnick-Shimmel 1985), whereas much higher concentrations are found in the phloem, *e.g.* 100 mM in rice (Hayashi and Chino 1990) and 60 to 140 mM in different sugar beet varieties (Lohaus et al. 1994).

Nitrogen Storage, Mobilisation, Remobilisation and Amino Acid Transport

Amino acid storage, remobilisation and transport allow the internal recycling of reduced nitrogen and maintenance of metabolism and growth, even when the surrounding becomes deprived of nitrogen. This storage and remobilisation depend closely on amino acid transport.

1. Laboratoire de Physiologie et Biochimie Végétales, ESA CNRS 6161, Bâtiment Botanique, 40 avenue du Recteur Pineau, 86022 Poitiers Cédex, France. *E-mail:* serge.delrot@univ-poitiers.fr
2. Laboratoire du Métabolisme, INRA, route de St-Cyr, 78026 Versailles Cédex, France. *E-mail:* rochatc@versailles.inra.fr
3. Plant Physiology, Zentrum für Molekularbiologie der Pflanzen (ZMBP), Eberhard-Karls-Universitat, Auf der Morgenstelle 1, D-72076 Tübingen, Germany. *E-mail:* Frommer@uni-tuebingen.de

Nitrogen storage includes short-term and long-term storage (Millard 1988). Short-term storage occurs in the form of nitrate, amino acids and vegetative storage proteins (VSP) of low molecular weight (12 to 42 kDa) in aerial parts (stems and leaves), when nitrogen uptake is in excess of the growth demand. Amino acids are also accumulated transiently in the vacuoles of leaf cells, mainly as glutamine and asparagine. VSPs accumulate in the wood parenchyma cells and in the cortex and the phloem parenchyma cells of the bark (Wetzel et al. 1989; Sauter and Van Cleve 1992). In this case, they are stored in protein bodies derived from vacuoles, and their accumulation is induced by short days and low temperatures (Van Cleve and Apel 1993). In poplar, when they are mobilised in spring (Sauter and Van Cleve 1992), the amino acids resulting from VSP hydrolysis are partly transformed into glutamine. This hydrolysis requires exposure of the plants for several weeks to low temperatures, and is decreased by bud removal, suggesting that it depends on nitrogen demand of the growing sinks (Coleman et al. 1993). Amino acids resulting from the hydrolysis of leaf VSPs are mobilised via the phloem, while those derived from roots and tubers are mobilised via the xylem.

Long-term storage occurs even in the case of nitrogen deprivation. Unlike short-term storage, long-term storage in leaves, seeds, tubers, storage roots and the bark of the trees competes with the nitrogen demand of growing organs (Heilmeier et al. 1994). Again, amino acids are involved in this long term storage, mainly as asparagine, glutamine, glutamate and proline (Tromp 1983; Sagisaka 1987). However, amino acids are generally incorporated into proteins which constitute the main form of long-term storage in leaves, bark of trees, cereal and legume seeds.

In leaves, RuBP carboxylase/oxygenase (Rubisco) may be considered as a long-term form of storage because it is remobilised more rapidly than the other proteins during senescence, and because it may, if nitrogen is present in surplus, be synthesised in excess, without increased carbon reduction (Millard 1988). Rubisco, which can account for over 50% of the soluble protein content of leaves of C_3 plants, is a major source of N for grain filling (Peoples et al. 1980).

Amino acid amides are often assumed to be the major amino acids exported in the phloem from senescing leaves, e.g. in *Lolium temulentum* (Thomas 1978). However, phloem exudates from wheat flag leaves contained only traces of asparagine and a small proportion of glutamine (Simpson and Dalling 1981). It has therefore been proposed that cellular pools of amino acids derived from protein hydrolysis may be metabolised in the leaf before being exported in the phloem in the form of amides or not, depending on the species.

In roots, the relative contribution made by amino acids and proteins in long-term storage depends widely on the species. Proteins are the most important in alfalfa and ray-grass roots (Corre et al. 1996), whereas amino acids account for 66% of the stored nitrogen in spurge roots (Cyr and Bewley 1989).

In seeds, nitrogen is accumulated as seed storage proteins which comprise several families whose importance varies with the plant group: gliadins and glutelins in cereals, globulins in legumes (see chap. 6). In the early stages of germination, seed proteins are hydrolysed into amino acids and small peptides, which are used for the growth of the young seedling. In addition to phloem loading by amino acid transporters, this process also involves peptide transporters in cereals. In barley grains, the peptide transport system rapidly develops in the scutellum at the beginning of germination, and the absorption of peptides is at least as important as the absorption of amino acids for the nitrogen nutrition of the embryo (Higgins and Payne 1978). The peptide transport system is specific for L-peptides up to four residues, and obeys Michaelis-Menten kinetics (Higgins and Payne 1977). A proton-peptide cotransport system has also been characterised in mature leaf tissues of broad bean (Jamai et al. 1994); its function, which may be redundant, as there are also numerous amino acid transporters present in leaf tissue, is still poorly understood. Several peptide transporters have been cloned from *Arabidopsis thaliana* (Steiner et al. 1994; Rentsch et al. 1995), which exhibit a classical topology with 12 putative transmembrane domains, and a wide specificity for various peptides, although the preferred substrate tested is Leu-Leu.

Physiological Aspects of Amino Acid Transport

Amino acids can move through the two long-transport systems of the plant, namely xylem and phloem, arranged together in vascular bundles (Esau 1965). The conducting elements of the xylem are the tracheids and wood vessels. The vessels, which are the most efficient elements in xylem sap conduction, are strongly lignified, elongated dead cells associated with the living parenchyma cells of the xylem. All of them are elongated cells with secondary cell walls that lack protoplasts at maturity. By contrast, the conducting elements of the phloem are living tissues, i.e. sieve elements arranged in sieve tubes and flanked (in angiosperms) by metabolically active, nucleated companion cells. Sieve elements and companion cells are connected to each other by branched plasmodesmata, which may allow transfer of assimilates, nucleotides and possibly nucleic acids and proteins.

The Xylem Sap

Throughout the plant life cycle, xylem transport provides the aerial parts with nitrogen taken up from the roots. This transport is mainly directed towards the actively transpiring leaves, whereas the young growing buds are less attractive. During spring, xylem transport also conveys the amino acids resulting from storage protein remobilisation from roots and tubers, in herbaceous biennial and perennial species, as well as in tree species. Depending on the species and the site of nitrogen reduction, amino acids account for variable proportions of the total xylem nitrogen (Fig. 1; Pate 1973). The xylem sap may contain inorganic nitrogen ureides, and amino acids, amongst which the amides glutamine and asparagine often constitute the major part, although frequently all amino acids are present at variable concentrations. In legumes, two groups may be distinguished with regard to the major nitrogenous compounds transported in the xylem sap, those which transport ureides (tropical, e.g. soybean) and those which transport amides (temperate, e.g. pea), mainly asparagine, especially when they are nodulated. In lupin, inhibiting nitrogen fixation by exposing nodules to oxygen dramatically decreased the proportion of asparagine in the xylem sap (Parsons and Baker 1996).

A comparison of the amino acid contents of xylem exudates from roots with that of the root tissue in *Ricinus* indicated that there were significant differences in the amino acid composition, which suggests that the uptake of amino acids in the xylem is a selective process (Schobert and Komor 1990).

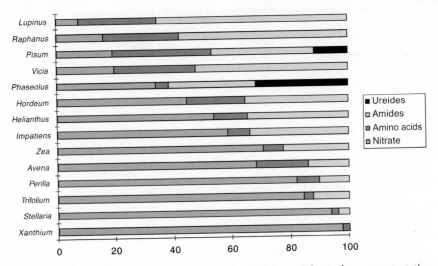

Fig. 1 : Relative proportions of nitrogen components of xylem sap from a large representation of monocot and dicot plants. (redrawn from Pate 1973). Plants were continously watered with a nutrient solution containing 140 ppm of nitrate.

The Phloem Sap

Content of Phloem Sap

The phloem sap conveys the amino acids from mature exporting leaves or from senescing leaves to the young growing leaves and fruits. It is, therefore, involved in the synthesis of structural and enzymatic proteins needed for cell construction, and in the synthesis of seed storage proteins. Amino acids usually account for 5 to 10% of total phloem solutes. As in the xylem, all amino acids may be present, but a major compound usually predominates, the nature of which depends on the species: asparagine in *Lupinus albus* (Pate et al. 1974), serine in *Yucca flaccida* (Tammes and Van Die 1964), allantoin and allantoic acid in *Acer pseudoplatanus* and citrulline in *Cucurbita ficifolia* (Eschrich 1963). The most common amino acids are asparagine/aspartate and glutamate/glutamine, serine, threonine, glycine, alanine, valine, leucine and isoleucine. The phloem concentration of proline, which is usually present in only trace amounts, is strongly increased by water stress (Hanson and Tully 1979; Girousse et al. 1996). Non-protein amino acids may aso be present, sometimes in large amounts, such as γ-aminobutyrate in the phloem sap of *Glycine max* (Housley et al. 1979), γ-methylene glutamine in peanut and homoserine in pea (Pate 1980). The amino acid composition of the phloem undergoes seasonal (Ziegler 1975) and diurnal (Pate 1976; Hocking 1980) variations, especially in trees, but the overall balance between the different amino acids remains quite constant (Pate 1976). In legumes, nodulation alters the amino acid composition of the phloem sap at the pod level in pea (Rochat and Boutin 1991). Phloem amino acids are either derived from photosynthesis and loaded after a short transport in the mesophyll cells, or retrieved from the xylem sap which irrigates the cell walls of the leaf cells, especially in the conducting bundles. In *Lupinus,* for example, phloem glutamine and asparagine mainly come from xylem-phloem exchanges, whereas alanine and glycine are mainly derived from photosynthesis (Pate 1976).

Phloem Loading

According to Pate (1976), the concentration of some amino acids is 3-to 47-fold higher in the phloem than in the xylem, which argues for active loading of the phloem. Parallel physiological studies have been done on sucrose uptake and phloem loading, and have relied on two main experimental systems, excised leaf discs of various species and *Ricinus* cotyledons. In soybean (Servaites et al. 1979) and in broad bean (Despeghel and Delrot 1983), the accumulation of amino acids in the phloem was inhibited by p-chloromercuribenzene sulphonic acid (PCMBS), a slowly permeable thiol reagent. This result indicates that, in these species, phloem loading is mediated by a protein carrier retrieving the amino acids from the apoplast. Similar conclusions were reached from experiments with castor bean cotyledons, which showed that exogenously supplied glutamine was much more efficiently loaded than glutamate, alanine or arginine (Schobert and Komor 1989). Comparative studies on the concentration of sucrose and amino acids in the cytoplasm and in the vacuole of mesophyll cells, as well as in the sieve tube sap, have been performed in spinach (Riens et al. 1991) and barley (Winter et al. 1992). The data show that the amino acid concentration of the sieve sap is similar to that of the cytoplasm, whereas sucrose is tenfold more concentrated in the sieve sap than in the cytoplasm of mesophyll cells. Phloem loading of amino acids therefore occurs against a small concentration gradient, if any. In barley, aspartate, glutamate, valine, lysine are slightly more concentrated in the phloem than in the cytoplasm of the mesophyll cell, while glutamine is less concentrated. The total amino acid concentration in the phloem increases from 186 mM in the light to 244 mM in the dark (Riens et al. 1991). The concentration of most amino acids undergoes diurnal variations in the phloem sap, which parallel the amino acid content of the leaf. One exception to this is the concentration of glutamate, which strongly increases in the phloem during the night, while the leaf concentration remains unaltered. In maize, similar approaches have led to the conclusion that asparagine and glutamine are much higher and glycine and aspartate much lower in the phloem sap than in the leaf extract (Weiner et al. 1991).

Altogether, these data show that phloem loading of amino acids is somewhat selective, and carrier-mediated. Measurements of proton fluxes accompanying amino acid uptake in leaf tissues and in cotyledons, as well as transient membrane depolarisations induced by amino acid uptake, indicate that uptake is mediated by proton/amino acid symporters. These symporters have been characterised by biochemical and molecular approaches (see below). Apoplastic loading of amino acids also implies that they are first leaked from the mesophyll cells into the apoplast. Little is known about this process, although it has been shown that apparent efflux of preloaded [^{14}C]-amino acids from *Commelina benghalensis* leaf discs is stimulated by light (Van Bel et al. 1986).

Phloem Unloading

In legume fruits, the phloem first enters the pod wall, then the seed coat through the funiculus, and ramification occurs in this organ. Metabolism and/or storage can occur all along this pathway. The pod wall serves as a transient reservoir for part of the nitrogen that will be finally stored in the seed. During the first stages of pod development, the pod wall uses the amino acids imported by the phloem to synthesise proteins which are later hydrolysed into peptides and amino acids, and transferred to the seeds at the latest stages of development (Kipps and Boulter 1974). Amino acid composition changes occur during the pathway from the fruit stalk to the seed funicle, due to the formation of threonine (probably from homoserine), and also in the seed coat due to production of glutamine, alanine and valine, which, together with threonine, are the major secreted amino acids from the seed coats (Rochat and Boutin 1991). Temporary storage of nitrogen also occurs in the seed coat, where arginine is detected during development (Rochat and Boutin 1992).

Embryos of developing legume seeds receive their nutrients from the surrounding seed coat, where the phloem terminates. The structural features of the conductive tissue in the fruits may differ among groups of species such as cereals or legumes (for review see Thorne 1985; Patrick 1997). However, some general properties of unloading pathways are shared: (1) unloading occurs entirely in maternal tissues, as no symplastic connection occurs between generations, (2) unloading occurs from vascular termini which are more or less reticulated depending on the species (3) the unloading site can include companion cells or vascular parenchyma cells. The mechanism by which sugars and amino acids are released by the seed coat is not fully understood. Removal of the developing seed in species producing large seeds has allowed the determination of the composition of metabolites supplied to the developing embryo (Thorne 1985). This technique, called the empty seed coat has also been very useful in determining factors that influence the unloading process. However, it must be stressed that the fluxes measured by this technique cover both phloem unloading *per se* and postphloem transfer in the seed coat cells. The osmotic environment of seed tissues seems to have a strong effect on assimilate transport into empty seeds (see reviews by Wolswinkel 1992; Patrick 1997). More attention has been paid in this area to sucrose transport. Recent studies have pointed out differences in release of sugars and amino acids from pea seed coats (de Jong and Wolswinkel 1995). Treatments with PCMBS inhibited both sugar and amino acid release, suggesting a carrier-mediated transport, but amino acid release was not affected by CCCP. More recently, it has been shown that the uptake of amino acids by pea seed coats is a passive, diffusion-like process (De Jong et al. 1997). The picture emerging from these data is that the plasma membrane of the seed coat parenchyma has pores through which molecules as large as amino acids (and sucrose) can diffuse freely.

Assimilates are then retrieved from the apoplastic cavity by the growing embryo. Amino acid absorption has been studied extensively with leguminous seeds such as soybean or pea (Bennett and Spanswick 1983; Lanfermeijer et al. 1990). These studies have demonstrated the presence of two different systems, a saturable and a non-saturable one. The saturable system seems to be an amino acid/H^+ symporter. Studies on valine absorption by pea cotyledons (Lanfermeijer et al. 1990) have shown that the two components may coexist during development. In early developmental stages, amino acids are taken up passively, the movements of amino acids being driven by their active metabolism in the cotyledon cells. At later developmental stages, passive transport is supplemented by a saturable and probably active transport system. Particular attention should be paid in the near future to the presence of specific amino acid transporters in the dermal cell layer of the embryo, as has been carried out for sucrose transport (Harrington et al. 1997).

Phloem/Xylem Exchanges

The use of nitrogenous compounds labelled with ^{15}N and ^{14}C has shown that organic nitrogen loaded in the phloem does not come exclusively from photosynthesis. Direct exchanges of amino acids between the xylem and the phloem (and *vice versa*) may also occur in the stem (Sharkey and Pate 1975; Van Bel 1984; Pate 1989). Quantitative measurements conducted in soybean indicate that the xylem/phloem exchanges account for 35 to 52% of the total nitrogen arriving in the fruit (Layzell and La Rue 1982). The precise mechanisms of these exchanges are still poorly known. Symplastic transport between the xylem and the phloem seems to occur via the wood rays, and involves three steps (1) transport across the xylem vessel/parenchyma interface; (2) radial transfer in the wood ray; (3) loading into the companion cell/sieve tube complex (Van Bel 1990). Also, in barley seedlings, the majority of the amino acids supplied to the roots by the phloem is immediately reexported towards the leaves via the xylem stream (Cooper and Clarkson 1989). The long-distance transport of amino acids therefore involves a constant recycling between the aerial parts of the plants and the roots, mediated by xylem/phloem exchanges. The amino acid phloem content might also control the rate of nitrogen uptake. Indeed, an artificial enrichment of the phloem sap in amino acids resulted in a decrease in nitrogen (Muller and Touraine 1992; Imsande and Touraine 1994).

Amino Acid Transporters:
Physiological and Biochemical Approaches

The long-distance transport and the compartmentation of amino acids into the different organs and plant cells finally depends on the distribution and activity of amino acid transporters located in the plasma membrane and in the various organelles, mainly the vacuole. During the past decade, considerable progress has been made concerning the biochemical and molecular characterisation of the plasma membrane amino acid transporters. In contrast, although there is some physiological information on transtonoplastic amino acid transport, no tonoplast amino acid transporter has yet been cloned.

Early physiological studies showed that amino acid transport across the plant plasma membrane is mediated by several transporters exhibiting a wide specificity, and energised by cotransport with protons and possibly other cations (Etherton and Rubinstein 1978; Kinraide and Etherton 1980; Kinraide 1981; Despeghel and Delrot 1983; Mounoury et al. 1984; Wyse and Komor 1984). Uptake in tissue samples exhibited complex kinetics due to the existence of several transporters and side effects that may have interfered with the actual uptake of labelled amino acids, i.e. diffusion into a tissue composed of several cell types, diffusion across the cell wall, metabolism or symplastic transport.

The ability to prepare highly purified plasma membrane vesicles (PMVs) with a reasonable yield, by phase partitioning of microsomal fractions into a mixture of polymers (Dextran T-500/polyethyleneglycol 3350) (Larsson et al. 1987) has allowed the study of transport across a single membrane. *In vivo*, the activity of the proton/amino acid symporters is powered by the proton-motive force created by the plasmalemma proton-pumping ATPase extruding protons from the cytoplasm to the cell wall, thus creating a pH gradient (ΔpH, about 2 pH units) and an electrical gradient ($\Delta\psi$, about -150 mV). Because the PMVs are mainly oriented right side out, uptake into PMVs must be energised by an artificial proton motive force (Fig. 2). This force is created by equilibrating the internal medium of the vesicles with an alkaline buffer containing potassium, and by resuspending them in an acidic buffer devoid of potassium, in the presence of the potassium ionophore valinomycin. The difference between the alkaline buffer inside and the acidic buffer outside the vesicles mimics the natural ΔpH, whereas the diffusion of potassium from the internal medium to the external buffer creates a transient electrical force, outside positive, mimicking the natural $\Delta\psi$. This experimental approach was developed by several groups and has allowed an extensive study of sugar (Buckhout 1989; Bush 1989; Lemoine and Delrot 1989) and amino acid (Bush and Langston-Unkefer 1988; Li and Bush 1990, 1991) uptake. The method offers several advantages for a precise study of the activity of the transporter: the substrate is not metabolised, it has direct access to the plasma membrane, and the proton-motive force is imposed externally, so that uptake is no longer dependent on the proton-pumping ATPase activity. The limitations of this method are due to the fact that the PMVs obtained are heterogeneous both in size (Lemoine et al. 1991) and in origin (several cell types), that the proton-motive force created is transient, mainly due to the lack of systems recycling the potassium, and that the substrate concentration rapidly builds up inside the vesicles. The transport activities due to PMVs from phloem cannot be detected in PMVs from exporting leaf tissues, because the PMVs from mesophyll cells largely predominate (Lemoine et al. 1996). Nevertheless, studies under initial uptake conditions allow a good appraisal of the activity, specificity, inhibitor sensitivity and kinetics of the transporters.

Initial evidence for the existence of proton-driven amino acid transport associated with PMVs came from cell fractionation studies carried out with zucchini hypocotyls (Bush and Langston-Unkefer 1988). These authors showed that alanine transport driven by an artificial proton-motive force, comigrated with markers of the plasma membrane when a microsomal fraction was fractionated on a sucrose density gradient. However, much of the work dealing with the biochemical characterisation of amino acid transporters has been run with PMVs purified from sugar beet leaves. These studies showed that both alanine (Li and Bush 1990) and valine (Gaillard et al. 1990) transport may be energised by either the ΔpH and/or the $\Delta\psi$ component of the proton-motive force. ΔpH-dependent transport of alanine, leucine, glutamine and glutamate exhibits simple saturable saturation kinetics, whereas isoleucine and arginine transport kinetics are biphasic (Li and Bush 1990). Transport of alanine is inhibited by diethylpyrocarbonate (DEPC), a chemical reagent binding covalently to the imidazole ring of histidine residues. Comparison of inactivation of alanine transport by DEPC in the presence and absence of amino acids showed that substrate binding protects the transporter against DEPC inhibition. This, together with the fact that histidine residues are often involved in rapidly reversible protonation reactions, suggests that a histidine residue is implicated in the reaction mechanism of the transporter (Bush 1993). Detailed studies on competitive inhibition of transport by various amino acids, indicated the presence of at least four transport systems, differing by their substrate specificity in PMVs from sugar beet leaves (Li and Bush 1990, 1991; Bush 1993). One transporter seems to be more specific for acidic amino acids, another for basic amino acids. There are two transport-

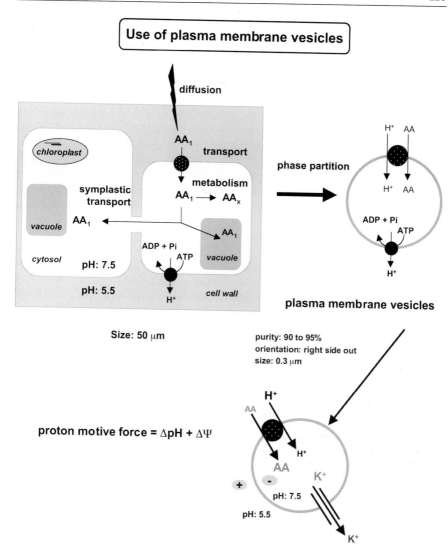

Fig. 2 : Diagram showing the value of using PMVs for the study of amino acid transport. Uptake studies in whole tissues are complicated by various side events that limit the interpretation of the fluxes measured. The purification of PMVs allows a direct study of the activity of the transporters without these interferences. Furthermore, the proton-motive force energising the transport is controlled by the experimenter, so that the transport activity depends neither on the activity of the H^+-ATPase nor on the metabolic status of the cell. The artificial proton-motive force is created by first equilibrating the PMVs in an alkaline K buffer. At the beginning of the incubation with the labelled amino acid, the PMVs are resuspended in an acidic buffer without potassium, and in the presence of the potassium ionophore valinomycin. The differences in the pHs of the buffer mimic the transmembrane pH gradient existing in the cell. Valinomycin allows the diffusion of the potassium preloaded in the PMVs. This creates an accumulation of positive charges outside the vesicles, which mimics the electrical gradient normally created by protons in the cells

ers specific for neutral amino acids; the neutral system I transports isoleucine, valine and threonine, whereas the neutral system II transports alanine and leucine. It has been suggested that branching at the β-carbon blocks the access of isoleucine, valine and threonine to the neutral system II. The positional relationship between the α-amino and carboxyl groups is important for substrate recognition, since D-isomers of amino acid do not compete with L-isomers. Although less restrictive, the molecular interactions associated with the distal end of the amino acid are also important, since the neutral amino acid transporters are able to discriminate between various neutral amino acids, and to exclude acidic and basic amino acids. Use of PMVs from *Ricinus communis* roots (Weston et al. 1994, 1995) and cotyledons (Williams et al. 1992), also demonstrated the multiplicity of amino acid transport systems in these organs. Two transport systems for neutral amino acids were characterised, which differed in their affinity for asparagine on one hand, and for isoleucine and valine on the other (Weston et al. 1995). In contrast to that which had been described for the sucrose transport activity, the transition from the importing, to the exporting stage of sugar beet leaf was not accompanied by an increase in amino acid transport measured in the PMVs (Lemoine et al. 1992).

Although studies with PMVs allow a better characterisation of the energetics and kinetics of amino acid transport, they do not yield unequivocal results concerning the number and specificity of the transporters. Indeed, PMV preparations may contain vesicles from different tissues differing by their transporter content, and a given amino acid may be transported simultaneously by several transporters with various affinities, yielding complex kinetics. In this context, the molecular cloning and the heterologous expression of plant amino acid transporters has revealed a complexity probably higher than was expected from biochemical studies.

Amino Acid Transporters:
Molecular Cloning and Heterologous Expression

Arabidopsis Thaliana

Complementation of yeast mutants deficient in amino acid transport allowed the cloning of the first plant amino acid transporter (*AtAAP1/Nat2*) from an *A. thaliana* cDNA library (Frommer et al. 1993; Hsu et al. 1993). AtAAP1 (amino acid permease, also called NAT2) is a 53-kDa polypeptide containing 486 amino acids. Computer analysis of the hydropathy profile suggests the existence of 10 to 12 putative transmembrane domains. Genetic engineering using a c-myc epitope on either the amino or carboxyl end of the protein, as well as immunofluorescence localisation of these termini in COS1-cells expressing AtAAP1, allowed Chang and Bush (1997) to present a topological model with 11 transmembrane domains. According to this model, the amino terminus of the protein is located in the cytoplasm, whereas the carboxyl terminus is outside the cell. Heterologous expression in a yeast mutant indicated that AtAAP1 transported neutral and acidic amino acids, and that its substrate specificity matched that of the neutral amino acid transporter II characterized in PMV studies (Hsu et al. 1993). Many other amino acid transporters have been cloned from *A. thaliana* (Kwart et al. 1993; Frommer et al. 1994, 1995; Fischer et al. 1995; Rentsch et al. 1996; Chen and Bush 1997). It is now clear that this plant possesses a wide range of different amino acid transporters differing by their specificity, their spatial and temporal expression, and their response to various stresses. This allows a fine tuning and control of nitrogen transport in a variety of physiological and environmental conditions. Based on sequence analysis, the amino acid transporters cloned so far may be classified in two major superfamilies, the ATFs (amino acid transporters) and the APCs (amino acid-polyamine-choline facilitators) (Fischer et al. 1998). The ATF superfamily comprises four families, i.e. AAPs, ProTs (proline transporters), LHT-related proteins (lysine and histidine transporters) and AUX1-related proteins (putative auxin transporters). The AAPs all exhibit strong sequence homologies, wide substrate specificity and are classified either as neutral and acidic amino acid transporters (AtAAP1, AtAAP2, AtAAP4 and AtAAP6), or as general amino acid transporters (AtAAP3 and AtAAP5) (Fischer et al. 1998). Northern analysis and GUS expression studies indicate that AtAAP1 is expressed in the endosperm and the cotyledons, whereas AtAAP2 is expressed in the phloem of the stem and in the veins supplying the seeds (Hirner et al. 1998).

ProT1 and ProT2 possess a strong selectivity for proline; *ProT1* is mainly expressed in roots, stems and flowers, whereas *ProT2* is expressed throughout the plant and its expression is induced by water or salt stress (Rentsch et al. 1996). *ProT1* and *ProT2* encode membrane proteins of 442 and 439 amino acids, each having 10 putative transmembrane domains. Interestingly, water stress induced the expression of *ProT2*, whilst at the same time it decreased that of the AAPs, showing that the expression of various amino acid transporters is strictly coordinated.

LHT1 contains 446 amino acids, 9 or 10 transmembrane domains, and exhibits a high affinity for lysine and histidine. It is expressed in all tissues, with the strongest expression in young leaves, flowers and siliques (Chen and Bush 1997).

Mutations in the gene encoding AUX1, whose sequence is similar to amino acid permeases, confer an auxin-resistant root growth phenotype, suggesting that AUX1 is an auxin transporter (Bennett et al. 1996). ART1, an aromatic amino acid transporter exhibiting sequence similarity with AUX1 and to tryptophan and tyrosine bacterial translocators, was recently identified by functional expression of an EST cDNA (Bush 1999). It was expressed at a very low level in all tissues, but more-specially on the surface of roots and leaves and in stomata.

The plant APCs include two main families, the CATs and proteins homologous to the yeast GABA permeases . AtCAT1, earlier called AAT1, is highly related to mammalian cationic amino acid transporters and recognises a broad spectrum of amino acids. *AtCAT1* is expressed in the vascular tissue (Frommer et al. 1995). An EST similar to the γ-aminobutyrate permease-related family has been identified by homology searches in the databases and is currently under investigation (Fischer et al. 1998).

Other Species

Little information is available on the molecular cloning of amino acid transporters in other species, especially in those where nitrogen transport has economical importance. RcAAP1, an amino acid transporter from *Ricinus communis,* transported a wide range of amino acids, but has a higher affinity for histidine, lysine and arginine. It did not recognize glycine, proline and citrulline (Marvier et al.

1998). A full-length clone (*VfAAP2*) and three partial amino acid transporter genes (VfAAPa, VfAAPb, VfAAPc) were also recently isolated from broad bean (*Vicia faba* L.). The function of VfAAP2 was tested by heterologous expression in a yeast mutant defective in proline transport (Montamat et al. 1999). VfAAP2 mediated proton-dependent proline uptake with an apparent K_m of about 1 mM. Aromatic amino acids, neutral aliphatic acids and citrulline were the best substrates of this transporter, whereas basic amino acids, ornithine and ureides were poorly recognised. Northern analysis indicates that all the four genes exhibit different pattern of expression. *VfAAP2* was most strongly expressed in the stem, at a lower level in sink leaves and pods. *VfAAPa, VfAAPb* and *VfAAPc* were most strongly expressed in the flowers, but their expression in the other organs differed.

Heterologous Expression

Heterologous expression of plant amino acid transporters is a convenient tool to assess their substrate specificity. Although the K_m measured in yeasts usually matches that observed in plant cells, some caution should always be taken. Indeed, different codons are used in yeasts and in plants, and it is assumed that this is one of the reasons why many cDNAs encoding plant membrane transporters cannot be successfully expressed in yeasts. Even when uptake activity is measured, it is likely that only a minor part of the transporter reaches the plasma membrane, due to the poor efficiency of the targeting process. Finally, the lipid environment of the transporter is different in a plant plasma membrane compared to the yeast plasma membrane. This might lead to some differences in transport activities.

Xenopus oocytes have also been widely used for heterologous expression of plant amino acid transporters, and have yielded interesting data on the reaction mechanism. Expression of *AAP1/NAT2* in *Xenopus* oocytes led to the conclusion that H^+/ amino acid cotransport mediated by this protein, occurs via a random simultaneous mechanism which does not depend on the amino acid (Boorer et al. 1996). Membrane voltage enhanced the maximal transport rate and the affinities for H^+ and amino acid. The transport velocity depends on the amino acid, and this may be due to differences in a controlling step in the transport cycle, possibly the translocation rate of the fully loaded transporter. The data also suggest that the transporter has more than one binding site for H^+ and amino acid. Stoichiometry experiments enabled Boorer and Fischer (1997) to show that AAP5 also transported anionic, cationic and neutral amino acids via the same mechanism, i.e. with a fixed amino acid coupling stoichiometry. Thus, *in planta*, the energy consumption for amino acid transport is independent of the net charge of the amino acid.

Regulation of Amino Acid Transport

Amino acid transport may be regulated indirectly via the proton-motive force developed by the plasma-membrane ATPase (Mounoury et al. 1984), and more directly at the transporter level by transcriptional, translational and posttranslational processes. Relatively little is known about the direct regulation of the amino acid transporters, because most of the studies published so far have concentrated on their expression estimated by Northern blot, i.e. on transcriptional control (Fig. 3).

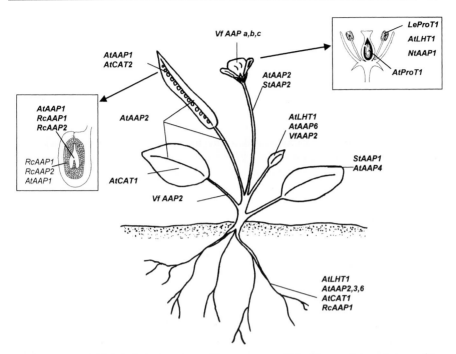

Fig. 3 : Tissue specificity of plant amino acid transporters. This diagram is based on results obtained either from Northern analyses or promoter-reporter gene fusions. (Fischer et al. 1998; Montamat et al. 1999 and references therein)

The expression of the various *A. thaliana* amino acid transporters has been reviewed recently (Fischer et al. 1998). Most of the transporters are expressed in source and sink organs, but some of them exhibit more specific expression: *AtAAP1* in the seed, *AtAAP3* in roots and seedlings, *AtAAP6* in root and sink leaves. At the tissue level within the seed, histochemical analysis with promoter-GUS fusions indicated that *AtAAP1* was expressed in the endosperm and in the embryo proper. This suggests a role in the import of amino acids in the endosperm, and subsequent uptake into the cotyledons of the developing embryo. *AtAAP2* was expressed in the phloem of stems and in the vascular strands of siliques and in funiculi (Hirner et al. 1998), which suggests a role in the uptake of amino acids into the seeds and in xylem/phloem exchanges along the path. The induction of *AtAAP1* and *AtAAP2* at the transcript level precedes the accumulation of 2S albumin transcripts encoding for storage protein, and both genes might be coregulated developmentally during silique maturation.

StAAP1 and *StAAP2* were expressed only in the source leaf and in the stem of potato (Fischer et al. 1998). In *Ricinus*, *RcAAP1* and *RcAAP2* were expressed abundantly in the cotyledon and roots, and at a lower level in the endosperm and hypocotyl. *In situ* hybridisation experiments further localised *RcAAP1* expression in the stele cells adjacent to the xylem poles by Bick et al. (1998). *AtLHT1*, a lysine/histidine transporter was present in all tissues, but more strongly expressed in sink organs such as young leaves, flowers and siliques. *In situ* hybridisation also localised the expression on the surface of roots in young seedlings and in pollen (Chen and Bush 1997). *AtProT1*, a proline transporter was expressed in all organs, but highest levels were found in roots, stems and flowers. *AtProT2* was expressed ubiquitously in the plant (Rentsch et al. 1996).

Some amino acid transporters seem more specifically expressed in reproductive cells. A transporter mediating the uptake of proline, glycine betaine and γ-aminobutyric acid (*LeProT1*) was expressed in mature and germinating pollen of tomato (Schwacke et al. 1999). It could serve for osmotic adjustment during pollen dehydration, and allow the uptake of material needed for the synthesis of OH-Pro-rich proteins during pollen elongation. *NsAAP1*, a putative amino acid transporter related to *AtLHT1* was expressed first during pollen mitosis and dramatically increased in mature pollen shortly before anthesis in *Nicotiana sylvestris* (Lalanne et al. 1997). So far, the only amino acid transporter which may have a specific role in ovule nutrition is *AtProT1*, whose expression in the flowers was downregulated after fertilisation (Rentsch et al. 1996).

Very few studies concerning the functional analysis of plant amino acid transporters promoters have been published so far. Although both *AtAAP1* and *AtAAP2* were expressed at the same time in the silique, no striking homologies could be found in their promoters by sequence analysis. Four ACGT-core-motifs supposed to be involved in many environmental responses were repeated around position -123 in *AtAAP1*. An E-box (also found in the β-phaseolin promoter) and a SEF3 motif were also found in this promoter. Since cell specificities of AtAAP1 and storage protein genes are similar, these elements might be involved in the developmental regulation of expression (Hirner et al. 1998). The 5' flanking region of *NsAAP1* (expressed in the pollen) contains long regions homologous to the promoter region of the tobacco pollen-specific eIF-4A8 translation factor (Lalanne et al. 1997).

The AtAAPs contain two conserved potential sites for phosphorylation (Thr[116] and Thr[248]), and okadaic acid also inhibits proton-driven amino acid transport in purified PMVs (Roblin, Bonmort and Delrot, submitted), which suggests a possible posttranslational regulation by phosphorylation/dephosphorylation. However, it has not yet been demonstrated unequivocally that amino acid transporters are phosphorylated. AtProT1 and AtProT2 contain one potential phosphorylatable site (Thr[247] in AtProt1). LeProT2 and LeProT3 both contain a putative 14-3-3 recognition site ($RSx_{1,2}SxP$), where the second site can be phosphorylated. This sequence is absent in Le ProT1, AtProT1 and AtProT2. Introduction of this sequence by site-directed mutagenesis into LeProT1 and its modification in LeProT2 did not change the activity of these proteins (Schwacke et al. 1999).

There is no evidence for N glycosylation of amino acid transporters of the plant plasma membrane (Chang and Bush 1997).

Amino Acid Transport Across the Tonoplast

Compartmentation studies show that plant vacuoles contain a considerable amount of amino acids (Martinoia et al. 1981; Alibert et al. 1982). This amount depends on the total concentration of amino acids, which, in turn, depends on the nutritional and metabolic status of the plant. The vacuolar content of amino acids increase with the free amino acid content of the cell (Martinoia and Ratajczak 1998), and the vacuolar amino acid concentrations fluctuate more strongly than the cytoplasmic concentrations. This implies that the vacuole may behave as a transient storage compartment for amino acids, and that reversible amino acid transport occurs across the tonoplast. In green tissues, chloroplasts are another compartment which may contain significant amounts of specific amino acids (i.e. aromatic amino acids). In *Chara* cells, it has been shown that the ratio of the different amino acids between the cytosol and the vacuole is quite constant, suggesting that there is no preferential uptake of amino acids across the tonoplast, at least in this organism (Martinoia and Ratajczak 1998). In the same organism, constant exchanges of amino acids between the vacuole and the cytosol have been demonstrated. Pre-loaded [^{14}C]labelled amino acids were exported from the vacuoles and metabolised into other amino acids that were reabsorbed and accumulated into the vacuoles. Under normal conditions, the vacuolar amino acid concentration is usually lower than the cytoplasmic one. Therefore, even if carrier-mediated amino acid uptake across the tonoplast may not require energy, the release of vacuolar amino acids should be energy-dependent.

In higher plants, amino acid transport across the tonoplast has been mostly studied with isolated vacuoles from barley leaves (Homeyer et al. 1989; Dietz et al. 1990; Martinoia et al. 1991; Goerlach and Willms-Hoff 1992). In the absence of ATP, the uptake of most amino acids occurred at a slow rate. Hydrophilic amino acids were taken up faster than hydrophobic ones. Uptake of phenylalanine into barley vacuoles was stimulated by both MgATP and PPi, and the accumulation was reversed by protonophores, suggesting that the observed effects may be due to the indirect activation of an amino acid/proton antiport via the ΔpH created by the V-ATPase and the PPase (Homeyer et al. 1989). Addition of ATP may also stimulate amino acid uptake by a direct effect on the amino acid transporter, since non-hydrolysable ATP analogues did not stimulate transport of several neutral and charged amino acids (glycine, arginine, aspartate). This shows that ATP hydrolysis by the V-type ATPase, or the amino acid transporter itself, are not involved in transport activation of these amino acids, unlike that which was observed for phenylalanine (Dietz et al. 1990; Martinoia et al. 1991; Goerlach and Willms-Hoff 1992). The activation is specific for ATP, since other nucleotides were without effect, and it exhibited a sigmoidal response as a function of ATP concentration. Depending on the amino acids, stimulation of transport was obtained with ATP only, or with either ATP or Mg-ATP. Maximal stimulation of transport was observed with ATP concentrations (3 to 5 mM), which exhibited a K_m for amino acids of about 2.5 mM. This transport system was inhibited by hydrophobic amino acids including phenylalanine, leucine and methionine. Its existence was also demonstrated in *Valerianella* and *Tulipa* (Thume and Dietz 1991). Interestingly, the efflux of amino acids from barley vacuoles exhibited features similar to the uptake, i.e. stimulation by ATP and inhibition by hydrophobic amino acids. This suggests that influx and efflux of amino acids across the tonoplast may be mediated by the same permease. Unlike the plasma membrane amino acid transporters which have been cloned (see above), no tonoplast amino acid transporter has yet been identified. Preliminary attemps to reconstitute and purify the ATP-dependent amino acid transporter led to the conclusion that transport activity was associated with a protein of about 210 kDa (Thume and Dietz 1991).

Despeghel JP, Delrot S (1983) Energetics of amino acid uptake by *Vicia faba* leaf tissue. Plant Physiol 71: 1-6

Dietz KJ, Jager R, Kaiser G, Martinoia E (1990) Amino acid transport across the tonoplast of vacuoles isolated from barley mesophyll protoplasts. Plant Physiol 92: 123-129

Esau K (1965) Anatomy of seed plants, 2nd ed. John Wiley, New York

Eschrich W (1963) Der Phloemsaft von *Cucurbita ficifolia*. Planta 60: 216-224

Etherton B, Rubinstein B (1978) Evidence for amino acid-H^+ co-transport in oat coleoptiles. Plant Physiol 61: 933-937

Feller U, Fischer A (1994) Nitrogen metabolism in senescing leaves. Crit Rev Plant Sci 13: 241-273

Fischer WN, Kwart M, Hummel S, Frommer WB (1995) Substrate specificity and expression profile of amino acid transporters (AAPs) in *Arabidopsis*. J Biol Chem 270: 16315-16320

Fischer WN, André B, Rentsch D, Krolkiewicz S, Tegeder M, Breitkreuz K, Frommer WB (1998) Amino acid transport in plants. Trends Plant Sci 3: 188-195

Frommer WB, Hummel S, Riesmeier JW (1993) Expression cloning in yeast of a cDNA encoding a broad specificity amino acid permease from *Arabidopsis thaliana*. Proc Natl Acad Sci USA 90: 5944-5948

Frommer WB, Hummel S, Rentsch D (1994) Cloning of an *Arabidopsis* histidine transporting protein related to nitrate and peptide transporters. FEBS Lett 347: 185-189

Frommer WB, Hummel S, Unseld M, Ninnemann O (1995) Seed and vascular expression of a high-affinity transporter for cationic amino acids in *Arabidopsis*. Proc Natl Acad Sci USA 92: 12036-12040

Gaillard C, Lemoine R, Delrot S (1990) Absorption de la L-valine par des vésicules de plasmalemme purifiées de feuilles de Betterave (*Beta vulgaris* L.). CR Acad Sci Serie D, Paris, Life sciences or sciences de la Vie, 311: 51-56

Girousse C, Bournoville R, Bonnemain JL (1996) Water-deficit-induced changes in concentrations in proline and some othe amino acids in the phloem sap of alfalfa. Plant Physiol 111: 109-113

Goerlach J, Willms-Hoff I (1992) Glycine uptake into barley mesophyll vacuoles is regulated but not energized by ATP. Plant Physiol 99: 134-139

Hanson AD, Tully RE (1979) Amino acids translocated from turgid and water-stressed barley leaves. II. Studies with ^{13}N and ^{14}C. Plant Physiol 64: 467-471

Harrington GN, Franceschi VR, Offler CE, Patrick JW, Tegeder M, Frommer WB, Harper JF, Hitz WD (1997) Cell specific expression of three genes involved in plasma membrane sucrose transport in developing *Vicia faba* seed. Protoplasma 197: 160-173

Hayashi H, Chino M (1990) Chemical composition of phloem sap form the uppermost internode of the rice plant. Plant Cell Physiol 31: 247-251

Heilmeier H, Freund M, Steinlein T, Schulze ED, Monson RK (1994) The influence of nitrogen availability on carbon and nitrogen storage in the biennial *Cirsium vulgare* (Savi) Ten. I. Storage capacity in relation to resource acquisition, allocation and recycling. Plant Cell Environ 17: 1125-1131

Hendershot KL, Volonec JJ (1993) Nitrogen pools in taproots of *Medicago sativa* L. after defoliation. J Plant Physiol 141: 129-135

Higgins CF, Payne JW (1977) Characterization of active dipeptide transport by germinating barley embryos: effects of pH and metabolic inhibitors. Planta 136: 71-76

Higgins CF, Payne JW (1978) Peptide transport by germinating barley embryos: uptake of physiological di- and oligopeptides. Planta 138: 211-215

Hirner B, Fischer WN, Rentsch D, Kwart M, Frommer WB (1998) Developmental control of H+/amino acid permease gene expression during seed development of *Arabidopsis*. Plant J 14: 535-544

Hocking PJ (1980) The composition of the phloem exudate and xylem sap from tree tobacco (*Nicotiana glauca*). Ann Bot 45: 633-640

Homeyer U, Litek K, Huchzermeyer B, Schultz G (1989) Uptake of phenylalanine into isolated barley vacuoles is driven by both adenosine triphosphatase and pyrophosphatase: evidence for a hydrophobic L-amino acid carrier system. Plant Physiol 89: 1388-1393

Housley TL, Schrader LE, Mill RM, Setter TL (1979) Partitioning of ^{14}C photosynthate and long distance translocation of amino acids in preflowering and flowering nodulated and nonnodulated soybeans. Plant Physiol 64: 94-98

Hsu LC, Chiou TJ, Chen L, Bush DR (1993) Cloning a plant amino acid transporter by functional complementation of a yeast amino acid transport mutant. Proc Natl Acad Sci USA 90: 7441-7445

Imsande J, Touraine B (1994) N demand and the regulation of nitrate uptake. Plant Physiol 105: 3-7

Jamai A, Chollet JF, Delrot S (1994) Proton-peptide co-transport in broad bean leaf tissues. Plant Physiol 106: 1023-1031

Kang SM, Titus JS (1980) Qualitative and quantitative changes in nitrogenous compounds in senescing leaf and bark tissues of the apple. Physiol Plant 50: 285-290

Kinraide TB (1981) Electrical evidence for different mechanisms of uptake for basic, neutral and acidic amino acids in oat coleoptiles. Plant Physiol 65: 1085-1089

Kinraide TB, Etherton B (1980) Interamino acid inhibition of transport in higher plants. Plant Physiol 68: 1327-1333

Kipps A, Boulter D (1974) Origin of the amino acids in pods and seeds of *Vicia faba* L. New Phytol 73: 675-684

Kwart M, Hirner B, Hummel S, Frommer WB (1993) Differential expression of two related amino acid transporters with differing substrate specificity in *Arabidopsis thaliana*. Plant J 4: 993-1002

Lalanne E, Mathieu C, Roche O, Vedel F, De Pape R (1997) Structure and specific expression of a Nicotiana sylvestris putative amino-acid transporter gene in mature and *in vitro* germinating pollen. Plant Mol Biol 35: 855-864

Lanfermeyer FC, Koerselmankooij J, Borstlap AC (1990) Changing kinetics of L-valine uptake by immature pea cotyledons during development. Planta 181: 576-582

Larsson C, Widell S, Kjellbohm P (1987) Preparation of high purity plasma membranes. Methods Enzymol 148: 558-568

Layzell DB, LaRue TA (1982) Modeling C and N transport to developing soybean fruits. Plant Physiol 70: 1290-1298

Lemoine R, Delrot S (1989) Proton motive force driven sucrose uptake in sugar beet plasma membrane vesicles. FEBS Lett 249: 129-133

Lemoine R, Bourquin S, Delrot S (1991) Active uptake of sucrose by plant plasma membrane vesicles: determination of some important physical and energetical parameters. Physiol Plant 82: 377-384

Lemoine R, Gallet O, Gaillard C, Frommer WB, Delrot S (1992) Plasma membrane vesicles from source and sink leaves. Changes in solute transport and polypeptide composition. Plant Physiol 100: 1150-1156

Lemoine R, Kühn C, Thiele N, Delrot S, Frommer WB (1996) Antisense inhibition of the sucrose transporter in potato: effects on amount and activity. Plant Cell Environ 19: 1124-1131

Li ZC, Bush DR (1990) ΔpH-dependent amino acid transport into plasma membrane vesicles isolated from sugar beet leaves: I. Evidence for carrier-mediated, electrogenic flux through multiple transport systems. Plant Physiol 94: 268-277

Li ZC, Bush DR (1991) ΔpH-dependent amino acid transport into plasma membrane vesicles isolated from sugar beet (*Beta vulgaris* L.) leaves. II. Evidence for multiple aliphatic, neutral amino acid symports. Plant Physiol 96: 1338-1344

Limami A, Dufossé C, Richard-Molard C, Fouldrin K, Roux L, Morot-Gaudry JF (1996) Effect of exogenous nitrogen ($^{15}NO_3$) on utilization of vegetative storage proteins (VSP) during regrowth in chicory (*Cichorium intybus* L.). J Plant Physiol 149: 564-572

Lohaus G, Burba M, Heldt HW (1994) Comparison of the contents of sucrose and amino acids in the leaves, phloem sap and taproots of high and low sugar-producing hybrids of sugar beet (*Beta vulgaris* L.). J Exp Bot 45: 1097-1101

Mae T, Makino A, Ohira K (1983) Changes in the amounts of ribulose bisphosphate carboxylase synthesized and degraded during the life span of rice leaf (*Oryza sativa*). Plant Cell Physiol 21: 1079-1081

Martinoia E , Ratajczak R (1998) Transport of organic molecules across the tonoplast. Adv Bot Res 25: 366-400

Martinoia E, Keck U, Wiemcken A (1981) Vacuoles as storage compartments for nitrate in barley leaves. Nature 289: 292-294

Martinoia E, Thume M, Vogt E, Rentsch D, Dietz KJ (1991) Transport of arginine and aspartic acid into isolated barley mesophyll vacuoles. Plant Physiol 97: 644-50

Marvier AC, Neelam A, Bick JA, Hall JL, Williams LE (1998) Cloning of a cDNA coding for an amino acid carrier from *Ricinus communis* (*RcAAP1*) by functional complementation in yeast: kinetic analysis, inhibitor sensitivity and substrate specificity. Biochim Biophys Acta 1373: 321-331

Millard P (1988) The accumulation and storage of nitrogen by herbaceous plants. Plant Cell Environ 11: 1-8

Montamat F, Maurousset L, Tegeder M, Frommer W, Delrot S (1999) Cloning and expression of amino acid transporters from broad bean. Plant Mol Biol 41: 259-268

Mounoury G, Delrot S, Bonnemain JL (1984) Energetics of threonine uptake by pod wall tissues of *Vicia faba* L. Planta 161: 178-185

Muller B, Touraine B (1992) Inhibition of NO_3^- uptake by various phloem-translocated amino acids in soybean seedlings. J Exp Bot 43: 617-623

Paiva E, Lister RM, Park WD (1983) Induction and accumulation of major tuber proteins of potato in stems and petioles. Plant Physiol 71: 161-168

Pate JS (1973) Uptake, assimilation and transport of nitrogen compounds by plants. Soil Biol Biochem 5: 109-119

Pate JS (1976) Nutrients and metabolites of fluids recovered from xylem and phloem: significance in relation to long distance transport in plants. In: Wardlaw IF, Passioura JB (eds) Transport and transfer processes in plants. CSIRO, Canberra, pp 253-345

Pate JS (1980) Transport and partitioning of nitrogenous solutes. Annu Rev Plant Physiol 31: 313-340

Pate JS (1989) Origin, destination and fate of phloem solutes in relation to organ and whole plant functioning. In: Baker DA, Milburn JA (eds) Transport of photoassimilates. Longman Scientific, Harlow, pp 138-166

Pate JS, Sharkey PJ, Lewis OAM (1974) Phloem bleeding from legume fruits: a technique for study of fruit nutrition. Planta 120: 229-243

Parsons R, Baker A (1996) Cycling of amino compounds in symbiotic lupin. J Exp Bot 47: 421-429

Patrick JW (1997) Phloem unloading: sieve element unloading and post-sieve element transport. Annu Rev Plant Physiol Plant Mol Biol 48: 191-222

Peoples MB, Beilharz VC, Waters SP, Simpson RJ, Dalling MJ (1980) Nitrogen redistribution during grain growth in wheat (*Triticum aestivum* L.). II. Chloroplast senescence and the degradation of ribulose-1,5-bisphosphate carboxylase. Planta 81: 494-500.

Rentsch D, Laloi M, Rouhara I, Schmelzer E, Delrot S, Frommer WB (1995) *NTR1* encodes a high affinity oligopeptide transporter in *Arabidopsis*. FEBS Lett 370: 264-268

Rentsch D, Hirner B, Schmelzer E, Frommer WB (1996) Salt stress-induced proline transporters and salt stress-repressed broad specificity amino acid permeases identified by suppression of a yeast amino acid permease-targeting mutant. Plant Cell 8: 1437-1446

Riens B, Lohaus G, Heineke D, Heldt HW (1991) Amino acid and sucrose content determined in the cytosolic, chloroplastic and vacuolar compartment and in the phloem sap of spinach leaves. Plant Physiol 97: 227-233

Rochat C, Boutin JP (1991) Metabolism of phloem-borne amino acids in maternal tissues of fruit of nodulated or nitrate-fed pea plants (*Pisum sativum* L.). J Exp Bot 42: 207-214

Rochat C, Boutin JP (1992) Temporary storage compounds and sucrose-starch metabolism in seed coats during pea seed development (*Pisum sativum*). Physiol Plant 85: 567-572

Rosnick-Shimmel I (1985) The influence of nitrogen nutrition on the accumulation of free amino acids in root tissue of *Urtica dioica* and their apical transport in xylem sap. Plant Cell Physiol 26: 215-219

Sagisaka S (1987) Amino acid pools in herbaceous plants at the wintering stage and at the beginning of growth. Plant Cell Physiol 28: 171-178

Sauter JJ , Van Cleve B (1992) Seasonal variation of amino acids in the xylem sap of "*Populus x canadensis*" and its relation to protein body mobilization. Trees 7: 26-32

Schobert C, Komor E (1989) The differential transport of amino acid into the phloem of *Ricinus communis* L. seedlings as shown by the analysis of sieve tube sap. Planta 177: 342-349

Schobert C, Komor E (1990) Transfer of amino acids and nitrate from the roots into the xylem of *Ricinus communis* seedlings. Planta181: 85-90

Schwacke R, Grallath S, Breitkreuz KE, Stransky E, Stransky H, Frommer WB, Rentsch D (1999) LeProT1, a transporter for proline, glycine betaine, and γ-aminobutyric acid in tomato pollen. Plant Cell 11: 377-391

Servaites JC, Schrader LE , Jung DM (1979) Energy-dependent loading of amino acids and sucrose in the phloem of soybean. Plant Physiol 64: 546-550

Sharkey PJ, Pate JS (1975) Selectivity in xylem to phloem transfer of amino acids in fruiting shoots of white lupin (*Lupinus albus* L.). Planta 127: 251-252

Simpson RJ , Dalling MJ (1981) Nitrogen redistribution during grain growth in wheat (*Triticum aestivum* L.). III. Enzymology and transport of amino acids from senescing flag leaves. Planta 151: 447-456

Staswick PE (1990) Novel regulation of vegetative storage protein genes. Plant Cell 2: 1-6

Steiner HY, Song W, Zhang L, Naider F, Becker JM, Stacey G (1994) An *Arabidopsis* peptide transporter is a member of a new class of membrane transport proteins. Plant Cell 6: 1289-1299

Tammes PML, Van Die J (1964) Studies on phloem exudation from *Yucca flaccida*. I. Some observations on the phenomenon of bleeding and the composition of the exudate. Acta Bot Neerl 13: 76-83

Thomas H (1978) Enzymes of nitrogen mobilization in detached leaves of *Lolium temulentum* during senescence. Planta 151: 447-456

Thorne JH (1985) Phloem unloading of C and N assimilates in developing seeds. Annu Rev Plant Physiol 36: 317-343

Thume M, Dietz KJ (1991) Reconstitution of the tonoplast amino acid carrier into liposomes: evidence for an ATP-regulated carrier in different species. Planta 185: 569-575

Tromp J (1983) Nutrient reserves in roots of fruit trees, in particular carbohydrates and nitrogen. Plant Soil 71: 401-413

Van Bel AJE (1984) Quantification of the xylem-to-phloem transfer of amino acids by use of inulin (^{14}C) carboxylic acid as xylem transport marker. Plant Sci Lett 35: 81-85

Van Bel AJE (1990) Xylem-phloem exchange via the rays: the under valued route of transport. J Exp Bot 41: 631-644

Van Bel AJE, Koops AJ, Dueck T (1986) Does light-promoted export from *Commelina benghalensis* leaves result from differential light-sensitivity of the cells in the mesophyll-to-sieve tube path? Physiol Plant 67: 227-234

Van Cleve B, Apel K (1993) Induction by nitrogen and low temperature of storage protein synthesis in poplar trees exposed to long days. Planta 189: 157-160

Weiner H, Blechschmidt-Schneider S, Mohme H, Eschrich W, Heldt HW (1991) Phloem transport of amino acids. Comparison of amino acid contents of maize leaves and of the sieve tube exudate. Plant Physiol Biochem 29: 19-23

Weston K, Hall JL, Williams LE (1994) Characterization of a glutamine/proton cotransporter from *Ricinus communis* roots using isolated plasma membrane vesicles. Physiol Plant 91: 623-630

Weston K, Hall JL, Williams LE (1995) Characterization of amino-acid transport in *Ricinus communis* roots using isolated membrane vesicles. Planta 196: 166-173

Wetzel S, Demmers C, Greenwood JS (1989) seasonally fluctuating bark proteins are a potential form of nitrogen storage in three temperate hardwoods. Planta 178: 275-281

Williams LE, Nelson SJ, Hall JL (1992) Characterisation of solute proton cotransport in plasma membrane vesicles from *Ricinus* cotyledons, and a comparison with other tissues. Planta 186: 541-550

Winter H, Lohaus G, Heldt HW (1992) Phloem transport of amino acids and sucrose in correlation to the corresponding metabolite levels in barley leaves. Plant Physiol 99: 996-1004

Wittenbach VA (1983) Purification and characterization of a soybean leaf storage glycoprotein. Plant Physiol 73: 125-129

Wolswinkel P (1992) Transport of nutrients into developing seeds: a review of physiological mechanisms. Seed Sci Res 2: 59-73

Wyse RE , Komor E (1984) Mechanism of amino acid uptake by sugarcane suspension cells. Plant Physiol 76: 865-870

Zhou JJ, Theodoulou F, Muldin I, Ingemarsson B, Miller AJ (1998) Cloning and functional characterization of a *Brassica napus* transporter that is able to transport nitrate and histidine. J Biol Chem 273: 12017-12023

Ziegler H (1975) Nature of transported substances. In: Zimmerman MN, Milburn JA (eds) Encyclopedia of plant physiology, new series, vol 1. Transport in plants. Springerg, Berlin Heidelberg New York, pp 59-100

Interactions Between Carbon and Nitrogen Metabolism

Christine H. Foyer[1], Sylvie Ferrario-Méry[2] and Graham Noctor[1]

Over the past two decades, many studies have revealed the interdependence of carbon and nitrogen assimilation. Primary carbon metabolism is dependent on nitrogen assimilation, most obviously because much of the nitrogen budget of the plant is invested in the proteins and chlorophyll of the photosynthetic apparatus. Conversely, nitrogen assimilation requires a continuous supply of energy and carbon skeletons. This means that photosynthetic products must be partitioned between carbohydrate synthesis and the synthesis of amino acids. Controls over this partitioning must be flexible, since both external nitrogen availability and internal nitrogen demand may be variable.

There is often substantial heterogeneity in the nature of the relationships between carbon and nitrogen assimilation in different plants. Interactions will vary considerably according to such factors as life cycle and longevity, habitat (which will determine, among other things, the types of nitrogen sources available) and the relative amounts of nitrogen assimilated in roots and leaves. In this chapter, our aim will be to place emphasis on unifying principles that are likely to govern how the assimilation of carbon and that of nitrogen interact in photosynthetic organisms, with particular reference to foliar metabolism. Three aspects of these interactions will be discussed: (1) the source of reductant and carbon skeletons for nitrogen assimilation; (2) a comparison of the effects of photosynthetic nitrogen assimilation and carbon fixation on cellular adenylate status and redox state; (3) metabolic and molecular cross-talk which controls key enzyme activities, allowing coordination of carbon and nitrogen assimilation as well as appropriate distribution of fixed carbon between carbohydrates and amino acids.

1. Department of Biochemistry and Physiology, IACR-Rothamsted, Harpenden, Herts AL5 2JQ, UK. *E-mail (Christine H. Foyer):* foyer@bbsrc.ac.uk. *E-mail (Graham Noctor):* graham.noctor@bbsrc.ac.uk.
2. Unité de Nutrition Azotée des Plantes, INRA, Route de Saint-Cyr, 78026 Versailles cedex. *E-mail:* ferrario@versailles.inra.fr

The Conversion of Nitrogen to Amino Acids: Reductant Supply

In higher plants, photosynthetic carbon assimilation occurs predominantly in the leaves and, apart from in a few specialised plant groups (e.g. CAM plants), can only be observed in the light. The reductive assimilation of nitrogen differs from that of carbon in two important respects. Firstly, significant amounts of nitrogen can be assimilated in darkened leaves or in heterotrophic tissues such as roots, depending on species and environment. In these cases, reductant must be supplied by reoxidation of carbon (Fig. 1). Secondly, the oxidation state of nitrogen sources is generally more variable than carbon, which in the vast majority of plants comes exclusively from CO_2. Depending on habitat and species, nitrogen sources available to plants (or their symbionts) include organic nitrogenous compounds, terrestrial ammonium, atmospheric ammonia, molecular dinitrogen, gaseous oxides of nitrogen and nitrate. In terms of reductant, inorganic N sources will require between two (ammonium) and ten (nitrate) electrons (Fig. 1). This reductant must ultimately be generated by the photosynthetic electron transport chain, whether it be directly through the photochemical reduction of ferredoxin, indirectly through the operation of redox shuttles, or more circuitously via the respiratory oxidation of fixed carbon (Fig. 1).

Both nitrite reductase (NiR) and the major foliar glutamate synthase activity (Fd-GOGAT) use ferredoxin (Fd) as reductant. In leaves, therefore, the photosynthetic electron transport chain will be the principal reductant source for nitrite reduction and the incorporation of ammonium. Although significant reduction of

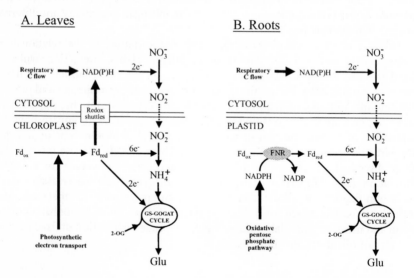

Fig. 1 A, B : Likely reductant sources for nitrate assimilation in illuminated leaves (**A**) and in roots (**B**). Redox shuttles are included in (**A**) to stress the possible contribution of the chloroplast to cytosolic nitrate reductase activity in the illuminated leaf. Exchange of reducing equivalents between the plastid and cytosol can also occur in roots. *Fd* Ferredoxin; *FNR* Ferredoxin-NADP oxidoreductase; *GS-GOGAT* Glutamine synthetase-glutamate synthase; *2-OG* 2-oxoglutarate

both nitrate and nitrite can occur in darkened leaves containing sufficient carbohydrate, rates are about two- to fivefold faster in the light (Aslam et al. 1979; Reed et al. 1983). In the dark, reductant is presumably generated by processes such as glycolysis and the oxidative pentose phosphate pathway. Generation of NADPH by the latter process is thought to be particularly important for N assimilation in roots, where specific forms of Fd and ferredoxin-NADP reductase may favour electron transfer from NADPH to Fd, the reverse of the direction of net flow during photosynthesis (Fig. 1B; Emes and Neuhaus 1997). In contrast to NiR and Fd-GOGAT, nitrate reductase (NR) is a cytosolic enzyme that predominantly uses NADH. This reductant must therefore be generated by carbon oxidation, either during respiratory carbon flows or via redox shuttles such as malate-oxaloacetate exchange between the chloroplast and cytosol (Fig. 1).

Even in the light, a considerable portion of the reductant for nitrogen assimilation in unicellular algae can come from respiration (Weger and Turpin 1989; Huppe et al. 1994). In illuminated leaves, 80% of the reductant necessary to assimilate nitrate into glutamate can be generated directly by the photosynthetic electron transport chain (Fig. 1A). Differences between unicellular algae and higher plants may be explained by interorgan specialisation in the latter. Unicellular algae must combine some of the metabolic properties of both leaves and roots within a single cell. They generally display higher capacities for respiration, relative to photosynthesis, than do leaves. Metabolic regulation in unicellular algae is geared to seize upon fortuitous upsurges in ambient nitrogen, whereas a leaf cell exists in a nutrient environment that is at least partly buffered by interorgan regulation. The response of the higher plant to environmental variables must be tempered by the homeostasis required by a complex life cycle. Although both respiration and photosynthesis can supply reductant for nitrogen assimilation in higher plants, their relative importance is probably organ-dependent, with the oxidative pentose phosphate pathway being crucial in roots (Bowsher et al. 1989) and the photosynthetic electron transport chain playing a dominant role in leaves (Fig. 1).

In non-chlorophyllous leaf cells or in the dark, reductant for foliar N assimilation must come from respiration. The marked stimulatory effect of anaerobiosis on nitrate reduction in darkened leaves suggests that one factor limiting dark N assimilation may be competition from the mitochondrial electron transport chain (Reed et al. 1983; Gray and Cresswell 1984). Conversely, nitrate-induced increases in the respiratory quotient (CO_2 evolved:O_2 consumed) in barley roots were mainly due to decreased O_2 consumption rather than accelerated CO_2 release (Bloom et al. 1992). This observation suggests that nitrate may compete with the mitochondrial electron transport chain for reductant generated from respiration and contrasts with data for unicellular algae, where nitrate supply increased the respiratory quotient by markedly stimulating CO_2 evolution (Weger and Turpin 1989).

What proportion of photosynthetic electron flow is linked to nitrogen assimilation in leaves? In Table 1 this parameter is calculated simply from measurements of foliar C:N contents, using a minimum number of necessary assumptions. For barley, pea and tobacco supplied with optimal amounts of nitrogen, the calculations predict that approximately 10% of photosynthetic electrons flow to nitrate (Table 1A). In contrast, this value is only about 1% in N-starved tobacco with a very high foliar C: N ratio. For plants supplied with optimal N, the calculated data are close to values derived from experiments with excised tobacco leaves supplied with nitrate (Table

1B; Morcuende et al. 1998). Conversely, calculating the foliar C:N ratio from the proportion of electron flow to nitrate yields a C: N value (6.3) that is similar to measured foliar values in tobacco at optimal N (7.0; Banks et al. 1999). Supplying tobacco leaves with sucrose, in concert with nitrate, yielded higher rates of N assimilation than when nitrate was supplied alone (Morcuende et al. 1998). In this case, nitrate assimilation accounted for about 25% of the total photosynthetic electron flow (Table 1B; Morcuende et al. 1998). The unphysiologically low C:N ratios derived from this value (Table 1B) suggest that, even if leaves can devote such a high proportion of their electron flow to nitrate, this condition would be exceptional.

Table 1. Predicted relationships between C: N assimilation ratios and the proportion of photosynthetic electron transport linked to nitrate assimilation

	Plant	Treatment	Foliar C:N contents	Calculated ratio of C:N assimilation	% electrons accounted for by N assimilation	Literature source of data in bold
A	Barley	Grown at optimal N	8.7	14.5	9	1
	Pea	Grown at optimal N	7.7	12.8	10	1
	Tobacco	Grown at optimal N	7.0	11.7	11	2
	Tobacco	Grown at low N	72.0	120	1.2	2
B	Tobacco	Excised leaves, fed nitrate + sucrose	2.6	4.3	25	3
	Tobacco	Excised leaves, fed nitrate	6.3	10.5	12	3
	Tobacco	Excised leaves, fed water	>20	>34	<4	3

In A, literature data for measured foliar C:N contents (in bold) are converted into mean C:N assimilation ratios by assuming that the leaves respire 40% of the C they assimilate, that the loss of assimilated N is 0, and that total export of C and N from the leaves occurs in proportion to their ratio of assimilation. The data are approximate, since (1) the ratio between total C assimilation and respiratory C loss will vary, notably as a function of leaf age, daylength and light capture; (2) leaves may assimilate N sources that are more reduced than nitrate, receive reduced N from the roots or assimilate some N in the dark; (3) at least small amounts of ammonia must escape from the leaf. The % of electron flow linked to N assimilation is calculated from C:N assimilation ratios by assuming that the number of electrons required per N and C assimilated is 10 and 7, respectively. All electrons are assumed to flow to C assimilation, N assimilation and photorespiration.

In B, literature data for the estimated proportion of electrons linked to N assimilation (in bold) are converted into C:N assimilation ratios and foliar C:N contents that would theoretically result from these ratios. Necessary assumptions as for A.

References: 1. De la Torre et al. (1991); 2. Banks et al. (1999); 3. Morcuende et al. (1998).

Since N assimilation can account for a significant part of photosynthetic electron flow, we might wonder whether carbon and nitrogen compete for reductant. Algae develop an increased capacity for N assimilation when grown at limiting N. Resupply of N to these cells brings about an inhibition of CO_2 fixation, which may reflect diversion of carbon from the Calvin cycle to respiratory pathways and/or utilisation of reductant by N assimilation (Elrifi and Turpin 1986). A study with spinach leaves failed to find evidence of significant competition between CO_2 fixation and nitrite assimilation (Robinson 1988). Simple competition for electrons should theoretically be more manifest at lower light intensities. Under many conditions, moderate to high light intensities require leaves to dissipate substantial amounts of excitation energy as heat, presumably an adaptation to conditions where the potential for electron transport outstrips electron demand. Although N sources stimulated light-dependent O_2 evolution in barley and pea, little associated inhibition of CO_2 fixation was observed (Bloom et al. 1989; de la Torre et al. 1991). In the C_4 plant maize, supplying nitrate to excised leaves produced a marked and rapid stimulation of amino acid contents and an associated partial inhibition of CO_2 fixation (Foyer et al. 1994a). Three observations suggest that the decreased CO_2 fixation in maize was the result of perturbed regulation rather than simple diversion of electrons to N assimilation. Firstly, the effect was transient; secondly, it was more pronounced at high than at low light; thirdly, it was associated with decreased photosystem II photochemical efficiency (Foyer et al. 1994a). Supplying nitrate and sucrose together to excised tobacco leaves at low light also caused rapid accumulation of amino acids, which was estimated to account for a substantial portion (25%) of total electron flow (Morcuende et al. 1998). The total electron flow itself (derived from chlorophyll fluorescence analysis) did not change, which the authors suggested might have been due to competition for electrons, with the N-linked electron flow producing a corresponding decrease in electron transport to CO_2 (Morcuende et al. 1998). Although further work may clarify the extent to which N and C compete for photosynthetic reducing equivalents, we can probably state with some confidence that competition in higher plants is a much less important phenomenon than in algae such as *Selenastrum*. This presumably reflects the differences discussed above.

An elegant effect found in *Chlamydomonas* is the nitrate-induced activation of the first enzyme of the oxidative pentose phosphate pathway, glucose-6-phosphate dehydrogenase (Huppe et al. 1994). As in higher plant leaves, this enzyme undergoes reductive light inactivation via the thioredoxin system. In algae, however, nitrate is apparently able to overcome the light inactivation, perhaps simply by competing with the thioredoxin system for electrons. Respiratory C flow is thereby engaged to supply carbon acceptors for amino acid synthesis, while the attendant oxidative inactivation of thiol-regulated Calvin cycle enzymes leads to lower photosynthetic CO_2 fixation (Huppe et al. 1994). This potentially self-regulating mechanism is likely to be of less significance in higher plants, where the homeostatic coordination of C and N metabolism probably requires more complex orchestration. Nevertheless, as discussed below, N-dependent stimulation of respiratory pathways is an important phenomenon in all plant species.

The Conversion of Nitrogen to Amino Acids: the Supply of Carbon

Amino acids are often divided into families, according to their primary amino acid or carbon skeleton precursor (see Chap. 4.1). In many plants, the major amino acids are derived from pyruvate (via alanine), oxaloacetate (via aspartate) and 2-oxoglutarate (2-OG: via glutamate), which is reflected by predominant aminotransferase activities (Ireland and Lea 1999). Important aminotransferase activities also catalyse the conversion of hydroxypyruvate to serine and glyoxylate to glycine, though these are involved in photorespiratory cycling rather than net N assimilation. The primary carbon acceptor of amino groups is 2-OG, used in the GS/GOGAT sequence (Fig. 1). In leaves from C3 species, even faster utilisation of 2-OG by GOGAT activity may occur during the photorespiratory nitrogen cycle (Keys et al. 1978). Unlike nitrogen assimilation, photorespiratory utilisation of 2-OG involves no net consumption of carbon skeletons, since glutamate:glyoxylate aminotransferase activity must ensure the stoichiometric regeneration of 2-OG.

The generation of 2-OG for nitrogen assimilation occurs through the oxidative decarboxylation of isocitrate, catalysed by isocitrate dehydrogenases (ICDH). As shown in Fig. 2, leaves contain isoforms of ICDH in the mitochondria and cytosol. Some activity is also associated with chloroplasts and peroxisomes (Elias and Givan 1977; Rasmusson and Møller 1990). The principal mitochondrial form is NAD-dependent and is a key enzyme in the TCA cycle, although mitochondria may also contain an NADP-linked enzyme of lower activity (Rasmusson and Møller 1990). The mitochondria have been considered the most important intracellular sites for production of 2-OG used in the chloroplastic GS-GOGAT cycle (Miflin and Lea 1982). On the other hand, it has been argued that net synthesis of 2-OG is unlikely to take place in the mitochondria, because 2-OG will rapidly be metabolised by 2-OG dehydrogenase, whose capacity is higher than mitochondrial ICDH (Chen and Gadal 1990). Studies with isolated mitochondria produced results consistent with the latter view, suggesting that a cytosolic NADP-linked ICDH may be more important in nitrogen assimilation (Hanning and Heldt 1993). In separate studies with tobacco, supplying nitrate was shown to induce transcripts for both the cytosolic NADP-linked isoform (Scheible et al. 1997a) and the mitochondrial NAD-dependent enzyme (Lancien et al. 1999). Expression studies in Scots pine provided little evidence that the activity of the cytosolic enzyme is tightly linked to nitrogen assimilation (Palomo et al. 1998). Moreover, drastic reduction of the cytosolic enzyme in potato by antisense technology resulted in no significant effects on growth and development, photosynthesis, or respiration (Kruse et al. 1998). It remains to be seen whether isocitrate dehydrogenases are yet another example of the apparent redundancy of plant enzymes. It is possible that 2-OG may be formed by several enzymes at different intracellular sites, each of which could be removed without affecting the overall rate of 2-OG production, due to compensatory increases in flux catalysed by the remaining enzymes.

Although much attention has focused on the site of 2-OG formation, nitrogen assimilation will pose similar demands on the pyruvate and oxaloacetate pools, and may also require organic acid synthesis to maintain acid-base balance (Raven and Smith 1976). Replenishment of the oxoacid pools (anapleurosis) will involve flux to the terminal reactions of glycolysis, as well as partial operation of the TCA cycle (Fig. 2). In addition to ICDH, candidate control points in governing flux to oxoacid formation are pyruvate kinase and phosphoenolpyruvate carboxylase (PEPC). In photosynthetically competent leaves, PEPC in particular will be crucial, since net formation of 2-OG without anapleurosis is not possible (PEPC regulation is discussed further below).

Bioenergetic Coordination in Photosynthetic Cells: Carbon, Nitrogen, and the Leaf ATP:Reductant Balance

While attention has been paid to possible competition between the reductive assimilations of C and N for electrons and metabolites, much less has focused on interactions at the level of ATP generation and turnover (for detailed analysis of this question, see Noctor and Foyer 1998). Nitrogen assimilation could affect chloroplastic adenylate status because it requires electrons and ATP at ratios that are different from the requirements of CO_2 fixation (Table 2). This question was addressed by Turpin and Bruce (1990), who showed that the distribution of excitation energy between photosystems I and II in algal cells was markedly influenced by

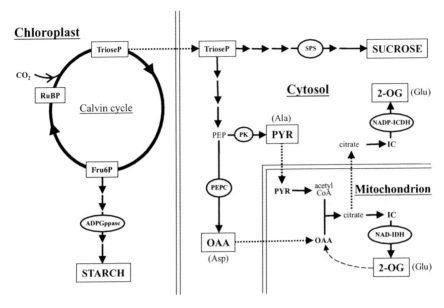

Fig. 2 : Major metabolic fates of carbon in the photosynthetic cell. Net products (*capital letters in boxes*) are starch, sucrose and oxoacids. Crucial enzyme activities, including those likely to be critical in controlling flux, are depicted in *ellipses*. Other respiratory pathways may contribute to oxoacid formation (for review, see Noctor and Foyer 1998). Amino acids resulting from transamination of oxoacids are shown in *brackets*. *Solid arrows* indicate chemical conversion; *dotted arrows* depict intracellular transport; *the dashed arrow* from mitochondrial 2-OG to OAA reflects possible complete turnover of the TCA cycle. ADPGppase, ADP-glucose pyrophosphorylase; *Fru6P* fructose-6-phosphate; *IC(DH)* or I(DH) isocitrate (dehydrogenase); *OAA* oxaloacetate; *2-OG,* 2-oxoglutarate; *PEP(C)* phosphoenolpyruvate (carboxylase); *PK* pyruvate kinase; *PYR* pyruvate; *RuBP* ribulose-1,5-bisphosphate; *SPS* sucrose-phosphate synthase; *trioseP* triose phosphate

the form in which N was supplied. The ATP:electron ratio required for the assimilation of nitrate is very different from that needed to assimilate ammonium (Table 2). Supplying ammonium, but not nitrate, induced the preferential excitation of photosystem I, suggesting enhanced engagement of cyclic electron transport to satisfy the relatively high ATP demand of N assimilation from ammonium (Turpin and Bruce 1990). In maize, however, a similar study provided less evidence of clearcut differential effects of nitrate and ammonium (Foyer et al. 1994a).

Over and above the impact of nitrogen-linked photosynthetic electron flow, cellular adenylate status could be affected by respiratory C flow to produce oxoacid acceptors. It is now accepted that some respiratory activity must occur in the light, to provide precursors for processes such as N assimilation. The extent to which this necessary oxidative C flow is linked to significant ATP production is less clear, since (1) multiple routes of C oxidation are possible; (2) NAD(P)H could be reoxidised by processes other than the mitochondrial electron transport chain (e.g. NR activity); (3) ATP yields associated with mitochondrial electron flow are variable (for further

Table 2. ATP:2 electron ratios required for carbon assimilation and the assimilation of nitrogen from sources with different reduction states

Substrate	Product	ATP:2 electron	
CO_2	Triose phosphate	1.5	2.5
	Sucrose	1.54	2.54
	Starch	1.58	2.58
NO_3^-	Glutamate	0.2	
	Glutamine	0.25	
NO_2^-	Glutamate	0.25	
	Glutamine	0.33	
NH_4^+	Glutamate	1	
	Glutamine	No electrons required	

For C assimilation, the two columns give the values for C_3 photosynthesis (left) and C_4 photosynthesis (decarboxylation via malic enzyme). In both cases, values are calculated for non-photorespiratory conditions. For N assimilation, values are calculated assuming that ammonium is integrated into amino acids via glutamine synthetase and glutamate synthase.

discussion see Noctor and Foyer 1998, 2000). Direct evidence in favour of a contribution of oxidative phosphorylation to cellular ATP supply in the light has come from studies with the ATPase inhibitor, oligomycin (Krömer et al. 1988). In algae, ammonium-induced respiratory oxygen uptake was shown to occur primarily through cytochrome oxidase rather than the alternative oxidase; i.e. respiratory flux was linked to mitochondrial pathways of high ATP yield (Weger et al. 1988, 1990). Increased foliar ATP levels in tobacco leaves supplied with nitrate and sucrose were explained in terms of sucrose-stimulated respiration to supply oxoacids (Morcuende et al. 1998). While there has been considerable discussion of the extent to which respiration (other than photorespiration) continues during photosynthesis, respiration linked to N assimilation should be seen as a minimal, necessary activity that is assimilatory rather than dissimilatory in nature. Dissimilatory respiration, involving full turnover of the TCA cycle, may occur in some leaves in the light, but is not necessarily required for N-linked respiration (Fig. 2).

N assimilation therefore engages two processes capable of generating excess ATP: photosynthetic photophosphorylation and ATP synthesis linked to respiratory C flow. Respiration in the light has been postulated to be necessary to support sucrose synthesis in the cytosol (Krömer et al. 1988). Carbon flux linked to N assimilation could also contribute more generally to the high ATP demands of carbon assimilation (Noctor and Foyer 1998). Any such contribution would entail either the suppression of those pathways that meet ATP demand in the absence of N assimilation or compensatory engagement of other ATP-consuming pathways. In the absence of a corresponding ATP sink, ATP formation linked to the resumption of N assimilation would rapidly result in dramatic perturbation of ATP/ADP ratios, due to the generally fast turnover of adenylate pools (Noctor and Foyer 2000). Consequently, the photosynthetic cell must have considerable in-built flexibility that allows rates of ATP consumption to be matched to production. As Table 1 shows, foliar C:N ratios change radically in response to changing N availability. During leaf development, the relative rates of C and N assimilation will vary, as will the relative capacities for photosynthesis and respiration. Shorter-term fluctuations in photorespiratory rates will also impinge on adenylate status. Future developments may shed light on mechanisms contributing to homeostasis in adenylate status against this backdrop of dynamic changes in metabolism.

Carbon and Nitrogen Assimilation: Cross-talk and Coordination

As discussed above, most evidence from short-term experiments suggests that increases in N assimilation do not bring about marked inhibition of CO_2 fixation in leaves from higher plants. On the contrary, most data highlight the homeostatic coordination of C and N assimilation so that C:N ratios remain relatively stable. Increases in CO_2 fixation with increasing irradiance are accompanied by enhanced N uptake (Gastal and Saugier 1989), while N assimilation is decreased at low CO_2 (Pace et al. 1990), or during depletion of carbohydrates in extended darkness (Rufty et al. 1989). The factors coordinating C and N assimilation are multiple and, at the whole-plant level, include effects on nitrate uptake by the roots, nitrate translocation in the xylem and nitrate reduction in the leaves.

Modifications in foliar nitrate reduction appear to be particularly significant (Rufty et al. 1989), effects that may partly reflect altered availability of C skeletons, reductant and ATP. These effects may also be mediated via changes in the activity of NR which, for example, is decreased at low CO_2 (Kaiser and Förster 1989; Pace et al. 1990). Quite apart from reductant supply or nitrate availability, the capacity for nitrate reduction (extractable NR activity) increases in the light. The activity of NR is under control at multiple levels, from expression of *nia* genes to covalent post-translational modification (for more detail, see Chap. 1.2). The most important intermediaries in light-regulation of nitrate reductase are nitrate, glutamine, and sugars. Expression of *nia* genes is induced by nitrate and sugars, and suppressed by glutamine (Hoff et al. 1994). However, neither *in vitro* NR activity nor *in vivo* nitrate reduction correlate tightly with *nia* transcript levels (Vincentz and Caboche 1991). One explanation may be effects of nitrate at the level of NR protein stability (Ferrario et al. 1995). A second could be the influence of phosphorylation on enzyme activity (Kaiser and Huber 1994). In itself, phosphorylation does not significantly affect enzyme activity, but it allows binding of a small protein belonging to the 14-3-3 class of inhibitor proteins found in all major classes of eukaryotes (Bachman et al. 1996). The light activation of NR activity presumably involves dephosphorylation followed by dissociation of enzyme and inhibitor (for more detailed discussion, see Foyer et al. 1999).

Notwithstanding downregulation by covalent modification, extractable NR activities are almost always in excess of the estimated *in vivo* flux, often markedly so (Pace et al. 1990; Foyer et al. 1994b; Morcuende et al. 1998). Either conditions are less favourable for NR activity *in vivo* than *in vitro* or other factors are more important in controlling flux from nitrate into amino acids. Constitutive expression of NR in tobacco gave extractable activities that were between two- and fivefold higher than wild-type tobacco but had no marked effect on chlorophyll contents, protein levels, photosynthesis or biomass, although glutamine levels were increased twofold (Foyer et al. 1994b; Ferrario et al. 1995). The last observation does suggest some effect of overexpression on flux from nitrate to glutamine: metabolism downstream of glutamine synthesis is obviously able to absorb this perturbation. Constitutive expression should theoretically remove the requirement for nitrate-induced activation of transcription and allow NR protein to remain high even at limiting nitrate supply. However, although nitrate does not influence the phosphorylation state of the enzyme, it appears to be necessary to prevent decreases in NR capacity due to protein turnover: the imposition of N deficiency caused extractable NR activities to decrease almost to zero in transformants and wild-type plants alike (Ferrario et al. 1995, 1996). Conversely, compensatory modifications at several levels (including transcription, protein turnover and phosphorylation status) dampened the impact of fewer *nia* genes on nitrate assimilation in tobacco (Scheible et al. 1997b). Several of these controls over NR activity seem to be affected by carbohydrate supply. As well as increasing *nia* transcripts, exogenous sugars influence the posttranslational regulation of NR (Kaiser and Huber 1994). Feeding sugars to tobacco leaves markedly stimulated nitrate reduction, an effect correlated both with greater stability of the NR protein and with an increase in NR activation state (Morcuende et al. 1998). Both increased stability and increases in activation state may be linked to sugar-induced decreases in NR phosphorylation status. Although further work is required to identify the carbon metabolite(s) most important in controlling NR activity at these levels, it seems that sugars and glutamine act antagonistically in matching the initial step of nitrate assimilation to foliar C and N status.

In addition to coordination through modulation of NR, the assimilation of C and N must be integrated at the level of ammonium assimilation. In C3 plants particularly, interactions at the level of GS and GOGAT are complicated by the fact that these enzymes are involved not only in primary N assimilation but also in the photorespiratory recycling of ammonium (Keys et al. 1978; Miflin and Lea 1982). At the level of ammonium incorporation, therefore, two distinct types of interaction between C and N metabolism are theoretically possible.

Much information has been generated through genetic manipulation of GS and Fd-GOGAT activity in plants (see Chap. 2.2). Photorespiratory mutants, so called because they are unable to grow in conditions favouring photorespiration, have been isolated for *Arabidopsis* and barley (Somerville and Ogren 1980; Kendall et al. 1986; Blackwell et al. 1987; Wallsgrove et al. 1987). On transfer to air, both GS and Fd-GOGAT mutants show a rapid inhibition of photosynthesis and a strong accumulation of ammonium (GS mutants: Blackwell et al. 1987; Wallsgrove et al. 1987), or an accumulation of ammonium, glutamine and malate twinned with decreases in glutamate, glycine and serine (Fd-GOGAT mutants: Somerville and Ogren 1980; Kendall et al. 1986). Rather than a direct effect of high ammonium levels, the inhibition of CO_2 fixation in air is probably principally due to slower recycling of amino donors for glyoxylate transamination, which leads to depletion of carbon metabolites in the Calvin cycle (Blackwell et al. 1987; Morris et al. 1989).

The inability of these mutants to photosynthesise and grow in air can be reversed by genetic crossing with wild-type plants to produce heterozygous mutants with intermediate activities (Hausler et al. 1994a, b). In these plants, significant metabolic perturbation was observed only under conditions of enhanced photorespiratory flux. Survival appeared to require about 40% of wild-type activity, suggesting that this residual activity was more than sufficient to satisfy the plant's requirements for primary N assimilation (Hausler et al. 1994a, b). It is conceivable that photorespiratory and assimilatory functions reside in different isoforms of the enzymes, e.g. cytosolic GS_1 or chloroplastic GS_2, although there is little evidence for a strict division of tasks. Of the two Fd-GOGAT genes found in *Arabidopsis*, one is expressed in a pattern consistent with a role in photorespiration, while the expression of the other suggests a function in primary N assimilation (Coschigano et al. 1998). Whether such a division of labour occurs at the biochemical level is, however, less clear, given the lack of evidence for pools of 2-OG or glutamine specific to either photorespiration or N assimilation. Indeed, when *Arabidopsis* mutants lacking the gene whose expression was linked preferentially with photorespiration were grown at high CO_2, a decreased capacity for N assimilation was observed (Coschigano et al. 1998).

Perhaps unsurprisingly, integration of foliar C and N metabolism at the level of ammonium incorporation appears to be inextricably linked to photorespiratory metabolism in C_3 plants. In tobacco transformed with a GS gene in the antisense orientation, decreased activities were associated with reduced activities of both phosphoeno*l*pyruvate carboxylase (PEPC) and hydroxypyruvate reductase (Temple et al. 1993). A more detailed study has been carried out with transformed tobacco with decreased Fd-GOGAT activities (Ferrario-Méry et al. 2000a, b). On transfer to air, these plants accumulate ammonium and glutamine in proportion to the decrease in enzyme activity (Ferrario-Méry et al. 2000a). Interestingly, lower Fd-GOGAT activities were also correlated with proportional increases in 2-OG (Ferrario-Méry et al. 2000a), which were associated with enhanced extractable PEPC activity (Ferrario-Méry unpublished). Higher PEPC activity may have resulted from increases in glutamine, which activates PEPC expression and is a positive effector of the enzyme (see next section), and can be viewed as an attempt to drive ammonium assimilation by the increased provision of C skeletons. Although this response was not sufficient to overcome the restriction on flux imposed by decreased Fd-GOGAT activity, it does suggest that photorespiratory conditions may stimulate pathways leading to net 2-OG synthesis under some circumstances. Under conditions where 2-OG utilisation is restricted, however, this stimulation could be partly counteracted by the negative effect of 2-OG on PEPC activity (see following section). An intriguing observation in tobacco with decreased Fd-GOGAT is an activation of glutamate dehydrogenase (GDH) activity (Foyer et al. 1999). Because of its relatively low affinity for ammonium, this enzyme is considered to catalyse glutamate oxidation rather than synthesis. Our working hypothesis is that GDH could make some contribution to N assimilation in the Fd-GOGAT antisense tobacco, where ammonium is increased.

Similarly to the positive effect of Fd-GOGAT underexpression on PEPC activity, NR transcripts were increased in the transformants, suggesting that high 2-OG levels in these plants may have overcome the predicted inhibitory effect of increased glutamine on NR gene expression (Ferrario-Méry et al. 2000b). In *Arabidopsis* Fd-GOGAT mutants transferred to photorespiratory conditions, no decrease in NR transcripts was observed, even though glutamine was observed to increase (Dzuibany et al. 1998). Although the authors interpreted this observation as evidence against a role for glutamine in regulating NR gene expression, they reported no data for 2-OG contents, which may have counteracted the increases in glutamine (Dzuibany et al. 1998). These observations highlight the possible contribution of the glutamine: 2-OG ratio, rather than glutamine concentration alone, to the coordination of C and N metabolism. As the initial organic product of N assimilation, glutamine levels should transmit information on N availability while 2-OG, as the initial C skeleton into which N is incorporated, will reflect the availability of oxoacid acceptors. In bacteria, 2-OG and glutamine concentrations are sensed by the PII signal transduction protein and the PII uridyltransferase-removing enzyme (Jiang et al. 1998). It remains to be seen whether this system operates in higher plants.

Control of Carbon Partitioning Between Carbohydrates and Amino Acids

Carbon fixed during photosynthesis can either be converted to starch in the chloroplast or exported to the cytosol, mainly in the form of triose phosphate. Once in the cytosol, triose phosphate can be converted to sucrose or oxidised in respiratory flow to produce energy, reductant and carbon skeletons for processes such as amino acid synthesis (Fig. 2). Reciprocal regulation of C partitioning between carbohydrates and amino acids has been revealed by perturbation of the C:N balance: leaf carbohydrate accumulates in response to low-N stress (Rufty et al. 1988) whereas resupply of ammonium to plant systems starved of N causes decreased flux into carbohydrates and enhanced flow of carbon to oxoacids and amino acids (Larsen et al. 1981).

Control over partitioning between carbohydrates and amino acids is probably assured by regulation of multiple enzyme activities, and involves a host of metabolite effectors as well as changes in gene expression. The activity of PEPC is likely to be a key factor in diverting carbon from carbohydrates to amino acids. In maize, the C_4-type PEPC (though not the C_3-type enzyme) appears to be regulated at both transcriptional and posttranscriptional levels by nitrate or its metabolites (e.g. glutamine: Suzuki et al. 1994, and references therein). Tobacco mutants deficient in NR activity, able to assimilate nitrate only very slowly, have been used to distinguish between signals mediated by nitrate itself and those arising from downstream metabolites (Scheible et al. 1997a). Supplying nitrate to these plants induced expression of several respiratory enzymes, including PEPC, pyruvate kinase, and citrate synthase, all necessary activities for anapleurotic respiratory activity (Fig. 2). Nitrate-induction of respiratory enzymes was accompanied by decreased expression of ADP-glucose pyrophosphorylase (Fig. 2), suggesting inverse regulation of starch formation and amino acid biosynthesis (Scheible et al. 1997a). The physiological significance of these elegant results must be interpreted with a certain degree of caution, due to the extreme phenotype of the mutants analysed.

Many studies have indicated inverse regulation of sucrose synthesis and oxoacid formation. Most attention has focused on coordinate regulation of cytosolic C flow, mediated through regulation of sucrose-phosphate synthase (SPS) and PEPC (Fig. 2). Both enzymes are subject to complex regulation by metabolite effectors (for review, Foyer et al. 1999). The most important effectors of SPS activity are glucose-6-phosphate (activating) and Pi (inhibiting). PEPC is also activated by glucose-6-phosphate. These effects allow simultaneous promotion of oxoacid formation and sucrose synthesis in response to high cytosolic sugar-phosphate levels. Specific integration of PEPC activity with the demands of N assimilation is ensured through inhibition of activity by organic acids (malate, 2-OG) and activation by glutamine (Schuller et al. 1990; Krömer et al. 1996).

The extractable foliar activities of SPS and PEPC also increase on illumination as a result of covalent modification, allowing coordination of sucrose synthesis and amino acid synthesis with photosynthesis. For both enzymes, light-dark regulation is due to changes in protein phosphorylation status, albeit in contrasting ways. Like NR, SPS is more active in the dephosphorylated form whereas the activity of PEPC is increased by phosphorylation (for review, Foyer et al. 1999). This light-dark regulation can be modulated by N status: feeding nitrate to wheat or maize leaves causes the phosphorylation status of both enzymes to increase, resulting in an inhibition of SPS and an activation of PEPC (Van Quy et al. 1991a, b; Foyer et al. 1994a). These observations led to the proposal that inverse regulation of these activities through phosphorylation status acts to achieve an appropriate allocation of C to sucrose and amino acids (Champigny and Foyer 1992).

Even if each of these controls makes a contribution to partitioning of C, it seems that a whole network of regulation operates, affecting numerous enzymes at multiple levels, in order to achieve proper allocation of C to amino acid synthesis. These controls ensure that C is directed towards oxoacid synthesis in amounts appropriate to the prevailing rate of N assimilation. This allows C to be partitioned between export and the *in situ* synthesis of nitrogenous cellular components (e.g. protein and chlorophyll) in young leaves, and may also be essential in ensuring appropriate rates of export of sucrose and amino acids from source leaves. The relative importance of each control may vary considerably. In the short term, direct control of enzyme activity by posttranslational modification and metabolite effectors will be influential, whereas developmental and adaptive changes are presumably orchestrated primarily through changes in gene expression which, ultimately, will determine changes in tissue C:N ratios. The role of sucrose and amino acids such as glutamine in signal transduction has been well characterised. The complex network of metabolic crosstalk cannot, however, be explained simply in terms of end-product modulation of gene expression. Other key signals, for example those arising from organic acid metabolism, serve to orchestrate gene expression in relation to supply and demand.

References

Aslam M, Huffaker RC, Rains DW, Rao KP (1979) Influence of light and ambient carbon dioxide concentration on nitrate assimilation by intact barley seedlings. Plant Physiol 63: 1205-1209

Bachmann M, Huber J, Liao P-C, Gage DA, Huber SC (1996) The inhibitor protein of phosphorylated nitrate reductase from spinach (*Spinacia oleracea*) leaves is a 14-3-3 protein. FEBS Lett 387: 127-131

Banks FM, Driscoll SP, Parry MAJ, Lawlor DW, Knight JS, Gray JC, Paul MJ (1999) Decrease in phosphoribulokinase activity by antisense RNA in transgenic tobacco: relationship between photosynthesis, growth, and allocation at contrasting nitrogen supplies. Plant Physiol 119: 1125-1136

Blackwell RD, Murray AJS, Lea PJ (1987) Inhibition of photosynthesis in barley with decreased levels of chloroplastic glutamine synthetase activity. J Exp Bot 38: 1799-1809

Bloom AJ, Caldwell RM, Finazzo J, Warner RL, Weissbart J (1989) Oxygen and carbon dioxide fluxes from barley shoots depend on nitrate assimilation. Plant Physiol 91: 352-356

Bloom AJ, Sukrapanna SS, Warner RL (1992) Root respiration associated with ammonium and nitrate absorption and assimilation by barley. Plant Physiol 99: 1294-1301

Bowsher CG, Hucklesby DP, Emes MJ (1989) Nitrite reduction and carbohydrate metabolism in plastids purified from roots of *Pisum sativum* L. Planta 177: 359-366

Champigny ML, Foyer CH (1992) Nitrate activation of cytosolic protein kinases diverts photosynthetic carbon from sucrose to amino acid biosynthesis. Basis for a new concept. Plant Physiol 100: 7-12

Chen R-D, Gadal P (1990) Do the mitochondria provide 2-oxoglutarate needed for glutamate synthesis in higher plant chloroplast? Plant Physiol Biochem 28: 141-145

Coschigano KT, Melo-Olivieira R, Lim J, Coruzzi GM (1998) Arabidopsis *gls* mutants and distinct ferredoxin-GOGAT genes: implications for photorespiration and primary nitrogen assimilation. Plant Cell 10: 741-752

De la Torre A, Delgado B, Lara C (1991) Nitrate-dependent O_2 evolution in intact leaves. Plant Physiol 96: 898-901

Dzuibany C., Haupt S., Fock H., Biehler K., Migge A, Becker T (1998) Regulation of nitrate reductase transcript level by glutamine accumulating in the leaves of a ferredoxin-dependent glutamate synthase-deficient *gluS* mutant of *Arabidopsis thaliana*, and by glutamine provided via the roots. Planta 206: 515-522

Elias BA, Givan CV (1977) Alpha-ketoglutarate supply for amino acid synthesis in higher plant chloroplasts. Intrachloroplastic localization of NADP-specific isocitrate dehydrogenase. Plant Physiol 59: 738-740

Elrifi IR, Turpin DH (1986) Nitrate and ammonium induced photosynthetic suppression in N-limited *Selenastrum minutum*. Plant Physiol 81: 273-279

Emes MJ, Neuhaus HE (1997) Metabolism and transport in non-photosynthetic plastids. J Exp Bot 48: 1995-2005

Ferrario S, Valadier MH, Morot-Gaudry JF, Foyer CH (1995) Effects of constitutive expression of nitrate reductase in transgenic *Nicotiana plumbaginofolia* L. in response to varying nitrogen supply. Planta 196: 288-294

Ferrario S, Valadier MH, Foyer CH (1996) Short-term modulation of nitrate reductase activity by exogenous nitrate in Nicotiana plumbaginofolia and Zea mays leaves. Planta 199: 366-371

Ferrario-Méry S, Suzuki A, Kunz C, Valadier MH, Roux Y, Hirel B, Foyer CH (2000a) Modulation of amino acid metabolism in transformed tobacco plants deficient in Fd-GOGAT. Plant and Soil 221: 67-79

Ferrario-Méry S, Masclaux C, Suzuki A, Valadier MH, Hirel B, Foyer CH (2000b) α-Ketoglutarate and glutamine are metabolic signals involved in NR gene transcription in untransformed tobacco plants deficient in Fd-GOGAT. Planta (submitted)

Foyer CH, Noctor G, Lelandais M, Lescure JC, Valadier MH, Boutin JP, Horton P (1994a) Short-term effects of nitrate, nitrite and ammonium assimilation on photosynthesis, carbon partitioning and protein phosphorylation in maize. Planta 192: 211-220

Foyer CH, Lescure JC, Lefebvre C, Morot-Gaudry JF, Vincentz M, Vaucheret H (1994b) Adaptations of photosynthetic electron transport, carbon assimilation, and carbon partitioning in transgenic *Nicotiana plumbaginofolia* plants to changes in nitrate reductase activity. Plant Physiol 104: 171-178

Foyer CH, Ferrario-Méry S, Huber SC (1999) Regulation of carbon fluxes in the cytosol. Co-ordination of sucrose synthesis, nitrate reduction and organic and amino acid biosynthesis. In: Leegood RC, Sharkey TD, Von Caemmerer S (eds) Photosynthesis: physiology and metabolism. Kluwer, Dordrecht, The Netherlands pp. 177-203

Gastal F, Saugier B (1989) Relationships between nitrogen uptake and carbon assimilation in whole plants of tall fescue. Plant Cell Environ 12: 407-418

Gray VM, Cresswell CF (1984) The effect of inhibitors of photosynthetic and respiratory electron transport on nitrate reduction and nitrite accumulation in excised *Z. mays* L. leaves. J Exp Bot 35: 1166-1176

Hanning I, Heldt HW (1993) On the function of mitochondrial metabolism during photosynthesis in spinach (*Spinacia oleracea* L.) leaves. Plant Physiol 103: 1147-1154

Häusler RE, Blackwell RD, Lea PJ, Leegood RC (1994a) Control of photosynthesis in barley mutants with reduced activities of glutamine synthetase and glutamate synthase. I. Plant characteristics and changes in nitrate, ammonium and amino acids. Planta 194: 406-417

Häusler RE, Lea PJ, Leegood RC (1994b) Control of photosynthesis in barley mutants with reduced activities of glutamine synthetase and glutamate synthase. II control of electron transport and CO_2 assimilation. Planta 194: 418-435

Hoff T, Truong H-N, Caboche M (1994) The use of mutants and transgenic plants to study nitrate assimilation. Plant Cell Environ 17: 489-506

Huppe HC, Farr TJ, Turpin DH (1994) Coordination of chloroplastic metabolism in N-limited *Chlamydomonas reinhardtii* by redox modulation. 2. Redox modulation activates the oxidative pentose phosphate pathway during photosynthetic nitrate assimilation. Plant Physiol 105: 1043-1048

Ireland RJ, Lea PJ (1999) The enzymes of glutamine, glutamate, asparagine and aspartate metabolism. In: Singh BJ (ed) Plant amino acids: biochemistry and biotechnology. Marcel Dekker, New York, pp 49-109

Jiang P, Peliska JA, Ninfa JA (1998) Reconstitution of the signal-transduction bicyclic cascade responsible for the regulation of the *Ntr* gene transcription in *Escherichia coli*. Biochemistry 37: 12795-12801

Kaiser WM, Förster J (1989) Low CO2 prevents nitrate reduction in leaves. Plant Physiol 91: 970-974

Kaiser WM, Huber SC (1994) Post-translational regulation of nitrate reductase in higher plants. Plant Physiol 106: 817-821

Kendall AC, Wallsgrove RM, Hall NP, Turner JC, Lea PJ (1986) Carbon and nitrogen metabolism in barley (*Hordeum vulgare* L.) mutants lacking ferredoxin-dependent glutamate synthase. Planta 168: 316-323

Keys AJ, Bird IF, Cornelius MJ, Lea PJ, Miflin BJ, Wallsgrove RM (1978) Photorespiratory nitrogen cycle. Nature 275: 741-743

Krömer S, Stitt M, Heldt HW (1988) Mitochondrial oxidative phosphorylation participating in photosynthetic metabolism of a leaf cell. FEBS Lett 226: 352-356

Krömer S, Gardeström P, Samuelsson G (1996) Regulation of the supply of cytosolic oxaloacetate for mitochondrial metabolism via phosph*enol*pyruvate carboxylase in barley leaf protoplasts. I. The effect of covalent modification on PEPc activity, pH response, and kinetic properties. Biochim Biophys Acta 1289: 343-350

Kruse A, Fieuw S, Heineke D, Müller-Röber B (1998) Antisense inhibition of cytosolic NADP-dependent isocitrate dehydrogenase in transgenic potato plants. Planta 205: 82-91

Lancien M, Ferrario-Méry S, Roux Y, Bismuth E, Masclaux C, Hirel B, Gadal P, Hodges M (1999) Simultaneous expression of NAD-dependent isocitrate dehydrogenase and other Krebs cycle genes after nitrate resupply to short-term nitrogen starved *Nicotiana tabacum*. Plant Physiol 120: 717-725

Larsen PO, Cornwell KL, Gee SL, Bassham JA (1981) Amino acid synthesis in photosynthesizing spinach cells. Effects of ammonia on pool sizes and rates of labeling from $^{14}CO_2$. Plant Physiol 68: 292-299

Miflin BJ, Lea P (1982) Ammonia assimilation. In: Miflin BJ (ed) The Biochemistry of plants, vol 5. Academic Press, New York, pp 169-202

Morcuende R, Krapp A, Hurry V, Stitt M (1998) Sucrose feeding leads to increased rates of nitrate assimilation, increased rates of α-oxoglutarate synthesis, and increased synthesis of a wide spectrum of amino acids in tobacco leaves. Planta 206: 394-409

Morris PF, Layzell DB, Canvin DT (1989) Photorespiratory ammonia does not inhibit photosynthesis in glutamate mutants of *Arabidopsis*. Plant Physiol 89: 498-500

Noctor G, Foyer CH (1998) A re-evaluation of the ATP:NADPH budget during C_3 photosynthesis. A contribution from nitrate assimilation and its associated respiratory activity? J Exp Bot 49: 1895-1908

Noctor G, Foyer CH (2000) Homeostasis of adenylate status during photosynthesis in a fluctuating environment. J Exp Bot 51: 347-356

Pace GH, Volk RJ, Jackson WA (1990) Nitrate reduction in response to CO_2-limited photosynthesis. Relationship to carbohydrate supply and nitrate reductase activity in maize seedlings. Plant Physiol 92: 286-292

Palomo J, Gallardo F, Suarez MF, Canovas FM (1998) Purification and characterization of NADP$^+$-linked isocitrate dehydrogenase from Scots pine – evidence for different physiological roles of the enzyme in primary development. Plant Physiol 118: 617-626

Rasmusson AG, Møller IM (1990) NADP-utilizing enzymes in the matrix of plant mitochondria. Plant Physiol 94: 1012-1018

Raven JA, Smith FA (1976) Nitrogen assimilation and transport in vascular land plants in relation to intracellular pH regulation. New Phytol 76: 415-431

Reed AJ, Canvin DT, Sherrard JH, Hageman RH (1983) Assimilation of [^{15}N]nitrate and of [^{15}N]nitrite in leaves of five plant species under light and dark conditions. Plant Physiol 71: 291-294

Robinson JM (1988) Spinach leaf chloroplast CO_2 and NO_2^- photoassimilations do not compete for photogenerated reductant: manipulation of reductant levels by quantum flux density titrations. Plant Physiol 88: 1373-1380

Rufty TW, Huber SC, Volk RJ (1988) Alterations in leaf carbohydrate metabolism in response to nitrogen stress. Plant Physiol 88: 725-730

Rufty TW, MacKown CT, Volk RJ (1989) Effects of altered carbohydrate availability on whole plant assimilation of $^{15}NO_3^-$. Plant Physiol 89: 457-463

Scheible W-R, Gonzáles-Fontes A, Lauerer M, Röber BM, Caboche M, Stitt M (1997a) Nitrate acts as a signal to induce organic acid metabolism and repress starch metabolism in tobacco. Plant Cell 9: 783-798

Scheible W-R, Gonzáles-Fontes A, Morcuende R, Lauerer M, Geiger M, Glaab J, Gojon A, Schulze E-D, Caboche M, Stitt M (1997b) Tobacco mutants with a decreased number of functional *nia* genes compensate by modifying the diurnal regulation of transcription, post-translational modification and turnover of nitrate reductase. Planta 203: 304-319

Schuller JA, Turpin DH, Plaxton WC (1990) Metabolite regulation of partially purified soybean nodule phosphoenolpyruvate carboxylase. Plant Physiol 94: 1429-1435

Somerville CR, Ogren WL (1980) Inhibition of photosynthesis in *Arabidopsis* mutans lacking leaf glutamate synthase activity. Nature 286: 257-259

Suzuki I, Crétin C, Omata T, Sugiyama T (1994) Transcriptional and posttranscriptional regulation of nitrogen-responding expression of phosphoenolpyruvate carboxylase gene in maize. Plant Physiol 105: 1223-1229

Temple SJ, Knight TJ, Unkefer PJ, Sengupta-Gopalan C (1993) Modulation of glutamine synthetase gene expression in tobacco by the introduction of an alfalfa glutamine synthetase gene in sense and antisense orientation : molecular and biochemical analysis. Mol Gen Genet 236: 315-325.

Turpin DH, Bruce D (1990) Regulation of photosynthetic light harvesting by nitrogen assimilation in the green alga *Selenastrum minutum*. FEBS Lett 263: 99-103

Van Quy L, Lamaze T, Champigny ML (1991a) Short-term effects of nitrate on sucrose synthesis in wheat leaves. Planta 185: 53-57

Van Quy L, Foyer CH, Champigny ML (1991b) Effect of light and nitrate on wheat leaf phosphoenolpyruvate carboxylase activity. Plant Physiol 97: 1476-1482

Vincentz M, Caboche M (1991) Constitutive expression of nitrate reductase allows normal growth and development of *Nicotiana plumbaginofolia* plants. EMBO J 10: 1027-1035

Wallsgrove RM, Turner JC, Hall NP, Kendall AC, Bright SWJ (1987) Barley mutants lacking chloroplast glutamine synthetase – biochemical and genetic analysis. Plant Physiol 83: 155-158

Weger HG, Turpin DH (1989) Mitochondrial respiration can support NO_3^- and NO_2^- reduction during photosynthesis. Plant Physiol 89: 409-415

Weger HG, Birch DG, Elrifi IR, Turpin DH (1988) Ammonium assimilation requires mitochondrial respiration in the light. A study with the green alga *Selenastrum minutum*. Plant Physiol 86: 688-692

Weger HG, Chadderton AR, Lin M, Guy RD, Turpin DH (1990) Cytochrome and alternative pathway respiration during transient ammonium assimilation by N-limited *Chlamydomonas reinhardtii*. Plant Physiol 94: 1131-1136

Nitrogen Traffic During Plant Growth and Development

Alain Ourry[1], James H. Macduff[2], Jeffrey J. Volenec[3] and Jean Pierre Gaudillere[4]

The supply of nitrogen in a plant-available form is one of the main factors controlling plant productivity in both natural and agricultural ecosystems. Higher plants have therefore evolved a variety of strategies to increase the effectiveness by which N is retained and utilised for growth and development, following its acquisition from the environment by root uptake or N_2 fixation. The concept of nitrogen use efficiency (NUE) has been widely employed in plant physiology and ecology in the analysis of these strategies. NUE has been defined in many ways, but there are strong theoretical grounds for regarding it as the ratio between the biomass produced by, and the flux of nitrogen through, a system over a given time interval (Garnier and Aronson 1998), typically expressed in units of kg biomass mol^{-1} N. Following Berendse and Aerts (1987), this definition can be expressed as: NUE = nitrogen productivity x mean residence time of N in the plant (MRT). The nitrogen productivity term (Ingestad 1979) provides a measure of the efficiency with which internal N is used to provide biomass, and MRT provides a measure of N retention within the plant.

   Nitrogen traffic can be defined rather loosely as the temporal and spatial pattern of N transfer and allocation occurring within the plant throughout its life cycle, and is subject to a combination of ontogenetic and environmental determination. At the whole-plant level, these transfers encompass both the initial allocation of newly acquired (from the soil/atmosphere) N to sink organs for growth or storage, and the recycling or reallocation of N from senescing and temporary storage tissues. Both primary N allocation and reallocation strategies affect NUE. The subject of this chapter is the N traffic associated with reallocation, particularly that which occurs to

1. UA INRA 950, Physiologie et Biochimie Végétales, Institut de Recherche en Biologie Appliquée, Université, 14032 Caen Cedex, France. *E-mail:* ourry@criuc.unicaen.fr
2. Institute of Grassland and Environmental Research, Plas Gogerddan, Aberystwyth, Ceredigion SY23 3EB, UK. *E-mail:* james.mcduff@bbsrc.ac.uk
3. Department of Agronomy, Purdue University, West Lafayette, IN 47907-1150, USA. *E-mail:* jvolenec@purdue.edu
4. Station de Physiologie Végétale, Centre INRA de Bordeaux, BP 81, 33 883 Villenave d'Ornon Cedex, France. *E-mail:* Gaudillere@bordeaux.inra.fr

and from storage tissues. The factors determining initial allocation of N have been reviewed elsewhere (e.g. Wilson 1988; Aerts 1994; Cheeseman 1996 et al; Farrar and Gunn 1998). Several authors have discussed the distinctions between N storage and N accumulation, and the appropriate time-scale-dependent classification of different metabolic processes and classes of N compounds (Millard 1988; Heilmeier and Monson 1994). In the context of this chapter it is sufficient to define storage as "resources that build up in the plant and can be mobilised in the future to support biosynthesis" (Chapin et al. 1990), although this covers not only reserve compounds in *sensu strictu* but also the recycling of resources from functionally redundant tissues.

Depending on the time scale in question, the effectiveness with which plants are able to recycle and reallocate N between internal sources and sinks will influence both the N productivity and mean residence time components of NUE. Garnier and Aronson (1998) specified (1) the N resorption efficiency from senescent tissue and (2) the rate of biomass loss by the plant as the main determinants of the mean residence time of N within the plant. However, the temporary storage of N in non-senescent tissues and its subsequent remobilisation, a characteristic of many herbaceous and woody species, is likely to exert effects on both the N productivity and mean residence time components. The ability to recycle N within the plant allows the reuse of this element by different tissues and at different times during ontogeny. Intuitively, this should provide a degree of buffering against temporary shortfalls in either the supply of soil mineral N or in metabolic activity associated with N acquisition and reduction by roots. The importance of N mobilisation and subsequent reallocation during periods of reduced N acquisition in sustaining the growth of new tissues has been demonstrated for spring growth of trees (Habib 1984; Millard 1993; Gaudillère 1997), and spring growth or regrowth following defoliation in forage species (Volenec et al. 1996).

In reviewing current understanding of the nitrogen traffic associated with N storage and remobilisation in higher plants, we will draw on recent studies of woody and forage species, demonstrating both the involvement of specific proteins in vegetative N storage and providing some indication of how their pattern of accumulation/hydrolysis may be regulated. Although a comprehensive description of the molecular and physiological determinants of N storage and remobilisation continues to emerge, understanding how the different components of plant N metabolism (i.e. N uptake, N_2 fixation, assimilation, storage, remobilisation) act alone or in concert is a key to understanding N regulation at the whole plant level.

N Reserves and N Traffic Within the Plant: Significance for Yield and Spring Growth

Isotopic techniques based on ^{15}N-labelling have proved to be the most effective means of measuring N fluxes between different tissues or organs, and for evaluating the respective contributions of N remobilised from storage tissues or recently acquired by uptake and fixation from the environment. When N is mobilised from a source to an actively growing sink tissue, free amino acids are normally the first class of compounds to be translocated in the phloem or the xylem vessels (Fig. 1). The soluble protein fraction usually constitutes a larger storage pool of N, but

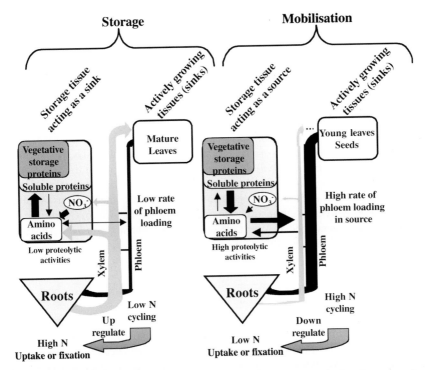

Fig. 1 : Simplified representation of N flow within the plant occurring during the accumulation of storage compounds (*left side*) or during the mobilization of N reserves to sink tissues (*right side*): young leaves (woody species during spring, forage species during spring or after shoot removal) or seeds (cereals)

requires proteolysis to amino acids to make it available for transport throughout the plant. The translocation of amino acids derived from this pool, therefore, tends to occur later than those from the free amino acid pool. For reasons of osmotic balance, plant cells may limit sucrose and amino acid concentrations. Hence, polymeric forms of carbon and nitrogen tend to accumulate, such as starch, fructans and soluble proteins.

Whilst the free amino acid pool is generally fully depleted under conditions typically associated with the mobilisation of N reserves, the soluble protein pool is usually only partially depleted; reducing by 40% in taproots or stolons of forage species (Avice et al. 1996; Corre et al. 1996), 50% in the bark of *Prunus persica* (Arora et al. 1992) and of *Malus pumila* (O'Kennedy and Titus 1979), 60% in tuberised roots of *Cychorium intybus* (Ameziane et al. 1997) and by as much as 90% in leaves and stems of *Triticum aestivum* (Waters et al. 1980) or *Oryza sativa* (Mae et al. 1983). The wide variation implies that different physiological processes are involved, depending on species and developmental stages. For example, the acute level of protein hydrolysis occurring in leaves of cereals during grain filling is related to irreversible and programmed foliar senescence. In contrast, the relatively modest hydrolysis found in woody and forage species is associated with recurring and reversible processes: the storage tissues or organs can accumulate N compounds on a regular basis. In a biennial species such as chicory (*Cychorium intybus*), mobilisation of N from tuberous roots occurs during the spring of the second year. Protein hydrolysis in this tissue is limited because the root must function for several months subsequent to the initial growth in Spring.

Deciduous trees remobilise N from roots and trunks to support the synthesis of new leaf tissue in the spring. In evergreen trees the previous year's foliage provides an additional source of N for translocation to the apical growing points during periods of flushing (Millard 1993). Similarly, early spring growth of leaves in many forage species requires the mobilisation of N from perennial tissues. This is also triggered by mechanical or herbivorous defoliation, which reduces soil N uptake and/or N_2 fixation (Volenec et al. 1996). In cereals, a large proportion of the N previously accumulated in leaves is mobilised to sustain grain filling when N uptake is limited by soil N availability and/or downregulation of the N uptake mechanisms. Additionally, N storage has been demonstrated to occur in tubers or roots of *Cychorium intybus* (Cyr and Bewley 1990a; Ameziane et al. 1997), *Euphorbia escula*, *Taraxacum officinale* (Cyr and Bewley 1990a and b), *Helianthus tuberosus* (Mussigmann and Ledoigt 1989), *Solanum tuberosum* (Paiva et al. 1983), *Ipomea batatas* (Li and Oba 1985), *Dioscorea rotundata* (Harvey and Boulter 1983) and *Vitis vinifera* (Conradie 1980). In these species the N reserves are mobilised primarily to support growth during the spring.

The dynamic nature of the balance between the accumulation and mobilisation of N reserves within the plant can be gauged from the changing pattern of source and sink interactions during its life cycle (Fig. 1). Synthesis and accumulation of reserve N compounds occur when a storage tissue is a net N importer (sink). This occurs when N acquisition by the plant exceeds that required to support synthesis of new tissue; for example when leaf growth is limited by low air temperature during autumn. In contrast, when N uptake and/or N_2 fixation are insufficient to meet the demands for growth, reserve N is mobilised, and storage organs behave as net exporters (source). When this occurs, amino acid concentrations in source tissues typically exhibit a pattern of transient decrease, increase and final drastic depletion. This reflects the changing balance between the flux of amino acids loaded into the phloem, and the flux generated by proteolysis of soluble proteins in the source tissue (Fig. 1). The regulation of proteolysis in this context remains uncertain, despite recent advances (see Chap. 5.2). Several studies have shown increased acid endoprotease activity (Waters et al. 1980; Dalling 1985; Feller and Keist 1986; Ourry et al. 1989) during N remobilisation from source to sink tissues. The transition from net sink to net source activity in these tissues is accompanied by changes in a number of enzyme activities (see Chap. 2.2), including transaminases, glutamine synthetase, glutamate synthase or glutamate dehydrogenase (Kang and Titus 1980; Simpson and Dalling 1981; Bigot and Boucaud 1992), but how these are coordinated remains unclear. This uncertainty also extends to the regulation of mineral N uptake and N_2 fixation by N remobilisation. It is somewhat paradoxical that, under certain conditions, the mobilisation of N reserves is regarded as compensating for low levels of N uptake from the soil. Yet, the resulting increase in amino acid cycling through the phloem (Fig. 1) may itself downregulate both N uptake (see Chaps. 1.1 and 4.2) and N_2 fixation (see Chap. 3.1).

The physiological and agronomic consequences of variation in N reserve status and in the capacity for N remobilisation have not been widely investigated. Ameziane et al. (1997) observed that the rate of accumulation of N reserves in *C. intybus* was affected by external N supply, but the final content of N reserves in the tuberous root was similar in N-replete or N-limited plants, indicating that reserve compounds are synthesised irrespective of a growth-limiting N supply. However, the N reserve status of lucerne (*Medicago sativa* L.) can be altered by limiting external N supply or by varying the height of cutting (Ourry et al. 1994). Both the yield and cumulative uptake of N between cuts by this species are linearly correlated with the initial level of available N reserves. Further, it has been argued that the competitive success of individual lucerne plants within dense stands is determined by N reserve status rather than by the availability of carbohydrates (Avice et al. 1997).

N reserve status exerts significant effects on the growth and yield of trees. On the basis of N fertiliser trials, Habib (1984) inferred that spring growth is more sensitive to the level of N reserves accumulated in the trunk and in the roots than to the availability of soil N. Similarly, N budget studies at the ecosystem level (Millard 1993) have indicated that enhanced soil fertility either has no effect, or merely decreases the proportion of leaf N content translocated to N storage tissues. So it would appear that both the extent of N remobilisation from storage tissues and its importance as the main source of N for leaf production by trees in the spring are affected by external N supply. There is convincing evidence that much of the annual increment of N allocated to reserves in woody species is provided by uptake of soil N during the autumn; resorption of N from senescing leaves appears to provide a relatively small proportion of the N subsequently remobilised during the spring. Consequently, N fertiliser strategies should be aimed at maximising N storage in autumn for use in the following spring.

Vegetative Storage Proteins (VSPs): Function and Diversity

Criteria for Identification of VSPs

The hydrolysable soluble protein content of many storage tissues includes a number of specific proteins thought to function as vegetative storage proteins (VSP). VSPs are identified by several criteria (Staswick 1990, 1994). They must be prominent polypeptides accounting for at least 5% of total soluble proteins, must be preferentially hydrolysed when compared with other soluble proteins, and be located specifically in vacuoles, progressively transformed into protein bodies. Traditionally, it was supposed that VSPs lack any catalytic activity, but several recent studies have shown many putative VSPs (Tables 1 and 2), have significant sequence homology with known enzymes (see below). An additional important criterion for classification as storage proteins, is that their patterns of accumulation/mobilisation should parallel the flux of N between source and sink tissues (Fig. 1), evaluated for example by ^{15}N labelling.

Occurrence of VSPs in Higher Plants

VSPs have been identified in herbaceous (Table 1) and woody plants, with the frequency of VSP occurrence increasing as more species are investigated. The characteristic molecular weights of the polypeptide subunits, estimated under denaturing conditions, range from 15 to 60 kDa, and dimeric structures have been reported in

Table 1. Summary of vegetative storage proteins found in herbaceous species, molecular weights, tissue or subcellular localisation, period of mobilisation, and conditions or treatments which up-regulate their accumulation.

Species	MW (kDa)	Localisation	Activity	Mobilisation	Upregulations	References
Glycine max	28[a], 31[a]	Vacuolar in PM, stem, leaves, seed pods,	Acid phosphatase	Seed filling	MeJa, wounding, N supply, sink limitation, soluble sugars, water deficit	Wittenbach (1983); Staswick (1989 1990); DeWald et al. (1990); Mason and Mullet (1990)
	94	Cytoplasmic or vacuolar in PM	Lipoxygenase		sink limitation	Tranberger et al. (1991); Stephenson et al. 1998
Arabidopsis thaliana	29[a], 30[a]	Flower, Buds	Acid phosphatase ?	Unknown	MeJa, wounding.	Berger et al. (1995)
Cychorium intybus	18[b]	Root	Unknown	Spring growth	?	Cyr and Bewley (1990)
Taraxacum officinale	17	Root	Unknown	Spring growth	N supply.	Ameziane et al. (1997)
Taraxacum officinale	18[b]	Root	Unknown	Spring growth	?	Cyr and Bewley (1990)
Helianthus tuberosus	16, 16.5, 18[b]	Tuber	Unknown	?	?	Mussigmann and Ledoigt (1989)
Ipomnoea batatas	25	Tuberous Root	Unknown	Spring growth	?	Li and Oba (1985)
Solanum tuberosum	41	Tuber	Unknown	?	?	Paiva et al. (1983)
Dioscorea rotundata	31	Tuber	Unknown			Harvey and Boulter (1983)
Euphorbia escula	26	Root	Unknown	Spring growth	?	Cyr and Bewley (1990)
Medicago sativa	15[c], 19[c], 32[c]	Vacuolar in Taproot,	Unknown	Spring growth and regrowth	MeJa, LT, SD	Hendershot and Volenec (1993)
	55.9	Taproot, Stem	ß-Amylase		?	Gana et al. (1998)
Trifolium repens	17.3	Stolon, Roots	Unknown	Spring growth and regrowth	LT, SD	Bouchart et al. (1998)
Brassica napus	23	Taproot	Unknown	Seed filling	MeJa, ABA	Ourry et al. (unpublished)

Abbreviations: Localisation: PM, paraveinal mesophyll. Upregulation of VSP accumulation: ABA, abscisic acid; LT, low temperature; MeJa, methyl jasmonate, SD short days. Molecular weight numbers sharing a similar letter indicate VSPs from different species that are immunologically related. After Volenec et al. 1996

Table 2. Summary of vegetative storage proteins found in woody species, molecular weights, tissue or subcellular localisation, period of mobilisation, and conditions or treatments which up-regulate their accumulation

Species	MW (kDa)	Localization	Mobilisation	Up-regulations	Reference
Gingko biloba	40, 45	Bark	Spring growth		Shim and Titus (1985)
Acer saccharum	16, 24	Bark	"		Wetzel et al. (1989)
Salix x smithiana	32 [a]	Bark	"		Wetzel et al. (1989)
Prunus persica	16, 19	Bark, wood	"		Arora et al. (1992)
Populus deltoides	32[a], 34 [a]	Bark	"	Short days	Wetzel et al. (1989)
				Wounding	Davies et al. (1993)
Populus euramericana	32[a], 34[a], 36[a], 38	Bark, wood	"		Stepien and Martin (1992)
Populus sp.	32[a] 36[a]	Bark, wood, roots PB		Short days	Langheinrich and Tischner (1991)
				N supply	Van Cleve and Apel (1993)
				Low temperature	
Larix decidua	25, 27, 32	Bark	"		Wetzel and Greenwood (1989)
Pinus strobus	15	Bark	"	?	Wetzel and Greenwood (1989)
Taxodium distichum	35[b]	Wood	"	?	Harms and Sauter (1991)
Metasequoia sp.	32, 34[b]	Wood	"	?	Harms and Sauter (1991)
Pseudotsuga sp.	30	Buds	"	?	Roberts et al. (1991)
Picea	20, 27	Buds, stems, roots	"	?	Roberts et al. (1991)

Abbreviations: Localisation: PB, protein bodies. Molecular weight numbers sharing a similar letter indicate VSPs from different species that are immunologically related.
After Stépien et al. 1994

poplar (Langheinrich and Tischner 1991; Stépien and Martin 1992). Most VSP sub-units show several isoforms, based on differences in their isoelectric points, revealed by two-dimensional-polyacrylamide gel electrophoresis. For example, analysis of *C. intybus* VSPs (Ameziane et al. 1997) by this method demonstrated up to seven isoforms. Excepting those identified in *Trifolium repens* (Bouchart et al. 1998), most VSPs appear to be glycosylated. It has been proposed that the oligosac-charide chains confer thermostability during winter as well as increasing the stored carbon content. Differing degrees of glycosylation may also explain the occurrence of different isoforms for a given VSP.

In biennial and perennial herbaceous plants (Table 1) the accumulation of VSPs occurs mainly during autumn, while they are mobilised during spring. Apart from providing N for shoot growth during spring, VSPs have also been shown to supply N to the regrowing shoots of forage species such as *M. sativa* (Volenec et al. 1996) and *T. repens* following severe defoliation when N uptake and/or fixation are severely inhibited. In annual species like *Glycine max* (Wittenbach 1983; Staswick 1992, 1994) and biennial species such as *Brassica napus* L. (A. Ourry et al., unpubl.), VSPs are mobilised during seed filling. In soybean, VSPs also accumulate in the cot-yledons during germination (Staswick 1994) and may provide a buffer of N between the hydrolysis of seed storage proteins and the release of amino acids required for seedling growth. In *Arabidopsis thaliana*, VSPs homologous to soybean VSPs have been characterised (Berger et al. 1995), and found mostly in flowers and buds, but their role is unknown. In trees (Table 2), overwintering VSPs play a well-established role in supporting spring leaf growth.

The mobilisation of N from leaves during grain filling in cereals and other crops has been studied in detail (Peoples et al. 1980; Waters et al. 1980; Simpson and Dalling 1981; Mae et al. 1983; Cliquet et al. 1990; Reilly 1990; Mackown et al. 1992). The hydrolysis of soluble proteins in leaves, resulting from increased proteolytic activity, coincides with the net export of free amino acids. Ribulose 1,5 bisphosphate carboxylase/oxygenase (Rubisco), which can account for over 50% of the soluble protein content of leaves of C_3 plants, is a major source of N for grain filling (Peoples et al. 1980; Mae et al. 1983). Its breakdown and use as a N source are described in detail in chapter 5.2. Despite the fact that remobilised N accounted for up to 83% of seed N in intact wheat plants, Mackown et al. (1992) did not identify VSPs serving as temporary storage for amino acids resulting from the disassembly of Rubisco in *T. aestivum* leaves. Irrespective of the proportion of the total sink capacity (spikelets) removed, Rubisco was hydrolysed continuously and amino acids accumulated in leaves without creation of VSP intermediates.

A number of fruitless searches for VSPs in forage species have been reported. For example, Li et al. (1996) analysed the SDS polyacrylamide gel elec-trophoresis patterns of soluble proteins extracted from different tissues of *Trifo-lium pratense*, *Lotus corniculatus* and *Melilotus officinalis*, sampled through an annual cycle, and could not detect any VSPs. It is probable that *Lolium perenne* also lacks VSPs, despite substantial mobilisation of N from sheath tissues to sup-port synthesis of new leaf tissue following severe defoliation. Loualhia et al. (1999) identified three candidate polypeptides in this species, all of which fol-lowed both the seasonal and defoliation-cycle patterns typical of VSPs, but were not sufficiently abundant to account for a significant fraction of the N mobilisa-tion measured after defoliation.

Sites of VSP Accumulation in Plants

The accumulation of VSPs in herbaceous species is generally confined to perennial tissues and organs such as roots, tubers and stolons (Table 1), although soybean and *A. thaliana* provide notable exceptions. In trees, the main organs of accumulation are in the bark and wood tissues of roots and the trunk (Table 2), consistent with the role of VSPs in spring growth. The location of VSPs in perennial tissues is seen as indirect evidence for their active role in promoting overwintering survival and recovery from defoliation. Soybean is unusual because VSPs are present almost exclusively in the above-ground vegetative tissues; relatively minor amounts are found in roots and seeds. Specific accumulation in the paraveinal mesophyll cells of soybean has been reported, and it has been argued that deposition of VSPs and other N storage compounds adjacent to xylem and phloem facilitates their subsequent mobilisation (Staswick 1994).

At the subcellular level, VSPs accumulate in protein bodies in *G. max* (Staswick 1992), poplar species (Van Cleve et al. 1988; Stépien et al. 1992) and *M. sativa* (Avice et al. 1996). Their location in protein-filled vacuoles may provide protection from hydrolysis. The pathway of VSP transport into the vacuole has been elucidated in soybean (Tranberger et al. 1991) and is similar to that for deposition of storage proteins in seeds (Muntz 1998). The 28- and 31-kDa VSPs are synthesised on membrane-bound ribosomes and are processed through the Golgi apparatus, before deposition in the vacuole. Some isoforms of the 94-kDa VSP accumulate in the cytoplasm and others in the vacuole (Stephenson et al. 1998). These are thought to be synthesised on free ribosomes, with deposition of some isoforms in vacuoles apparently occurring without involvement of the Golgi apparatus.

Polymorphism of VSPs

Some homology between the VSPs identified in different species, has been demonstrated (Tables 1 and 2). Antibodies raised against soybean VSPs cross-react with the VSPs of *A. thaliana*, and with the VSPs of other *Glycine* species (Staswick 1997). Similarly, VSPs of *C. intybus*, *T. officinale* and *H. tuberosus* seem to be immunologically related. The three VSP polypeptides observed in taproots of *M. sativa* are also abundant in roots of other perennial *Medicago* species, but were absent or present at much lower levels in roots of annual *Medicago* species (Cunningham and Volenec 1996). In trees, VSPs of different poplar species (Wetzel et al. 1989; Langheinrich and Tischner 1991; Stépien et al. 1992) and of *Salix* x *smithiana* (Wetzel et al. 1989) are also recognised by the same polyclonal antibody.

Absence of catalytic activity, one of the criteria for classification of proteins as functional VSPs, has been questioned based upon sequence homology data of several VSPs. In soybean, genes for the 28- and 31-kDa VSPs show significant homology with genes encoding an acid phosphatase. DeWald et al. (1992) also demonstrated that purified soybean VSP homodimers and heterodimers exhibited acid phosphatase activity. However, the significance of this activity has been challenged (Staswick et al. 1994), on the basis that the VSP fraction accounts for less than 0.1% of total acid phosphatase activity in soybean leaves. More recently, Penheiter et al. (1997) have identified an acid phosphatase in the nodules of soybean that possesses close homology with specific soybean VSPs. Further evidence for the occurrence of proteins having dual VSP/enzymatic functions in soybean can be inferred from the suggestion that a protein of 94 kDa with lipoxygenase activity may also be involved in N storage (Stephenson et al. 1998; Transbarger et al. 1991). Similarly, Gana et al. (1998) suggested that ß-amylase from *M. sativa* taproots functions as a VSP. This protein qualifies in terms of abundance (Boyce and Volenec 1992) and pattern of protein accumulation/hydrolysis (Volenec et al. 1996) and steady-state transcript levels (Gana et al. 1998). Its pattern of activity is also opposite to that of starch hydrolysis following shoot removal (Boyce et al. 1992), implying that it is not associated with starch breakdown.

There are no reports of VSPs from woody species being associated with enzymatic activity, to our knowledge. However, close relationships between poplar VSPs and wound-induced proteins belonging to the *Win* multigene family have been described (Clausen and Appel 1991; Coleman et al. 1992). Davis et al. (1993) reported a high-sequence homology between poplar VSPs and *Win-4*, a gene whose expression is wound-inducible; but while the poplar VSP occurs mainly in the stem and is upregulated by short daylength, *Win-4* is expressed in stems and leaves, and is not affected by shortening daylengths. More recently, it has been shown that the synthesis of *Win-4* mRNA and VSP mRNA both increase with increasing N supply (Laurence et al. 1997). Beardmore et al. (1996) reported a significant sequence homology between the genes encoding seed storage proteins and those for VSPs in poplar.

Taking the literature as whole, it is apparent that a wide range of proteins have been implicated as VSPs in higher plants. Despite their generally common subcellular vacuolar location, there appears to be little homology between putative VSPs from different species in terms of molecular weight and residual catalytic activities. This high degree of polymorphism is similar to the wide variation observed for seed storage proteins, characterised mainly in cereals (see Chap. 6). In both cases this would suggest that selective constraints were relatively low for such proteins during evolution, resulting in the persistence of many non-lethal, dysfunctional mutations. This is consistent with the high diversity in residual enzyme activity observed across VSPs. The appearance of VSPs may be fairly recent in evolutionary terms, although further evidence for this hypothesis is required. Amongst various explanations offered for the origin of VSPs, one of the most promising ascribes their appearance to mutations which partially or entirely deregulated the synthesis of enzymatic proteins, which were previously subject to posttranscriptional regulation of activity.

Regulation of N Traffic Through VSPs

Dependence on Plant N Status

Most of the information on the regulation of VSP metabolism has been provided by studies on soybean and poplar. The expression of genes encoding for VSPs appears to be highly regulated, although this needs further investigation. Given that these proteins function as N stores and tend to accumulate when the flux of N entering the plant exceeds the requirements for synthesis of new tissue, it is reasonable to suppose that VSP synthesis may be subject to regulation by an indicator of plant N status. This view is supported by work on soybean (Staswick 1992, 1997) and poplar (Van Cleve and Apel 1993), showing that the corresponding mRNA accumulated when plants were exposed to a high N supply, although the external concentrations of N were extremely high (25 mM N for poplar and 40 mM N for soybean). A physiologically more realistic high N supply (10 mM N), had no effect on VSP accumulation in non-nodulated lucerne (Kalengamaliro et al. 1997), probably because the consequent increase in shoot growth rate prevented a build up of surplus N within the plants. Similar behaviour was observed with B. napus L. (Ourry et al. unpubl.). The rate of VSP accumulation in C. intybus was unaffected over a range of different levels of N supply (Ameziane et al. 1997), suggesting that the genes for VSPs in this species are constitutively expressed rather than induced by changes in environmental or nutritional conditions as in poplar, soybean or A. thaliana. These authors also proposed that the flux of N resorbed from the leaves may regulate VSP synthesis in C. intybus, subject to some upper limit.

An alternative approach to manipulating plant N status and the availability of N for storage is excision of the actively growing vegetative or reproductive sinks. Removal of pods (Wittenbach 1982; Staswick 1989; Bunker et al. 1995; Stephenson et al. 1998) and buds (Stephenson et al. 1998) from soybean plants increased the accumulation of VSPs (28 and 31 kDa) and the synthesis of the corresponding mRNAs. This has also been reported for the four isoforms of the 94-kDa VSP exhibiting lipoxygenase activity, and its mRNA (Bunker et al. 1995; Stephenson et al. 1998). Both an increase in level of N supply and the removal of active sinks are likely to increase the size of available pools of amino acids, at least on a temporary basis. An increase in amino acid concentrations at the relevant sites might therefore appear to be a prime candidate for triggering or increasing synthesis of VSPs. However, Mason et al. (1992) failed to detect increases in mRNA coding for VSPs following the exogenous supply of free amino acids. Furthermore, mobilisation of N reserves, and particularly VSP hydrolysis, increases the flux of amino acids recycling within the plant, casting doubt on their role in inducing VSP accumulation.

The VSPs in poplar usually accumulate during autumn and winter, and removing the leaf buds (Coleman et al. 1993) can inhibit their hydrolysis in spring. The nearly complete defoliation frequently experienced by forage species induces mobilisation of N, and therefore hydrolysis of VSPs, in the storage tissues; the released N supporting synthesis of new leaves (Hendershot and Volenec 1993; Corre et al. 1996). Similar responses have been reported with other types of vegetation. For example, decapitation or defoliation of Euphorbia esula during autumn and winter reduced its capacity to accumulate a 26-kDa VSP (Cyr and Bewley 1990b).

Wounding and Methyl Jasmonate

Leaf wounding stimulates production of VSPs and associated mRNAs in soybean (Mason and Mullet 1990, Staswick 1990; 1994; Mason et al. 1992) and A. thaliana (Berger et al. 1995). Commonality between aspects of the wounding response and VSP synthesis in poplar has been inferred from sequence homology, and from the common induction by wounding and high N supply seen for VSPs and wound-induced proteins encoded by the Win multigene family (Coleman et al. 1992; Davis et al. 1993; Beardmore et al. 1996; Laurence et al. 1997). Several authors have suggested that synthesis of storage proteins is an adaptive response conferring tolerance to wounding and pathogen attack, by way of recycling nutrients from damaged areas (Staswick 1994; Creelman and Mullet 1997). Jasmonic acid and its derivative methyl esters, notably methyl jasmonate, are widely distributed compounds in the plant kingdom, and function both as promoters of senescence and regulators of plant growth (Creelman and Mullet 1997). They activate genes involved in pathogen and insect resistance and genes encoding VSPs in soybean, but repress genes encoding proteins involved in photosynthesis. Foliar spraying of methyl jasmonate on soybean (Anderson 1988; Mason and Mullet 1990; Staswick 1990) and application to the roots of *A. thaliana* (Berger et al. 1995), even at low concentrations, has been shown to increase the synthesis of VSP mRNA. The VSP accumulation is elicited in all cell types under these conditions, compared with the limited expression pattern observed in control plants. Treatment with methyl jasmonate also increased levels of VSPs in lucerne and oilseed rape (A. Ourry et al. unpubl.). Methyl jasmonate-induced expression of VSP genes in soybean is stimulated synergistically by sugars (Mason et al. 1992) and inhibited by high phosphate supply and auxin application (DeWald et al. 1994). In their review, Creelman and Mullet (1997) concluded that jasmonate and methyl jasmonate accumulate systematically in response to localized wounding, following the release of an induction signal from the wound site and its translocation through the phloem (Anderson 1985) to other plant tissues. There appears to be sufficient evidence to ascribe a regulatory role to methyl jasmonate in the expression of VSP genes following localised wounding. Whether this extends to the regulation of VSP synthesis under other conditions remains an open question.

Dependence on Photoperiod and Temperature

The seasonal fluctuations in VSP levels observed with many species, and particularly the tendency for their accumulation during autumn, prompt the question of whether low temperature and/or shortening daylength are prerequisites. In poplar (Langheinrich and Tischner 1991, Coleman et al. 1991; 1993; Van Cleve and Apel 1993), exposure to short days induced a substantial increase in the level of the 32-kDa VSP and the corresponding transcript, as shown by *in vitro* translations (Coleman et al. 1991). The same response was obtained following exposure to long days and low temperatures (Van Cleve and Apel 1993). Growth under short days also stimulates VSP accumulation in white clover (Bouchart et al. 1998) and lucerne (A. Ourry et al. unpubl.). Exposure to low root temperature had a synergistic effect with a short photoperiod in white clover but an antagonistic effect in lucerne. The involvement of phytochrome in the response to short days has been proposed (Langheinrich and Tischner 1991), as have a number of alternative explanations. For example, Coleman et al. (1993) argued that the effect is a direct consequence of modified source-sink relationships for N within poplar. These authors observed that 28 days under a short photoperiod, followed by exposure to long days, produced a pattern of VSP accumulation followed by mobilisation, but when exposure to short days was extended beyond 28 days, the buds remained dormant, and VSPs were not hydrolysed when a long photoperiod was introduced. Likewise, when buds were removed under long photoperiod conditions, VSP mobilisation was inhibited. These observations are consistent with the hypothesis that bud dormancy in woody and herbaceous perennials during late autumn and winter induces VSP accumulation due to a large reduction in the requirement for shoot N. The N absorbed from soil exceeds that required for growth and, as a consequence, excess

N is partitioned to root storage tissues (Bouchart et al. 1998). However, Cunningham and Volenec (1998) recently demonstrated that VSP accumulation in roots of non-dormant alfalfa cultivars was similar to that of dormant cultivars. VSP gene expression could therefore be affected by environmental factors (photoperiod, temperature, soil N availability), which modify the source-sink relationships within the plant, and hence alter the growth-related N demand of the aerial sink tissues, differently from N uptake or fixation.

Conclusions and Perspectives

Nitrogen storage is a widespread phenomenon in higher plants, occurring in different tissues and for a number of purposes. Mobilisation of N reserves provides N for tissue synthesis during the rapid growth of many woody and perennial herbaceous species in the early spring, after winter dormancy. Mobilisation of N can also compensate for low rates of N uptake from the environment resulting from agronomic practices (e.g. defoliation) or physiological changes in the plant (e.g. reduced N uptake during grain filling in cereals). These N reserves are usually in the form of free amino acids, readily available for mobilisation, and soluble proteins. The soluble protein fraction is the largest storage pool of N, but requires hydrolysis to amino-N before translocation to actively growing tissues.

Further molecular-level research directed at the genes encoding VSPs and enzymes of the jasmonate pathway, together with the relevant signal transduction systems will undoubtedly add to our understanding of mechanism controlling N cycling within plants. However, there are also a number of questions that require attention at the physiological and agronomic levels of analysis. Firstly, relatively little is known about the general regulation of N mobilisation and the specific regulation of VSP proteolysis. It is not clear how N and VSP mobilisation from different tissues is coordinated or exactly how changes in environmental conditions initiate mobilisation. This uncertainty extends to the dependence of mobilisation on ontogeny, and the causes of interspecific variation in this linkage. A second area of concern centres on the trophic and agronomic significance of N storage. This needs to be investigated more comprehensively and in a wider range of species, including plants inhabiting natural and agricultural ecosystems. Quantitative assessments of the impact on NUE are required, and in the case of crop species, of the potential for genetic improvement through selective breeding. Much of the work aimed at elucidating the molecular basis of the physiological functioning of VSPs has been conducted with the model species of soybean and poplar. However, the range of VSPs identified across the plant kingdom exhibits a high degree of polymorphism in terms of molecular characteristics and physiological functions. This may limit the relevance of studies with model species to other economically important plants, including forages.

Amongst the strategies proposed for increasing the efficiency of N uptake, retention and utilisation, overexpression of root N transporters (see Chap. 1.1) has been championed, especially in low N input conditions. However, benefits from this approach remain to be demonstrated, because N transporter activity is regulated in a number of ways, so that N acquisition matches N demand (see Chaps. 1.1 and 1.2). Manipulating N storage characteristics such as capacity, location and turnover rates holds promise for enhancing N acquisition, especially when there is little or no N demand associated with vegetative or reproductive growth. This approach could improve seasonal NUE when growth is restricted by the environment, and provide N reserves when conditions are suitable for growth return.

Much of the genetic improvement in crop productivity over the past 50 years has been conducted under high N input conditions. Selection in this environment has reduced the N storage capacity of grassland species. Thornton et al. (1994) found that species associated with environments in which levels of available N are generally low rely more on N storage than species usually found in N-rich environments. Likewise, shoot regrowth of *Lolium temulentum*, which is a ruderal species, appears to be limited by the availability of N reserves (Ourry et al. 1996). In contrast, *Lolium perenne* can compensate for low N reserve status by rapidly increasing root N uptake (Louahlia et al. 1999). It has also been argued from a wider ecological perspective (Chapin et al. 1990) that the most efficient mechanisms for N recycling and storage are likely to be found in species adapted to low N fertility environments. This has significant consequences for identifying genetic material for breeding programmes aimed specifically at enhancing NUE, and more widely at improving the sustainability of agriculture as a whole.

References

Aerts R (1994) The effect of nitrogen supply on the partitioning of biomass and nitrogen in plant species from heathlands and fens: alternatives in plant functioning and scientific approach. In: Roy J, Garnier E (eds) A whole plant perspective on carbon-nitrogen interactions. SPB Academic, The Hague, pp 247-265

Améziane R, Richard-Molard C, Deléens E, Morot-Gaudry JF, Limami AM (1997) Nitrate ($^{15}NO_3$) limitation affects nitrogen partitioning between metabolic and storage sinks and nitrogen reserve accumulation in chicory (*Cichorium intybus* L.). Planta 202: 303-312

Anderson JM (1985) Evidence for phloem transport of jasmonic acid. Plant Physiol 77: 75

Anderson JM (1988) Jasmonic acid-dependant increases in the level of specific polypeptides in soybean suspension cultures and seedlings. J Plant Growth Regul 7: 203-211

Arora R, Wisniewski ME, Scorza R (1992) Cold acclimation in genetically related (sibling) deciduous and evergreen peach [*Prunus persica* (L.) Batsch]. Plant Physiol 99: 1562-1568

Avice JC, Ourry A, Volenec JJ, Lemaire G, Boucaud J (1996) Defoliation-induced changes in abundance and immunolocalization of vegetative storage proteins in taproots of *Medicago sativa* L. Plant Physiol Biochem 34: 561-570

Avice JC, Ourry A, Lemaire G, Volenec JJ, Boucaud J (1997) Root protein and vegetative storage protein are key organic nutrients for alfalfa shoot regrowth. Crop Sci 37: 1187-1193

Beardmore T, Wetzel S, Burgess D, Cherest PJ (1996) Characterization of seed storage proteins in populus and their homology with populus vegetative storage proteins. Tree 16: 833-840

Berendse F, Aerts R (1987) Nitrogen-use efficiency: a biologically meaningful definition? Funct Ecol 1: 293-296

Berger S, Bell E, Sadka A, Mullet JE (1995) *Arabidopsis thaliana atvsp* is homologous to soybean *VspA* and *VspB*, genes encoding vegetative storage protein acid phosphatases, and is regulated similarly by methyl jasmonate, wounding, sugars, light and phosphate. Plant Mol Biol 27: 933-942

Bigot J, Boucaud J (1992) Activities of enzymes involved in amino acids metabolism in perennial ryegrass ; changes with regrowth. Phytochem istry 31: 4071-4074

Bouchart V, Macduff JH, Ourry A, Svenning MM, Gay AP, Simon JC, Boucaud J (1998) Seasonal pattern of accumulation and effects of low temperatures on storage compounds in *Trifolium repens* L. Physiol Plant 104: 65-74

Boyce PJ, Volenec JJ (1992) ß-Amylase from taproots of alfalfa. Phytochem istry 31: 427-431

Boyce PJ, Penaloza E, Volenec JJ (1992) Amylase activity in taproots of *Medicago sativa* L. and *Lotus corniculata* L. following defoliation. J Exp Bot 43: 1053-1059.

Bunker TW, Koetje DS, Stephenson LC, Creelman RA, Mullet JE, Grimes HD (1995) Sink limitation induces the expression of multiple soybean vegetative lipoxygenase mRNAs while the endogenous jasmonic acid level remains low. Plant Cell 7: 1319-1331

Chapin FS III, Schulez ED, Mooney HA (1990) The ecology and economics of storage in plants. Annu Rev Ecol Syst 21: 423-447

Cheeseman JM, Barreiro R, Lexa M (1996) Plant growth modelling and the integration of shoot and root activities without communicating messengers: opinion. Plant Soil 185: 51-64

Clausen S, Appel K (1991) Seasonal changes in the concentration of the major storage protein and its mRNA in xylem ray cells of poplar trees. Plant Mol Biol 2: 365

Cliquet JB, Déléens E, Mariotti A (1990) C and N mobilization from stalk and leaves during kernel filling by ^{13}C and ^{15}N tracing in *Zea mays* L. Plant Physiol 94: 1547-1553

Coleman GD, Chen THH, Ernst SG, Fichigami L (1991) Photoperiod control of poplar bark storage protein accumulation. Plant Physiol 96: 686-692

Coleman GD, Chen THH, Ernst SG, Fichigami L (1992) Complementary DNA cloning of poplar bark storage protein and control of its expression by photoperiod. Plant Physiol 98: 687-693

Coleman GD, Englert JM, Chen THH, Fuchigami LH (1993) Physiological and environmental requirements for poplar (*Populus deltoides*) bark storage protein degradation. Plant Physiol 102: 53-59

Conradie WJ (1980) Seasonal uptake of nutrients by chenin blanc in sand culture: I. Nitrogen. S Afr J Enol Vitic 1: 59-65

Corre N, Bouchart V, Ourry A, Boucaud J (1996) Mobilization of nitrogen reserves during regrowth of defoliated *Trifolium repens* L. and identification of vegetative storage proteins. J Exp Bot 301: 1111-1118

Cunningham SM, Volenec JJ (1996) Purification and characterization of vegetative storage proteins from alfalfa (*Medicago sativa* L.) taproots. J Plant Physiol 147: 625-632

Cunningham SM, Volenec JJ (1998) Seasonal carbohydrate and nitrogen metabolism in roots of contrasting alfalfa (*Medicago sativa* L.) cultivars. J Plant Physiol 153: 220-225

Creelman RA, Mullet JE (1997) Biosynthesis and action of jasmonates in plants. Annu Rev Plant Physiol Plant Mol Biol 48: 355-381

Cyr DR, Bewley JD (1990a) Proteins in the roots of the perennial weeds chicory (*Cichorium intybus* L.) and dandelion (*Taraxacum officinale* Weber) are associated with overwintering. Planta 182: 370-374

Cyr DR, Bewley JD (1990b) Seasonal variation in nitrogen storage reserves in the roots of leafy spurge (*Euphorbia escula*) and responses to decapitation and defoliation. Physiol Plant 78: 361-366

Dalling MJ (1985) Proteolytic enzymes and leaf senescence. In: Thompson WW, Nothnagel EA, Huffaker RC (eds) Plant senescence: iIts biochemistry and physiology. American Society of Plant Physiologists, Rockville, Maryland, pp 54-60

Davis JM, Coleman GD, Chen TH Gordon MP (1993) *Win4* a family of wound-induced genes in poplar. Plant Physiol 102S: 25

DeWald DB, Mason HS, Mullet JE (1992) The soybean vegetative storage proteins *VSPa* and *VSPb* are acid phosphatases active on polyphosphates. J Biol Chem 267: 15958-15964

DeWald DB, Sadka A, Mullet JE (1994) Sucrose modulation of soybean *Vsp* gene expression is inhibited by auxin. Plant Physiol 104: 439-444

Farrar J, Gunn S (1998) Allocation: allometry, acclimation – and alchemy? In: Lambers H, Poorter H, Van Vuuren MMI (eds) Inherent variation in plant growth. Physiological mechanisms and ecological consequences. Backhuys, Leiden, pp 183-198

Feller U, Keist M (1986) Senescence and nitrogen metabolism in annual plants. In: Lambers H, Neetson JJ, Stulen I (eds, Fundamental, ecological and agricultural aspects of nitrogen metabolism. Nijhoff, Dordrecht, pp 219-234

Gana JA, Kalengamaliro NE, Cunningham SM, Volenec JJ (1998) Expression of ß-amylase from alfalfa taproots. Plant Physiol 118: 1495-1506

Garnier E, Aronson J (1998) Nitrogen-use efficiency from leaf to stand level: clarifying the concept. In: Lambers H., Poorter H., Van Vuuren MMI (eds) Inherent variation in plant growth. Physiological mechanisms and ecological consequences. Backhuys, Leiden, pp 515-538

Gaudillère JP (1997) Gestion de l'azote chez les espèces ligneuses. In: Morot-Gaudry JF (ed) Assimilation de l'azote chez les plantes: aspects physiologique, biochimique, moléculaire. INRA, Paris, pp 295-305

Habib R (1984) La formation des réserves azotées chez les arbres fruitiers. Fruits 39: 623-635

Harms U, Sauter JJ (1991) Storage proteins in the wood of Taxodiaceae and of *Taxus*. J Plant Physiol 138: 497-499

Harvey PJ, Boulter D (1983) Isolation and characterization of the storage protein of yam tubers (*Dioscorea rotundata*). Phytochemistry 22: 1687-1693

Heilmeyer H, Monson RK (1994) Carbon and nitrogen storage in herbaceous plants. In: Roy J, Garnier E (eds) A whole-plant perspective on carbon-nitrogen interactions. SPB Academic, The Hague, pp 149-171

Hendershot KL, Volenec JJ (1993) Nitrogen pools in taproots of *Medicago sativa* L. after defoliation. J Plant Physiol 141: 129-135

Ingestad T (1979) Nitrogen stress in birch seedlings II. N, K, P, Ca and Mg nutrition. Physiol Plant 45: 149-157

Kalengamaliro NE, Volenec JJ, Cunningham SM, Joern BC (1997) Seedling development and deposition of starch and storage proteins in alfalfa roots. Crop Sci 37: 1194-1200

Kang S, Titus JS (1980a) Activity profiles of enzymes involved in glutamine and glutamate metabolism in the apple during autumnal senescence. Physiol Plant 50: 291-297

Kang S, Titus JS (1980b) Qualitative and quantitative changes in nitrogenous compounds in senescing leaf and bark tissues of apple. Physiol Plant 50: 285-290

Langheinrich U, Tischner R (1991) Vegetative storage protein in poplar: induction and characterization of a 32- and a 36-kD polypeptide. Plant Physiol 97: 1017-1025

Lawrence SD, Greenwood JS, Korhnak TE, Davis JM (1997) A vegetative storage protein homolog is expressed in the growing shoot apex of hybrid poplar. Planta 203: 237-244

Li H, Oba K (1985) Major soluble proteins of sweet potato roots and changes in proteins after cutting, infection, or storage. Agric Biol Chem 49: 737-744

Li R, Volenec JJ, Joern BC, Cunningham SM (1996) Seasonal changes in nonstructural carbohydrates, protein, and macronutrients in taproots of alfalfa, red clover, sweetclover, and birdsfoot trefoil. Crop Sci 36: 617-623

Loualhia S, Macduff JH, Ourry A, Humphreys M, Boucaud J (1999) N reserve status affects the dynamics of nitrogen remobilization and mineral nitrogen uptake during recovery from defoliation by contrasting cultivars of *Lolium perenne*. from defolaition. New Phytol 142, 451-462

MacKown CT, Van Sanford DA, Zhang N (1992) Wheat vegetative nitrogen compositional changes in response to reduced reproductive sink strength. Plant Physiol 99: 1469-1474

Mae T, Makino A, Ohira K (1983) Changes in the amounts of ribulose bisphosphate carboxylase synthesized and degraded during the life span of rice leaf (*Oryza sativa* L.). Plant Cell 24: 1079-1086

Mason HS, Mullet JE (1990) Expression of two vegetative storage protein genes during development and in response to water deficit, wounding and jasmonic acid. Plant Cell 2: 569-579

Mason HS, Dewald DB, Creelman RA, Mullet JE (1992) Coregulation of soybean vegetative storage protein gene expression by methyljasmonate and soluble sugars. Plant Physiol 98: 859-867

Millard P (1988) The accumulation and storage of nitrogen by herbaceous plants. Plant Cell Environ 11: 1-8

Millard P (1993) A review of internal cycling of nitrogen within trees in relation to soil fertility. In: Fragoso MAC, van Beusichen ML (eds) Optimization of plant nutrition. Kluwer Dordrecht, pp 623-628

Muntz K (1998) Deposition of storage proteins. Plant Mol Biol 38: 77-99

Mussigmann C, Ledoigt G (1989) Major storage proteins in Jerusalem artichoke tubers. Plant Physiol Biochem 27: 81-86

O'Kennedy BT, Titus JS (1979) Isolation and mobilization of storage proteins from apple shoot bark. Physiol Plant 45: 419-424

Ourry A, Bigot J, Boucaud J (1989) Protein mobilization from stubble and roots and proteolytic activities during post-clipping re-growth of perennial ryegrass. J Plant Physiol 134 : 298-303

Ourry A, Kim TH, Boucaud J (1994) Nitrogen reserve mobilization during regrowth of Medicago sativa L.: relationships between their availability and regrowth yield. Plant Physiol 105: 831-837

Ourry A, Macduff JH, Ougham HJ (1996) The relationship between mobilization of N reserves and changes in translatable messages following defoliation in Lolium temulentum L. and Lolium perenne L. J Exp Bot 47: 739-747

Paiva E, Lister RM, Park WD (1983) Induction and accumulation of major tuber proteins of potato in stem and petioles. Plant Physiol 71: 161-168

Penheiter AR, Duff SM, Sarath G (1997) Soybean root nodule acid phosphatase. Plant Physiol 114: 597-604

Peoples MB, Beilharz VC, Waters SP, Simpson RJ, Dalling MJ (1980) Nitrogen redistribution during grain growth in Wheat Triticum aestivum L. Planta 149: 241-251

Reilly ML (1990) Nitrate assimilation and grain yield. In: Abrol YP (ed), Nitrogen in higher plants. Research Studies, Milton Keynes, pp 335-367

Roberts LS, Toivonen P, McInnis SM (1991) Discrete proteins associated with overwintering in interior spruce and douglas-fir seedlings. Can J Bot 69, 437-441

Shim KK, Titus JS (1985) Accumulation and mobilization of storage proteins in gingko shoot bark. J Kor Soc Hortic Sci 26: 350-360

Simpson RJ, Dalling MJ (1981) Nitrogen redistribution during grain growth in wheat (Triticum aestivum L.) Planta 151: 447-456

Staswick PE (1989) Developmental regulation and the influence of plant sinks on vegetative storage protein gene expression in soybean leaves. Plant Physiol 89: 309-315

Staswick PE (1990) Novel regulation of vegetative storage protein genes. Plant Cell 2: 1-6

Staswick PE (1992) Jasmonate, genes, and fragrant signals. Plant Physiol 99: 804-807

Staswick PE (1994) Storage proteins of vegetative plant tissues. Annu Rev Plant Physiol Plant Mol Biol 45: 303-322

Staswick PE (1997) The occurrence and gene expression of vegetative storage proteins and a rubisco complex protein in several perennial soybean species. J Exp Bot 48: 2031-2036

Staswick PE, Huang J, Rhee Y (1991) Nitrogen and methyl jasmonate induction of soybean vegetative storage protein genes. Plant Physiol 96: 130-136

Staswick PE, Papa C, Huang JF, Rhee Y (1994) Purification of the major soybean leaf acid phosphatase that is increased by seed-pod removal. Plant Physiol 104: 49-57

Stephenson LC, Bunker TW, Wesley ED, Grimes HD (1998) Specific soybean lipoxygenases localize to discrete subcellular compartments and their mRNAs are differentially regulated by source-sink status. Plant Physiol 116: 923-933

Stépien V, Martin F (1992) Purification, characterization and localization of the bark storage proteins of poplar. Plant Physiol Biochem 30: 399-407

Stépien V, Sauter JJ, Martin F (1992) Structural and immunological homologies between storage proteins in the wood and the bark of poplar. J Plant Physiol 140: 247-250

Stépien V, Sauter JJ, Martin F (1994) Vegetative storage proteins in woody plants. Plant Physiol Biochem 32: 185-192

Thornton B, Millard P, Duff EI (1994) Effects of nitrogen supply on the source of nitrogen used for regrowth of laminae after defoliation of four grass species. New Phytol 128: 615-620

Tranbarger TJ, Franceschi VR, Hildebrand DF, Grimes HD (1991) The soybean 94-kilodalton vegetative storage protein is a lipoxygenase that is localized in paraveinal mesophyll cell vacuoles. Plant Cell 3: 973-987

Van Cleve B, Apel K (1993) Induction by nitrogen and low temperature of storage protein synthesis in poplar trees exposed to long days. Planta 189: 157-160

Van Cleve B, Clausen S, Sauter JJ (1988) Immunochemical localization of a storage protein in poplar wood. J Plant Physiol 133: 3146-3153

Volenec JJ, Ourry A, Joern BC (1996) A role for nitrogen reserves in forage regrowth and stress tolerance. Physiol Plant 97: 185-193

Waters SP, Peoples MB, Simpson RJ, Dalling MJ (1980) Nitrogen redistribution during grain growth in wheat Triticum aestivum L. Planta 148: 422-428

Wetzel S, Demmers C, Greenwood JS (1989) Seasonally fluctuating bark proteins are a potential form of nitrogen storage in three temperate hardwoods. Planta 178: 275-281

Wilson JB (1988) A review of evidence on the control of shoot:root ratio, in relation to models. Ann Bot 61: 433-449

Wittenbach VA (1982) Effect of pod removal on leaf senescence in soybeans. Plant Physiol 70: 1544-48

Wittenbach VA (1983) Purification and characterization of a soybean leaf storage glycoprotein. Plant Physiol 73: 125-129

Protein Hydrolysis and Nitrogen Remobilisation in Plant Life and Senescence

Renaud Brouquisse[1], Céline Masclaux[2], Urs Feller[3] and Philippe Raymond[1]

Proteolysis in cell life

In plant cells, as in all other cells, proteins are submitted to permanent turnover, and the intracellular content of a given protein depends on its rate of both synthesis and degradation. The life time of most proteins is shorter than that of the cell. Thus, in young leaves of *Lemna minor*, the average half-life of protein was estimated to be 7 days, and it was shorter under stress conditions (Davies 1982). Such observations mean that nitrogen and amino acid fluxes are both cylic and permanent. Although protein turnover may appear wasteful, in terms of energy, numerous studies have shown that proteolysis provides multiple functions in cell physiology, and is an essential regulatory mechanism of cell metabolism and development.

Proteolysis plays an essential role in the processing of proteins, such as transit peptide hydrolysis or zymogen maturation, and in the regulation of metabolic pathways via the degradation of enzymes or regulatory proteins (Vierstra 1996). For instance, in eukaryotic cells, the selective degradation of positive or negative regulators is essential in the progression through the cell cycle (Hershko and Ciechanover 1998). Similarly, throughout plant life, cell growth or differentiation, plastid differentiation, or peroxisome to glyoxisome conversion are some of the many events which involve considerable modification of metabolism and require the intervention of specific proteases. Environmental or nutritional stresses cause protein denaturation and degradation (Davies 1982), and acclimation mechanisms often involve the induction of specific proteolytic processes (Vierstra 1996). In such cases, proteolysis plays a housekeeping role by removing abnormal, inactivated or denatured

1. Unité de Physiologie et de Biotechnologie végétales, Centre INRA de Bordeaux Aquitaine, BP 81, 33883 Villenave-d'Ornon Cedex, France. *E-mail:* brouquis@bordeaux.inra.fr et raymond@bordeaux.inra.fr
2. Unité de Nutrition Azotée des Plantes, INRA-Versailles, route de Saint-Cyr, 78026 Versailles, France. *E-mail:* masclaux@versailles.inra.fr
3. University of Department of Plant Biology, Altenbergrain 21, Bern, CH-3013, Switzerland. *E-mail:* urs.feller@pfp.unibe.ch

proteins, the accumulation of which could be toxic for the cell. During germination, the degradation of seed storage proteins supplies amino acids for protein synthesis in growing organs (Mitsuhashi and Oaks 1996; Schmid et al. 1998). Seed storage proteins themselves are synthesised with amino acids derived from the degradation of proteins in senescing leaves, stems or tubers (Staswick 1994). Leaf senescence is associated with increased proteolysis (Peoples and Dalling 1988; Smart 1994; Feller and Fischer 1994). In such cases, proteolysis plays an essential role in the redistribution of organic nitrogen within the whole plant.

The mechanisms and the regulation of cellular proteolysis are still poorly known. This is due to both the very large number of potential substrates, since each protein is likely to be hydrolysed during the cell life, and to the large number of different proteases together with the complexity of the proteolytic mechanisms: for intance, more than 100 different proteins are involved in the ubiquitin-dependent proteolysis alone. Furthermore, the fact that short- and long-life proteins coexist in the same compartment suggests that the proteolytic machinery is able to discriminate between substrates. However, whereas most of the purified proteases hydrolyse peptide bonds at particular sites, they are rarely specific for one given substrate protein. Thus, the occurrence of different proteolytic systems, and their respective compartmentation, are important elements in the global regulation of cell metabolism.

Within the scope of this book, the question of the role of proteolysis in relation to nitrogen metabolism and mobilisation is raised. At the level of the cell, what are the different proteolytic systems involved? How are they regulated? How are the proteins selected for degradation? When, how, and in which tissues are proteolytic processes important during plant development? In the present chapter, we first describe general features of proteases and proteolytic systems in plant cells. We then focus on the role of proteolysis during plant development, through three particular situations, namely germination, response to environmental and nutritional stresses, and senescence.

Protease classes

Proteases, also known as peptidases or peptide hydrolases, may be classified according to three major criteria: (1) the nature of the hydrolysed site of the substrate protein, (2) the chemical nature of the catalytic site of the protease, and (3) the homologies of peptidic sequences between proteases. Such criteria have allowed the definition of protease families, subfamilies and clans, whose characteristics have been detailed in reference books such as Barrett et al. (1998). The main features commonly used to distinguish the different types of proteases are summarised below.

Proteases may be divided into two subclasses comprising the exopeptidases and the endopeptidases. Exopeptidases hydrolyse peptide bonds near the end of polypeptidic chains. Those acting at the N-terminal extremity release a single amino acid, a dipeptide or a tripeptide (they are respectively named aminopeptidase, dipeptidylpeptidase and tripeptidylpeptidase) and those acting at the C-terminal extremity release a single amino acid or a dipeptide (respectively, carboxypeptidase and peptidyldipeptidase). Other exopeptidases are specific for dipeptides (dipeptidase), tripeptides (tripeptidase) or release terminal residues that are substituted, cyclised or linked by isopeptidic bonds (ω-peptidase). Endopeptidases act inside peptidic chains and the presence of free α-amino or α-carboxyl groups generally inhibits their activity. The specificity of the exo- as well as endopeptidases depends on the nature of the amino acids near the hydrolysis site. For instance, chymotrypsin- and trypsin-type proteases cleave a peptide bond at the carboxyl side of aromatic or basic amino acids, respectively.

Proteases may also be characterised according to the nature of their catalytic site, usually defined by their sensitivity to more or less specific inhibitors. Five protease types have been defined on this basis. Serine-type proteases have a serine residue in their active site (most serine proteases possess a catalytic triad, involving Ser, His and Asp residues in the catalytic process). The inhibitors most commonly used to characterise serine proteases are 3,4-dichloroisocoumarin (3,4-DCI), diisopropyl fluorophosphate (DFP) and phenylmethylsulfonyl fluoride (PMSF). Cysteine-type proteases have a cysteine residue in their active centre, and they are inhibited by the epoxide inhibitor E-64, iodoacetamide, iodoacetate or N-ethylmaleimide. Some peptide aldehydes of bacterial origin, such as leupeptin or chymostatin are useful serine and cysteine protease inhibitors. The aspartic-type peptidases (all of which are endopeptidases) possess two aspartic acid residues in their active site. The best identified inhibitor for aspartic proteases is pepstatin. The metallo-proteases use a metal ion (commonly zinc) in the catalytic mechanism. They may be inhibited by metal chelators, the most useful of them being 1,10 phenanthroline. The threonine-type proteases possess a threonine residue in their active site. The proteasome and its relatives are the only known threonine type peptidases (they possess a catalytic tetrad: Thr, Glu/Asp, Lys and the N-terminal amine), but there are homologues of trypsin in which the catalytic Ser is replaced by Thr. Proteasome is inhibited by peptides like lactacystin, and different leupeptin analogues like MG132 or MG262. Finally, there are a number of proteases for which the catalytic type remains to be established. Those for which the amino acid sequences are known can be grouped into families, which have been given names that begin with the letter U to signify peptidase unclassified with regard to catalytic mechanism.

Protein remobilisation and proteolytic enzymes

Proteases have been found in almost all cell compartments (Fig. 1).

Vacuolar Proteolysis

Vacuoles are the lytic compartment of plant cells, and contain a number of low-specificity hydrolases such as endo- and carboxypeptidases, RNAase, DNAase, acid phosphatase and glycosidases. It seems that these enzymes may function in differentiating cells through a process of autophagy where portions of the cytoplasm are transported into the vacuole (Matile 1997). Three main vacuole types have been recognised in plant cells: protein storage vacuoles (PSV) that contain seed-type storage proteins, delta vacuoles that contain vegetative storage proteins and pigments, and lytic vacuoles. In many cases, two separate types, and in some cases three types coexist in the same cell (Paris et al. 1996), and possess different proteolytic activities (Fig. 1). Thus, some observations indicate that in barley aleurone, as in pea and barley roots, PSV contain a phytepsin-type aspartic protease, whereas vacuoles of another type contain the cysteine protease aleurain (Okita and Rogers 1996).

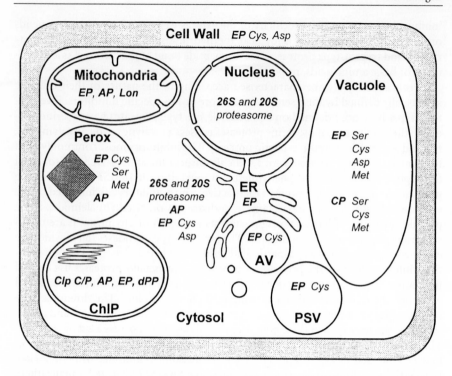

Fig. 1 : Protease distribution in plant cells. AV Autophagic vacuole; ChlP chloroplast; ER endoplasmic reticulum; Perox peroxisome; PSV protein storage vacuole. Protease classes are indicated as follows: *AP* aminopeptidase; *CP* carboxypeptidase; *EP* endopeptidase; *dPP*
dipeptidylpeptidase. *Ser, Cys, Met* and *Asp* refer respectively to serine, cysteine, metallo and
aspartic protease types. The mitochondrial protease *Lon* (not referred to in the text) is a homologue of the Lon, or La protease present in bacteria, yeast and mammal cells. It catalyses the
ATP-dependent degradation of proteins in the mitochondrial matrix (Barakat et al. 1998).

Most of the protease-related cDNAs that have been isolated from senescing
plant organs encode cysteine proteases. Although none of their encoded products
has yet been localised, biochemical data strongly suggest that the main proteolytic
events occurring during senescence take place in vacuoles, probably lytic vacuoles.
Indeed, most of the cysteine proteases have an acidic pH optimum *in vitro*, suggesting that they are localised in the vacuole *in vivo* (Callis 1995). Endopeptidases with
a pH optimum close to 5 and carboxypeptidases have also been localised in the vacuole (Peoples and Dalling 1988). The predominant endopeptidase in senescent
leaves of *Oryza sativa* and *Lolium temulentum* (Morris et al. 1996) was maximally
active at pH 5 at the beginning of the degradative process and at pH 8 during the late
stage of senescence. In *L. temulentum*, it was shown to cross-react with an antibody
raised against the non-glycosylated cysteine-endopeptidase papain (Morris et al.
1996). The appearance of new proteolytic activities with a higher pH optimum at the
late phase of senescence suggests the involvement of either extravacuolar endopeptidases or endopeptidases in late-differentiated neutral vacuoles, in the degradative
process of the cell. Vacuolar endopeptidases are separated by several membranes

(tonoplast, chloroplast envelope) from Rubisco, the predominant leaf protein located in the chloroplast stroma. Changes in compartmentation (e.g. transfer of polypeptides accross membranes, membrane rupture or inclusion of plastids into vacuoles) are therefore a prerequisite for the degradation of Rubisco by vacuolar enzymes. It is still open as to what extent vacuolar proteases contribute to the degradation of chloroplast proteins during senescence.

In carbon-starved plant cells, the remobilisation of proteins is associated with cellular autophagy, a process whereby parts of the cytosol and organelles are included inside double-membrane-bound vesicles that are expelled into the vacuole, where their lipid and protein constituents are hydrolysed (Aubert et al. 1996). The major endopeptidase present in carbon-starved root tips of maize (James et al. 1996) is not a cysteine but a serine endopeptidase, located in the vacuole. The low specificity of this enzyme makes it an adequate protease for the hydrolysis of the proteins that would be trapped by an autophagic process.

Cytosolic and Nuclear Proteolysis by the Ubiquitin-Proteasome System

Ubiquitin is a small (8-kDa) protein, and the 26S proteasome is a large 1500-kDa proteolytic complex, that are present in the cytosol and nucleus of eukaryotic cells, including plants (Fu et al. 1999, and references therein). Proteins to be degraded in the ubiquitin-proteasome pathway are targeted for degradation by conjugation with chains of multiple ubiquitins that serve as recognition signals for the 26S proteasome. The selective degradation of ubiquitinated proteins by the proteasome system implies the presence in target proteins of signals for ubiquitin ligation (Hershko and Ciechanover 1998). Among degradation signals in primary protein structure the best characterised is the N-end rule system, in which the ubiquitination and degradation of a protein is determined by the nature of its terminal amino acid residue. In mitotic cyclins and other cell-cycle regulators, the signal is constituted by a 9-amino acid sequence motif (named destruction box) usually located 40-50 amino acid residues from the N terminus of the protein. Alternatively, many proteins containing PEST elements (regions enriched in Pro, Glu, Ser and Thr residues) are targeted for ubiquitination by phosphorylation. This is the case for certain transcriptional regulators, β-catenin and G-type cyclins. However, phosphorylation of some other proteins (c-Mos, c-Fos and c-Jun protooncogenes) prevents their ubiquitination and degradation. In other eukaryotes, the proteasome has also been implicated in the degradation of abnormal proteins.

The importance of this system in the regulation of plant development is illustrated by its role in the control of cell division and in the response to auxin and light. The degradation of mitotic cyclins, which determines the exit from mitosis, has recently been demonstrated to depend on the ubiquitin-proteasome system (Genschik et al. 1998), by showing that chaemeric proteins bearing the cyclin destruction box fluctuated in a cell-cycle specific manner, but remained stable in the presence of a specific proteasome inhibitor. Concerning the active form of phytochrome A, PFR, further evidence for its degradation by the ubiquitin-proteasome pathway has been obtained by showing that modified PFR proteins that were not ubiquitinated were also not degraded (Clough et al. 1999). Genetic analyses of A. thaliana showed that the action of auxin requires functional ubiquitinating systems, probably involved in the degradation of protein repressors of the auxin responses (del Pozo and Estelle 1999)

The ubiquitin-proteasome system is also modulated in response to development and stress. The expression of particular polyubiquitin genes is increased in response to stress and during senescence, and the expression of an E2-type ubiquitin carrier encoding gene (*UBC4*) was shown to be induced in senescing leaves of both *A.thaliana* and *Nicotiana sylvestris* (Genschik et al. 1994) and in spinach cotyledons. The expression of genes encoding the 20S catalytic core and regulatory complex of the cytosolic proteasome was increased during senescence (Ito et al. 1997). All these data suggest that the cytosolic ubiquitin-proteasome pathway contributes to protein remobilisation in response to stress, and possibly, senescence.

The ubiquitin-proteasome pathway is also present in the nucleus. A cDNA (*NtPSA1*) that encodes a putative nuclear-located α-type subunit of the 26S proteasome was characterised recently (Bahrami and Gray 1999). The level of the related mRNA declined during senescence of both leaves and flowers of *A. thaliana*. This suggested that the role of proteasomes in plant nuclei is different from that in the cytosol, and may be to regulate developmental events in developing and proliferating tissues rather than to degrade proteins in senescing ones.

Peroxisomal Proteolysis

Peroxisomes, or microbodies, are invoved in a number of catabolic processes, including the β-oxidation of fatty acids, and more specialised functions like the glyoxylic acid cycle, for gluconeogenesis in glyoxisomes of growing seedlings, the oxidation of glycolate, for photorespiration in leaf peroxisomes, etc. All peroxisome proteins are encoded in the nucleus, and peroxisome proteases are involved in processing of imported proteins (Schmid et al. 1998).

During senescence, peroxisomes keep their integrity longer than chloroplasts (Aubert et al. 1996), but senescence-induced proteolysis in peroxisomes was recently reported (Distefano et al. 1997). Peroxisomes purified from young pea leaves were found to contain three endopeptidases, and four additional endopeptidases were present in senescing leaves, all of which are neutral proteases. By using inhibitors of different protease classes, three serine proteases, two cysteine proteases, and a metallo-protease were identified. The serine proteases represent approximatively 70% of the total endoproteolytic activity. In senescing leaves, an overall decrease in the number of polypeptides of peroxisomal matrices was observed, and this self-proteolytic effect was abolished by PMSF, a serine proteases inhibitor, thus demonstrating the role of serine endoproteases in peroxisomal protein turnover in senescencing tissue.

Chloroplastic Proteolysis

Several ATP-dependent and ATP-independent peptide hydrolases have been identified in plastids (Shanklin et al. 1995, Desimone et al. 1998). Among these, the ATP-dependent Clp protease (Shanklin et al. 1995) represents an interesting enzyme containing a ClpP subunit with proteolytic activity, encoded in the plastome, and a ClpC subunit with ATPase activity, encoded in the nucleus. ClpP has a molecular mass of about 23 kDa and ClpC of about 100 kDa. Similar to other self-compartmentalised proteases like the proteasome, subunits of ClpP can form a barrel-shaped structure with a central pore (Lupas et al. 1997). This architecture may separate the proteolytic domains (most likely oriented toward the central pore) from native substrate proteins in the stroma. The regulatory subunits (ClpC) may force unfolded proteins into the central pore. The conformation of the substrate protein and of the ClpP subunits might be equally important for the degradation rate (accessibility of susceptible peptide bonds in the substrate for the proteolytic domain of the protease). Both subunits of Clp have been detected on immunoblots of expanding (highest levels of Clp), mature and senescing (lowest levels of Clp) *A. thaliana* leaves. These results indicate that Clp is expressed in all stages of leaf development and is not a senescence-specific protease. The nuclear *clpC* and the plastidial *clpP* genes are expressed in expanding, mature and senescing bean leaves. However, a nuclear gene encoding an analogue of the regulatory subunit of Clp has been reported to be upregulated in *A. thaliana* during natural senescence, dark-induced senescence or dehydration (Nakashima et al. 1997).

It appears likely that metallo-endopeptidase activities are important for proteolysis in chloroplasts (Roulin and Feller 1997). A developmental-stage-dependent protease in the thylakoids of bean plastids has been reported very recently (Tziveleka and Argyroudi-Akoyunoglou 1998). The authors suggested that the stability of membrane proteins in the presence of this protease may depend on interactions of the apoprotein with other chloroplast constituents (e.g. LHCII with chlorophyll). Final conclusions concerning the functions of the various peptide hydrolases in the plastids are not yet possible. It must be borne in mind that the cleavage of peptide bonds is not restricted to net protein remobilisation during senescence, since protein turnover, the removal of the transit peptide from nuclear encoded stroma proteins or of the lumen targeting signal, and the degradation of damaged proteins depend on peptide hydrolases throughout the life cycle of a leaf (Vierstra 1996).

Proteolysis in plastids, as in other cell compartments, may be regulated by the type and activity of the peptide hydrolases present in the same compartment, by the susceptibility of the substrate proteins or by the actual solute composition (Feller and Fischer 1994). Therefore an increase in the total, or in a particular proteolytic activity is only one possible mechanism to accelerate protein degradation. Modifications in the substrate proteins may affect their susceptibility to proteolytic attack. Interactions with solutes (e.g. substrates) or covalent modifications can alter the conformation of a native protein and, as a consequence, its susceptibility to proteolysis. Active oxygen species and their effect on the catabolism of chloroplast proteins have been a major issue during the past few years (Stieger and Feller 1997; Desimone et al. 1998). In this context it must be considered that dithiothreitol (DTT), which is a widely used compound in chloroplast isolation and incubation media, inhibits at least two enzymes involved in the detoxification of active oxygen species in chloroplasts (Neubauer 1993). Metal-catalysed reactions may cause a site-specific damage in a protein by active oxygen species and make these proteins susceptible for proteolysis (Stadtman 1992). In intact pea chloroplasts, high light or the addition of methylviologen at low light, accelerated the degradation of glutamine synthetase, phosphoglycolate phosphatase and the large subunit of Rubisco (Stieger and Feller 1997). These enzymes possess binding sites for divalent cations and may therefore be more easily damaged than other enzymes by radicals produced under these incubation conditions. A radical-dependent tagging mechanism was considered for the regulation of proteolysis in chloroplasts, since under certain conditions an oxidative modification may precede protein degradation.

The remobilisation of nitrogen from senescing leaves contributes to an efficient use of nitrogen by annual crop plants. Reduced nitrogen in mesophyll cells of C_3 plants is present mainly (up to 75% of the total nitrogen in these cells) in the chloroplasts (Peoples and Dalling 1988). Rubisco is the most abundant stromal protein in these plants, while the light-harvesting chlorophyll a/b-binding polypeptide (LHCII) is predominant in the thylakoid membranes. Chlorophyll contributes only a minor percentage to the total leaf nitrogen. It is widely accepted that the first steps in the degradation of chlorophyll and of chloroplast proteins occur in the intact organelles (Desimone et al. 1998). Chlorophyll degradation products can be transported from the chloroplast to the vacuole and accumulate there (Tommasini et al. 1998). Therefore chlorophyll degradation products cannot be considered as nitrogen sources for growing plant parts. As compared to chlorophyll catabolism, little is known about the degradation of chloroplast proteins and the compartmentation of this important process. In intact chloroplasts isolated from pea leaves, nuclear-encoded stromal proteins are degraded to fragments no longer detected on immunoblots, indicating that the proteolytic enzymes required, are present inside these organelles (Roulin and Feller 1997). Rubisco, glutamine synthetase and several other stromal enzymes are degraded in intact pea chloroplasts, while these catabolic steps were no longer observed after lysis of the isolated chloroplasts. It is still open to debate whether the dilution of soluble constituents by chloroplast rupture, an inadequate composition of the medium or the loss of structural prerequisites are responsible for this effect. The incubation conditions for intact chloroplasts (e.g. illumination, temperature, solutes in the medium) influence the velocity of degradation as well as the pattern of accumulating fragments from Rubisco (Feller and Fischer 1994; Roulin and Feller 1997; Stieger and Feller 1997). The occurrence of the first proteolytic cleavages in the intact chloroplasts does not necessarily mean that the polypeptides are completely degraded to free amino acids inside these organelles. Oligopeptide translocators have been identified in the plasmalemma of barley scutella and of bean leaf cells (Jamai et al. 1994). It appears possible that oligopeptide translocators are also present in other membranes (e.g. chloroplast envelope), allowing the exchange of peptides between different subcellular compartments, i.e. the cytosol or vacuole, where they can be degraded to amino acids, which can be loaded into the phloem and translocated to sinks. The involvement of extraplastidial peptide hydrolases in the degradation of Rubisco and other chloroplast proteins could, under certain conditions, also be achieved by an altered subcellular distribution caused by membrane rupture or membrane fusion. The compartmentation of chloroplast protein degradation and the enzymes involved in this process may depend on the actual physiological status (e.g. cause of senescence, energy supply).

Proteolysis in Germination and Seedling Growth

During germination, the storage proteins synthesised throughout seed maturation are degraded into peptides and amino acids to be exported to the growing parts of the seedling (Mitsuhashi and Oaks 1996). Besides storage proteins, quiescent seeds contain alkaline proteases, carboxypeptidases and a few endopeptidases. The degradation pathways of storage proteins are approximately the same in both mono- and dicotyledonous seedlings and may be schematised according to two scenarios.

Newly synthesised proteases (mainly endopeptidases) are imported into protein bodies via the secretory pathway, and initiate the hydrolysis of the storage proteins, thus making them sensitive to other, already present or newly imported, endopeptidases and carboxypeptidases, which continue the degradation process. Peptides formed during proteolysis are then transported to the cytosol, where they are degraded to amino acids by alkaline proteases. Cysteine proteases are particularly involved in this processes. For example, during barley germination, the aleurone and scutellar epithelial cells secrete two types of cysteine endopeptidases, EP-A and EP-B, into the endosperm. These proteases catalyse the first step of B- and D-type hordein hydrolysis (Callis 1995). Similarly, in maize seed endosperm, the sequential disappearence of α, β, γ and δ-type zeins was correlated to the synthesis of three different cysteine endopeptidases (Mitsuhashi and Oaks 1996). Such germination-induced cysteine proteases, many of which are related to papain, named GerCP (for germination cysteine proteinases), have been described in monocots like wheat (*Triticum aestivum*), barley (*Hordeum vulgare*), corn (*Zea mays*) and rice (*Oryza sativa*) and in legumes like soyabean (*Glycine max*), common bean (*Phaseolus vulgaris*), or black gram (*Vigna mungo*); for a review see Granell et al. (1998). They are characterised by small molecular size, generally acidic pI and pH optimum, broad substrate specificity and sensitivity to cysteine protease inhibitors. Many GerCPs are normally synthesised in one cell type to initiate the degradation of the storage proteins in another cell type; they are therefore secreted. GerCPs have been reported to carry out limited proteolysis, initiating the hydrolysis of seed storage proteins. Their activity seems to be regulated both at the transcriptional and posttranslational levels by limited proteolysis of an inactive precursor by an asparaginyl endopeptidase, called vacuolar processing enzyme (VPE) or legumain, that is coordinately regulated (Hara-Nishimura et al. 1998). The regulation of GerCP-specific expression during germination seems to be partially conserved through different species, as demonstrated by the ability of the promoter of GerCP from legumes to drive specific expression during germination in transgenic tobacco. In some plant species, GerCPs (endo- and carboxypeptidases) seem to be hormonally induced by gibberellins and repressed by abscisic acid, during seed germination. Indeed, abscisic acid is present in the dry seed and its concentration decreases at the onset of germination, while gibberellins are synthesised by the embryo and act on the aleurone layer. Moreover, in chickpea (*Cicer arietinum*), ethylene seems to modulate GerCP expression. The promoter region of *GerCP* genes also appears to contain hormone regulation elements. In addition, it has been hypothesised that GerCPs could participate in the activation of β-amylase, primarily responsible for the mobilisation of sugars (Granell et al. 1998).

In some species, like buckwheat (*Fagopyrum esculentum*) or rice, the proteases involved in the degradation of storage proteins are synthesised during seed maturation but remain inactive, either because of the presence of protease inhibitors, or because they need an additional maturation step to become active. In the dry seeds of buckwheat, the protein bodies contain the main storage protein (13S globulin), together with a zinc metallo-peptidase. The premature hydrolysis of the globulin is avoided because of the reduced availability of zinc necessary for protease activity and a high concentration of protease inhibitor. At the onset of germination, the zinc concentration increases, the inhibitor-protease complex breaks up, and the globulin is hydrolysed by the protease. Processing of the zymogen to a mature form and/or changes in vacuolar pH are other known mechanisms for protease activation during germination (Okita and Rogers 1996). Aspartic peptidases, known as phytepsins, could possibly be involved in this type of regulation. Phytepsin was found to be present in a large variety of monocots and dicots, and in several grain tissues including the embryo, scutellum and aleurone layer (Kervinen 1998). Barley grains contain a prominent phytepsin activity which has been suggested to initiate the hydrolysis of grain storage proteins at the onset of germination. However, despite the presence of phytepsin in developing, resting and germinating grains, its role in the hydrolysis of storage proteins remains to be confirmed.

The involvement of ubiquitin-dependent proteolysis during germination has also been studied. The first plant 20S proteasome to be purified was extracted from dry pea seeds, and significant changes in ubiquitin conjugates occur during germination (Agustini et al. 1996). Recently, the expression of 20S and 26S proteasome subunits was shown to be induced in germinating spinach seeds (Ito et al. 1997). Therefore, it is likely that ubiquitin-dependent proteolysis is involved in the major metabolic disruptions associated with the outbreak of germination.

Proteolysis and Nitrogen Remobilisation in Senescence

Senescence is commonly defined as the sequence of biochemical and physiological events comprising the final stage of development and leading to cell death (Smart 1994; Noodén et al. 1997). In both annual and perennial plants, it has been macroscopically defined as the progressive yellowing of the leaves, which is associated with the disruption of most cellular components, and is followed by leaf death and abscission. However, it is now clear that senescence is distinct from chlorophyll degradation, and would be better characterised as the interruption of growth and decrease in the efficiency of cellular processes (Bleecker 1998) that in leaves are concomitant with the mature stage (Crafts-Brandner et al. 1998).

Senescence is genetically controlled, e.g. it will occur at a given time in the life of the leaf, even under optimal growth conditions: this is often called natural senescence. However, senescence is also triggered by a number of unfavourable environmental conditions or by nutrient deprivation. The possibility that senescence might result from a genetic program only, with no metabolic or environmental stimulus, has been discussed critically (Bleecker 1998). In any case, the major process associated with senescence in plants is the ordered degradation of cell constituents, and the features discussed below apply to senescence, whether natural, or induced by environmental or nutritional constraints.

At the onset of senescence, a net protein degradation is initiated. Originally, only vacuolar proteolysis was thought to be responsible for protein degradation in senescing plant tissues, although the presence of proteolytic activities had also been demonstrated in other cell compartments such as the chloroplast, cell wall, microsomes, mitochondria, cytosol, Golgi apparatus, nucleus and peroxisomes (Distefano et al. 1997, and references therein). It is now clearly established that senescence-associated proteases are present in the vacuoles, plastids, peroxisomes and cytosol. Most of them are cysteine proteinases, related to papain, and were named SenCP for senescence cysteine proteinases (Granell et al. 1998).

Studies on senescence and proteolysis have been performed mainly on flowers and leaves. Marked changes in exo- and endoproteolytic activities have been observed in many different plant species during senescence. While total aminopeptidase activities are high in young tissues and diminish early during senescence, carboxypeptidases are maintained longer and endopeptidases activities increase dramatically up to the late phases of senescence (Peoples and Dalling 1988).

The remobilisation of nutrients from senescing to growing plant parts is essential in plant development. The major source of nitrogen is the chloroplasts which contain the photosynthetic machinery of the cell, carry out major biosyntheses and store most of the leaf proteins. Cytosolic proteins (e.g. nitrate reductase) are also degraded early in senescence, but they represent a minor part of the nitrogen source for remobilisation. Protein breakdown is usually associated with a specific metabolism in which both carbon and nitrogen are mobilised into transportable organic compounds, mostly represented by amino acids such as glutamine and asparagine. It has been hypothesised that during senescence the massive release of ammonia is a consequence of amino acid deamination and catabolism of nucleic acids. The parallel induction of cytosolic glutamine synthetase (GS_1) and the concomitant decrease of its plastidic counterpart (GS_2) in senescing leaf tissues (Kamachi et al. 1992; Pérez-Rodriguez and Valpuesta 1996) indicates that ammonia assimilation is shifted from the chloroplast to the leaf cytosol. Moreover, the GS_1-encoding gene appears to be specifically induced in leaf tissues during ageing (Kamachi et al. 1992; Watanabe et al. 1994; Buchanan-Wollaston 1997). GS_1 isoenzymes would be involved in the formation of the amide glutamine, which is then exported *via* the phloem to various sink tissues such as shoots, floral meristems and developing seeds and fruits. In the different leaf stages of vegetative tobacco plants, glutamate dehydrogenase (GDH) follows a pattern of expression similar to that of GS_1 at both the protein and mRNA levels, and both are correlated with the increase in endoproteolytic activity (Masclaux et al. 2000). Whether GDH is also involved in the reassimilation of ammonia or participates in the catabolism of amino acids *via* the deamination of glutamate is still controversial (Lea and Ireland 1999). However, it is clear that the process of protein breakdown is tightly linked to the expression and activity of both GDH and GS_1. Since prolonging photosynthesis during grain filling is recognised as critical to improve yield (Matile 1997), manipulating the expression or function of endoproteolytic activities, GS_1 or GDH appears as a means to achieve this goal.

Leaf or flower senescence is an actively regulated process that involves coordinated expression of specific genes (Buchanan-Wollaston 1997; Noodén et al. 1997), involving the perception of specific signals. Signals involved in the expression of senescence-induced proteases are currently under investigation.

Gibberellic acid (GA) is effective in retarding leaf senescence (Smart 1994). However, the mechanism of action of GA on senescence has yet to be elucidated. GA has been shown to be involved in many physiological processes, some of them, such as seed germination, involving the mobilisation of stored proteins (Gan and Amasino 1997). As such, GA was put forward as a potential factor triggering protease gene expression. Exogenous GA application to cut flowers increased transcript levels of both *SEN11* and *SEN102* genes that encode two senescence-induced cysteine proteases (Guerrero et al. 1998). In contrast, during pea ovary senescence, a cysteine protease encoding gene was downregulated by gibberellin.

Physiological studies have shown that in many species, cytokinins can inhibit leaf senescence. Transgenic plants expressing an autoregulated, senescence induced, gene for cytokinin biosynthesis exhibit a delayed senescence (Gan and Amasino 1997). Conversely, the endogenous cytokinin level drops with the progression of leaf senescence, and lowering cytokinin during callus growth of *Petunia hybrida* induced the expression of the *P21* and *P17* cysteine protease encoding genes (Tournaire et al. 1996). Homologous genes were also induced in senescent leaves of pea. Lowering cytokinin levels in senescing leaves has been suggested as one of the factor triggering induction of *P21* and *P17* during leaf senescence.

Ethylene is known to accelerate many of the physiological changes associated with leaf senescence. It was proposed that ethylene hastens the progress of senescence by activating senescence-associated genes. The genes *pDCCP1*, *SENU2* and *SENU3* which encode potential cysteine proteases were induced in carnation petals (Jones et al. 1995) and tomato leaves (Drake et al. 1996) during senescence, and have been shown to be upregulated by ethylene.

The role of the carbohydrate status in protease induction in senescing tissues is discussed below.

Proteolysis and Senescence in Nutritional Stresses

Similar proteases are induced according to changes in developmental stages, or in response to diverse environmental or nutritional constraints. It is interesting to observe that proteases are also induced as stress is released, as shown by the induction of the serine protease that degrades the light-stress protein, in plants that are transferred to low-light conditions (Adamska et al. 1996). This is consistent with the idea that proteases degrade regulatory proteins, or damaged proteins, or proteins that have become useless because changing developmental stages or environmental conditions have led to distinct cell activities. Since plants are constantly exposed to a changing environment, it is likely that such proteases are frequently expressed, although at varying levels, and that their activity contributes to cell fitness. However, in many cases, stress conditions or nutrient deprivation (Fig. 2) induce senescence, thus indicating that cell acclimation was not successful, and the constraint progressively, or suddenly, leads to cell death.

The detailed study of the sequence of events that occur in sugar-starved roots of maize gives an example of a smooth transition from an acclimation to a senescence process that eventually leads to cell death (Fig. 2). In the first 4 h after being placed in a sugar free medium, the level of amino acids, including asparagine, remains steady, thus indicating that no net catabolism of protein occurs (Brouquisse et al. 1992). However, many changes are already taking place: for example, the amount of cell division-related genes begins to decrease (Chevalier et al. 1996b), and fatty acids become the major respiratory substrate (Dieuaide-Noubhani et al. 1997). Therefore, despite the absence of net breakdown of protein it is likely that, at this stage, the protein turnover has already increased to respond to the required changes in cell activities and metabolism. Net protein breakdown and asparagine accumulation become obvious after 8 h and last for the following 30 h (Brouquisse et al. 1992). At the same time, various proteolytic activities increase, particularly that of a vacuolar serine endopeptidase (James et al. 1996). The function of many of the overexpressed genes is related to the catabolism of proteins and amino acids, including proteases, protease inhibitors, metallothioneins, asparagine synthetase etc. (Chevalier et al.

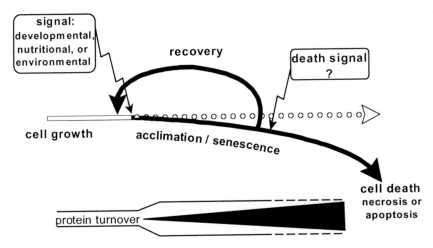

Fig. 2 : Protein remobilisation in plant cell life and death. A number of signals may induce an interruption of growth and initiate the remobilisation of proteins, and other cellular constituents as well, in processes which contribute to cell acclimation although they are very similar to senescence. The increase in protein turnover may, or may not, be associated with protein losses. Recovery occurs if acclimation succeeds, or conditions become favourable before cells are disorganised, or before a possible death signal is given

1996a). This senescence-like process can be reversed by the supply of metabolisable sugars, i.e. glucose or sucrose, but in the absence of sugars it leads to cell death. Overall, however, the decrease in cell activity together with the changes in metabolism and in gene expression appeared adequate to help the cells to survive in the absence of sugars (Brouquisse et al. 1992; Dieuaide-Noubhani et al. 1997). Some of these symptoms also occur in various organs of whole plants submitted to prolonged darkness (Brouquisse et al. 1998).

It may be noted that the process induced by sugar starvation in the roots, with cessation of cell growth and remobilisation of proteins, exhibits all the characteristics of a senescence process and shares many similarities with the dark-induced as well as the natural senescence of leaves (Buchanan-Wollaston 1997). It is likely that the dark-induced senescence of leaves is a consequence of sugar starvation (Kleber-Janke and Krupinska 1997; Brouquisse et al. 1998). However, the analysis of leaf senescence is complex, because sugar supply can induce senescence phenomena in two opposite ways: one is through sugar depletion as described above, and the other is through sugar excess: indeed, an increase of different protease activities was demonstrated in leaf segments of *Cyperus rotundus* in the presence of high, as well as of low carbohydrate levels (Fischer et al. 1998). According to nitrogen remobilization markers (GDH and GS_1), the sink to source transition of tobacco leaves is characterised by an accumulation of sugars, and endoproteolytic activities were induced at the same time as hexose concentrations in leaf blades reached their maximum (Masclaux et al. 2000). In non-senescent leaves, sugar accumulation leads to a decline in chlorophyll and photosynthetic proteins and represses the transcription of photosynthetic genes (Gan and Amasino 1997; Wingler et al. 1998, and references therein), which may be a first step in the induction of senescence. However, it remains to be determined whether hormonal and sugar signals are part of a more general signalling pathway, tuned in response to leaf natural or induced senescence, or are inducing a set of genes specifically involved in proteolysis and N remobilisation.

It may be surprising that very different proteases, like the very specific proteasome system (Ito et al. 1997, see above) or cysteine proteases, are induced in senescing tissues. This may correspond to an overlap of distinct phases of the senescence process where specific and unspecific proteolytic processes may be induced successively.

Conclusion: from Protein Turnover in Cells to Nitrogen Remobilisation in Plants

Proteolysis and nitrogen remobilisation are constantly associated with both plant responses to stresses and plant senescence processes. The senescence associated with fruit maturation or seed filling is a dramatic event in the biology of plants (Peoples and Dalling 1988), and its importance in crop yield, particularly in monocarpic plants, has been emphasised. How have plants evolved such a process? The major two hypotheses about senescence induction in relation to flower or fruit production (see references in Noodén et al. 1997; Wilson 1997), either through the production of a death hormone or through nutrient withdrawal, share the common view that the function of senescence is remobilisation of resources for reproductive development, and is therefore whole-plant-oriented. The classical concept of one single senescence mechanism was consistent with this starting point. However, present evidence suggests that the senescence phenomena can have multiple causes that may converge to common or similar pathways from a number of different situations or signals (Feller and Fischer 1994; Noodén et al. 1997; Bleecker 1998). This recent view suggests possible links between the function of proteolysis in normal cell life, including responses to developmental or stress signals, and that in protein remobilisation during senescence.

The case of a sugar starvation-induced senescence in maize root tips provides an interesting perspective, because although the signal in the whole plant is nutrient withdrawal (Brouquisse et al. 1998, and references therein), plant-oriented remobilisation can be excluded: indeed, the root tip cells have no neighbouring xylem or phloem vessels, so that remobilisation to other plant organs, if any, is limited. Therefore, the effect of protein remobilisation could be interpreted as a means to improve the fitness of root tip under sugar starvation. In the first hours of sugar starvation, this was by providing preexisting amino acids to build up the needed proteins, and, after a few hours, by supplying substrates for respiration. Similarly, the transient accumulation of asparagine was considered within the economy of the root cells, as a transient storage of reduced nitrogen to be used for recovery (Fig. 2), rather than for remobilisation to other plant parts.

The commonalities between sugar starvation-induced events in roots and other senescence processes, most often described in leaves, suggest that a large part of the remobilisation phenomena may have evolved as a response to constraints at the cell level. In this scheme, the function of protein remobilization is cell acclimation, although it progressively wastes cell resources and ultimately leads to cell death. A death signal may be emitted at some time during such senescence processes (Fig. 2), since nucleosomal DNA fragmentation, a sign of apoptosis, has been detected in a number of cases. It is possible that, building upon these cell-oriented processes, plants have evolved additional mechanisms that allow them to take advantage of the degradation of senescing organs to built up their seeds, fruits and other sinks. These additional whole-plant-oriented mechanisms may include the choice of metabolic intermediates for nitrogen transport (asparagine or glutamine), the "decision" to form a glyoxylic acid cycle to export the sugars made from lipids, as well as transport capacities through the plant, importation and utilisation in sink cells, etc. A key point in cell survival may be the time when cell metabolism switches from protein mobilisation for acclimation, to mobilisation for exportation.

References

Adamska I, Lindahl M, Roobolboza M, Andersson B (1996) Degradation of the light-stress protein is mediated by an ATP-independent, serine-type protease under low-light conditions. Eur J Biochem 236: 591-599

Agustini V, McIntosh T, Malek L (1996) Ubiquitination and ATP levels in garden pea seeds. Physiol Plant 97: 463-468

Aubert S, Gout E, Bligny R, Marty-Mazars D, Barrieu F, Alabouvette J, Marty F, Douce R (1996) Ultrastructural and biochemical characterization of autophagy in higher plant cells subjected to carbon deprivation: control by the supply of mitochondria with respiratory substrates. J Cell Biol 133: 1251-1263

Bahrami AR, Gray JE (1999) Expression of a proteasome alphatype subunit gene during tobacco development and senescence. Plant Mol Biol 39: 325-333

Barakat S, Pearce DA, Sherman F, Rapp WD (1998) Maize contains a Lon protease gene that can partially complement a yeast pim-1deletion mutant. Plant Mol Biol 37: 141-154

Barrett AJ, Rawlings ND, Woessner JF (1998) Handbook of proteolytic enzymes. Academic Press, New York

Bleecker AB (1998) The evolutionary basis of leaf senescence : method to the madness? Curr Opin Plant Biol 1: 73-78

Brouquisse R., James F, Pradet A, Raymond P (1992) Asparagine metabolism and nitrogen distribution during protein degradation in sugar-starved maize root tips. Planta 384-395

Brouquisse R, Gaudillère JP, Raymond P (1998) Induction of a carbon-starvation-related proteolysis in whole maize plants submitted to light/dark cycles and to extended darkness. Plant Physiol 117: 1281-1291

Buchanan-Wollaston V (1997) The molecular biology of leaf senescence. J Exp Bot 48: 181-199

Callis J (1995) Regulation of protein degradation. Plant Cell 7: 845-857

Chevalier C, Bourgeois E, Just D, Raymond P (1996a) Metabolic regulation of asparagine synthetase gene expression in maize (Zea mays L.) root tips. Plant J 9: 1-11

Chevalier C, Lequerrec F, Raymond P (1996b) Sugar levels regulate the expression of ribosomal protein genes encoding protein S28 and ubiquitin-fused protein S27a in maize primary root tips. Plant Sci 117: 95-105

Clough RC, Jordan-Beebe ET, Lohman KN, Marita JM, Walker JM, Gatz C, Vierstra RD (1999) Sequences within both the N- and C-terminal domains of phytochrome A are required for PFR ubiquitination and degradation. Plant J 17: 155-167

Crafts-Brandner SJ, Holzer R, Feller U (1998) Influence of nitrogen deficiency on senescence and the amounts of RNA and proteins in wheat leaves. Physiol Plant 102: 192-200

Davies DD (1982) Physiological aspects of protein turnover. In: Boulter D, Parthier B (eds) Nucleic acids and proteins in plants. I. Structure, biochemistry and physiology of proteins. Encyclopedia of Plant Physiology. New series. Berlin Heidelberg New York, pp 189-228

del Pozo JC, Estelle M (1999) Function of the ubiquitin-proteosome pathway in auxin response. Trends Plant Sci 4: 107-112

Desimone M, Wagner E, Johanningmeier U (1998) Degradation of active-oxygen-modified ribulose-1,5-bisphosphate carboxylase/oxygenase by chloroplastic proteases requires ATP-hydrolysis. Planta 205: 459-466

Dieuaide Noubhani M, Canioni P, Raymond P (1997) Sugar-starvation-induced changes of carbon metabolism in excised maize root tips. Plant Physiol 115: 1505-1513

Distefano S, Palma JM, Gomez M, Delrio LA (1997) Characterization of endoproteases from plant peroxisomes. Biochem J 327: 399-405

Drake R, John I, Farrell A, Cooper W, Schuch W, Grierson D (1996) Isolation and analysis of cDNAs encoding tomato cysteine proteases expressed during leaf senescence. Plant Mol Biol 30: 755-767

Feller U, Fischer A (1994) Nitrogen metabolism in senescing leaves. Crit Rev Plant Sci 13: 241-273

Fischer A, Brouquisse R, Raymond P (1998) Influence of senescence and of carbohydrate levels on the pattern of leaf proteases in purple nutsedge (Cyperus rotundus). Physiol Plant 102:385-395

Fu H, Doelling J, Rubin D, Vierstra R (1999) Structural and functional analysis of the six regulatory particle triple-A ATPase subunits from the Arabidopsis 26S proteasome. Plant J 18:529-539

Gan SS, Amasino RM (1997) Making sense of senescence – molecular genetic regulation and manipulation of leaf senescence. Plant Physiol 113: 313-319

Genschik P, Marbach J, Uze M, Feuerman M, Plesse B, Fleck J (1994) Structure and promoter activity of a stress and developmentally regulated polyubiquitin-encoding gene of Nicotiana tabacum. Gene 148: 195-202

Genschik P, Criqui MC, Parmentier Y, Derevier A, Fleck J (1998) Cell cycle-dependent proteolysis in plants: identification of the destruction box pathway and metaphase arrest produced by the proteasome inhibitor MG132. Plant Cell 10: 2063-2075

Glathe S, Kervinen J, Nimtz M, Li GH, Tobin GJ, Copeland TD, Ashford DA, Wlodawer A, Costa J (1998) Transport and activation of the vacuolar aspartic proteinase phytepsin in barley (Hordeum vulgare L.). J Biol Chem 273: 31230-31236

Granell A, Cercós M, Carbonell J (1998) Plant cysteine proteinases in germination and senescence. In: Barrett AJ, Rawlings ND, Woesner JF (eds) Handbook of proteolytic enzymes. Academic Press, London, CD-rom, chap 199

Guerrero C, Delacalle M, Reid MS, Valpuesta V (1998) Analysis of the expression of two thiolprotease genes from daylily (*Hemerocallis* spp.) during flower senescence. Plant Mol Biol 36: 565-571

Hara-Nishimura I, Kinoshita T, Hiraiwa N, Nishimura M (1998) Vacuolar processing enzymes in protein-storage vacuoles and lytic vacuoles. J Plant Physiol 152: 668-674

Hershko A, Ciechanover A (1998) The ubiquitin system. Annu Rev Biochem 67: 425-479

Ito N, Tomizawa K, Tanaka K, Matsui M, Kendrick RE, Sato T, Nakagawa H (1997) Characterization of 26S proteasome alpha- and beta-type and ATPase subunits from spinach and their expression during early stages of seedling development. Plant Mol Biol 34: 307-316

Jamai A, Chollet J, Delrot S (1994) Proton-peptide co-transport in broad bean leaf tissues. Plant Physiol 106: 1023-1031

James F, Brouquisse R, Suire C, Pradet A, Raymond P (1996) Purification and biochemical characterization of a vacuolar serine endopeptidase induced by glucose starvation in maize roots. Biochem J 320: 283-292

Jones ML, Larsen PB, Woodson WR (1995) Ethylene-regulated expression of a carnation cysteine proteinase during flower petal senescence. Plant Mol Biol 28: 505-512

Kamachi K, Yamaya T, Hayakawa T, Mae T. Ojima K (1992) Changes in cytosolic glutamine synthetase polypeptide and its mRNA in a leaf blade of rice plants during natural senescence. Plant Physiol 98: 1323-1329

Kervinen J (1998) Plant cysteine proteinases in germination and senescence. In: Barrett AJ, Rawlings ND, Woesner JF (eds) Handbook of proteolytic enzymes. Academic Press, London, CD-rom chap 278

Kleber-Janke T, Krupinska K (1997) Isolation of cDNA clones for genes showing enhanced expression in barley leaves during dark-induced senescence as well as during senescence under field conditions. Planta 203: 332-340

Lea PJ, Ireland RJ (1999) Nitrogen metabolism in plants. In: Singh BK (ed) Plant amino acids. Marcel Dekker, New York, pp 1-47

Lupas A, Flanagan JM, Tamura T, Baumeister W (1997) Self-compartmentalizing proteases. Trends Biochem Sci 22:399-404

Masclaux C, Valadier MH, Brugière N, Morot-Gaudry JF, Hirel B (2000) Characterization of the sink/source transition in tobacco (*Nicotiana tabacum, L.*) shoots in relation to nitrogen management and leaf senescence. Planta, in press

Matile P (1997) The vacuole and cell senescence. *In:* Leigh RA, Sanders D, (eds) Advances in botanical research incorporating advances in plant pathology, vol 25. Academic Press, San Diego, pp 87-112

Mitsuhashi W, Oaks A (1996) Localization of major endopeptidase activities in maize endosperms. J Exp Bot 47: 749-754

Morris K, Thomas H, Rogers LJ (1996) Endopeptidases during the development and senescence of *Lolium temulentum* leaves. Phytochemistry 41:377-384

Nakashima K, Kiyosue T, Yamaguchi Shinozaki K, Shinozaki K (1997) A nuclear gene, *erd1* encoding a chloroplast-targeted Clp protease regulatory subunit homolog is not only induced by water stress but also developmentally up-regulated during senescence in *A. thaliana*. Plant J 12: 851-861

Neubauer C (1993) Multiple effects of dithiothreitol on nonphotochemical fluorescence quenching in intact chloroplasts. Influence on violaxanthin de-epoxidase and ascorbate peroxidase activity. Plant Physiol 103: 575-583

Noodén LD, Guiamét JJ, John I (1997) Senescence mechanisms. Physiol Plant 101: 746-753

Okita TW, Rogers JC (1996) Compartmentation of proteins in the endomembrane system of plant cells. Annu Rev Plant Physiol Plant Mol Biol 47:327-350

Paris N, Stanley CM, Jones RL, Rogers JC (1996) Plant cells contain two functionally distinct vacuolar compartments. Cell 85: 563-572

Peoples M, Dalling M (1988) The interplay between proteolysis and amino acid metabolism during senescence and nitrogen reallocation. *In*: Noodén L, Leopold AC (eds, Senescence and aging in plants. Academic Press, San Diego, pp 181-217

Pérez- Rodriguez J, Valpuesta V (1996) Expression of glutamine synthetase genes during natural senescence of tomato leaves. Physiol Plant 97: 576-582

Roulin S, Feller U (1997) Light-induced proteolysis of stromal proteins in pea (*Pisum sativum L*) chloroplasts: requirement for intact organelles. Plant Sci 128: 31-41

Schmid M, Simpson D, Kalousek F, Gietl C (1998) A cysteine endopeptidase with a C-terminal KDEL motif isolated from castor bean endosperm is a marker enzyme for the ricinosome, a putative lytic compartment. Planta 206: 466-475

Shanklin J, Dewitt ND, Flanagan JM (1995) The stroma of higher plant plastids contain ClpP and ClpC, functional homologs of *Escherichia coli* ClpP and ClpA: an archetypal two-component ATP-dependent protease. Plant Cell 7: 1713-1722

Smart CM (1994) Gene expression during leaf senescence. New Phytol 126: 419-448

Stadtman E (1992) Protein oxidation and aging. Science 257: 1220-1224

Staswick (1994) Storage proteins of vegetative plant tissues. Annu Rev Plant Physiol Plant Mol Biol 44: 303-322

Stieger PA, Feller U (1997) Degradation of stromal proteins in pea (*Pisum sativum* L.) chloroplasts under oxidising conditions. J Plant Physiol 151: 556-562

Tommasini R, Vogt E, Fromenteau M, Hortensteiner S, Matile P, Amrhein N, Martinoia E (1998) An ABC-transporter of *Arabidopsis thaliana* has both glutathione-conjugate and chlorophyll catabolite transport activity. Plant J 13: 773-780

Tournaire C, Kushnir S, Bauw G, Inze D, Teyssendier de la Serve B, Renaudin JP (1996) A thiol protease and an anionic peroxidase are induced by lowering cytokinins during callus growth in *Petunia*. Plant Physiol 111: 159-168

Tziveleka L-A, Argyroudi-Akoyunoglou JH (1998) Implications of a developmental-stage-dependent thylakoid-bound protease in the stabilization of the light-harvesting pigment-protein complex serving photosystem II during thylakoid biogenesis in red kidney bean. Plant Physiol 117: 961-970

Vierstra RD (1996) Proteolysis in plants: mechanisms and functions. Plant Mol Biol 32: 275-302

Watanabe Akio, Hamada K, Yokoi H, Watanabe Akira (1994) Biphasic and differential expression of cytosolic glutamine synthetase genes of radish during seed germination and senescence of cotyledons. Plant Mol Biol 26: 1807-1817

Wilson JB (1997) An evolutionary perspective on the "death hormone" hypothesis in plants. Physiol Plant 99: 511-516

Wingler A, Vonschaewen A, Leegood RC, Lea PJ, Quick WP (1998) Regulation of leaf senescence by cytokinin, sugars, and light – Effects on NADH-dependent hydroxypyruvate reductase. Plant Physiol 116: 329-335

The Biochemistry and Molecular Biology of Seed Storage Proteins

Jean-Claude AUTRAN[1] , Nigel G. HALFORD[2] and Peter R. SHEWRY[2]

Introduction

Economic importance of seed storage proteins

Most plants synthesise proteins in their organs of reproduction and propagation, such as seeds of gymnosperms and angiosperms. Storage proteins are usually located in two tissues. In dicotyledonous plants they may be located in the diploid cotyledons (exalbuminous), in the triploid endosperm (albuminous) or, occasionally, in both tissues. In monocotyledonous cereals they are primarily located in the triploid endosperm tissue. They are deposited in high amounts in the seed, in discrete deposits (protein bodies) and survive desiccation for long periods of time. In most cases, storage proteins lack any other biological activity and simply provide a source of nitrogen, sulphur and carbon skeletons for the developing seedling (Shotwell and Larkins 1989; Shewry 1995).

From the human point of view, seeds represent the most important plant tissue that is harvested and consumed, consequently, the economic importance of seed proteins is considerable. Seed storage proteins form the most important source of dietary proteins for humans as about 70% of the total intake comes directly from this source. In addition, seed proteins provide the major component of the diet of non-ruminant farm animals. Although proteins make up a relatively small proportion of the cereal grain (usually 7-15%, compared with up to 40% in legumes), cereals are the dominant world crops in terms of both dry matter production and protein pro-

1. INRA, Unité de Technologie des Céréales et des Agropolymères,
2, Place Viala
34060 Montpelier Cedex 01
France
E-mail: autran@ensam.inra.fr
2. IACR-Long Ashton Research Station
Bristol
BS41 9AF
UK
E-mail: peter.shewry@bbsrc.ac.uk

duction. For instance, the annual yields of the eight most important species (wheat, maize, rice, barley, sorghum, oats, millets and rye in order of decreasing importance) exceeded 1700 million metric tons in the 1995-1997 period, which corresponds to about 200 million tons of proteins, *i.e.* in theory a sufficient amount to meet the requirements of mankind. There has therefore been a considerable economic stimulus to the study of cereal proteins and, in particular, the storage proteins that account for 50% of the total (Shewry et al. 1994b). Of the remaining plant proteins (about 70 million tons), nearly all come from dicotyledonous seeds, especially the legumes (soybean, pea, peanut, bean, faba bean, lentil, chickpea, lupin etc.) and oilseeds (cottonseed, sunflower, oilseed rape etc.).

Because of their abundance, the storage proteins are largely responsible for the nutritional quality and technological properties of the seed. These aspects have, therefore, been the subject of considerable research since 1745, when Beccari is credited with having isolated gluten from wheat.

Nutritional Quality

Cereals and legumes are not only the major crops used to provide energy in food and feed, but they also supply most of the proteins consumed by humans and used for animal production. Cereals have some advantages in containing very few types of antinutritional factors, but their storage proteins have low nutritional value, as they are limiting in lysine besides having extremely high levels of non-essential amino acids (*e.g.* glutamine and proline). If the nutritionally inferior storage proteins of the two most important crops in the world, wheat and maize, could be converted into proteins with better nutritional value, it would certainly have a great impact on human nutrition in many areas as well as on animal production (Doll 1984). In contrast, the storage proteins of legume seeds have a much better balance of essential amino acids, although still limiting in methionine and cysteine. They often, however, also contain various types of antinutritional factors (*e.g.* trypsin inhibitors, phytohaemagglutin, α-galactosides, glucosinolates, alkaloids) which may be only partially removed by processing or plant breeding.

Much research has been devoted to increasing the amounts of seed proteins, and their contents of essential amino acids, to improve the nutritional quality of seeds. At first, this seemed possible because seed storage proteins can undergo major changes, as indicated by their high level of biochemical heterogeneity and genetic polymorphism. They may vary between wide limits (as the result of glycosylation, posttranslational processing, gene mutations and environmental effects), with such changes in protein composition being tolerated by the developing seed. However, it must be noted that seed storage proteins also possess certain essential properties that enable them to fulfil their physiological role (Spencer 1984). There may, therefore, be less flexibility than expected to tailor storage protein composition with a view to improve end-use quality. For instance, storage proteins are sequestered in protein bodies where they are not exposed to the proteinases responsible for the breakdown of metabolic proteins, and their structure is likely to contain information that determines their selective transport from the site of synthesis to the site of accumulation. Such constraints on seed storage proteins are also common to all secretory proteins made on the ER and processed in the endomembrane system, and the variation observed is restricted to very specific regions. Although we know the role of the typical leader sequence that directs the transport of the nascent polypeptide through the membrane of the endoplasmic reticulum and into the lumen, we understand very little of the sequence requirements that specify the subsequent steps in the transport of storage proteins to, and their deposition in, the protein bodies (see below; Spencer 1984).

Use in the Food Industry

Setting aside nutritional considerations, proteins are used as food ingredients for their functional properties, *i.e.* to provide a certain specific function in the product. The proteins of most concern in the food industry are the storage proteins, although this is not to deny that some enzymes may also be important (Miflin et al. 1983). Most functional properties influence the sensory characteristics of food, especially texture, but they can also affect the physical behaviour of foods or food ingredients during processing (*e.g.* mixing, extrusion, fermentation, heating, drying, cooking) and storage. These properties are discussed in more detail in a later section.

Cereal seeds provide the raw material for two of mankind's oldest technologies: the baking of bread and the fermentation of alcoholic beverages. One of the questions that has challenged cereal chemists for a long time is why wheat protein is unique among cereals and other plant proteins in forming a dough with viscoelastic properties ideally suited to making leavened bread (Bushuk and MacRitchie 1989). Today, the basis of wheat "protein quality" remains poorly understood in detail, although most scientists agree on the fact that the ability to form a viscoelastic gluten depends on the capacity of wheat storage proteins to interact and to form polydisperse aggregates in an appropriate balance. In contrast, the storage proteins of barley tend to have, when in excess, a negative influence on endosperm disaggregation during the malting process and on brewing properties.

The functional properties of legume proteins relate mainly to their ability to stabilize emulsions or foams, and to impart textural attributes (Wright and Bumstead 1984). Proteins of legume seeds are often refined using dry (air-classification) or wet (alkaline extraction followed by acid precipitation) methods with selective removal or destruction of undesirable components. Concentrates or isolates are then processed to make meat substitutes or functional agents for the food industry using texturisation.

Although recent research has increased our knowledge of the components of storage proteins so that we are now in a better position to relate protein composition to functionality for specific end uses, the basic mechanisms that determine the functional properties are seldom clearly understood because of the complexity of the various food systems and of the processes by which the raw materials are transformed into end-products (Cheftel et al. 1985).

Interest in Seed Storage Proteins

Seed storage proteins, and especially wheat storage proteins, have been the subject of considerable research during several decades with the use of increasingly sophisticated analytical methods, leading to detailed knowledge of their structures and properties and to impacts on quality improvement through breeding, varietal identification and better control of technological processes. However, studies of cereal proteins have tended to lag behind studies of other more fashionable proteins such as enzymes. Interestingly, there has been a great resurgence of interest in seed storage proteins over the past two decades. This is partly because the unusual features of storage proteins, such as their synthesis in large amounts in specific tissues at precise stages of development, have made them attractive for studies of cDNA cloning and gene regulation. In addition, the availability of complete amino acid sequences of many plant storage proteins and the recent development of transient expression and transformation systems have stimulated renewed interest in their biophysics and cell biology (Shewry 1995).

This Chapter Reflects All of These Interests

It aims to review our current knowledge on seed storage proteins, focusing on their biochemistry and molecular biology, including classification, structures, evolution, synthesis and deposition, biophysical properties, genetic manipulation and their impact on technological utilization. It is necessary to be selective in order to keep the chapter down to a reasonable size, but the aim is to give a both broad and up-to-date account of seed storage proteins.

Classification of Seed Proteins

Classification is an artificial process reflecting the purpose of the classifier (Boulter and Derbyshire 1978). Seed proteins may be therefore classified in a variety of ways (*e.g.* chemical structure, mechanisms of action, biological function, location, genetic relationships and the separation procedures employed in purification). The ideal classification may be based on the mechanism of protein action but we do not yet know enough about the detailed three-dimensional structures of plant proteins. Other systems have therefore been used, based mainly on separation procedures, biological function (storage, metabolic or structural proteins), physicochemical properties (electrophoretic mobility, contents of sulphur-containing amino acids) or genetics (gene location, duplication and divergence).

In fact, the classification of plant proteins was historically based on the work carried out around the turn of the century by Osborne (1907), who classified plant proteins into groups (called Osborne groups) on the basis of their extraction in a series of solvents: water (albumins), dilute salt solution (globulins), aqueous ethanol (prolamins) and dilute alkali or acid (glutelins). The two former groups essentially included metabolic (*e.g.* enzymatic) and storage proteins, whereas the two latter were mainly storage proteins. Prolamins are usually given trivial names based on the Latin generic name of the cereal, for example secalins in rye (*Secale cereale*), hordeins in barley (*Hordeum vulgare*) and zeins in maize (*Zea mays*). In wheat, the prolamin storage proteins are usually classified into two groups, gliadins and glutenins, which together form gluten. Whereas gliadins are monomeric proteins, glutenins are polymers consisting of disulphide-bonded polypeptides, so-called subunits.

Despite the paucity of knowledge of protein structure in Osborne's time, his classification has proved to be remarkably durable and still provides a framework for modern studies of cereal proteins (Shewry 1995). Although this classification is simple in concept, it has led to a considerable amount of confusion and dispute, as discussed by Shewry and Miflin (1985). The basis of these problems is that the extractability of proteins is affected by many factors including the physiological state of the tissue. Coupled with this is the fact that many modifications to Osborne's original extraction procedures have been introduced by different workers, without always monitoring the fractions for purity and cross-contamination using electrophoresis and amino acid analysis (Shewry and Miflin 1985). Consequently, in the 1980s, following the proposal of Field et al. (1983), a majority of cereal protein chemists agreed to take into account physicochemical, molecular and genetic properties to redefine prolamins to include cereal storage proteins previously defined as both prolamins and glutelins (*i.e.* gliadins and glutenins of wheat). These proteins are discussed in detail below. In contrast, the glutelins were redefined to include only non-storage proteins (mainly enzymic and structural) and will not be discussed further.

The prolamins are unusual in being restricted to only one family of plants, the grasses, which include the cultivated cereals. This contrasts with the globulin and albumin storage proteins which have wider distributions (Table 1).

Table 1. Major groups of seed storage proteins and their distributions. (Shewry 1995)

Type	2S Albumins	Prolamins	7S Globulins	11S Globulins
Solubility	Water	Aq. alcohols (± reducing agents)	Dilute saline	Dilute saline
Major components	Brassicas Sunflower Castor bean Brazil nut	Cereal endosperms (wheat, barley, rye, maize)	Legumes Cottonseed	Most legumes Cucurbits Brassicas Endosperms of oats and rice Castor bean
Minor components		Oats and rice endosperm	Cereal embryos and aleurones	Wheat endosperm

The globulin storage proteins of seeds were historically separated from pea, soybean and faba bean by cryoprecipitation, differential salt solubility and heat coagulation. Two broad types, called legumins and vicilins, were recognized by Osborne (1924). An important technical advance was the introduction of ultracentrifugation (Danielsson 1949), which allowed characterisation of the main storage proteins, legumins and vicilins, with sedimentation coefficients (S_{20w}) of about 11S-12S and 7S-8S, respectively. These 7S and 11S storage globulins have similar characteristics, but vary widely in their relative amounts depending on the species. The 11S globulins are the most widely distributed (Table 1) whereas the 7S globulins are more restricted in distribution, being present in some legumes, cottonseed and in cereal embryos and aleurones. Finally, the 2S albumins represent a fourth major group of storage proteins, occurring in a range of dicotyledonous species, including oilseeds such as rapeseed, sunflower and castor bean (Youle and Huang 1981).

Structures and Characteristics of Storage Proteins

Storage proteins are probably ubiquitous in seeds. In the vast majority of cases they have no known function apart from providing nutrition (carbon, nitrogen and sulphur) to the developing seedling. It is probable, therefore, that the structures of storage proteins are not as highly constrained as those of many other proteins, such as enzymes, although there is clearly a requirement that the protein should be efficiently synthesised, packaged, stored and mobilized during germination. Consequently, it is not surprising that storage proteins exhibit great variation in their structures and properties. Nevertheless, almost all of the storage proteins present in the seeds of major crops fall into four groups which derive from only two gene superfamilies.

Globulin Storage Proteins

Globulins are the most widespread type of storage protein and may well prove to be present in all angiosperm seeds, although varying in amount, properties and tissue distribution. Two types are recognised, with sedimentation coefficients of 7-8S and 11-12S. Both have been studied in most detail from legume seeds and are often called legumins (11S) and vicilins (7S), based on the taxonomy of the species from which they were first derived (family Leguminosae, tribe Viciae). These names are currently used for fractions from *Vicia faba* (field bean) and *Pisum sativum* (pea) but different names are used for *Phaseolus* (7S phaseolin) and soybean (11S glycinin, 7S β-conglycinin) globulins. Similarly, specific trivial names are often used for fractions from other plant groups, notably for 11S globulins, which are more widely distributed than 7S. These include cruciferin and helianthinin for 11S globulins of crucifers and sunflower, respectively (Casey 1999).

A typical 11S globulin has an M_r of about 300 000-400 000 and is a hexamer of six subunits (M_r about 60,000) associated by non-covalent forces. Each subunit is posttranslationally "nicked" by a specific proteinase to give acidic and basic polypeptides (M_r about 40 000 and 20 000, respectively), which are linked by a single disulphide bond. Thus, native legumins are broken down into six acidic and six basic polypeptides when treated with reducing agent under denaturing conditions (Casey 1999; Casey and Domoney 1999).

The 7S globulins differ from the 11S in being trimeric, with M_r typically about 150 000-190 000. The subunit M_r is typically about 50 000, but proteolytic processing can lead in some species to the generation of smaller polypeptides while further polymorphism can result from glycosylation. Thus, in pea, proteolysis of M_r 50 000 precursors at one or two sites gives rise to polypeptides ranging in M_r from about 12 000-33 000, some of which may be glycosylated, in addition to uncleaved precursor (Casey 1999; Casey and Domoney 1999). Unlike 11S globulins, the 7S globulins contain no disulphide bonds.

Although the 7S and 11S globulin subunits have little or no amino acid sequence identity, sophisticated alignments and structural predictions indicate that they are indeed related and this is supported by analyses of their three-dimensional structures. Thus, the acidic and basic polypeptides of the 11S globulin subunits appear to correspond to the *N*- and *C*-terminal regions, respectively, of the 7S globulin subunits (Argos et al. 1985; Lawrence et al. 1994). High-resolution 3-D structures have been determined for two 7S globulins, phaseolin (Lawrence *et al.* 1990, 1994) and canavalin from jack bean (Ko et al. 1993). Each subunit comprises two structurally similar units, each consisting of a β-barrel of antiparallel β-strands followed by α-helices which form loops (Fig. 1). The three subunits form trimers of dimensions 90 x 90 x 35 Å (phaseolin) and 80 x 80 x 40 Å (canavalin). Although the structures of 11S glob-

Fig. 1 : Schematic ("backbone-worm") representation of the phaseolin trimer, based on the X-ray structure of Lawrence et al. (1994). The *N* and *C* termini of each polypeptide are labelled. The location of the threefold axis perpendicular to the plane of the figure is indicated by the *white triangle*, whilst the locations of the pseudo-twofold axes are indicated by *white lines*. The latter axes lie in the plane of the paper and occur both as intrasubunit axes (relating the *N*- and *C*-terminal modules of the same polypeptide) and as inter-subunit axes (relating *N*- and *C*-terminal modules of neighbouring polypeptides). The polypeptide link between helix 4 and *C*-terminal strand A is absent in the structure. (Lawrence 1999)

ulins have not been determined in such detail, preliminary studies of the trimeric proglycinin expressed in *E. coli* show a similar structure to that of 7S globulins (Utsumi et al. 1993; 1996), with the backbones of the two protein types being readily superimposed. The trimeric proglycinin also has similar dimensions (93 x 93 x 36 Å) to 7S globulins with two trimers being assembled within the vacuole to give a hexamer of about 110 x 110 x 80 Å (Badley et al. 1975).

The similar structures of the 7S and 11S globulins will presumably facilitate their regular packing within protein bodies, the two protein groups occurring together in many species. It also implies that the two groups have evolved from a common ancestor. The presence of two structurally similar units within each 7S/11S globulin subunit indicates that a short ancestral domain may have initially been duplicated to give two domains corresponding to the 7S N-terminus/11S acid chain and 7S C-terminus/11S basic chain and that this ancestral protein was then duplicated to give the ancestral 7S and 11S globulin subunits, as shown in Fig. 2 (Lawrence 1999).

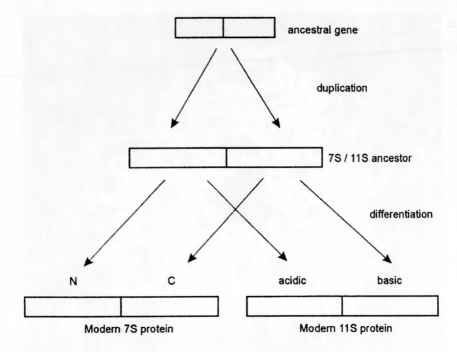

Fig. 2 : Possible evolutionary pathway for 7S and 11S globulins, based on an ancestral gene duplication (Gibbs et al. 1989; Lawrence et al. 1994; Shutov et al., 1995; Lawrence 1999).

In fact, more recent studies have shown that 7S and 11S globulins belong to an even more extensive superfamily of functionally diverse proteins found in plants and microbes. They include "germin" (oxalate oxidase) of germinating wheat embryos, spherulation-specific spherulins of myxomycetes (slime moulds), plant auxin binding proteins and various enzymes. Dunwell (1998) has coined the name "cupins" (latin for small barrel) for this superfamily to reflect the presence of a core β-barrel structure.

2S Albumins

Albumin storage proteins were initially defined as a group in 1981 when Youle and Huang (1981) isolated and characterized 2S albumin fractions from seeds of 12 botanically diverse species including two legumes but only one monocot (*Yucca*, Liliaceae). Detailed characterization has since been reported for 2S albumins from a range of dicotyledonous plants allowing the basic structure to be defined, but the presence of homologous proteins in monocots remains to be confirmed.

A typical 2S albumin (*e.g.* the napins of oilseed rape and other crucifers) comprises two subunits of about 30-40 residues and 60-90 residues, respectively, with two interchain disulphide bonds and two intrachain bonds within the large subunit. It is synthesised as a single precursor protein with posttranslational proteolysis resulting in the loss of an *N*-terminal prosequence and a linker peptide between the two subunits. However, there is considerable variation on this basic structure (see Fig. 3; Shewry and Pandya 1999).

1. Pro and/or linker sequences are absent from some albumins.

2. In castor bean two heterodimeric proteins are released by proteolysis of a single precursor protein.

3. In sunflower the mature protein consists of a single subunit (*i.e.* cleavage into large and small subunits does not occur) with either one or two albumins being released from a single precursor protein.

4. Most albumins have a conserved cysteine skeleton of eight residues, which form four disulphide bonds as discussed above. However, an additional unpaired cysteine is present in conglutin δ of lupin while the pea albumin subunits PA1a/PA1b contain a total of ten cysteine residues which do not apparently form interchain disulphide bonds.

Of particular interest is the presence in some species of methionine-rich components, the most widely studied being in Brazil nut (19 methionines out of 101 residues) and sunflower (16 methionines, 103 residues). The main interest in these proteins has been in relation to improving sulphur-poor forage and legume crops by genetic engineering. Although work on the Brazil nut protein was discontinued when it was shown to be allergenic, Molvig et al. (1997) reported that the expression of SFA8 in lupin seeds resulted in a 98% increase in seed methionine. However, the

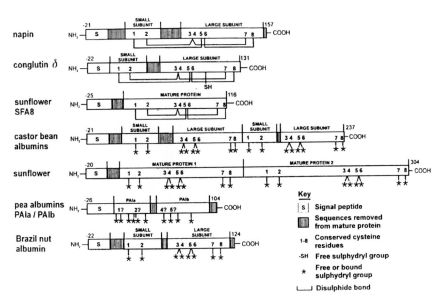

Fig. 3 : Schematic depictions of various types of 2S albumin, indicating their origins from precursor proteins and their disulphide structures. Conserved cysteine residues are numbered *1-8* and free sulphydryl groups shown as *-SH*. Cysteine residues whose behaviour is unknown are indicated by *. The precise correspondence between the cysteines in the pea albumins PA1a and PA1b and those in the other albumins is not known and potentially conserved residues are indicated by the *number* and ? The pea albumins PA1a and PA1b differ from the dimeric albumins in that the two subunits do not remain associated by interchain disulphide bonds. (Shewry and Pandya 1999)

total contents of sulphur and nitrogen in the seeds remained constant and the increase in methionine was at the expense of free sulphate and, to a lesser extent, cysteine (reduced by about 12%).

Although the small size of the 2S albumins would be expected to facilitate 3 D structure analysis, only one structure has so far been determined, for a 2S napin from oilseed rape by NMR spectroscopy (Rico et al. 1996). It shows five α-helices and a C-terminal loop in a right handed spiral (Fig. 4) with a global fold similar to other S-rich low M_r seed proteins (see below).

Although the major function of 2S albumins is undoubtedly storage, some components have been shown to exhibit biological activity. These include napins from kohlrabi (*Brassica napus* var *rapifera*), charlock (*Sinapis arverse*) and black mustard (*B. nigra*), all of which are inhibitors of serine proteinases (Svendsen et al. 1989, 1994; Genov et al. 1997) and may therefore play a role in defence. Similarly, napins from radish (*Raphanus sativus*) inhibit the growth of a range of plant pathogenic

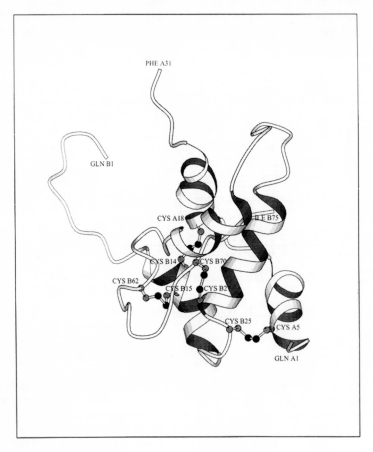

Fig. 4 : Ribbon diagram showing the 3-D structure of the *B. napus* napin Bn1b determined by NMR spectroscopy (Rico *et al.* 1996). The positions of cysteine residues and the *N*- and *C*-terminal residues of the polypeptide chains are indicated. (Rico et al. 1996)

fungi, particularly when in the presence of thionins (Terras et al. 1993). In this case, the mechanism probably involves membrane permeabilisation. A different type of biological activity, as allergens, has been referred to above in relation to the Brazil nut methionine-rich protein. This property is shared by albumins from other species such as castor bean, yellow mustard and oriental mustard.

Prolamins

The prolamin storage proteins differ from albumins and globulins in being restricted to the seeds of only one family of plants, the Gramineae (grasses,) which include the cereals. Because cereals form a major source of proteins for animal nutrition and food processing, their protein components have been studied in some detail. This has resulted in the availability of amino acid sequences and, in some cases, also structural data, for "typical" prolamins from all the major cereals, allowing their structural and evolutionary relationships to be determined.

The range of variation in the structures and properties of prolamins is vast, both within and between species. It is therefore not possible to provide a full account within the size limits of the present chapter. We will therefore focus on only two species, wheat and maize, referring the reader to recent review articles for more detailed accounts of these and other species (see, for example, Shewry (1995) and Shewry and Casey (1999)).

The prolamins of wheat correspond broadly to the gluten proteins (see p 317) and are classically divided into two groups on the basis of their solubility (gliadins) or insolubility (glutenins) in alcohol/water mixtures. The gliadins comprise monomeric proteins which interact in gluten by non-covalent forces (principally hydrogen bonds) and are further divided on the basis of their electrophoretic mobility at low pH into α-gliadins (fast), β-gliadins, γ-gliadins and ω-gliadins (slow) (Fig. 5).

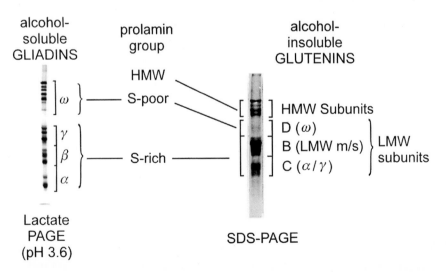

Fig. 5: The classification and nomenclature of wheat gluten proteins separated by SDS-PAGE and electrophoresis at low pH. The D group of LMW subunits are only minor components and are not clearly resolved in the separation shown. (Shewry et al.. 1999)

In contrast, the glutenins consist of high M_r polymers stabilized by interchain disulphide bonds. Once these bonds are reduced the component subunits are soluble in alcohol/water mixtures and it is therefore usual to define both gliadins and glutenins as prolamins. The reduced glutenin subunits can be separated by SDS-PAGE into two major groups called high molecular weight (HMW) and LMW, the latter being further divided into B, C and D groups (Fig. 5).

Although the gliadin/glutenin classification is still routinely used, comparison of amino acid sequences shows that it is possible to divide the whole range of wheat prolamins into three broad groups and into subgroups within these (Fig. 5). Similar groups are also present in other members of the tribe Triticeae (barley, rye), confirming the validity of this classification (see Shewry 1995; Shewry et al. 1999).

The HMW prolamins comprise the HMW subunits of wheat glutenin. Either three, four or five individual HMW subunit proteins are present in cultivars of hexaploid bread wheat, each accounting for about 2% of the total grain protein (Halford et al. 1992). The availability of the complete amino acid sequences of a number of subunits, derived from genomic DNA sequences, shows that they have an organisation similar to that shown in Fig. 6. They comprise between 627 and 827 res-

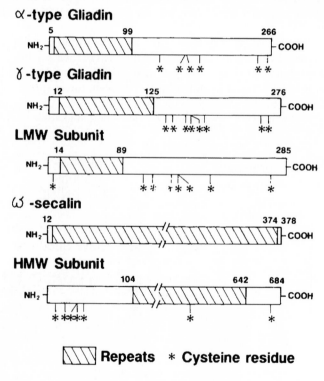

Fig. 6: Schematic structures of typical wheat gluten proteins, based on published amino acid sequences. The repeated sequences are based on the following consensus sequences: α-type gliadin, PQPQPFP + PQQPY; γ-type gliadin, PQQPFPQ; LMW subunit, PQQPPFS + QQQQPCL; ω-secalin, PQQPFPQQ; HMW subunit, GYYPTSLQQ + PGQGQQ. Full references are given in Shewry and Tatham (1990) and Shewry et al. (1999). (Shewry et al. 1994a)

idues with M_r ranging from about 67 500 to 88 100. They also have a clear domain structure with an extensive repetitive domain flanked by shorter non-repetitive domains at the N- and C-termini (of 81-104 and 42 residues, respectively). The repeats are based on two (PGQGQQ + GYYPTSP or LQQ) or three (also GQQ) motifs which appear to form a loose spiral supersecondary structure resulting in an extended conformation for the whole molecule. Cysteine residues are largely restricted to N-terminal (three or five residues) and C-terminal domains (one residue), providing cross-linking sites for polymer formation.

The S-poor prolamins of wheat comprise the ω-gliadins and the D group of LMW subunits, which together account for about 10-20% of the total prolamin fraction. The ω-gliadins contain no cysteine residues but high contents of glutamine (40-50 mol%), proline (20-30 mol%) and phenylalanine (8-9 mol%) (Kasarda et al. 1983). No complete amino acid sequences of ω-gliadins have so far been reported but studies of the homologous ω-secalins of rye (Fig. 6) and C hordeins of barley show that they consist almost entirely of repeated sequences based on an octapeptide motif (consensus PQQPFPQQ). Although there is no homology between this motif and the repeated sequences present in the HMW subunits, the repeated sequences in the S-poor prolamins appear to form a similar loose spiral structure. The D group of LMW subunits are only minor components and appear to be derived from the ω-gliadins by the addition of a single cysteine residue, allowing polymer formation.

The final group of prolamins, called S-rich, forms about 60-80% of the total fraction and comprises the α-type gliadins (α/β-gliadins), the γ-type gliadins and the B and C groups of LMW subunits. These all have similar sequences, with repetitive N-terminal and non-repetitive C-terminal domains (Fig. 6). The C-terminal domains of the α-type and γ-type gliadins contain six and eight cysteine residues, which form three or four intra-chain bonds, respectively. The C-type LMW subunits appear to correspond to α-type and γ-type gliadins with the addition of unpaired cysteines, which allow the formation of polymers. In contrast, the B-type LMW subunits form a discrete group with three intra-chain disulphide bonds and one or two additional unpaired cysteines. The repetitive domains of the S-rich prolamins are rich in proline and glutamine and are based on either one or two short consensus motifs.

Despite the wide variation in sequence and structure within the prolamins of wheat and other members of the Triticeae, two lines of evidence indicate that they have evolved from a common ancestral gene (Kreis and Shewry 1989; Shewry and Tatham 1990). Firstly, it is possible to recognise three short conserved sequences, each comprising about 30 residues and labelled A, B and C in Fig. 7, within the C-terminal domains of S-rich prolamins. Related regions are also present in the HMW prolamins but in this case they are separated in the N-terminal (region A) and C-terminal (regions B, C) domains. These regions also show homology with each other, suggesting that they arose from the triplication of a short ancestral sequence. Insertion of further non-repetitive sequences between these regions and of repeated sequences may subsequently have given rise to the S-rich and HMW prolamins. The S-poor prolamins do not contain regions A, B and C but the repeated sequences, which comprise most of the proteins, are clearly related to those in the S-rich prolamins. It can therefore be proposed that they originated from the same ancestral protein as the S-rich prolamins by further amplification of the repeated sequences followed by loss of most of the non-repetitive sequences including regions A, B and C.

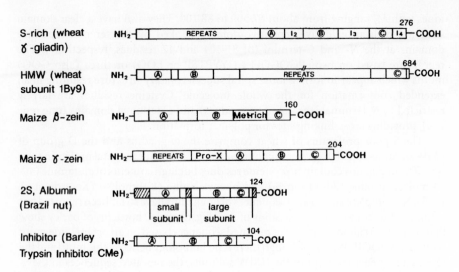

Fig. 7 : Schematic structures of various members of the prolamin superfamily. (After Shewry 1995)

Whereas the prolamins of wheat and other members of the Triticeae range in M_r from about 30 000-90 000, all of the maize prolamins (termed zeins) have M_r below 30 000. SDS-PAGE (Fig. 8) resolves two major bands of apparent M_r about 19 000 and 22 000 (α-zeins) with minor bands of apparent M_r about 27 000, 16 000 (γ-zeins), 14 000 (β-zeins) and 10 000 (δ-zeins). The Z19 and Z22 α-zeins actually comprise about 210-220 residues and 240-245 residues, with true molecular masses of about 23 000-24 000 and 26 500-27 000, respectively. They have similar structures (Fig. 7), with short N-terminal (36-37 residues) and C-terminal (10 residues) domains flanking either nine or ten blocks of degenerate repeats each comprising about 20 residues. These repeats are rich in non-polar amino acids (leucine, alanine) and each has been proposed to form an α-helix.

There is no evidence that the α-zeins are related to any of the prolamins of the Triticeae, or indeed to any other proteins except α-type prolamins in related panicoid cereals (sorghum, millets). In contrast, the other zein groups all appear to be related to the prolamins of the Triticeae (Fig. 7). The β-zeins and γ-zeins both contain regions corresponding to A, B and C in the Triticeae but have little other sequence homology with each other. In the γ-zeins either two (M_r 16 000) or eight (M_r 27 000) tandem repeats of the peptide PPPVHL are present, followed by a 22-residue Pro-X region in which almost every other residue is proline. In contrast, the β-zeins do not contain repeats but have a methionine-rich region close to the C-terminus. The δ-zeins are also methionine-rich but contain no sequence repeats or other distinctive features. Nevertheless, the identification of structural similarities to the methionine-rich 2S albumin of Brazil nut suggests that the δ-zeins may be related to other prolamins, as discussed below.

The α-zeins contain little or no methionine and only one or two cysteine residues, the latter resulting in their presence as monomers or oligomers. Higher levels of cysteine are present in the β-, γ- and δ-zeins, which is consistent with their presence in high M_r polymers. As in wheat, the reduced subunits are soluble in alcohol/water mixtures although the polymers may be soluble. The reduced γ-zein subunits are also readily soluble in water, being unique among prolamins in this respect.

The Prolamin Superfamily

Sequences related to those of regions A, B and C can also be identified in a range of other seed proteins, including prolamins from rice and oats, the 2S albumins and the cereal proteinase/amylase inhibitors (Kreis et al. 1985; Kreis and Shewry 1989). In the heterodimeric 2S albumins, such as the methionine-rich albumins of Brazil nut (Fig. 8), the site for proteolysis is located between regions A and B, with region A being present in the small subunit and regions B and C in the large subunit.

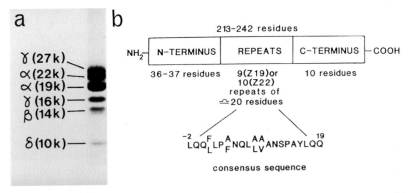

Fig. 8 : a. One-dimensional SDS-PAGE of total zeins of maize. b. Schematic structures of the M_r 19 000 (Z19) and M_r 22 000 (Z22) α-zeins of maize (After Shewry and Tatham 1990)

The cereal inhibitors have M_r ranging from about 12 000-16 000 and may be present as monomers or in dimeric or tetrameric complexes. The individual subunits may be inactive or exhibit activity against trypsin, exogenous α-amylases (notably from insect larvae), or both enzymes. They appear to be located in protein bodies within the mature cereal grain and are presumed to play a role in conferring broad-spectrum resistance to insect predators. They also have impacts on grain utilization, contributing to the development of respiratory allergy (baker's asthma) in workers in the milling and baking industry and affecting the stickness of pasta products made from durum wheat.

In addition to the 2S albumins and inhibitors, cereal grains contain several other groups of low M_r cysteine-rich proteins. These include the puroindolines, which are also rich in tryptophan and play a role in determining grain texture (*i.e.* hardness or softness) (Giroux and Morris 1998), and the non-specific lipid transfer proteins which are able to transport phospholipids *in vitro* (Breu et al. 1989). Both groups of proteins may also confer resistance to fungal pathogens *in vivo* (Dubreil et al. 1998; Terras et al. 1992). Both of these groups of proteins have highly conserved skeletons of cysteine residues, including Cys Cys and Cys Xaa Cys motifs, which are similar but not identical to those in the inhibitors (Gautier et al. 1994) and 2S albumins (Egorov et al. 1996). Not surprisingly, all of these low-M_r S-rich proteins also appear to have similar 3-D structures, resembling the soybean hydrophobic protein in having four α-helices forming a right-handed superhelix (the α-type proteins) (Baud et al. 1993). Similar structures have been determined (by NMR or X-ray analysis) or predicted for 2S albumins from oilseed rape (Rico et al. 1996; Fig. 4) and sunflower (Shewry and Pandya 1999); a bifunctional α-amylase/trypsin inhibitor from ragi (Indian finger millet) (Strobl et al. 1995), non-specific lipid binding proteins from various species including wheat and maize (Gincel et al. 1994; Gomar et al. 1996) and wheat puroindolines (reviewed in Shewry et al. 2000).

Structure and Expression of Storage Protein Genes

Storage proteins of all the major crop species are encoded by multigene families clustered at different loci in the genome. This presents a very complicated picture, but there is considerable evidence to suggest that storage protein genes have evolved from a small number of ancestral genes. The prolamins of the *Triticeae*, for example, are encoded by genes in complex loci on the homoeologous group 1 chromosomes (*Glu-1, Sec3, Hor3; Gli-1, Sec1, Hor1; Glu-3, Hor2, Hor5*), minor remote loci on the same chromosomes (*Gli-3,4,5; Sec4, Hor4*) and additional major loci on chromosomes 6A, B and D of wheat and 2R of rye (see Shewry et al. 1999 for a more detailed review). This is consistent with their origin from single ancestral loci present on the group 1 chromosomes, and the translocation of some loci to other chromosomes in wheat and rye. This hypothesis is supported by the fact that all of the prolamin genes contain regions of some homology and share the unusual characteristic of containing no introns. The diversity seen today may have arisen from the multiplication of a single ancestral gene, the insertion of sequences and the duplication of repetitive regions (Kreis et al. 1985).

The α-zein genes of maize are encoded by a multigene family that is at least as complex as those in the Triticeae, with between 70 and 100 members (Wilson and Larkins 1984) present at several loci on chromosomes 1, 4, 7 and 10 (Shen et al. 1994). The β-δ- and γ-zeins, in contrast, are located at single loci (Das and Messing 1987). The β- and δ-zein genes are single copy, while there are either one or two γ-zein genes.

Legume globulin gene families are also quite large. In pea, for example, there are at least 18 genes encoding vicilin subunits (Casey et al.1986), divided into three small, related gene families (Casey et al. 1988), two genes encoding convicilin (Domoney and Casey 1985) and more than 10 genes encoding legumins (Domoney and Casey 1985; Domoney et al. 1986). As with the prolamin genes of the *Triticeae*, legume globulin genes show similarities in sequence, structure and organization that suggest that they may have evolved from a common ancestor. Sequence similarities *within* the 7S (vicilin-type) and 11S (legumin-type) globulin gene families from different species are greater than *between* the 7S and 11S globulins from any single species, but it is clear that the 7S and 11S families are related (Gibbs et al. 1989).

The expression of seed protein genes is subject to tissue-specific and developmental regulation. Prolamin genes of the *Triticeae* and maize are expressed exclusively in the starchy endosperm during mid- and late development, although there are some differences in spatial distribution of the different protein classes within the endosperm and in the exact timing of expression during development. They are also subject to nutritional regulation, responding to the availability of nitrogen and sulphur in the grain (Giese and Hopp 1984; Duffus and Cochrane 1992). Control of gene expression is exerted primarily at the transcriptional level (Bartels and Thompson 1986; Sörensen et al. 1989). Legume globulin genes are expressed in the parenchyma cells of the cotyledons and in the endosperm, where this is present.

Genes that share a common ancestry and show similar patterns of expression would be expected to have regulatory sequences in common, and several regulatory elements have been identified and characterized in seed storage protein gene promoters. The first to be identified was a sequence of approximately 30 bp positioned around 300 bp upstream of the transcription start site of several gliadin and hordein genes of wheat and barley (Forde et al. 1985). It was first termed the -300 element, subsequently the prolamin box or endosperm element, and one or more copies of it are present in the promoters of all the S-poor and S-rich storage protein genes of wheat, barley and rye characterized so far. The consensus sequence for the element is 5'- TGACATGTAA AGTGAATAAG ATGAGTCATG but it contains two, separate conserved motifs, TGTAAAGT and G(A/G)TGAGTCAT, with a more variable region in between. The former has been called the endosperm motif (Hammond-Kosack et al. 1993), the latter the GCN4-like motif (GLM), nitrogen element or N motif (Hammond-Kosack et al. 1993; Müller and Knudsen 1993). The N motif is similar to the binding site of the yeast transcription factor, GCN4, which is involved in nitrogen signalling. It is inverted with respect to the E motif in S-poor prolamin genes (Shewry et al. 1999).

Detailed functional analyses have been performed on versions of this element from C hordein and LMW subunit genes. Müller and Knudsen (1993) used particle bombardment of cultured barley endosperms with C hordein promoter/β-glucuronidase (GUS) constructs to confirm a regulatory role for the prolamin box and to show that the E and N motifs were separate elements. The N motif acted as a negative element at low nitrogen levels and interacted with the E motif and other upstream elements to give high expression when nitrogen levels were adequate. Hammond-Kosack et al.(1993) used *in vivo* footprinting and gel retardation assays to show that E motifs within the prolamin box and further upstream in the promoter of the wheat LMW subunit gene, *LMWG-1D1*, bound a putative transcription factor, ESBF-1, during early grain development, whereas no binding could be detected in other tissues. A second putative transcription factor, ESBF-II, bound the N motif prior to maximum expression of the gene. This study was followed by a functional analysis of the prolamin box of this gene in transgenic tobacco, which showed that both motifs were required for seed-specific expression, and the cloning of SPA, a bZIP transcriptional activator which recognized the N motif (Albani et al. 1997).

The E motif is also present in the promoters of zein genes and has been shown to act as a tissue-specific enhancer (Quayle and Faix 1992). It is not accompanied by an adjacent N motif, although motifs similar to the N motif are present elsewhere in the promoters of some zein genes (de Freitas et al. 1994). A transcription factor, encoded by the *Opaque2* gene, has been shown to bind a sequence close to the E motif and to regulate expression of α-zein genes (Schmidt et al. 1987, 1992; Lohmer et al. 1991). Another DNA-binding protein, OHP-1, has been shown to interact with the Opaque2 protein, but its exact role is not understood.

Perhaps surprisingly, the prolamin box is not present in the HMW prolamin gene promoters, despite the fact that HMW prolamin genes are subject to tissue-specific and developmental regulation similar to S-poor and S-rich prolamin genes. This may be because they are expressed at higher levels and require a more powerful enhancer element. The major regulatory element of the HMW prolamin promoters is a 38-bp sequence, GTTTTGCAAA GCTCCAATTG CTCCTTGCTT ATCCAGCT, first identified by Thomas and Flavell (1990) in the *Glu-1D-2* gene from the cultivar Chinese Spring at position -186 to -148. All of the HMW prolamin genes characterized so far contain this sequence and its position is tightly conserved (Shewry et al. 1999).

As with cereal genes, a regulatory element in legume storage protein genes was first identified by sequence comparisons. This revealed the presence of a conserved element of 28 bp, known as the legumin box, in the promoters of 11S globulin genes (Bäumlein et al. 1986). Functionality of this sequence has been tested in transgenic plants and the central core motif of CATGCAT shown to be required for high levels of gene expression (Bäumlein et al. 1992; Lelievre et al. 1992). A related sequence is present in the promoters of 7S globulins and has been termed the vicilin box. The legumin box is present in a β-phaseolin gene promoter within an upstream activator sequence (UAS1) that was identified by deletion analysis of the promoter using reporter gene expression in transgenic tobacco (Bustos et al. 1989). UAS1 was found to drive seed-specific expression at 80% of the level of the longest promoter sequence (795 bp) used in the study (Bustos et al. 1991). However, several other regulatory elements that do not resemble the legumin box were identified, including two other activating sequences, UAS2 and UAS3, and two elements, NRS1 and 2, that downregulated expression. Fine control of reporter gene expression in the transgenic plants required all of these elements (reviewed by Hall et al. 1999).

Storage Protein Synthesis and Deposition

Seed storage proteins are products of the secretory pathway which takes place within the endomembrane system of the cell. The proteins are initially synthesised on ribosomes attached to the rough endoplasmic reticulum (ER) and are directed into the lumen of the ER by an *N*-terminal signal peptide which is removed by proteolysis. The subsequent folding, assembly, processing and deposition of the proteins occur in one or more compartments of the endomembrane system: the ER lumen, the Golgi apparatus and the vacuole.

Protein folding and disulphide bond formation are thought to occur within the ER lumen. These processes may be assisted by proteins resident in the ER lumen, molecular chaperones such as a binding protein (BiP) and enzymes such as protein disulphide isomerase (PDI) and prolyl peptidyl *cis trans* isomerase (PPI). It is difficult to prove that such "helper" proteins are essential for storage protein folding and assembly, although the levels of BiP do increase in some seeds during the period of storage protein accumulation (Xia and Kermode 1999). In addition, the levels of BiP are increased in maize mutants in which storage protein synthesis is impaired and protein body morphology affected (Boston et al.1991; Fontes et al. 1991).

The assembly of proteins into oligomers and polymers stabilized by interchain disulphide bonds or non-covalent interactions is also considered to occur, or at least to be initiated, within the ER lumen. In the case of cereal prolamins this may include the formation of extensive disulphide-stabilized polymers such as the glutenins, which are important in determining the functional properties of wheat. Assembly of 7S globulins into their mature trimeric form and of 11S globulins into an intermediate trimeric form of about 9S also occurs in the ER lumen.

In cereal prolamins no further modifications of the proteins occur but protein body formation occurs by two different routes. In maize and rice the prolamins accumulate within the lumen of the ER, giving rise to ER-derived protein bodies. In wheat and barley the situation is less clear but it is probable that two populations of protein bodies occur. Some prolamins, particularly the polymeric glutenins of wheat and polymeric hordeins of barley, appear to accumulate within the ER lumen as in rice and maize, but others (notably monomeric hordeins and gliadins) are transported via the Golgi apparatus to the vacuole, forming a second population of protein bodies (Levanony et al. 1992; Rubin et al. 1992). It is probable that the relative amounts of protein trafficking through these two pathways varies depending on the environmental conditions and developmental stage, while the two populations of protein body also fuse during the later stages of development to give a continuous proteinaceous matrix in the cells of the mature barley or wheat grain (Parker 1980).

The trafficking of proteins and their retention within specific compartments of the secretory pathway is usually determined by the presence of specific signals on the proteins, which may be cleavable or non-cleavable peptides at the N- or C-termini or surface patches on the folded protein, and their recognition by specific receptors. Proteins which do not contain such signals are thought to pass through the ER and Golgi to the cell membrane where they are secreted as a default destination. Retention within the ER is generally determined by the presence of C-terminal tetrapeptides (either KDEL or HDEL), which are present on the lumenal "helper proteins" discussed above, while vacuolar targeting signals vary considerably in their location and sequence.

No retention signals have yet been identified on prolamins and the retention of some components within the ER may be determined by their solubility properties which result in the formation of insoluble aggregates which are not readily transported through the endomembrane system. Similarly, no targeting signals have yet been identified on the prolamins destined for vacuolar storage.

The 2S albumins, 7S globulins and 11S globulins are all transported via the Golgi apparatus to the vacuole where proteolytic processing occurs to produce the mature proteins. Cleavage of the 11S globulin subunits into α/β chains also appears to be a prerequisite for assembly of the intermediate trimers formed in the ER into the mature hexamers in soybean but not in pea (Dickinson et al. 1989; Kermode and Bewley 1999).

The extent of glycosylation of 7S globulins varies between species and individual subunits whereas 11S globulin subunits are glycosylated only rarely (*e.g.* in lupin) (Casey 1999). Glycosylation of asparagine residues (N glycosylation) may occur during translocation across the ER membrane with further modifications occurring in the ER lumen and in the Golgi. In contrast, the glycosylation of serine and threonine residues (O glycosylation) occurs only in the Golgi. As with prolamins, the mechanisms that determine the targeting of 2S albumins and globulins to the vacuole are still unclear, but there is some evidence that specific sequences which are not subsequently cleaved are recognised. The vacuoles containing deposits of 2S albumin and globulin storage proteins then appear to divide to give discrete protein bodies.

Manipulation of Seed Storage Proteins

Classical and Mutation Breeding

Because the storage proteins are responsible for many aspects of the functional properties of seeds, their amount and composition have doubtless been manipulated by many generations of plant breeders whose aim has included the improvement of end-use properties. However, only rarely, and in recent years, have plant breeders been able to select for individual proteins with known quality attributes. The best-known example of this is the HMW subunits of wheat glutenin and breadmaking quality. The identification of specific allelic subunits associated with good or poor breadmaking performance (see Payne 1987, and below) and the ease with which these proteins can be followed in plant breeding programmes by SDS-PAGE of single grain extracts allowed breeders in the 1980s and a 90s to routinely select for quality associated subunits in their breeding programmes. There is no doubt that this has made an important contribution to quality improvement of European and other wheats. However, in most cases, it has not been possible to associate major quality traits with specific protein components.

Plant breeding has also been used to manipulate the total protein content of the seeds of some crops, exploiting variation which occurs between genetically determined limits for each species. Thus soybean varieties with between about 35 and 50% protein have been developed although there appear to be negative correlations between protein content and yield and between protein content and oil content (Salunkhe et al. 1992).

The ability to manipulate seed protein composition depends on the availability of genetic variation in the characters of interest, either within the crop itself or in related species with which it can be crossed. In some cases, such variation exists either within normal lines or spontaneous mutants. However, there is not sufficient variation available in many characters and further variation must be generated by mutagenesis or transformation.

Mutagenesis, using either chemical or physical mutagens, was extensively used in the 1960s and a 70s to generate variation in a range of characteristics. In barley, a major target was to generate high lysine mutants with improved nutritional quality, similar to the spontaneous mutants identified in maize (Misra et al. 1975; Bright and Shewry 1983). However, in both cereals there has been limited success in separating the high lysine character from deleterious associated effects to produce high-yielding lines with acceptability to farmers and end-users.

The high lysine mutants of maize were initially identified by visual inspection, due to associated changes in grain texture. About 12 mutant lines with lesions in different genes have been identified, with lysine contents ranging from about 110 to over 200% of those in the control non-mutant lines (Bright and Shewry 1983). The most widely studied, and most successful in terms of plant breeding, is the *opaque-2* mutation, which was also the first to be identified in 1964 (Mertz et al. 1964). The *Opaque-2* gene encodes a transcriptional activator which regulates the expression of the M_r 22 000 α-zein genes. Thus in *opaque-2* mutants the amount of M_r 22 000 α-zeins is drastically reduced but there are also decreases in other zein classes, probably due to pleiotropic "knock on" effects (Coleman and Larkins 1999). Although *opaque-2* maize contains higher levels of lysine and tryptophan the kernal is also softer making it more liable to damage and infection. In addition, the yield may be lower than that of normal maize lines. Nevertheless, it has proved possible to produce hard-textured *opaque-2* lines suitable for commercial cultivation by using genetic modifiers. The resulting material, called quality protein maize (QPM), has only slightly lower quality than normal *opaque-2* lines. The mechanism of endosperm texture modification is still not completely understood but the lines contain elevated levels (two- to threefold) of γ-zein which result from increased steady state levels of transcripts (Coleman and Larkins 1999).

Several other high lysine mutants of maize have been studied in detail including *floury 2* and *opaque-15* which specifically affect M_r 19 000 α-zeins and M_r 27 000 γ-zeins, respectively. However, none of these mutants has been successfully used to produce commercial varieties.

In barley a number of induced high lysine mutants and one spontaneous mutant (Hiproly) are available, with increases in lysine of up to 40% (Bright and Shewry 1983). Despite intensive breeding efforts with Hiproly and Risø mutant 1508 (see Munck 1992), no commercially successful lines have been produced.

Genetic Transformation

Transformation provides an attractive opportunity to manipulate the protein composition of seeds as it allows single defined sequences to be inserted. It is, therefore, possible to add additional copies of endogenous genes and to regulate their levels and their temporal and spatial patterns of expression using specific promoters. It is also possible to use antisense or other technologies to downregulate the levels of expression of endogenous genes and to add completely new genes from other plants, microbes or animals. It is not surprising, therefore, that seed protein composition has been a major target for genetic engineering experiments.

Two strategies have been used in order to increase the levels of nutritionally limiting amino acids in seeds. The first is to increase the levels of free lysine based on manipulation of its biosynthetic pathway. Two key regulatory enzymes in the lysine biosynthetic pathway (aspartate kinase and dihydrodipicolinate synthase) are normally feedback-regulated by lysine, resulting in a low amount of free amino acid. Transformation with feedback-insensitive forms of these enzymes from bacteria resulted in two- and fivefold increases in total lysine in seeds of canola (oilseed rape) and soybean, respectively, (Falco et al. 1995) but no impact on total lysine was reported in transgenic barley (Brinch-Pedersen at al.1996).

The second approach is to transform plants with additional genes encoding proteins rich in essential amino acids. In legumes the focus has been on sulphur amino acid content, using genes for methionine-rich 2S albumins. The 2S albumin from Brazil nut contains about 26 mol% methionine and has been used to increase total seed methionine by up to 30% in tobacco and oilseed rape (Altenbach et al. 1989, 1992) and up to three-fold in narbon bean (*Vicia narbonensis*) (Saalbach et al. 1995). However, this protein is now known to be allergenic to humans limiting commercial development. The 2S albumin of sunflower contains 16 methionines and 8 cysteines in a protein of 103 residues and does not appear to be allergenic. Expression in lupin resulted in a 94% increase in total seed methionine although there was a small decrease in cysteine and no increase in total seed sulphur (Molvig et al. 1997).

Although lysine-rich proteins have been characterized from plants (for example, the barley chymotrypsin inhibitors CI-1 and CI-2 which contain about 9.5 and 11.5 g% lysine, respectively) they have not so far been used for genetic engineering. Instead, Keeler et al. (1997) adopted a different approach, by designing and expressing a completely new protein containing 31% lysine and 20% methionine. This resulted in a significant increase in total lysine in seeds of transgenic tobacco but no significant effect on the methionine content.

Nutritional quality is readily defined in terms of the contents of nutritionally essential amino acids and is therefore also relatively easy to manipulate. In contrast, the functional properties that determine processing quality in food systems are more difficult to define in molecular terms and may involve multicomponent interactions (*i.e.* protein:protein, protein: starch, protein:lipid) which may change during the processing itself. Nevertheless, progress has been made in using genetic engineering to improve some aspects of seed protein functionality.

In wheat an obvious target for manipulation is the high molecular weight subunits of glutenin, due to the association between allelic variation in their number and composition and breadmaking quality (see above). Four laboratories have reported the transformation of bread wheat with additional genes for HMW subunits, in order to increase the number of expressed subunits and total amount of HMW subunit protein (Altpeter et al. 1996; Blechl and Anderson 1996; Barro et al. 1997; Alvarez et al., 2000), resulting in one case in increased dough strength (Barro et al. 1997). In one line the total amount of HMW subunit protein was increased from 12.7 to 20.5% of the total extractable protein, resulting in a highly elastic dough which appeared to be too strong to be mixed under normal conditions (Rooke et al. 1999).

Soybean proteins are extensively used in the food industry to confer functional properties, including gelation to form tofu (a traditional food in the Far East) and emulsification. Utsumi and colleagues have, therefore, focused on understanding and manipulating these properties using protein engineering of soybean 11S globulin (glycinin) subunits. They showed that several mutations resulted in increased gel hardness and improved emulsification properties, including deletion of short sequences at the N-terminus and the insertion of short methionine-rich sequences into variable domains (Utsumi 1992). The latter are of particular interest as they may allow the simultaneous improvement of functional properties and nutritional quality. Expression of one of the methionine-enhanced glycinins in transgenic tobacco resulted in accumulation up to about 4% of the total proteins but about half of the protein was partially degraded and not correctly assembled. Current work includes transformation of rice with the same constructs in order to improve its nutritional quality and confer novel functional properties.

These studies with wheat and soybean demonstrate therefore, that it should be possible to use genetic transformation to improve the functional properties of commercial crops, based on a detailed understanding of the molecular basis for protein functionality (see following section).

Biophysical Properties and Impact on Utilization

Seed storage proteins are used either as raw materials or ingredients for their functional properties. The latter are physicochemical properties that contribute to a specific function in the product, for example determining the desired processing and product characteristics (sensory, physical, textural, etc.). Functional properties are usually classified into three main groups:

– hydration properties depend on protein-water interactions and include water absorption, solubility, viscosity and adhesion,
– properties based on protein-protein interactions which include precipitation, gelatinisation, dough formation and shaping, viscoelastic properties,
– surface properties involve superficial strain and include emulsifying and foaming characteristics.

The objective of food scientists is to find a mechanistic explanation of functional behaviour. Obviously, elucidating molecular mechanisms should help us to optimize industrial processes and improve end-use qualities as well as to develop methods to screen for high quality genotypes in plant breeding programs or to test grain at harvest. However, our current state of knowledge and the complexity of the various food systems do not yet allow us to clearly understand how a given structure will determine, for example, the texture of a food product. The main difficulty lies in the fact that native structures are often modified during processing (*e.g.* mixing, extrusion, fermentation, heating, drying, cooking,) of raw materials or protein ingredients into the final complex food (Cheftel et al. 1985).

General Aspects of the Molecular Basis of Functionality

In order to understand the relationship between the structure of a protein and its functional properties, researchers have endeavoured to relate the latter to the protein structure as measured by a range of physicochemical parameters. For example, the size as given by the molecular weight, the charge and polarity as given by the electrophoretic mobility, isoelectric point and average hydrophobicity, and, finally, the intramolecular forces as given by the solubilization or dissociation behaviour in various solvents and stability to denaturation. At an intuitive level it is possible to select those physicochemical properties which are likely to contribute to specific functional properties. For example, a functional property such as solubility might be influenced by the molecular weight, hydrophobicity and net surface charge of the protein. Because, at a fundamental level, these properties can all be regarded as functions of the amino acid composition, primary sequence and structure of the protein, it follows that similarities in structural characteristics between proteins can be assumed to imply a corresponding similarity in both their physicochemical and functional properties (Wright 1983). To what extent this assumption is valid for seed storage proteins will be discussed below using cereals, mainly wheat, and legumes, mainly soybean, as examples.

Functionality of Cereal Proteins

Gluten proteins contribute unique technological characteristics to doughs, making wheat the most widely used cereal. Thus, in a developed wheat flour dough, the protein forms a continuous network which gives rise to the viscoelastic properties essential for sheeting and gas retention. However, the molecular explanation for this uniqueness depends on differences which must be quite subtle, since cereals which are related genetically to wheat and show superficial similarities in protein composition do not form dough with similar viscoelastic properties. To understand the viscoelasticity of wheat gluten, a property which is inherent to polymers and associated with flexible thread-like molecules, it is necessary to know the structures of the individual proteins and how they interact in the developing and mature seed, as well as in food systems. Gaining this knowledge has required analyses at several levels. The amino acid sequences of individual monomers and subunits were originally determined by direct analysis of isolated proteins, but are now usually deduced from the nucleotide sequences of cloned cDNAs or genes encoding the proteins. This approach has been used to determine the sequences of a number of zeins, gliadins, glutenins and hordeins. Knowledge of the locations of intrachain disulphide bonds in monomers and interchain bonds in polymers determined by sequence analysis of peptides isolated from unreduced protein preparations has also proved to be extremely valuable (Müller et al. 1998).

The conformations and potential interactions of the proteins are important aspects to study. However, in contrast to legume proteins in which water or salt solvents extract almost all of the storage proteins, conformational analyses of cereal proteins are limited by the solubility properties of prolamins and studies of their structures using a range of spectroscopic or hydrodynamic procedures can be only performed in a limited range of solvents. Polymeric prolamins (*e.g.* the HMW subunits of glutenin or whole gluten) cannot be dissolved without reduction of disulphide bonds, which makes it difficult to relate the results to the native state, or restricts the choice of analytical procedures to those that can be used on material in the solid state (scanning probe microscopy and NMR and FT-IR spectroscopy) (Tatham et al. 1990) or to computer modelling approaches (Kasarda et al. 1994).

Functionality in Legume Proteins

Proteins of legume seeds are often refined using dry (air-classification) or wet (alkaline extraction followed by acid precipitation) procedures with selective removal or destruction of undesirable components. The resulting concentrates or isolates are then processed to make meat substitutes using texturization or functional agents for the food industry. Consequently, in legume proteins, functional properties mainly relate to an ability to stabilize emulsions or foams and impart textural attributes. Emulsions (*e.g.* margarine, salad cream) and foams (*e.g.* whipped desserts, toppings) represent disperse systems in which one phase (air or oil) is dispersed throughout a continuous phase (water). Proteins provide the stabilizing force that prevents these systems reverting to two separate phases. Proteins migrate to the air:water or to the oil:water interface and, on unfolding, form an interface layer with consequent alteration to the surface properties. It is generally accepted that the balance of hydrophobic to hydrophilic regions (and therefore the amino acid composition and sequence) of a given protein has a significant effect on its ability to stabilize such dispersions (Wright and Bumstead 1984). On the other hand, the textural properties of legume proteins originate in their capacity to heat-set and form a stable matrix or gel, *i.e.* in cakes. Gelling involves the denaturation of proteins, basically the rupture of intramolecular bonds and unfolding of polypeptide chains, followed by the formation of intermolecular crosslinks between newly exposed residues of the denatured protein. The ability of a protein to form a gel depends upon its size, the structure and the nature of internal bonding, the previous history of the protein and also on extrinsic parameters such as solvent characteristics (Wright and Bumstead 1984).

Gluten Proteins and Breadmaking, Biscuitmaking and Pastamaking

Genetics of Protein Monomers and Protein Quality

Wheat is processed in the food industry into a range of products, including bread, other baked goods (cakes, biscuits), noodles and pasta. The ability to make these products is determined, to a great extent, by the gluten proteins which confer cohesive and viscoelastic properties to doughs. The concept of protein quality was born several decades ago, when it was realised that different wheat varieties had different baking performance. In the early 1970s it was demonstrated that the electrophoretic pattern of gliadins was a fingerprint of the wheat variety. At about the same time, wheat breeders in several countries began to develop varieties that had extremely high yield potential, but many of these varieties had unacceptable baking quality. Electrophoretic analysis of gliadins was quickly adopted for detecting the presence of admixtures in official grades of wheats or the presence of undesirable varieties in deliveries to the flour mill (Wrigley 1995). While widespread use has been made of the polymorphism of the gliadin proteins in wheat variety identification, it is research on glutenin subunits that has contributed significantly to a better assessment of protein quality and prediction in breeding programmes.

Glutenin subunits consist mainly of two types, the high molecular weight glutenin subunits (HMW subunits) and the low molecular weight glutenin subunits (LMW subunits). The HMW subunits (which account for about 6-12% of the total protein fraction) are so designated because they form a slower-moving group of components when reduced glutenin subunits are separated by SDS-PAGE with apparent M_r from about 80 000-120 000 (Fig. 9). Most wheat varieties have four or five HMW subunits, which frequently differ in their electrophoretic mobility among varieties. These are encoded on the long arms of chromosomes 1A, 1B and 1D and can be divided into x-type and y-type (Payne et al. 1984).

In the 1980s, researchers at the Plant Breeding Institute, Cambridge, UK, showed that the presence of certain subunits (e.g. 1Ax1, 1Ax2*, 1Dx5+1Dy10, or 1Bx7+1By9) were correlated with high breadmaking quality (Payne et al. 1987; Fig. 10). Scientists in several countries took advantage of this relationship between the genetics and quality of wheat proteins to develop new varieties more adapted to the modern baking technologies that require higher dough strength (e.g. the Chorleywood Bread Process, fast-food breads, rolls, buns, frozen doughs). Not all baking technologies, however, could benefit from this relationship. For example, for wholemeal bread, protein quantity is generally more important than protein quality. On the other hand, in southwestern Europe, breadmaking technologies are quite different from those commonly used in North America or in northern Europe. For example, in France, typical breads are made of essentially four ingredients, flour, water, yeast, and salt, with little or no additives, and they are normally baked on the oven hearth rather than in a pan. In this case, doughs with very high strength and tenacity are detrimental to baking score or loaf volume and a high extensibility of dough is required (Autran et al. 1997). To better understand the physicochemical bases of

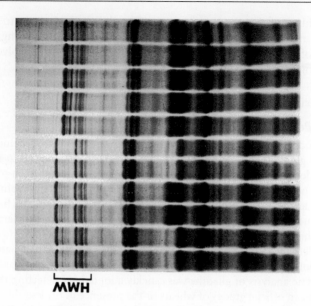

Fig. 9: Sodium dodecyl sulphate polyacrylamide gel electrophoresis (SDS-PAGE) of wheat storage proteins, with indication of the high molecular weight (HMW) subunits of glutenin

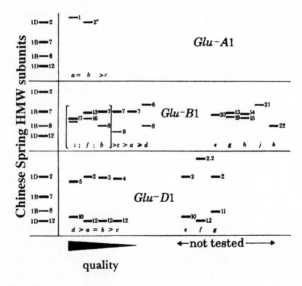

Fig. 10: Allelic variation in the high molecular weight (HMW) subunits of glutenin and their relation to bread-making quality (Payne et al. 1984). On the *left hand side* are the standard HMW subunits from cultivar Chinese Spring. The subunits have been split into three groups according to whether their genes are controlled by chromosome 1A, 1B or 1D. The *letter beneath each allelic group* refers to the international HMW allele nomenclature. The subunits that associate most strongly with good quality have been placed on the *left hand side* of each group

dough extensibility and to facilitate breeding of new types of wheats with a satisfactory balance between dough strength and extensibility, it has been necessary to study protein fractions other than the HMW subunits of glutenin. Recent reports have emphasized the possible role of LMW subunits of glutenin, that are more abundant than the HMW subunits and are encoded by genes that are genetically linked to some of the gliadin genes on the short arms of chromosomes 1A and 1B. New protein markers in wheat breeding have been proposed. For example, combining the chromosome 1D-encoded HMW subunit (*Glu-D1*) alleles that impart a high dough tenacity (*e.g.* subunits 1Dx5+1Dy10) with the chromosome 1A-encoded (*Glu-A3*) LMW subunit alleles *o* and *n* that impart high extensibility should result in the development of new wheats adapted to the baking technologies of southwestern Europe. When aiming at breeding of biscuit-type wheats, it could be recommended to screen lines to select seeds containing both the *Glu-D1* allele 1Dx2+1Dy12 and the *Glu-B3* allele *III*. Moreover, because many of the food products made from soft milling wheats require little or no elasticity (*e.g.* to form doughs or batters) they can be made from a new type of wheat that has been produced by transferring null alleles at *Glu-D1* and *Glu-A1* loci into the soft wheat cultivar Galahad. One line, called Galahad-7 contains only one HMW subunit (subunit 1Bx7) and produces extremely extensible doughs (Payne and Seekings 1996).

A similar strategy had been used for improving pasta products before being used for breadmaking. Durum wheat (*Triticum durum* Desf.) is widely considered to be the best type of wheat for pasta making due to its excellent amber color and superior cooking quality. Differences in cooking quality (*i.e.* high firmness and good surface condition of cooked pasta) were attributed to the protein content and composition of the grain. A major breakthrough in our understanding of the biochemical and genetic basis of pasta quality was made by Damidaux et al. (1978), with the discovery of a clear-cut relationship between the electrophoretic pattern of γ-gliadins and gluten strength, an indicator of pasta firmness. The γ-45 gliadin was associated with strong gluten, whereas the allelic γ-42 gliadin was associated with weak gluten. In fact, the positive effect of the γ-45 gliadins (*Gli-B1*) locus arises from its genetic linkage with the *Glu-B1* locus encoding LMW subunits of glutenin (Pogna et al.1988) and probably results from differences in the amounts of expressed LMW subunits between the two allelic types.

However, new specifications for durum wheat proteins have recently been introduced as a result of the use of higher drying temperatures in the pasta industry. While the roles played by protein content and protein composition have almost the same importance when pasta is dried at low temperature (55 °C), when using 70-90 °C drying the protein content becomes most important. The question that is presently challenging researchers is, therefore, how to make it possible to increase of the protein content of the grain without simultaneous increases in the brownness and ash content of the semolina.

Functional Properties of Glutenin Polymers

To evaluate the functionality of wheat proteins and to manipulate them in breeding and during food processing, it is necessary to investigate the molecular basis of the above-mentioned correlations used by breeders. This shows that studies based only on electrophoresis are not adequate because functionality is primarily determined by the presence of large protein aggregates rather than protein monomers.

In fact, the gluten proteins include monomeric and polymeric prolamins. In the monomeric gliadins disulphide bonds are either intra-molecular (in α- β- and γ- gliadins) or are absent (ω-gliadins). Purified hydrated gliadins have little elasticity and cohesion and contribute mainly to the extensibility of the dough system. The aggregated glutenins, in contrast to the gliadins, have intermolecular disulphide bonds in addition to intramolecular ones (Kasarda 1989). Hydrated glutenin polymers, free of gliadins, are highly cohesive and elastic and contribute elasticity to the dough system. It seemed reasonable, therefore, to attribute this elasticity largely to the crosslinked nature of the glutenin subunits.

The aggregated prolamins comprise two groups of polypeptides that can be separated under reducing conditions: the HMW subunits (apparent M_r range 80 000-140 000) and LMW subunits (apparent M_r range 50 000-80 000). Although the HMW subunits account for about 6-12% of the total protein of wheat, they appear to be of particular importance in determining the viscoelastic properties of wheat gluten and in turn determine the functionality of wheat doughs in various food systems, specifically breadmaking.

Molecular and biophysical studies have revealed details of HMW subunit structure that may relate to their role in viscoelastic polymers. As reviewed by Shewry et al. (1998) and discussed above, analysis of genes encoding nine different HMW subunits (including allelic and homeoallelic forms derived from the A, B and D genomes) demonstrates that the proteins have similar structures. All consist of three domains, with short N-terminal and C-terminal domains (both containing one or several cysteine residues) flanking a central repetitive domain based on hexapeptide and nonapeptide motifs with tripeptide motifs present in x-type subunits only. Previous studies have suggested that these repeated sequences form an unusual supersecondary structure, a loose spiral based on β-reverse turns. This β-spiral is of special interest since it appears to be unique among proteins (although a similar structure has been demonstrated for a synthetic polypeptide based on a repeat motif present in elastin, an elastomeric protein of mammals). Such a structure may contribute to the mechanism of gluten viscoelasticity, via intrinsic elasticity and/or the formation of extensive hydrogen bonds with adjacent proteins, the latter being facilitated by the high content of glutamine residues ($\cong 40$ mol%). However, another important aspect of HMW subunit structure is the presence of cysteine residues near the ends of the polypeptide chain, a feature that can facilitate a linear extension of the glutenin polymer into huge molecules consisting of polypeptides attached to one another by disulphide bonds, with entanglement regions, a system that may give a rubber-like elasticity.

The other family of glutenin subunits, the LMW subunits, is one of the most abundant storage protein groups in wheat endosperm. However, they are much less well characterised than HMW subunits and the effects of individual components have not been studied (Sissons *et al.* 1998). Also, functional and structural studies of the LMW subunits have always been limited by the difficulty of preparing adequate amounts of single homogeneous polypeptides. This is because the LMW subunits are somewhat insoluble after reduction of intermolecular disulphide bonds, which is necessary for their purification, but which also causes the exposure of buried hydrophobic regions. The LMW subunits also derive from many more genes than the HMW subunits, and our knowledge is based on a limited number of complete DNA sequences, some of which may not encode major components.

The LMW subunits were initially classified into B and C types according to their mobility in SDS-PAGE. However, there is growing evidence of similarity of groupings of LMW polypeptides based on sequence studies. Lew et al. (1992) suggested that it would be more valid to classify LMW subunits into classes based on sequence rather than on their electrophoretic mobility. Six main sequence types were defined on the basis of N-terminal sequences. Three of these can be considered to form LMW subunits proper (N-terminal sequences: SHIPGLERPSGL-, METSHIPGL-, METS(R)CIPGL-) while the others closely resemble the α- γ-, and ω-type gliadins (Sissons et al. 1998). At first, the LMW subunits were assumed to be chain-terminators in the growth of the glutenin polymer as their cysteine residues were only present in the C-terminal part of the molecule (Kasarda 1989). More recently, LMW subunits were found with a cysteine residue near their N terminus, making such subunits also available as chain extenders. However, despite reported correlations between the allelic composition of LMW subunits and dough extensibility, there is still little evidence as to how individual LMW subunits affect dough quality.

In addition to the individual glutenin subunits, the structures of the polymeric glutenins that they form have also been studied. Pioneering studies relating the molecular weight distribution of such glutenins to breadmaking quality were based on solubility methods or on conventional chromatography and hence suffered from many disadvantages: they were tedious, lengthy and difficult to reproduce or to quantify. The advent of HPLC techniques that have capabilities of automation, reproducibility and quantitation has allowed studies of larger series of samples. In contrast with studies based on reversed-phase HPLC (RP-HPLC), which are generally aimed at fingerprinting varieties based on gliadins or reduced glutenin subunits, size-exclusion HPLC (SE-HPLC) has the potential to retain large aggregates in a quasi-native state, and to provide information on aggregate size. For example, SE-HPLC of unreduced phosphate-SDS extracts proved to be a powerful tool for studying the physico-chemical and structural basis of wheat quality, and one that is applicable to the rapid examination of size differences of glutenins from wheat flours and industrial glutens (Autran 1994; Fig. 11).

Recently, large-size protein aggregates have been investigated in a more dynamic way. For example, the amount and rheological characteristics of the SDS-insoluble gel protein fraction decreased during dough mixing, whilst the amount increased again during dough resting. In this way, differences in reactivity were shown for the various HMW glutenin subunits. In particular, subunits 1By9, 1Dy10 and 1Dy12 were incorporated into polymeric glutenin at a faster rate and to a higher level than the subunits 1Dx2, 1Dx5, 1Bx7, 1Ax1 and 1Ax2*, which can be of importance when blending flours of different subunit composition (Weegels et al. 1997). Also, the presence of subunits 1Dx5+1Dy10 appeared to be associated with higher polymer molecular weight distribution than the presence of subunits 1Dx2+1Dy12 (MacRitchie 1998). Popineau et al. (1993) used gluten extraction and fractionation by a sequential procedure that preserved functionality, as well as dynamic measurements in shear to investigate large glutenin polymers of various isogenic lines of the cultivar Sicco with different HMW subunit compositions, and found a strong correlation between the amount of large glutenin polymers and the viscoelasticity of gluten subfractions. They concluded that both the quantity of HMW subunits and subunit composition influenced gluten viscoelasticity by modifying the polymerisation state of gluten proteins. Also, glutens differing in their LMW and HMW subunit

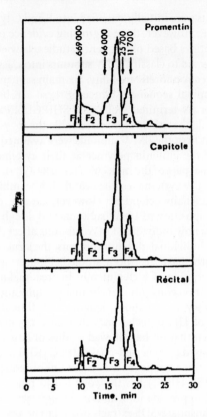

Fig. 11 : Elution profiles obtained by size-exclusion high performance liquid chromatography (SE-HPLC) of unreduced flour proteins extracted with phosphate-SDS from three bread wheat cultivars differing in baking strength (Alveograph W indices were 105, 110 and 200, respectively). Elution positions of molecular weight standards are indicated. The elution curve is divided into four regions: F_1 (high molecular size aggregates), F_2 (intermediate size aggregates), F_3 (gliadins), F_4 (albumins and globulins)

compositions were analysed to determine their size distribution and their rheology in the dynamic regime and by ESR spin-labelling, demonstrating a relationship between their aggregative properties, segmental flexibility and viscoelastic behaviour, with the proportion of rigid polypeptide segments related to the height of the elastic plateau (Hargreaves et al.1996).

    Complementary to physicochemical techniques, the potential of immunochemical methods, especially those based on monoclonal antibodies, has been exploited to recognise protein conformation, to yield information on functionally important groups, and to quantify specific flour polypeptides. For example, Andrews and Skerritt (1994) reported positive correlations between antibody binding to chromosome 1D-encoded HMW glutenin subunits (*i.e.* 1Dx5+1Dy10 or 1Dx2+1Dy12) and dough strength.

    With the advent of molecular genetics, it has become possible to produce whole wheat proteins and protein domains via heterologous expression in *E. coli* in sufficient amounts for detailed spectroscopic analyses and to express wild-type and mutant subunits in transgenic wheat plants. These new developments, together with molecular modelling, should provide information on the relationship between the protein primary structure and functionality and on the molecular basis for gluten viscoelasticity and will aid the development of strategies to improve the functional properties of gluten and wheat for a range of end uses, as discussed by Ciaffi et al. (1998), Shewry (1998), Barro et al. (1997) and Rooke et al.(1999).

Effect of Environment on Functional Properties and End-Uses

The intrinsic processing quality of wheat cultivars is changed significantly by cultural practices (*e.g.* amount, type and application dates of nitrogen fertilisation; sulphur availability) and climate (grain filling duration, temperature and relative humidity during grain filling) via modification of flour protein content and composition. For example, a high amount of nitrogen fertiliser generally leads to a significant increase in the total protein content, but this increase affects mainly the gluten proteins with little effect on the albumins and globulins. Several reports have shown that a high amount of N fertiliser, especially when applied at late stages of plant growth, results in an increased gliadin to glutenin ratio (although the change in this ratio may not be consistent for certain genotypes) and an increased ratio of HMW to LMW subunits, but no change in the ratio of x-type: y-type HMW subunits (Pechanek et al. 1997). Whereas the proportions of the various groups of gluten protein may be affected by N fertiliser, the polypeptide composition within each group proved to be constant with respect to growing conditions, so that gliadin or glutenin patterns can be used for fingerprinting genotypes using techniques such as electrophoresis or RP-HPLC. The exception to this general rule relates to the increased synthesis of sulphur-poor prolamins (*i.e.* ω-gliadins) when sulphur is deficient (Wrigley et al. 1984; Fig. 12).

In recent years, interest in the effects of environmental factors has been stimulated by studies of heat stress during grain filling in wheat. High temperatures (>32 °C) for as few as 1 or 2 days during grain filling were shown to result in decreases in the protein quality and dough properties (Randall and Moss 1990), with a change in the protein composition generally resulting in a decreased dough strength (Ciaffi et al. 1995). Blumenthal et al. (1998) recently reviewed the main hypotheses that have been advanced to account for the observed changes
1. A decrease in the ratio of glutenin: gliadin results from gliadin synthesis continuing during heat stress whereas glutenin synthesis is greatly decreased. This effect was explained by the presence of heat-stress elements (HSE) in the upstream regions of some gliadin genes, but not in the published sequences of glutenin genes.
2. A decrease of the size of glutenin polymers in the mature grain under heat stress, resulting in weakening of the resulting dough.
3. The synthesis of a M_r 70 000 heat-shock protein (HSP 70) as a reaction to heat stress, resulting (if still present in the mature grain) in a loss of dough quality.

In fact, the most recent investigations of Blumenthal et al. (1998) showed that the amount of HSP 70 in mature grain was not correlated with most indicators of dough strength, while incorporation of purified HSP 70 into dough showed no dramatic effect on dough properties. Also, sequencing the upstream regions of HMW subunit genes failed to show the presence of heat-shock promoters even in widely different genotypes. Consequently, research is presently focused on the degree of

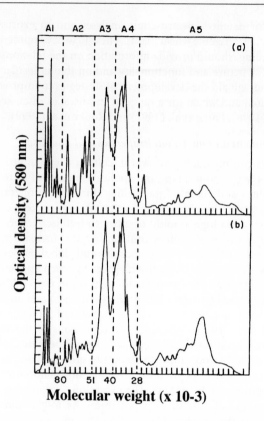

Fig. 12 : Densitometer scans of SDS gel separations of polypeptides extracted with SDS + 2-mercaptoethanol from sulphur deficient flour (a 0.083% S, 1.84% N) and from normal flour (b 0.161 S, 1.96% N). The peaks in particular regions of the densitometer scans are divided into five groups: *A1* (HMW subunits of glutenin); *A2* (mainly ω-gliadins); *A3* (mainly LMW subunits of glutenin); *A4* (mainly α, β- and γ-gliadins); *A5* (albumins and globulins). (Wrigley et al. 1984)

polymerization of the glutenin chains and on the roles of HSP and chaperones in the developing grain (Blumenthal et al. 1998). This is because it is considered likely that HSP modifies the folding and aggregation of gluten proteins *in situ* during grain filling, especially during stress situations, thereby altering their dough-forming potential. HSPs have been implicated in such processes in other organisms.

Legume Seed Proteins

Legume seeds constitute the basic protein source in the diets of many developing countries. In developed countries they are used mainly as protein-rich food in intensive animal production, but they are also of importance (especially in the case of soybean products that are the most important in trade) as meat substitutes or functional agents in the food industry. Because it is not possible to grow soybean in many parts of the world, other leguminous species have been studied as vegetable protein sources (Gueguen 1983). For example, in Europe many studies have been carried out on faba bean, pea and lupin in order to provide alternatives to meat by developing meat-like foods and to develop novel protein-rich foods as a complement to cereals. More recently, an interest has developed in using legume proteins for non-food markets. Various oilseeds have also been used to produce protein for processing. However, in oilseeds the yield and oil content rather than the protein properties have determined the choice of species.

This section will be concerned with the factors affecting the functional properties of legume storage proteins, with special reference to those of soybean, although most aspects may apply to proteins of other seeds such as pea, lupin or cruciferous oilseeds. Three principal aspects will be considered; processing, functionality and the possibilities available to manipulate functional properties.

Processing Legume Seed Proteins

Unlike many plant materials destined for manufacturing into foods, legume seeds are very rarely available in a form that is immediately usable by the food industry. Various types of processes are mandatory or desirable to purify the protein constituents and to transform them into ingredients suitable for the food processor. These processes may vary from one species to another, but all have implications with regard to the subsequent use of the material.

Refining Processes

An important constraint of refining is its cost, which favours simple processes with few steps, low energy consumption and a stable supply of raw material in large tonnages (the latter being readily fulfilled by seed proteins). However, processes may be mandatory, such as the removal of antinutritional factors (trypsin inhibitors, phytohaemagglutinin, goitrogen, saponin), desirable (removal of indigestible sugars) or optional (protein isolation, fractionation or specific modification) (Lillford 1981). Historically, a variety of procedures have been used to eliminate toxic substances and antinutrients. The processing steps generally included dehulling, boiling or cooking, grinding, toasting, puffing and fermentation (Deshpande and Damodaran 1990). Refinement processes can also be classified as dry (mechanical separation, air-classification) or wet (solvent extraction and washing, precipitation by pH adjustment, centrifugation). A typical dry process, pin-milling of legume seeds, leads to flours containing two populations of particles differentiated by both size and density. Using pin-milling, protein bodies can be detached from the surface of the starch granules so that, after air-classification, the heavy or coarse fraction (the starch fraction) can be separated from the light or fine fraction (the protein concentrate). However, only partial fractionation of protein and starch can be achieved, as even after repeated pin-milling and air-classification of pea flour, the lightest fraction still contains 8% starch in addition to 60% protein (this fraction is also enriched in lipids and ash) whereas the heavest fraction contains 5% protein in addition to 78% starch (Gueguen 1983).

Wet processes are recommended in order to prepare highly purified protein fractions (Fig. 13). For example, to prepare protein concentrates from defatted soy flour, the protein is generally immobilized by a choice of several treatments to enable removal of soluble sugars by washing with aqueous alcohol or dilute acid. However, these treatments are likely to leave the functionality of the proteins somewhat impaired (Wright and Bumstead 1984). On the other hand, isoelectric precipitation

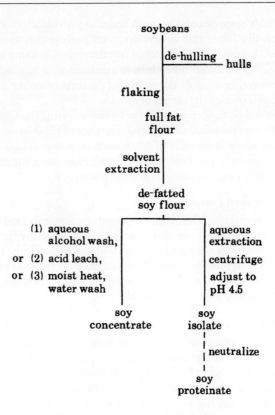

soybeans

de-hulling ———— hulls

flaking

full fat
flour

solvent
extraction

de-fatted
soy flour

(1) aqueous   aqueous
    alcohol wash,   extraction

or (2) acid leach,   centrifuge

or (3) moist heat,   adjust to
    water wash   pH 4.5

soy   soy
concentrate   isolate

neutralize

soy
proteinate

Fig. 13 : Processing of oilseeds (soybean). (Wright and Bumstead 1984)

at pH 4.5 is generally used to prepare protein isolates. However, even using mild conditions, precipitation can lead to loss of solubility and final neutralisation of the protein isolate before drying can pose a problem in subsequent processing and formulation because of residual salt. Following application of such processes to a defatted soy flour (containing 56% protein, 33% carbohydrates), typical protein concentrates can contain about 70% protein and 17% carbohydrates while protein isolates can contain as much as 95% protein with less than 1% carbohydrates.

Texturization of Seed Proteins:

As noted by Giddey (1983), the procedure used in the manufacture of concentrates or isolates is appropriate only if most of the protein remains in a native state. For example, concentrates obtained by a water/alcohol leaching process generally do not meet this requirement as partly denatured proteins cannot participate in the same way in stabilizing the artificial structures that are required for texturized end-products.

The primary task of texturization is to orientate the individual protein molecules so as to confer, after setting, a directional structure and thus an anisotropic resistance to the food. The texturization mechanisms therefore consist of protein insolubilization, including thermal (reversible or non-reversible) coagulation and denaturation.

Numerous patented processes of texturization have been described, which were divided by Giddey (1983) into nine classes: wet spinning, cooking extrusion, gel texturization, tear texturization, melt spinning, solvent texturization, texturization by surface deformation, texturization by freezing and biological texturization.

Processing and Functionality

Processes aim to confer desirable functional properties. These, in turn, relate mainly to conferring greater interfacial properties compared with most low-molecular-weight surfactants and emulsifiers, resulting in greater ability to stabilize emulsions or foams, and to impart textural attributes (Gueguen and Popineau 1998). Because of the complexity of food systems, information on how a protein will behave and on the relationships between microstructure and functional properties are extremely difficult to obtain, so that studies of functional properties generally start from simpler model systems before being extrapolated to mild processes and then to commercially available products. In addition, appropriate and relevant tools for functional characterization (water-binding capacity, hydrophobicity, charge and polarity, emulsifying and foaming characteristics, state of aggregation etc.) must be available and critically evaluated.

Functional properties are a manifestation of the inherent composition and structure of the protein, i.e. its amino acid composition, primary sequence and, finally, the organisation of the polypeptide chains and subunits in the native protein. For legume proteins, and specifically for soy proteins, the intrinsic properties of 7S and 11S globulins have been investigated in detail, including the amino acid compositions and sequences of subunits and their unfolding and association-dissociation upon heating, the latter using size separation and differential scanning calorimetry.

7S and 11S globulins have significantly different functional properties for emulsifying or gelation. Applying the same processing (e.g. to form tofu) to raw materials having different ratios of 7S and 11S globulins gives totally different product properties. For example, 7S globulins have higher emulsifying properties whereas gels made from 11S globulins are generally firmer with higher water holding capacities than their 7S analogues, the differences being ascribed to the contribution of disulphide bonds in the 11S globulins to the gel matrix. Thus, when a 7S protein solution is heated up to 100 °C at pH 7-8 and low ionic strength, the molecule dissociates into its three subunits, without further reaggregation, whereas reaggregation occurs if the protein solution is slightly concentrated (1%). As reviewed by Cheftel et al. (1985), the mechanism of denaturation during heating of 11S glycinin consists of the following steps:

1. The prevalence of hydrophobic interactions over electrostatic repulsion at the isoelectric point brings together the basic subunits,

2. Binding of the basic subunits together through disulphide bonds, leading to an oligomeric structure, followed by exchange of sulphydryl groups resulting in a release of the acid subunits which remain soluble,

3. Aggregation and precipitation of the basic oligomers as the result of hydrophobic interactions, leading to high M_r (10^7) through new exchanges of disulphide bonds, which may, in turn, initiate a three-dimensional protein network under specific conditions.

In 7S proteins (e.g. glycinin), the gelation mechanisms are different, as they involve more electrostatic interactions. The gelation mechanisms are also different and more complex upon heating of a solution containing both 7S (emulsion-enhancing) and 11S (gel-enhancing) protein fractions (Fig. 14).

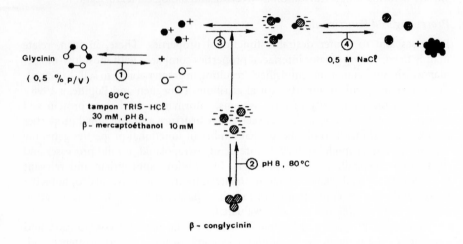

Glycinin

(0,5 % P/V)

80°C
tampon TRIS–HCℓ
30 mM , pH 8,
β– mercaptoéthanol 10 mM

0,5 M NaCℓ

② pH 8 , 80 °C

β – conglycinin

Fig. 14 : Hypothetical sequence of the dissociation-aggregation reactions during heating up to 80 ° C of a solution containing 11S glycinin + 7S β-conglycinin at pH 8.0.
○ acid (A) subunits of glycinin; ● basic (B) subunits of glycinin; ○ subunits of β-conglycinin. Glycinin ○–● is made of both A subunits ○ and B subunits ●. (Damodaran and Kinsella 1982)

In soybean the relationships between structural, chemical and functional properties have been largely substantiated. Because of the broad similarity with other seed proteins that have been studied, substitution of one protein for a related one generally presents few problems in food processing. However, large differences in the properties of 7S and 11S storage proteins will inevitably mean that the ratio of 7S and 11S proteins will play a significant role in determining the overall properties of the material (Wright 1983). In addition, other non-protein components present in the seeds may interact with proteins and thereby alter their basic properties. Finally, the structure/function relationships are based mainly on qualitative comparisons. According to Wright (1983), unless some quantitative relationships are established between functional and physicochemical properties, it will not be possible to predict with confidence the functional behaviour of any "protein X".

Manipulation of Functional Properties

Because the balance of 7S and 11S proteins is likely to play a significant role in determining the overall properties of the processed material, a number of ways to alter this ratio have been suggested. For example, this has been achieved during processing through the exploitation of intrinsic differences in the properties of the two protein types. For example, differences in the pH solubility profiles of the 7S and 11S globulins (Fig. 15) or differential effects of divalent cations on precipitation have been used in patented preparation procedures for 7S- and 11S-enriched fractions of legumes (Wright and Bumstead 1984). Technological approaches to quality improvement were also extended so as to include the elimination of unde-

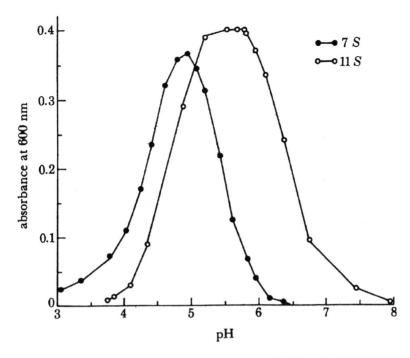

Fig. 15 : Effect of pH on precipitation of 7S and 11S soybean globulins in 0.06 M Tris-HCl buffer. (Wright and Bumstead 1984)

sirable components, including inactivation of trypsin inhibitors of soybean through heat treatments and high-shear extrusion to arrest lipoxygenase activity and thus eliminate "beany" flavours of soybean products that showed also have good shelf life against fat oxidation (Lusas 1998).

Another option for manipulating the ratio of 7S and 11S globulins is to exploit the natural variation in germplasm (*e.g.* the ratios vary from 0.6-2.0 in soybean, 0.7-5.0 in pea and 0.3-0.6 in faba bean) and, based on our knowledge of the inheritance of these proteins, to enhance the level of variation through breeding of high 7S or high 11S lines. Such a genetic approach has been successfully used to alter the ratios of 7S (emulsion-enhancing) and 11S (gel-enhancing) protein fractions. Soybeans with higher sulphur-amino acid contents (resulting from increased methionine) are also being developed (Lusas 1998).

A direct approach to modifying the functional properties is through the application of enzymic hydrolysis or chemical derivatization of specific residues to alter the structure or composition of the protein. Although the degree of hydrolysis is a highly critical parameter that must be carefully controlled to minimise the production of bitter-tasting peptides, peptide hydrolysates of soybean protein were obtained which had increased solubility and enhanced foaming and emulsifying properties (Wilde 1998).

Acknowledgement. IACR receives grant-aided support from the Biotechnology and Biological Sciences Research Council of the United Kingdom.

References

Albani D, Hammond-Kosack MCU, Smith C, Conlan S, Colot V, Holdsworth M Bevan MW (1997) The wheat transcriptional activator SPA: seed-specific bZIP protein that recognizes the GCN4-like motif in the bifactorial endosperm box of prolamin genes. Plant Cell 9: 171-184

Altenbach SB, Pearson KW, Meecker G, Staraci LC, Sun SSM (1989) Enhancement of the methionine content of seed proteins by the expression of a chimeric gene encoding a methionine-rich protein in transgenic plants. Plant Mol Biol 13: 513-522

Altenbach SB, Kuo C-C, Staraci LC, Pearson KW, Wainwright C, Georgescu A, Townsend J (1992) Accumulation of a Brazil nut albumin in seeds of transgenic canola results in enhanced levels of seed protein methionine. Plant Mol Biol 18: 235-245

Altpeter F, Vasil V, Srivastava V, Vasil K (1996) Integration and expression of the high-molecular-weight glutenin subunit *1Ax1* gene into wheat. Nat Biotechnoe 14 1155-1159

Alvarez ML, Guelman S, Halford NG, Lustig S, Reggiardo MI, Rybushkina N, Shewry PR, Stein J, Vallejos RH (2000) Silencing of HMW glutenins in transgenic wheat expressing extra HMW subunits. Theor Appl genet 100: 319-327

Andrews JL, Skerritt JH (1994) Quality-related epitopes of high Mr subunits of wheat glutenin. J Cell Sci 19: 219-230

Argos P, Narayana SVL, Nielsen NC (1985) Structural similarity between legumin and vicilin storage proteins from legumes. EMBO J 4: 1111-1117

Autran JC (1994) Size-exclusion high-performance liquid chromatography for rapid examination of size differences of wheat and cereal proteins. In: Kruger JE, Bietz JA (eds) HPLC of cereal and legume proteins. American Association of Cereal Chemists, St Paul, MN, pp 326-372

Autran JC, Hamer RJ, Plijter JJ Pogna NE (1997) To explore and improve the industrial use of wheats. Cereal Foods World 42: 216-227

Badley RA, Atkinson D, Hauser H, Oldani D, Green JP, Stubbs JM (1975) The structure, physical and chemical properties of the soybean protein glycinin. Biochim Biophys Acta 412: 214-228

Barro F, Rooke L, Békés F, Gras P, Tatham AS, Fido R, Lazzeri PA, Shewry PR, Barcelo P (1997) Transformation of wheat with high molecular weight subunit genes results in improved functional properties. Nat Biotechnol 15: 1295-1299

Bartels D,. Thompson RD (1986) Synthesis of messenger RNAs coding for abundant endosperm proteins during wheat grain development. Plant Sci 46: 117-125

Baud F, Pebay-Peyroula E, Cohen-Addad C, Odani S, Lehmann MS (1993) Crystal structure of a hydrophobic protein from soybean; a member of a new cysteine-rich family. J Mol Biol 231: 77-887

Bäumlein H, Wobus U, Pustell J, Kafatos FC (1986) The legumin gene family: structure of a B type gene of *Vicia faba* and a possible legumin gene specific regulatory element. Nucleic Acids Res 14: 2707-2720

Bäumlein H, Nagy I, Villaroel R, Inze D, Wobus U (1992) Cis-analysis of a seed protein gene promoter: The conservative RY repeat CATGCAT within the legumin box is essential for tissue-specific expression of a legumin gene. Plant J 2: 233-239

Blechl AE, Anderson OD (1996) Expression of a novel high-molecular-weight glutenin subunit gene in transgenic wheat. Nat Biotechnol 14: 875-879

Blumenthal C, Stone PJ, Gras PW, Békés F, Clarke B, Barlow EWR, Appels R, Wrigley CW (1998) Heat-shock protein 70 and dough-quality changes resulting from heat stress during grain filling in wheat. Cereal Chem 75: 43-50

Boston RS, Fontes EBP, Shank BB, Wrobel RL (1991) Increased expression of the maize immunoglobulin binding protein homologue b-70 in three zein regulatory mutants. Plant Cell 3: 497-505

Boulter D, Derbyshire E (1978) The general properties, classification and distribution of plant proteins. In: Norton G (ed) Plant proteins. Butterworths, London, pp 3-24

Breu V, Guerbette F, Kader JC, Kannangara CG, Svensson B, Von Wettstein-Knowles P. (1989) A 10 kD barley basic protein transfers phosphatidylcholine from liposomes to mitochondria. Carlsberg Res Commun 54: 81-84

Bright SWJ, Shewry PR (1983) Improvement of protein quality in cereals. CRC Crit Rev Plant Sci 1: 49-93

Brinch-Pedersen H, Galili G, Knudsen S, Holm PB (1996) Engineering of the aspartate family biosynthetic pathway in barley (Hordeum vulgare L.) by transformation with heterologous genes encoding feed-back-insensitive aspartate kinase and dihydrodipicolinate synthase. Plant Mol Biol 32: 611-620

Bushuk W, MacRitchie F (1989) Wheat proteins: aspects of structure that determine breadmaking quality. In: Dixon-Phillips R, Finley JW (eds) Protein quality and the effects of processing. Marcel Dekker, New York, pp 345-369

Bustos MM, Guiltinan MJ, Jordano J, Begum D, Kalkan FA, Hall TC (1989) Regulation of β-glucuronidase expression in transgenic tobacco by an A/T-rich, cis-acting sequence found upstream of a French bean β-phaseolin gene. Plant Cell 1: 839-853

Bustos MM, Begum D, Kalkan FA, Battraw MJ, Hall TC (1991) Positive and negative cis-acting DNA domains are required for spatial and temporal regulation of gene expression by a seed storage protein promoter. EMBO J 10: 1469-1479

Casey R (1999) Distribution and some properties of seed globulins. In: Shewry PR, Casey R (eds) Seed proteins. Kluwer Dordrecht, The Netherlands, pp 159-169

Casey R and Domoney C (1999) Pea globulins. In: Shewry PR, Casey R (eds) Seed proteins. Kluwer Dordre cht, The Netherlands, pp 171-208

Casey R, Domoney C, Ellis N (1986) Legume storage proteins and their genes. Oxford Surv Plant Mol Cell Biol 3:1-95

Casey R, Domoney C, Ellis N, Turner S (1988) The structure, expression and arrangement of legumin genes in pea. Biochem Physiol Pflanz 183: 173-180

Cheftel J.-C, Cuq J-L., Lorient D (1985) Proteines alimentaires.. Technique et Documentation, Lavoisier-APRIA, Paris, 309 pp

Ciaffi M, Margiotta B, Colaprico G, de Stefanis E, Sgrulletta D, Lafiandra D (1995) Effect of high temperatures during grain filling on the amount of insoluble proteins in durum wheat J Genet Breed 49: 285-296

Ciaffi M, Lee YK, Tamas L, Gupta R, Appels R (1998) Molecular analysis of low Mr glutenin genes in *Triticum durum*. In: Gueguen J, Popineau (eds) Plant proteins from European crops – food and non-food applications. Springer, Berlin Heidelberg New York, pp 58-63

Coleman CE, Larkins BA (1999) Prolamins of maize. In: Shewry PR, Casey R (eds) Seed proteins. Kluwer, Dordrecht, The Netherlands, pp 109-139.

Damidaux R, Autran JC, Grignac P, Feillet P (1978) Mise en évidence de relations applicables en sélection entre l'électrophorégramme des gliadines et les propriétés viscoélastiques du gluten de *Triticum durum* Desf. C R Acad Sci Paris Sér D 287: 701-704

Damodaran S, Kinsella JE (1982) Effects of conglycinin on the thermal aggregation of glycinin. J Agric Food Chem 30: 812-817

Danielsson CE (1949) Seed globulins of the Gramineae and Leguminoseae. Biochem J 44: 387-400

Das OP, Messing JW (1987) Allelic variation and differential expression at the 27-kilodalton zein locus in maize. Mol Cell Biol 7: 4490-4497

de Freitas FA, Yunes JA, da Silva MJ, Arruda P, Leite A (1994) Structural characterization and promoter activity analysis of the γ-kafirin gene family from sorghum. Mol Gen Genet 245: 177-186

Deshpande SS, Damodaran S (1990) Food legumes: Chemistry and technology. In: Pomeranz Y (ed) Advances in cereal science and technology, vol 10. American Association of Cereal Chemists, St Paul, Minnesota, pp 147-241

Dickinson CD, Hussein EHA, Nielsen NC (1989) Role of posttranslational cleavage in glycinin assembly. Plant Cell 1:459-469

Doll H (1984) Nutritional aspects of cereal proteins and approaches to overcome their deficiences. Philos Trans R Soc Ser B 304: 373-380

Domoney C, Casey R (1985) Measurement of gene number for seed storage proteins in Pisum. Nucleic Acids Res 13: 687-699

Domoney C, Ellis THN, Davies DR (1986) Organization and mapping of legumin genes in *Pisum*. Mol Gen Genet 202: 280-285

Dubreil L, Gaborit T, Bouchet B, Gallant DJ, Broekaert WF, Quillien L, Marion D (1998) Spatial and temporal distibution of the major isoforms of puroindolines (puroindoline-a and puroindoline-b) and non-pecific lipid transfer protein (ns-LTPle₁) of *Triticum aestivum* seeds. Relationships with their in vitro antifungal properties. Plant Sci 138: 121-135

Duffus CM, Cochrane MP (1992) Grain structure and composition. In: Shewry PR (ed) Barley: genetics, biochemistry, molecular biology and biotechnology. CAB International, Wallingford, pp 291-317

Dunwell JM (1998) Cupins: A new superfamily of functionally diverse proteins that include germins and plant storage proteins. Biotechnol Genet Eng Rev 15: 1-32

Egorov TA, Odintsova TI, Musolyamov A Kh, Fido R., Tatham AS, Shewry PR (1996) Disulphide structure of a sunflower seed albumin: conserved and variant disulphide bonds in the cereal prolamin superfamily. FEBS Lett 396: 285-288

Falco SC, Guida T, Locke M, Mauvais J, Sanders C, Ward RT, Webber P (1995) Transgenic canola and soybean seeds with increased lysine. Bio/Technology 13: 577-582

Field JM, Shewry PR, Miflin BJ (1983) Solubilisation and characterization of wheat gluten proteins: correlations between the amount of aggregates proteins and baking quality. J Sci Food Agric 34: 370-377

Fontes EBP, Shank BB, Wrobel RL, Moose SP, O'Brian GR, Wurtzel ET, Boston RS (1991) Characterization of an immunoglobulin binding protein homologue in the maize floury-2 endosperm mutant. Plant Cell 3: 483-496

Forde BG, Heyworth A, Pywell J, Kreis M (1985) Nucleotide sequence of a B1 hordein gene and the identification of possible upstream regulatory elements in endosperm storage protein genes from barley, wheat and maize. Nucleic Acids Res 13: 7327-7339

Gautier M-F, Aleman ME, Guirao A, Marion D, Joudrier P (1994) Triticum aestivum puroindolines, two basic cysteine-rich seed proteins: cDNA sequence analysis and developmental gene expression. Plant Mol Biol 25: 43-57

Genov N, Goshev I, Nikolova D, Georgieva DN, Filippi B, Svendsen I (1997) A novel thermostable inhibitor of trypsin and subtilisin from the seeds of Brassica nigra: amino acid sequence, inhibitory and spectroscopic properties and thermostability. Biochim Biophys Acta 1341: 157-164

Gibbs PEM, Strongin KB, McPherson A (1989) Evolution of legume seed storage proteins – domain common to legumins and vicilins is duplicated in vicilins. Mol Biol Evol 6: 614-623

Giddey C (1983) Phenomena involved in the "texturization" of vegetable proteins and various technological processes used. In: Bodwell CE, Petit L (eds) Plant proteins for human foods. Martinus Nijhoff/Dr W. Junk, The Hague, pp 221-233

Giese H, Hopp E (1984) Influence of nitrogen nutrition on the amount of hordein, protein Z and β-amylase messenger RNA in developing endosperms of barley. Carlsberg Res Commun 49: 365-383

Gincel E, Simorre J-P, Caille A, Marion D, Ptak M, Vovelle F (1994) Three-dimensional structure in solution of a wheat lipid-transfer protein from multidimensional ^{1}H-NMR data. A new folding for lipid carriers. Eur J Biochem 226: 413-422

Giroux MJ, Morris CJ (1998) Wheat grain hardness results from highly conserved mutations in the friabilin components of puroindoline-a and b. Proc Natl Acad Sci USA 95: 6262-6266

Gomar J, Petit MC, Sodano P, Sy D, Marion D, Kader JC, Vovelle F, Ptak M (1996) Solution structure and lipid binding of a nonspecific lipid transfer protein extracted from maize seeds. Protein Sci 5: 565-577

Gueguen J (1983) Legume seed protein extraction, processing, and end product characteristics. In: Bodwell CE, Petit L (eds) Plant proteins for human foods. Martinus Nijhoff/Dr W Junk, The Hague, pp 63-99

Gueguen J, Popineau Y (1998) Plant proteins from European crops – food and non-food applications. Springer, Berlin Heidelberg New York, 339 pp

Halford NG, Field JM, Blair H, Urwin P, Moore K, Robert L, Thompson R, Flavell RB Tatham AS, Shewry PR (1992) Analysis of HMW glutenin subunits encoded by chromosome 1A of bread wheat (*Triticum aestivum* L.) indicates quantitative effects on grain quality. Theor Appl Genet 83: 373-378

Hall TC, Chandrasekharan MB Guofu L (1999) Phaseolin, its past, properties, regulation and future. In: Shewry PR, Casey R (eds) Seed proteins. Kluwer, Dordrecht, The Netherlands, pp 209-240

Hammond-Kosack MCU, Holdsworth MJ, Bevan MW (1993) *In vivo* footprinting of a low molecular weight glutenin gene (LMWG-1D1) in wheat endosperm. EMBO J 12: 545-554

Hargreaves J, Popineau Y, Cornec M, Lefebvre J (1996) Relationships between aggregative, viscoelastic and molecular properties in gluten from genetic variants of bread wheat. Int J Biol Macromol 18: 69-75

Kasarda DD (1989) Glutenin structure in relation to wheat quality. In: Pomeranz Y (ed) Wheat is unique: structure, composition, processing, end-use properties, and products. American Association of Cereal Chemists, St Paul, Minnesota, pp 277-302

Kasarda DD, Autran J-C, Lew E.J-L, Nimmo CC, Shewry PR (1983) N-terminal amino acid sequences of ω-gliadins and ω-secalins: implications for the evolution of prolamin genes. Biochim Biphys Acta 747: 138-150

Kasarda DD, King G, Kumosinski TF (1994) Computer molecular modeling of HMW-glutenin subunits. In: Molecular modeling. ACS Symposium Series No 576, American Chemical Society, Washington, DC, pp 209-220

Keeler SJ, Maloney CL, Webber PY, Patterson C, Hirata LT, Falco SC, Rice JA (1997) Expression of *de novo* high-lysine α-helical coiled-coil proteins may significantly increase the accumulated levels of lysine in mature seeds of transgenic tobacco plants. Plant Mol Biol 34: 15-29

Kermode AR, Bewley JD (1999) Synthesis, processing and deposition of seed proteins: the pathway of protein synthesis and deposition in the cell. In: Shewry PR, Casey R (eds) Seed proteins. Kluwer, Dordrecht, The Netherlands, pp 807-841

Ko T-P, Ng JD, McPherson A (1993) The three-dimenstional structure of canavalin from jack bean (*Canavalia ensiformis*). Plant Physiol 101: 729-744

Kreis M, Shewry PR (1989) Unusual features of seed protein structure and evolution. Bio-Essay 10: 201-207

Kreis M, Forde BG, Rahman S, Miflin BJ, Shewry PR (1985) Molecular evolution of the seed storage proteins of barley, rye and wheat. J Mol Biol 183: 499-502

Lawrence MC (1999) Structural relationships of 7S and 11S globulins. In: Shewry PR, Casey R (eds) Seed proteins. Kluwer, Dordrecht, The Netherlands, pp 517-541

Lawrence MC, Suzuki E, Varghese JN, Davis PC, Van Donkelaar A, Tuloch PA. Colman PM (1990) The three-dimensional structure of the seed storage protein phaseolin at 3 Å resolution. EMBO J 9: 9-15

Lawrence MC, Izard T, Beuchat RJ, Blagrove M, Coleman P (1994) Structure of phaseolin at 2.2 Å resolution. Implications for a common vicilin/legumin structure and the genetic engineering of seed storage proteins. J Mol Biol 238: 748-776

Lelievre JM, Oliveira LO, Nielsen NC (1992) 5'-CATGCAT-3' elements modulate the expression of glycinin genes. Plant Physiol 98: 387-391

Levanony H, Rubin R, Altschuler Y, Galili G (1992) Evidence for a novel route of wheat storage proteins to vacuoles. J Cell Biol 119: 1117-1128

Lew E J-L, Kuzmicky DD, Kasarda DD (1992) Characterization of low-molecular-weight glutenin subunits by reversed-phase high-performance liquid chromatography, sodium dodecyl sulfate-polyacrylamide gel electrophoresis, and N-terminal amino acid sequencing. Cereal Chem 69: 508-515

Lillford PJ (1981) Extraction processes and their effect on protein functionality. In: Bodwell CE, Petit L (eds) Plant proteins for human food. Martinus Nijhoff/Dr W. Junk, The Hague, pp 199-205

Lohmer S, Maddaloni M, Motto M, Di Fonzo N, Hartings H, Salamini F, Thompson RD (1991) The maize regulatory locus Opaque-2 encodes a DNA binding protein which activates the transcription of the b-32 gene. EMBO J 10: 617-624

Lusas E (1998) Achievements, status and challenges in food protein processing. In: Guegen J, Popineau Y (eds) Plant proteins from European crops – food and non-food applications. Springer, Berlin Heidelberg New York, pp 257-264

MacRitchie F (1998) Protein composition and physical properties of wheat flour doughs. In: Guegen J, Popineau Y (eds) Plant proteins from European crops – food and non-food applications. Springer, Berlin Heidelberg New York, pp 113-119

Mertz ET, Bates LS, Nelson OE (1964) Mutant gene that changes protein composition and increases lysine content of maize endosperm. Science 145: 279-280

Miflin BJ, Field JM, Shewry PR (1983) Cereal storage proteins and their effect on technological properties. In: Daussant J, Mossé J,. Vaughan J (eds) Seed proteins. Phytochemical Society of Europe Symposia, Series n° 20. Academic Press, London, pp 255-319

Misra PS, Mertz ET, Glover DV (1975) Studies on corn proteins. VI. Endosperm protein changes in single and double endosperm mutants of maize. Cereal Chem 52: 161

Molvig L, Tabe LM, Eggum BO, Moore AE, Craig S, Spencer D, Higgins TJV (1997) Enhanced methionine levels and increased nutritive value of seeds of transgenic lupins (Lupinus angustifolius L.) expressing a sunflower seed albumin gene. Proc Natl Acad Sci USA 94: 8393-8398

Müller M, Knudsen S (1993) The nitrogen response of a barley C-hordein promoter is controlled by positive and negative regulation of the GCN4 and endosperm box. Plant J 4: 343-355

Müller S, Vensel WH, Kasarda DD, Köhler P Wieser H (1998) Disulphide bonds of adjacent cysteine residues in low molecular weight subunits of wheat glutenin. J Cereal Sci 27: 109-116

Munck L (1992) The case of high-lysine barley breeding. In: Shewry PR (ed) Barley genetics, biochemistry, molecular biology and biotechnology. CAB International, Wallingford, pp 573-601

Osborne TB (1907) The proteins of the wheat kernel. Carnegie Inst. Washington, Washington DC, Publ No 84

Osborne TB (1924) The vegetable proteins. Longmans, Green, London

Parker ML (1980) Protein body inclusion in developing wheat endosperm. Ann Bot 46: 29-36

Payne PI (1987) Genetics of wheat storage proteins and the effect of allelic variation on breadmaking quality. Annu Rev Plant Physiol 38: 141-153

Payne PI, Seekings JA (1996) Characterisation of Galahad-6, Galahad-7 and Galahad-8, isogenic lines that contain only one HMW glutenin subunit. In: Wrigley CW (ed) Gluten '96, Proc 6th Int Gluten Workshop, 1-6 September, Sydney, Australia, pp 14-17

Payne PI, Jackson EA, Holt LM, Law CN (1984) Genetic linkage between endosperm storage protein genes on each of the short arms of chromosomes 1A and 1B in wheat. Theor Appl Genet 67: 235-243

Payne PI, Nightingale MA, Krattiger AF, Holt LM (1987) The relationship between HMW glutenin subunit composition and the bread-making quality of British-grown wheat varieties. J Sci Food Agric 40: 51-65

Pechanek U, Karger A, Groger S, Charvat B, Schöggl G, Lelley T (1997) Effect of nitrogen fertilization on quantity of flour protein components, dough properties, and breadmaking quality of wheat. Cereal Chem 74: 800-805

Pogna N, Lafiandra D, Feillet P and Autran J-C (1988) Evidence for a direct causal effect of low molecular weight subunits of glutenins on durum viscoelasticity in durum wheats. J Cereal Sci, 7: 211-214

Popineau Y, Cornec M, Lefebvre J, Marchylo B (1993) Influence of HMW glutenin subunits on glutenin polymers and rheological properties of glutens and gluten subfractions of near-isogenic lines of wheat Sicco. J Cereal Sci 19: 231-241

Quayle T, Feix G (1992) Functional analysis of the -300 region of maize zein genes. Mol Gen Genet 23: 369-374

Randall PJ, Moss HJ (1990) Somme effects of temperature regime during grain filling on wheat quality. Aust J Agric Res 41: 603-617

Rico M, Bruix M, González C, Monsalve RI, Rodríguez R. (1996) ^{1}H NMR assignment and global fold of napin BnIb, a representative 2S albumin seed protein. Biochemistry 35: 15672-15682

Rooke L, Békés F, Fido R, Barro F, Gras P, Tatham AS, Barcelo P, Lazzeri P, Shewry PR (1999) Overexpression of a gluten protein in transgenic wheat results in highly elastic dough. J Cereal Sci 30: 115-120

Rubin R, Levanony H, Galili G (1992) Evidence for the presence of two different types of protein bodies in wheat endosperm. Plant Physiol 99: 718-724

Saalbach I, Waddell D, Pickardt T, Schieder O, Muntz K (1995) Stable expression of the sulphur-rich 2S albumin gene in transgenic *Vicia narbonensis* increases the methionine content of seeds. J Plant Physiol 145: 674-681

Salunkhe DK, Chavan JK, Kadam SS (1992) World oilseeds – chemistry, technology, and utilisation. Van Nostrand Reinhold, New York, 554 pp

Schmidt RJ, Burr FA, Burr B (1987) Transposon tagging and molecular analysis of the maize regulatory locus *Opaque-2*. Science 238: 960-963

Schmidt RJ, Ketudat M, Aukerman M, Hoschek G (1992) *Opaque-2* is a transcriptional activator that recognizes a specific target site in 22-kDa zein genes. Plant Cell 4: 689-700

Shen B, Carneiro N, Torres-Jerez I, Stevenson B, McCreery T, Helentjaris T, Baysdorfer C, Almira E, Ferl RJ, Habben JE, Larkins BA (1994) Partial sequencing and mapping of clones from two maize cDNA libraries. Plant Mol Biol 26: 1085-1101

Shewry PR (1995) Plant storage proteins. Biol Rev 70: 375-426

Shewry PR (1998) The HMW subunits and their role in determining the functional of wheat gluten and dough. In: Workshop on Biopolymer Science – Food and Non Food Applications, 28-30 September, Montpellier (France)

Shewry PR, Casey R (eds) (1999) Seed proteins. Kluwer, Dordrecht, The Netherlands

Shewry PR, Miflin BJ (1985) Seed storage proteins of economically important cereals. In:Pomeranz Y (ed) Advances in cereal science and technology, vol 7. American Association of Cereal Chemists, St Paul, Minnesota, USA, pp 1-83

Shewry PR, Pandya M J (1999) The 2S albumins storage proteins. In Shewry PR, . Casey R (eds) Seed proteins. Kluwer Dordrecht, The Netherlands, pp 563-586

Shewry PR, Tatham AS (1990) The prolamin storage proteins of cereal seeds: structure and evolution. Biochem J 267: 1-12

Shewry PR, Halford NG, Tatham AS (1994a) Analysis of wheat proteins that determine breadmaking quality. Food Sci Technol Today 8: 31-36

Shewry PR, Miles MJ, Tatham AS (1994b) The prolamin storage proteins of wheat and related species. Prog Biophys Mol Biol 61: 37-59

Shewry PR, Greenfield J, Buonocore F, Wellner N, Belton PS, Parchment O, Osguthorpe D, Tatham AS (1998) Conformational studies of the repetitive sequences of HMW subunits of wheat glutenin. In: Gueguen J, Popineau Y (eds) Plant proteins from European crops – food and non-food applications. Springer, Berlin Heidelberg New York, pp 120-125

Shewry PR, Tatham AS, Halford NG (1999) The prolamins of the Triticeae. In: Shewry PR, Casey R (eds) Seed proteins. Kluwer Dordrecht, The Netherlands, pp 35-78

Shewry PR, Tatham AS, Popineau Y (2000) The chemical basis of wheat grain quality. In: Autran JC, Hamer RJ (eds) Wheat Science and Technology in the European Union. IRTAC, Paris. In press

Shotwell MA, Larkins BA (1989) The biochemistry and molecular biology of seed storage proteins. In: Marcus A (ed) The biochemistry of plants, vol 5. Academic Press, San Diego, pp 297-345

Shutov AD, Kakhovskaya IA, Braun H, Bäumlein H, Muntz K (1995) Legumin-like and vicilin-like seed storage proteins: evidence for a common single-domain ancestral gene. J Mol Evol 41: 1057-1069

Sissons MJ, Békés F, Skerritt JH (1998) Isolation and functionality testing of low molecular weight glutenin subunits. Cereal Chem 75:30-36

Sörensen MB, Cameron-Mills V Brandt A (1989) Transcriptional and post-transcriptional regulation of gene expression in developing barley endosperm. Mol Gen Genet 217: 195-201

Spencer D (1984) The physiological role of storage proteins in seeds. Philos Trans R Soc Ser B 304: 275-285

Strobl S, Mühlhahn P, Bernstein R, Wilscheck R, Maskos K, Wunderlich M, Huber R, Glockhuber R, Holak TA (1995) Determination of the three-dimensional structure of the bifunctional α-amylase/trypsin inhibitor from ragi seeds by NMR spectroscopy. Biochemistry 34: 8281-8293

Svendsen I, Nicolova D, Goshev I, Genov N (1989) Isolation and characterization of a trypsin inhibitor from the seeds of kohlrabi (*Brassica napus* var. *Rapifera*) belonging to the napin family of storage proteins. Carlsberg Res Commun 54: 231-239

Svendsen I, Nicolova D, Goshev I, Genov N (1994) Primary structure, spectroscopic and inhibitory properties of a two-chain trypsin inhibitor from the seeds of charlock (*Sinapis arvensis* L), a member of the napin protein family. Int J Pept Protein Res 43: 425-430

Tatham AS, Shewry PR, Belton PS (1990) Structural studies of cereal prolamins, including wheat gluten. In: Pomeranz Y (ed) Advances in cereal science and technology, vol 10. American Association of Cereal Chemists, St Paul, Minn., USA, pp 1-78

Terras FRG, Goderis IJ, Van Leuven F, Vanderleyden J, Cammue BPA, Broekaert WF. (1992) *In vitro* antifungal activity of a radish (*Raphanus sativus* L.) seed protein homologous to nonspecific lipid transfer proteins. Plant Physiol 100: 1055-1058

Terras FRG, Schoofs HME, Thevissen K, Osborn RW, Vanderleyden J, Cammue BPA, Broekaert WF (1993) Synergistic enhancement of the antifungal activity of wheat and barley thionins by radish and oilseed rape 2S albumins and by barley trypsin inhibitors. Plant Physiol 103: 1311-1319

Thomas MS, Flavell RB (1990) Identification of an enhancer element for the endosperm-specific expression of high molecular weight glutenin. Plant Cell 2: 1171-1180

Utsumi S (1992) Plant food protein engineering. Adv Food Nutr Res 36: 82-208

Utsumi S, Gidamis AB, Mikami B, Kito M (1993) Crystallization and preliminary X-ray crystallographic analysis of the soybean proglycinin expressed in *Escherichia coli*. J Mol Biol 233: 177-178

Utsumi S, Gidamis AB, Takenaka Y, Maruyama N, Adachi M, Mikami B (1996) Crystallization and X-ray analysis of normal and modified recombinant soybean proglycinins: three-dimensional structure of normal proglycinin at 6 Å resolution. In: Parris N, Kato A, Creamer LK, Pearce J (eds) Macromolecular interactions in food technology. American Chemical Society, Washington, DC, pp 257-270

Weegels PL, Hamer RJ Schofield JD (1997) Depolymerisation and re-polymerisation of wheat glutenin during dough processing. II. Changes in composition. J Cereal Sci 25: 155-163

Wilde PI (1998) Modifying the interfacial behavior and functional characteristics of proteins. In: Guegen J, Popineau Y (eds) Plant proteins from European crops – food and non-food applications. Springer, Berlin Heidelberg New York, pp 105-112

Wilson DR, Larkins BA (1984). Zein gene organization in maize and related grasses. J Mol Evol 29: 330-340

Wright DJ (1983) Comparative physical and chemical aspects of vegetable protein functionality. Qual Plant Plant Foods Hum Nutr 32: 389-400

Wright DJ. Bumstead MR (1984) Legume proteins in food technology. Philos Trans R Soc B 304: 381-393

Wrigley CW (1995) Identification of food-grain varieties. American Association of Cereal Chemists, St Paul, Minnesota, 283 pp

Wrigley CW, du Cros DL, Fullington JG, Kasarda DD (1984) Changes in polypeptide composition and grain quality due to sulphur deficiency in wheat.J Cereal Sci 2:15-24

Xia J.-H, Kermode AR (1999) Analyses to determine the role of embryo immaturity in dormancy maintenance of yellow-cedar (*Chamaecyparis nootkatensis*) seeds: synthesis and accumulation of storage proteins and proteins implicated in desiccation tolerance. J Exp Bot 50: 107-118

Youle RJ, Huang AHC (1981) Occurrence of low molecular weight and high cysteine containing albumin storage proteins in oilseeds of diverse species. Am J Bot 68: 44-48

Nitrogen, Plant Growth and Crop Yield

David W. LAWLOR[1], Gilles LEMAIRE[2] and François GASTAL[2]

Nitrogen Use in World Agriculture

Nitrogen, Population and Food Production

The use of fertilisers in agriculture and horticulture is the key to production of sufficient food (including the fodder for animals) to maintain the global human population (currently 6 billion; Evans 1998) and to permit its continued rapid growth to the expected 10 or even 12 billion (Bumb 1995). Nitrogen in a form which can be used by plants is essential to crop production, and application of N fertilisers, produced industrially by chemical reduction of atmospheric (gaseous) nitrogen, has enabled the enormous and unprecedented expansion of the world's human population and the food supply (Bacon 1995; Evans 1998). The increased nitrogen supply is probably a consequence of population driven technological advances, in a complex interaction which is poorly understood (Evans 1998). Phosphate and potassium are also essential elements whose supply may not be sustainable in the long-term, as they are mined. However, the role of N in crop production is a critical aspect of crop production. Understanding the mechanisms by which crops respond to nitrogen is the key to maintaining and improving crop growth and yield, and the efficiency with which N is used and other resources also (Sinclair and Horie 1989; Bock and Hergert 1991; Lawlor et al. 1989; Grindlay 1997). This review analyses crop responses to N supply and integrates our knowledge of subcellular, cellular and organ processes to clarify the needs for N by plants and problems of quantification.

Global Aspects of Crop Production and N Fertilisers

To meet the dietary needs of the human population requires corresponding agricultural production. The scale of current global food production may be gauged by the three major grain cereals: wheat (600 Mt), maize (580 Mt) and rice (520 Mt), a total of 1700 Mt (Khush and Peng 1996). A commensurate fertiliser supply is

1. Biochemistry and Physiology Department, AFRC Institute of Arable Crops Research, Rothamsted Experimental Station, Harpenden, Herts AL5 2JQ, UK. *E-mail:* David.Lawlor@bbsrc.ac.uk
2. Station d'Écologie des Plantes Fourragères, INRA, 86600 Lusignan, Fr.
E-mail: lemaire@lusignan.inra.fr et gastal@lusignan.inra.fr

required. A simple calculation indicates just how much N is needed: with 2% N in grain dry matter (i.e. 0.02 Mt per Mt dry matter) and 1% in the non-grain component, which may be of equal mass to the grain, some 51 Mt of N is needed for annual grain production. This is the N removed from the soil by the crop. Grain is consumed and provides protein N. The N in straw may reenter the soil if it is ploughed in, but straw has value as fodder in much of world agriculture, even if not in developed agriculture. With 75% efficiency of removal of N by the crop compared to that applied, the total N required is about 68 Mt for the three staple crops alone. However, relatively little N is applied to cereals in subsistence farming or in areas prone to drought, for example, so part of this N is derived from the soil, rainfall etc. Crops differ in their N content and, therefore, in N requirement. Leaf vegetables generally have larger N content than cereals and root vegetables smaller contents. Although the total production of these crops is much smaller than the staple cereals, they constitute a substantial demand for N. Grasslands in developed agriculture do receive N fertilisers (Bélanger et al. 1992a + b), although rangelands in much of world agriculture do not. In developed agriculture, amounts of N fertiliser applied per unit land area are generally very large, although dependent on the type of crop. Vegetables, for example, often receive over 500 kg ha^{-1}, but even arable crops are heavily fertilised. Maize in the USA in the late 1980s received an average of 150 kg N ha^{-1} (Follett et al. 1991; Bumb 1995) and winter wheat in the UK 180 kg ha^{-1} N at the peak of N use (Addiscott *et al* 1991). In addition, considerable applications of N fertilisers are also made to crops in developing economies, e.g. tea in Sri Lanka (J. Mahotti, pers. comm).

For much of human history, N supply limited crop production, and still does in much subsistence and low-input farming. Shifting agriculture, fallowing, use of legumes (with symbiotic N$_2$ fixation: also the role of blue-green algae and *Azolla* in symbiosis in rice cultivation should be noted) and application of animal manures supplied the demand in part (Evans 1998). Later, concentrated forms of N-containing fertilisers of natural origin (e.g. guano) provided some of the N required for crops in the industrial economies. Since the beginning of the 20th century, the use of chemically fixed N has increased massively in importance. Between 1900 and 1960 N-fertiliser use increased from 1 to 10 Mt, then to 15 Mt in 1966 and to 50 Mt in 1977, reaching 80 Mt in 1990 (Bock and Hergert 1991; Bumb 1995), and has helped to keep the human population in the main, adequately fed, although many people in the developing world are undernourished, particularly with respect to proteins. Fertiliser use increased from 1966 to 1977 by 70% in developed (industrial) agricultural countries but by 200% in developing (less industrial) countries. The lavish use of N in the 1970s was related to the relative decrease in the cost of fertiliser N compared to the economic benefit. After 1979/1980, total N fertiliser use either stabilised or increased slightly (North America) or decreased considerably (Western Europe). This was particularly so after 1989/1990 in Eastern Europe and the former Soviet Union, largely associated with social changes and economic decline, and in Asia following economic recession (Bumb 1995). The increased cost of fertiliser, combined with substantial overproduction of agricultural products and decreased prices, reduced applications of N per unit land area and in total. There are other reasons for the changes in N use. In more developed economies, the suspicion that nitrate and nitrite might adversely affect human health (Follett et al. 1991; Bacon 1995) and the realisation that inappropriate use of large amounts of N fertilisers was contributing to environmental pollution (Addiscott *et al.* 1991) was a factor. Indeed, 80 Mt of industrial N is 70-80% of total terrestrial biological N fixation and has implications for altering ecosystem processes and the biosphere as a whole, (e.g. N pollution by runoff and drainage affecting surface, ground-waters and estuaries; Bacon 1995). Emphasis has shifted to increasing the efficiency with which N is used and to avoiding waste, with its attendant environmental problems. To achieve these goals, better use of our current understanding of the role of N in crops and, in particular, how the processes interact is required (Greenwood et al. 1991). Also, the effects of environmental conditions on the processes which determine its use by plants are crucially important.

N Supply and Crop N Demand

A complex chain of events determines the uptake of N, its use in plant production and the efficiency with which it is used. N is rapidly and efficiently absorbed by plants when their root systems come into contact with the required forms (nitrate ions, NO_3^-, and ammonium ions, NH_4^+), even in the micromolar range of concentrations in the soil solution (Chaps. 1.1 and 2.2). An extensive, yet dense, root system enables efficient absorption of the amounts of N required by the plant if the external concentrations are adequate. However, under some conditions such as cold or very dry or waterlogged soils, absorption is much less efficient. Also, if very large amounts of N are applied to soils, even if crops have well-developed root systems and efficient mechanisms for absorption, the plant may become saturated. Efficiency of N absorption, as well as of N utilisation, therefore, decreases with increasing amount of N applied. Thus, availability of N to, and also the uptake of N by, plants in the field depend on many factors, so the efficiency with which N is taken up by plants varies (Bock and Hergert 1991).

Analysis of crop requirement for N is illuminating, showing the magnitude of the N fluxes. Consider a wheat crop grown with ample water and nutrients, free from pests and diseases in the cool, moist conditions of northwest maritime Europe (at Rothamsted, Harpenden, UK, D.W. Lawlor and JE Leach, unpubl.). To achieve a grain yield of 10 t dry matter ha^{-1} and straw production of 10 t ha^{-1} [i.e. a ratio of grain to straw biomass or harvest index (HI) of 50%] with an average of 1.5% N in dry matter, the total N needed is 300 kg ha^{-1}. A grass crop, used as fodder for animals for meat and milk production in feed lots, may produce 20 t dry matter ha^{-1} containing 4% N in a similar environment. Thus, 800 kg N ha^{-1} may be removed from the soil (and is then largely emitted as animal waste, which has to be utilised). Incidentally, this illustrates the demands placed on N from plants by animal production. The N is supplied from the soil and, particularly, from fertiliser applications. Where N supply is less than the amount of N removed (e.g. the average UK application of 180 kg N ha^{-1} to wheat is less than the calculated amount of N removed), the deficiency may be supplied by N from the soil. This depends on factors such as N release from the previous crop, soil organic matter and soil water content, temperature and N in rainfall (which may give some 50 kg N ha^{-1} $year^{-1}$ in industrial areas) and in irrigation water. However, loss of N also occurs, e.g. excess rain causes leaching, warmth may increase gaseous losses etc. The supply of soil N is, therefore, very variable (Addiscott *et al.* 1991). For a crop growing on soil with a small N content (e.g. with little organic matter, or after removal of N by a large crop), or where applied N is lost (e.g. by leaching and gaseous emissions from soils and perhaps from plants), correspondingly more N must be supplied if potential yields are to be obtained.

The amount of N taken up by a crop and the efficiency with which is used in crop production (e.g. defined as yield/N supplied) is not a constant, but changes with environmental factors and N supply. With no (or very small) N fertiliser application, uptake of N is small but efficiency large compared to large N applications. For example, maize in the USA in 1950 (when very little N was applied) removed from soils about 2.5 times more N than was applied. By 1988, only 40-50% of the N applied was removed in the crop. The remainder was left in the soil or returned as crop residues and N-use efficiency decreased by a factor of 4 (Follett et al. 1991). Excess N may remain in the soil for later absorption, although this increases the chance of leaching and breakdown, with release of gaseous forms of N which are potential pollutants (Bacon 1995). There are differences in the requirements for N and in the efficiency of N use by different crops; although the same qualitative mechanisms operate, the quantitative relationships are much harder to establish. It is these relationships which determine the economic returns from N applications to crops and also the risks for environment. Thus they are of the greatest importance.

The Biochemical and Physiological Role of N in Crops

In plants, N is principally required for the synthesis of proteins, both structural and enzymatic. Enzymes are responsible for the synthesis not only of other proteins (life is an autocatalytic process!) but all metabolic intermediates, components of cellular structure and storage including carbohydrates, fats and pigments. These varied compounds constitute the plant body and are required for the growth of cells and organs, including the production of yield components (Lemaire et al. 1992; Lawlor 1995). Proteins, and particularly some of the amino acids they contain, are dietary imperatives for humans and other animals which are unable to synthesise them. Growth of crops, and the accumulation of proteins, carbohydrates, lipids etc. is determined by the characteristics of the organs of the plant and, therefore, of the cellular and subcellular processes which constitute that organ (Evans 1983; Jeuffroy and Meynard 1997). Hence, we address the cellular aspects as the key to understanding crop N interactions.

N in Cells and Organs

Proteins are present as enzymatic and structural storage components in cell walls and vacuoles but by far the greatest concentrations are in organelles, such as nuclei, mitochondria and chloroplasts, and in the cytosol. Active metabolism is associated generally with large numbers and often amounts of different proteins. Enzymes may differ greatly in activity per unit of protein (specific activity), so the relation between amount (mass) of a protein and its role in metabolism depends on the type of protein. There is large variation in the composition of different types of cells within a plant and from different plants. Cells which store carbohydrates, or are support and water-transport cells (xylem, for example), have little or no protein or other components containing N, but have a large proportion of cellulose, lignin etc. Therefore, their N content as a proportion of dry matter is low. In contrast, metabolically active cells, such as meristems, which are growing and dividing, and photosynthetic cells, contain considerable quantities of proteins, pigments etc. and relatively little cellulose or storage carbohydrates, so they have a large N content per unit dry matter (Evans 1983; Lawlor et al. 1988). Large amounts and concentrations of N may also occur in cells which store protein (e.g. in seeds of many species: see Chap. 6) although they may have low metabolic activity. Transport proteins of many types move ions, electrons and metabolites between compartments and cells, and may predominate in organs such as roots, but as they are relatively efficient only a small mass of these proteins may be involved. Organs which store starch, e.g. many root tubers and grains, may also have rather small concentrations of N and low rates of metabolism when mature. The proportion of different types of cells in an organ determines its composition. There may be great heterogeneity in an organ. In seeds, for example, the small embryo contains cells with rapid (potential) metabolic activity and large N concentrations, and others (e.g. in the endosperm) which store protein and starch. The link between biochemical composition and function of an organ helps to explain much about the nutrient composition and also the nutrient requirements of the plant (Evans 1983; Lawlor et al. 1989). We have examined this for leaves, focussing on those of cereals.

Composition of the Photosynthetic System

The main function of the mature, active leaves of higher plants is the assimilation of CO_2 by photosynthesis, producing the large mass of carbohydrates, particularly sucrose, which is exported to the rest of the plant. Synthesis of carbohydrates is a function of the gross photosynthetic rate of the leaf minus respiration, both dark- and photorespiration, which are intimately involved in the utilisation of assimilates. Leaves also reduce NO_3^- and synthesise amino acids, which are used in protein synthesis (see Chaps. 1.1, 2.1, 4.1). The complex metabolism involved in all these processes requires a large number of different enzymes, pigments and proteins (see Lawlor 1993).

The components of the photosynthetic system consist of the light reactions, the CO_2 and NO_3^- assimilating machinery, and the mechanisms for transport of assimilates and for synthesis of the photosynthetic components (Evans 1989; Lawlor 1993). The light-harvesting structures of photosynthesis, which capture the energy of photons, are the chlorophyll-protein complexes (LHCP). Electron transport components transfer the electrons to electron acceptors (ferredoxin and NADPH) which ultimately are consumed in the assimilation of CO_2 and NO_3^-. ATP is also required for CO_2 and NO_3^- assimilation and many cellular processes and is synthesised by ATP synthase (coupling factor, CF) in chloroplasts and mitochondria. ATP and NADPH are used by the photosynthetic carbon reduction (Calvin) cycle to synthesise ribulose bisphosphate (RuBP) which is the substrate for CO_2 assimilation catalysed by RuBP carboxylase-oxygenase (Rubisco). In leaves, Rubisco is the most important protein in terms of mass. It is a large (550 kDa) protein of eight large and eight small subunits, with eight CO_2 assimilation sites per molecule. The enzymes from different species of higher plants have similar properties (Makino et al. 1988) and are particularly slow in catalysis, with a turnover rate of 2-3 s per site. In the current atmospheric CO_2 and O_2 concentrations, the carboxylase activity of the enzyme is not saturated with CO_2 and there is substantial oxygenase activity, which is responsible for the production of phosphoglycolate. This is metabolised with the release of CO_2 as photorespiration, which decreases the efficiency with which Rubisco assimilates CO_2. Thus, the total amount of assimilated carbon, as sucrose and starch, which is available to the plant, is less than potentially available from the gross photosynthetic rate. Photorespiration is exacerbated by low CO_2 concentrations and low CO_2/O_2 ratio and by warmer temperatures. To achieve large rates of CO_2 assimilation in the current atmosphere requires a substantial investment of N in the light-harvesting system and the enzymes of the Calvin cycle, and especially in Rubisco (Evans and Seemann 1984, 1989). About 40-50% of the soluble protein of the leaf is in Rubisco alone and another 25% in the light harvesting and electron transport components. That corresponds to about 25% of total leaf N in Rubisco (Fig. 1). Chloroplasts, where Rubisco and the other photosynthetic components are located, contain a large proportion of the total leaf N. With ample N supply, the flag leaf of wheat contains about 2.5 g N m^{-2}, some 14 g m^{-2} total soluble protein (tsp) and 7 g m^{-2} of Rubisco protein. This corresponds to about 3.5-4% of N in dry matter (Lawlor et al. 1989). In an actively growing crop, the amount of N as amino acids, nitrate and nitrite and other small molecules is generally small relative to proteins, pigments, nucleic acids and other macromolecules. However, with very large N supply, the concentrations of such small molecules may increase considerably (Addiscott *et al.* 1991).

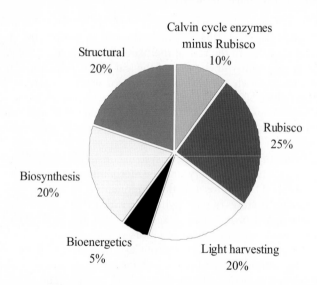

Fig. 1 : The percentage of total N in C_3 leaves in the main groups of N-containing components. (After Evans and Seemann 1989)

Chloroplast Numbers and Sizes with N Supply

The effects of N supply on the chloroplasts during leaf growth have been analysed frequently (Kutik et al. 1995). Ample N increases the numbers of chloroplasts per mesophyll cell and their cross-sectional area and length; compared with deficient N, the chloroplasts have fewer starch granules (because assimilates are used in growth more effectively) and slightly more thylakoid membrane (both stromal and granal) but less stromal volume. Also the density of protein (predominantly Rubisco) in the stroma is greater with large N supply (Kutik et al. 1995). A somewhat simplified view of the impact of N on chloroplasts is that increased availability of substrate for protein synthesis allows the development of more and larger chloroplasts, with a more extensive thylakoid system and, particularly, larger stromal volume containing more Rubisco and other proteins associated with CO_2 assimilation.

Leaf Composition

The composition of leaves varies with developmental stage and environmental conditions, and hence, in the field, with the season. In northwestern Europe, crops sown in autumn or early spring develop under cool, moist conditions, in dim light and short days. Leaves grown under such conditions with ample N have large protein content and relatively little cell wall, support tissue or carbohydrate. Later insertions of leaves, developed and functioning under warmer, brighter and drier conditions have relatively more structural and storage carbohydrate and so a smaller proportion of N in dry matter (Greenwood et al. 1990; Lemaire and Gastal 1997), although they may have larger protein and N content per unit area. The small N% is, in part, a consequence of the basis for expression of N content (Leigh and Johnston 1987) and the changes in proportions of cellular and organ components. The developmental factors determining and regulating leaf formation are still rather poorly understood, as is the role of environment (e.g. temperature and water supply) in the processes determining the composition. Clearly, light quality and intensity have major effects on leaf growth and development via regulation of gene expression and production of assimilates (Evans 1989). Leaves of cereals have the greatest concentrations of N-containing components at full expansion and are then most active photosynthetically. After a period, the amounts of components decrease and also metabolic activity as senescence progresses (Lawlor et al. 1988, 1989). Remobilisation of the components occurs: amino acids from proteolysis are transported to younger developing leaves, reproductive organs etc. (see Chap. 4.2).

Pigments are also broken down, often in parallel with the loss of proteins but not always, e.g. some stay-green mutants of plants lose protein from leaves but retain chlorophyll, as enzymes which are required for pigment breakdown are absent (Thomas and Howarth, 2000). The rate of leaf senescence is under genetic control (although modified by environment), so general slowing of senescence may prolong the period of photosynthesis and improve yield. Another consequence of ageing and senescence is that older leaves in the lower canopy, where light intensity and quality are unfavourable for rapid photosynthesis, contribute N to the growth of leaves at the top of the canopy (Lemaire et al. 1991). These develop with a greater content of metabolic machinery as a consequence, thus tending to optimise photosynthesis in relation to N supply and light (Werger and Hirose 1991; Lemaire et al. 1991). In cereals, some three-quarters of the N required for grain development and growth is from the vegetative organs, particularly leaves, so Rubisco serves as a store of N.

Rubisco as N Store

Although the role of Rubisco in crops is primarily to assimilate CO_2, it is often underutilised and a storage function has long been mooted (Millard 1988; Lawlor et al. 1988, 1989). Remobilisation of protein and utilisation of the liberated amino acids for synthesis of new proteins is particularly important for the synthesis of storage proteins in grains of cereals (see Chap. 6). There is a logic to this, which is often overlooked. Wheat originated from areas with moisture, and thus soil nitrate, available in winter and early spring when vegetative growth takes place. Use of the excess N to make Rubisco increases the capacity for photosynthetic CO_2 fixation and storage of N. Remobilisation of N for grain growth occurs during the dry conditions under which the crop matures, when water stress decreases N availability and also inhibits photosynthesis. Under such conditions, the use of Rubisco as a storage protein would be of the greatest importance to grain growth and production. Indeed, under such conditions, decreasing the amount of Rubisco, as has been suggested to increase N use efficiency and decrease the amount of N required for crop growth, would probably be a retrograde step. The advantage to crops of the ability to accumulate excess N in the form of an enzyme with potentially useful function (e.g. in periods of very bright light or when CO_2 concentration is large within the canopy) has been addressed (Lemaire and Millard 1999). Accumulation of nitrate ions and synthesis of amino acids might allow the crop to store N very rapidly, at a low energy cost (see Chaps. 2.2 and 4.1). However, the need to regulate ionic and osmotic concentrations might increase the energy costs, compared to the synthesis of proteins and particularly Rubisco, especially when the return on energy invested in terms of additional photosynthesis and availability of N are considered. Such detailed calculations of the mass and energy balances of N metabolism of crops in the long term have not been made. In most cases we may regard it as having a biologically sensible dual function.

Photosynthetic Rate, Leaf Composition and N

From the preceding discussion of the composition of leaves, it is clear why leaves contain a large concentration of Rubisco and other components and, hence, a large concentration of N. Particularly large contents are necessary in bright light in the current atmospheric CO_2 concentration (Evans, 1983, Evans and Seeman 1984; Lawlor 1993). There is a strong relationship between N supply, the concentration of N and the amounts of components per unit leaf area and the rate of CO_2 assimilation (Fig. 2), as shown in many studies (Evans 1983; Makino et al. 1988; Evans and Seemann 1989; Lawlor et al. 1989; Nakano et al. 1997; Theobald et al. 1998). This correlation applies in the range of deficient to sufficient N supply. However, with larger N supply, the content of components, particularly Rubisco, may increase, often substantially, without stimulating the photosynthetic rate, although in much brighter light and high CO_2 concentrations, greater amounts of components are required for large rates of photosynthesis (Nakano et al. 1997; Theobald et al. 1998). The N-use efficiency for carbon assimilation (mol C assimilated/mol N used) is generally greatest with deficient N and smallest with abundant N (Evans and Seemann 1989). The relationship between the components of the photosynthetic system may change over the range of N content, reflecting adaptation of the photosynthetic system. With abundant N compared to deficient N supply, there is more Rubisco relative to other leaf components; 26% compared to 21% in a field study with winter wheat in the UK (Lawlor et al. 1989). As the N of spring wheat leaves increased from 0.5 to 2.5 g m^{-2}, Rubisco increased some 30-fold but chlorophyll and ATP synthase only increased 4-fold and the chlorophyll a/b ratio was unchanged (Fig. 2; Theobald et al. 1998). This supports the role of Rubisco as a

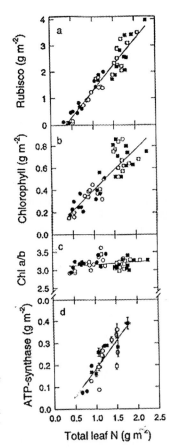

Fig. 2 : Amounts of ribulose bisphosphate carboxylase-oxygenase and ATP synthase protein and chlorophyll m^{-2} leaf of spring wheat and the chlorophyll a/b ratio as a function of N content m^{-2} leaf for spring wheat. Plants were grown in controlled environments with deficient (*circles*) or sufficient (*squares*) N applications and at 36 Pa CO$_2$ (*open symbols*) or 70 Pa CO$_2$ (*closed symbols*). (After Theobald et al. 1998)

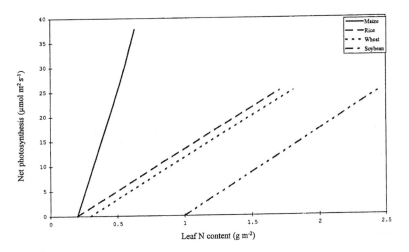

Fig. 3 : Relation between net photosynthetic carbon dioxide assimilation and N content of leaves in maize (C$_4$ plant), and rice, wheat and soybean (after various authors)

protein/N store mentioned earlier. Under the expected future increase in atmospheric CO_2, which will tend to saturate Rubisco, C_3 plants, such as wheat and rice, may have very large excess of carboxylation capacity. Indeed, Rubisco carboxylation capacity may exceed that for ATP synthesis, particularly in leaves in relatively low light with large N contents. This suggests that plant breeding or genetic engineering might be used to decrease the amount of Rubisco, thus decreasing the requirement for N fertilisers (Nakano et al. 1997) as there is little evidence of natural acclimation due to high CO_2, at least in cereals (Theobald et al. 1998). However, the risk of such a reduction in Rubisco content could be a limitation of N storage capacity in leaves (see above).

This discussion has centred on C_3 plants. Those with C_4 photosynthesis have different response of photosynthesis to N (Sage et al. 1987). C_4 plants assimilate CO_2 initially by the phosphoenol-pyruvate carboxylase (PEPC) reaction which has a high affinity for HCO_3^- (which is in equilibrium with atmospheric CO_2) and no oxygenase reaction (see Lawlor 1993). The oxaloacetic acid produced is converted to malate or aspartate (depending on the species) which is transported to the bundle sheath, where Rubisco is located, and CO_2 released there. The PEPC reaction acts as a mechanism to increases the concentration of CO_2 around Rubisco, thereby substantially decreasing the oxygenase reaction and the rate of photorespiration. Consequently, C_4 plants have faster rates of photosynthesis in the current atmosphere than C_3, particularly under warm conditions in bright light. Also, C_4 plants contain less Rubisco protein per unit area (although more of other enzymes) and thus have less N per unit leaf area and are more efficient users of N (Sage et al. 1987). However, both C_3 and C_4 plants require N and their rates of photosynthesis respond similarly to changing N supply with an approximately hyperbolic relationship, although the rates of photosynthesis for a given N content in C_4 leaves are greater than in C_3 (Fig. 3).

Respiration and N

Respiratory loss of assimilated carbon is often regarded as wasteful, decreasing net photosynthetic CO_2 assimilation, despite the central role of respiration in all aspects of cellular metabolism, including the synthesis of ATP and the production of metabolic intermediates (Ferrario et al. 1997). In leaves with low N content, respiration may be a larger proportion of CO_2 assimilation than in leaves with high N content, although the effect on overall carbon balance is probably small. However, considering the whole crop, the relative inefficiency of assimilate production per unit of N with an ample N supply may be exacerbated by the greater respiration rates. The relation between respiration and photosynthesis is also very dependent upon other environmental conditions. Effects of temperature on respiration, for example, are an important consideration for the balance between assimilate production and utilization. Respiration has a Q_{10} of about 2 but for photosynthesis it is generally smaller, so that at warmer temperatures relatively more respiration occurs than assimilation. The relationship depends on the species, with C_3 plants having a lower threshold than C_4 plants. Also, if there is a large proportion of non-photosynthetic, respiring tissue, then the demand for assimilates will increase relative to assimilation. This may be the case for crops grown with ample N. Decreasing the amount of protein synthesised (e.g. Rubisco), to avoid accumulation of protein in excess of the needs of photosynthesis, should decrease respiration and CO_2 loss, slowing consumption of carbohydrates and allowing more energy for other processes in the plant and greater accumulation of storage (harvestable) products. However, the value of such optimisation has not been rigorously tested (Lemaire and Millard 1999) and, in any case, it is unlikely that the savings from reduced respiration would offset the losses resulting from decreased photosynthetic rate under fluctuating light, and the benefits of Rubisco as N store.

Plant Growth and Development

Nitrogen supply affects the growth of plants greatly, in terms not only of amount of biomass produced but the size and proportion of organs and their structure (Pearman et al. 1977; Sinclair and Horie 1989; Greenwood et al. 1991; Lemaire et al. 1992; Jeuffroy and Meynard 1997). Also plant development may be affected, in part because the growth and size of the plant may affect developmental regulation. Brief consideration of the way in which N affects organ growth and size is important, for these are essential features of crop production and greatly dependent on N supply.

Cell Division and Expansion

N supply has very marked effects on the growth of leaves, stems etc., so the processes determining their production must be affected. Regulation of cell division in meristems, cell development and expansion in primordia and young organs and growth of organs with respect to N supply are not well understood or quantified (Gastal and Nelson 1994). A certain minimum size and composition (proteins etc) of meristematic cell must be required for division, as well as for expansion. Cell expansion is related to (and possibly regulated by) wall proteins, which are integral components of the wall matrix, and responsible for the synthesis and structure of the cellulose microfibrils and other components and their links. Cytoskeletal proteins may be important in division and growth. A large N supply increases the numbers of cells per leaf and their size, and decreases the proportion of structural protein but increases the synthesis of non-structural proteins in larger organelles. In old wheat leaves with N deficiency, 3 g m^{-2} of protein could be regarded as structural, estimated from the residual N content, equivalent to about 60-70% of the maximum content of total soluble protein in younger N-deficient leaves. In contrast, old leaves grown with a very large N content have a residual protein which is about 20-30% greater but is only about 20% of the total soluble protein formed (Lawlor et al. 1989). This suggests preferential synthesis of structural protein associated with cell growth/expansion, with other cellular proteins made as N becomes more available. This would allow cells to expand, and thus leaf area to increase, even with N deficiency, but with a much smaller amount of metabolic protein per unit leaf area. This is consistent with the decrease in N and protein content per unit leaf area, as a consequence of N deficiency. It should be noted that if all proteins were synthesised in strict proportion with changing N, and if only cells with the full complement of components developed, then fewer cells would be made. Only the size of the leaf would change, not the amounts of components per unit leaf area. Clearly, this is not the case, as discussed earlier. There is considerable flexibility in the organisation and regulation of cell number and composition in response to N.

Organ Numbers

Abundant N supply increases the number of meristems produced by plants and their growth, thus encouraging shoot (branches and tillers) formation and growth in most plants (Lawlor et al. 1988, 1989). Arguably, this is the single most important process affected by N, for it is the way in which plants increase their competitive ability in vegetation and the number of fruits and seeds. In a growing crop, there is a hierarchy of responses to N supply. During early growth of the seedling in cereals, for example, the main stem and the first leaves and seminal root system utilises the N stored in the grain together with N from the soil. As additional N becomes available, primary tillers, with their leaves, grow from the meristems made in the axils of leaves on the main stem and coleoptile. Additional (secondary) tillers grow from the meristems in these primary tillers. If the supply of N is adequate, the plant forms large numbers of tillers and leaves (Pearman et al. 1977). The number of tillers formed in forage grasses and cereals is proportional to the N supply, reaching a maximum value (depending on the crop and cultivar), above which further N does not stimulate tillering, because light interception is saturated and because of the regulation of tiller appearance by light quality (red / far red ratio) (Deregibus et al. 1983). There are complex interactions between the light environment and N supply, which determine the structure of cereal and grass crops. In dense crop canopies, high N supply exacerbates competition for light (both total energy and spectral quality of the radiation) that determines whether new tillers are produced and die or survive (Simon and Lemaire 1987; Werger and Hirose 1991).

The increase in organs is a very important aspect of production of biomass and yield in all crops, including grasslands for grazing and grain production in cereals. More reproductive organs are formed with a large N supply, e.g. ear and grain, thus increasing yield potential. However, there is often a loss of size and quality associated with large numbers of seeds, fruits etc. (see earlier discussion on assimilate supply and demand).

Leaf Area

As N-supply has such a profound effect on the development of individual leaves, both on main and secondary branches and tillers (Pearman 1977; Lawlor et al. 1988, 1989; Jeuffroy and Meynard 1997), it is not surprising that the primary effect of N on crop production is via the formation of leaf area (Sinclair and Horie 1989; Van Keulen et al. 1989; Bélanger et al. 1992a). The area of individual laminae increases as the N supply increases, until a maximum size is reached (Lawlor et al. 1988, 1989; Grindlay 1997) which will depend on other environmental factors, e.g. water, temperature (Gastal et al. 1992). Production of both more and larger cells plays a role. With deficient N, a 50% decrease in area per leaf was related to about 40% decrease in number of cells per leaf and a 30% decrease in cell volume (Lawlor et al. 1989), suggesting greater sensitivity of meristematic cell formation than of cell expansion.

Leaf Area Index and Light Interception

As the consequence of an increasing N supply, LAI (leaf area/ground area) increases substantially (Fig. 4; Bélanger et al. 1992a; Lawlor 1995; Lemaire and Millard 1999). This may be seen for a wheat crop sown in autumn and growing slowly through the

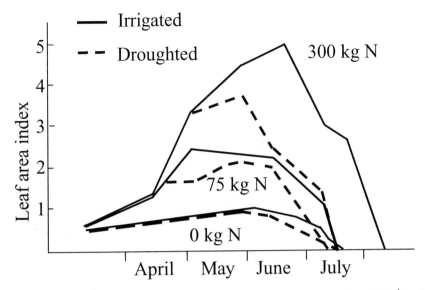

Fig. 4: Effects of three different total amounts of nitrogen fertiliser (0, 75 and 300 kg ha^{-1}) and of irrigation on the growth of leaf area in winter wheat over the season in the field at Rothamsted, UK

winter in north-temperate Europe. LAI increases slowly to perhaps 0.5 in early March and N-supply has little effect. As temperatures increase, so does LAI, peaking at 5-6 by mid-June. An even greater LAI can develop if crops receive ample water together with N, particularly under warm conditions. At an LAI of 6, the corresponding leaf dry matter is some 2.5 t ha^{-1} with 3.5% N, equivalent to about 88 kg N ha^{-1}, which must be met if maximum light interception and photosynthetic rates are to be achieved. With less N available, LAI is decreased, and on very N-deficient soils (after removal of N by a previous N-deficient crop, for example) and without N application it may not reach 1.

Solar radiation incident on the soil surface is intercepted by the leaves, which photosynthesise and form the assimilates used in growth. Interception of energy is proportional to LAI to a value of about 4, depending on crop architecture, e.g. erect leaves allow more radiation to reach lower leaves than lax (horizontal) leaves, thus increasing light interception at an equivalent LAI. As most of the incident radiation is intercepted by an LAI of about 4, there is little or no extra radiation intercepted by a greater LAI. Total above-ground dry mass (biomass) of a crop is proportional to total radiation intercepted over the growing season (Monteith 1977), with inadequate N decreasing the leaf area and shortening its duration, thus decreasing light interception and yield (Fig. 5). Also, a deficiency of N decreases the efficiency with which biomass is produced per unit of intercepted radiation (Table 1), as expected from the reduction in photosynthesis caused by N deficiency (Sinclair and Horie 1989; Gastal and Bélanger 1993; Lawlor 1995). As Table 1 shows, N deficiency also increases the proportion of assimilates used in root growth, leading to a decreasing shoot/root ratio. Generally, there is a strong relationship between yield and aboveground biomass, although this is not fixed; yield often forms a smaller fraction of the

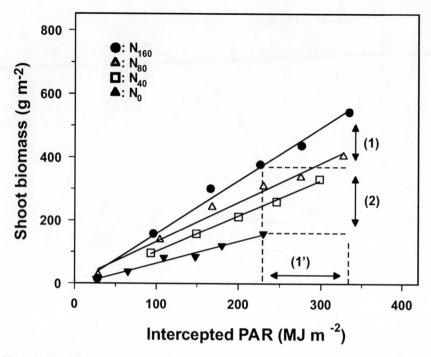

Fig. 5 : Relationship between the above-ground biomass accumulation and the quantity of photosynthetically active radiation (PAR) intercepted by a grass sward (tall fescue) related to the N applied during regrowth after cutting. *1* shows the difference in biomass due to effect of N on leaf area growth and *1'* the consequences for PAR interception. *2* shows the difference in biomass due to the effect of N on the radiation use efficiency. (After Bélanger et al. 1992a)

Table 1. Effect of N on the efficiency of canopy gross photosynthesis (E_p, g m^{-2} MJ^{-1}) and on the efficiency of the conversion of light to total biomass (E_t), and for conversion to above-ground biomass (E_a) for a tall fescue sward in spring after different applications of N

Nitrogen applied kg ha^{-1}	E_p	E_t	E_t/E_p	E_a	E_a/E_t
0	4.82	1.78	0.37	1.12	0.63
60	5.03	2.28	0.45	1.73	0.76
120	5.05	2.56	0.51	2.05	0.81
180	5.62	2.65	0.47	2.21	0.83

total dry matter (smaller harvest index) as biomass increases. This may be due to competition for assimilates between vegetative and yield bearing organs (see earlier discussion). Hence, it is important for the production of biomass and yield, that a LAI of 4 or somewhat greater should be formed but that excessive leaf area should be avoided because it has costs, e.g. for energy and N, which may detract from any additional yield. Nevertheless, such an extra LAI, even if it represents a cost, has to be also considered as beneficial in term of N storage to provide further N supply by remobilisation during grain filling, as demonstrated for maize (Plenet and Lemaire 1999) Thus, the N supply must be adequate to achieve the desired LAI and match the rates of development (determined by genotype, temperature etc.). If the N supply is inadequate, then LAI does not develop at the potential rate and light interception is decreased. Also, as discussed, the N deficiency slows the potential photosynthetic rate (decreasing photosynthetic efficiency per unit of radiation intercepted), so decreasing production (Fig. 5). Generally the change in LAI has a greater impact on total production and yield than the change in photosynthetic efficiency (Lawlor 1995).

Nitrogen Effects on Yield and Its Components

It is the ability of N to improve crop production, quality and the productivity of farming, that justifies interest in N and plants. Earlier sections have shown that the effects of N are achieved at all scales, from the biochemical to whole plant. For a crop, the total yield depends on the number of yield components and their size.

Yield Components

Yield of crops, total fresh or dry mass per unit land area, depends on the often repeated relationship:
number of plants ha^{-1} x number of yield bearing organs $plant^{-1}$ x mass yield-bearing $organ^{-1}$

In cereals, the number of tillers $plant^{-1}$ is generally greater than the number of ears formed: also the number of grain/ear and the dry weight (mass) $grain^{-1}$ are important components of yield. Planting density determines the number of plants $area^{-1}$ and the tillering of individual plants determines the number of potential ears, but a number of factors (light, developmental stage etc.) affect ear development and determine the number of ears produced. Within each ear a potential number of grains is formed which often greatly exceeds the number of grains finally established, possibly because of competition for assimilates between grain. Potential grain size is genetically determined but the actual size achieved is greatly influenced by environmental conditions, particularly those which affect the availability of carbon assimilates and protein production. Nitrogen supply affects all the components of grain yield, as discussed, via the effects on the number of tillers, ears and grains formed (Lawlor 1995; Grindlay 1997; Jeuffroy and Meynard 1997). The number of ears is largely determined during early growth, with grain number and potential size determined later via carbon and nitrogen assimilate supply, which, of course, greatly affect the final grain size. However, N applied early to cereal crops stimulates vegetative growth relatively more than the formation of reproductive structures. So numbers of grain, although massively increased by nitrogen, are affected less than tillering, which produces stem and leaf growth, often without ear formation in the case of late or low-ranking tillers. Thus, harvest index decreases as N supply increases. Overproduction of straw in much industrial agriculture is unwanted (but not in less developed agriculture, where it is valuable as fodder): there are problems of disposal and it is an inefficient use of N.

There are important interactions between environmental conditions and plant processes, which affect the production and quality of grain. For example, an increased N supply increases photosynthesis unit^{-1} leaf area and of the total crop (Evans and Seemann 1989; Sage et al. 1987), as we have seen, and this may increase the mass grain^{-1} but decrease the N%. In some cases this is desirable, in malting barley for example, but for feed and bread-making wheats, the opposite may be true. Inadequate N may cause shrunken grain, but large amounts of N may also have this effect by stimulating vegetative growth, causing excessive water use, drought stress and haying-off. Thus, there is need to manipulate the supply of N to the plant relative to carbon assimilation at critical times, such as grain filling, and with regard to other environmental conditions to achieve the desired quality characteristics. This is not simply a question of changing the fertiliser application at particular times. To achieve a low N% in cereals, for example, decreasing N results in remobilisation of proteins from leaves to grain and also decreases photosynthesis and carbohydrate production and yet may have little effect on the N% of grain. Genetic characteristics of the plant then become important (Chap. 6), with the rate and extent of remobilisation of photosynthetic components affecting grain quality. In all optimisations, there is need to balance the production of total yield and quality (grain size, N%) which are major economic considerations. Poor understanding of the factors regulating the processes which determine N content is hampering improvements in N-use efficiency for total production and in attainment of quality goals and their optimisation.

Other environmental factors may substantially alter the balance between processes within the plant, affecting quality and yield, and substantially altering the N responses and requirements. Temperature has a major impact, as already mentioned. Warm conditions increase the rate of development but decrease the duration of periods of growth, generally advancing maturity, senescence and ripening. Difference in the behaviour of crops in different years is due (in part) to the alterations in the balance between biochemical and physiological processes driven by temperature and the supply of N. For example, early growth of cereals under warm, wet conditions with abundant N results in considerable vegetative growth which competes for assimilates during grain filling and may result in small mass per grain with relatively large N%, but yet large numbers of grain. If later grain filling takes place during hot dry conditions, then the rapid loss of leaf area may substantially decrease grain size and production but with relatively large N%; both scenarios give a low harvest index. Water supply also affects plant growth, vegetative and reproductive processes, N availability etc. and thus affects crop dry matter production and yield. Nitrogen supply has a dominant effect on yield production because agriculture is an extractive process, with removal of a crop also removing N from the soil, so very marked deficits can show quickly. Other factors, e.g. rainfall, are more variable and therefore, less easily established. However, when N supply can be regulated with relative ease and deficits can be avoided, the impact of other factors cannot be ignored.

Dry Matter and Yield

Despite the very large differences in dry matter production and yield caused by N supply and the variability introduced by other environmental conditions (Lawlor 1995), there is a relatively uniform response of dry matter and yield production to total N accumulated in crops, such as winter wheat (Fig. 6). This shows an expo-

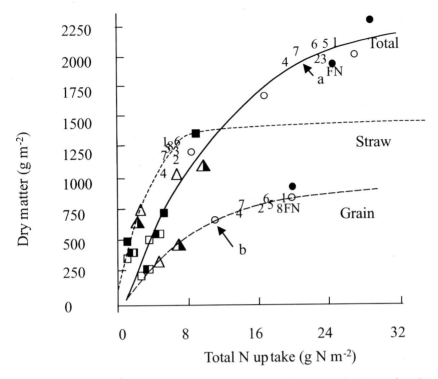

Fig. 6 : Dry matter (g m^{-2}) in total above-ground crop and in the straw and grain as a function of the N uptake in each for winter wheat crops grown with different N applications, in different years, one with irrigation (D.W. Lawlor and J.E. Leach, unpubl)

nential response of straw, grain and total dry matter to total N accumulated by a crop grown with different N amounts, with and without water deficits, in different years. Close adherence of yield to a single response indicates that the availability of N is the main factor determining production. As N uptake increases, production also rises but with abundant N, dry matter and grain production does not increase further: growth is N-saturated.

Dynamics and Potential of Crop Growth and Efficiency of N Use

Understanding the dynamics of growth and N supply is crucial in achieving maximum production in crops (van Keulen et al. 1989; Grindlay 1997; Lemaire and Gastal 1997; Lemaire et al. 1997). It is necessary to consider the potential total demand of the crop for N (kg N ha^{-1}) and the rate of demand (kg N ha^{-1} day^{-1}). Demand must then be considered relative to the supply of N, which is also affected by the environment but in a different way, so the interaction between the processes is very dynamic and variable. However, for field crops, general relationships between N supply and plant yield (N response curves) are apparent. The finer responses, which determine quality of yield, are still not clarified.

Potential Crop Growth and N

Potential crop growth rate, defined as the rate of dry matter production over a given period (Lawlor 1991), is a useful concept to understand the requirement for, and plant responses to N. A similar concept is the potential total production over a period, e.g. growing season. Potential growth rate and duration of growth depend on the genetic composition of the plant interacting with environmental conditions. They are expressed under ideal environmental conditions where all constraints are removed. Potential growth rate cannot be determined *a priori* but may be approximated (but with some difficulty) from experiments with combinations of conditions – temperature, light, water, nutrients etc. In the field, the potential is measured with temperatures and radiation of that location and season as the maximum rate of dry matter accumulation that the crop will achieve with inputs (such as fertilisers, water, control of pests and diseases) optimised, i.e. ideally neither limiting nor in excess. In practice, the potential growth of the crop and its demand for N is assessed over a range of N supply which includes the maximum production (Lemaire and Millard 1999). The N required to maintain the potential growth rate (N requirement) over a period is given by the potential crop growth rate and the N concentration in dry matter. Total N demand then depends on crop size, which is related to the duration of growth. Potential growth rate and, therefore, N requirement, are temperature-dependent, warmer temperatures than the optimum increasing and cooler slowing them. However, the duration of growth is shorter under warm and longer under cool conditions, so the effects on total production and demand for N may be rather similar. Compensating changes in processes determining crop production are frequent, but often not completely, matched.

If the supply of N from soil and storage in the plant is less than needed to sustain the potential growth rate, for the biochemical and physiological reasons already discussed, then N becomes limiting and the actual growth rate is smaller than the potential. Therefore, additional N fertilisers are needed, but applications may not be appropriate in quantity or timing to meet a demand (Follet et al. 1991). Under agricultural conditions, very large amounts may be applied before the onset of rapid growth (to anticipate the demand and avoid restricting crop responses). However, this can be very inefficient: the storage capacity of the plant is limited and the rates of protein synthesis and growth processes are relatively slow, so the fertiliser is left in the soil and may be lost (Addiscott et al. 1991). This problem may be greatly exacerbated if the weather cools, slowing growth. When rapid growth resumes, the crop may have an inadequate supply of N to ensure growth at the potential rate. Dry soils may also decrease the rate of N supply to the plant. As N application is increased and the frequency decreases (and water supply, temperature etc. become less favourable) the ratio of N uptake/N supply decreases greatly. Applications of small amounts of fertiliser at intervals related to weather conditions substantially reduce the losses and provide the plant with adequate N for growth. Thus, it is likely that the current efficiency of N use in the field may be greatly improved by adopting appropriate methods of supplying N to crops, with frequent, small amounts applied taking into account N in the soil, weather conditions etc. With very small, frequent N applications the ratio of N uptake/N supply may approach 1, or even exceed it as N is removed from the soil. Fertiliser application in irrigation water (fertigation) is particularly efficient (D.W. Lawlor and J.E. Leach, unpubl.).

N Concentration and critical N concentration

As shown in Fig. 7, the total N content of the above-ground biomass increases with increasing biomass, but not linearly because the leaf/stem ratio of a crop (e.g. lucerne) decreases progressively as crop mass increases, and because stem N% is much lower than leaf N%. The ratio of stem to leaf increases as the crop grows; so a higher proportion of N accumulates in the stems despite the lower concentration of N.

The nitrogen content in crop dry matter (N%) when growth rate is optimal is regarded as a critical value, $N\%_{critical}$(Fig. 8), which may be used in agronomy to calculate fertiliser applications to achieve the best production (Gastal and Lemaire, 1997; Lemaire and Millard 1999). The $N\%_{critical}$ changes with age of the crop, proportion of different organs etc. However, $N\%_{critical}$ is relatively constant for a crop at a particular growth stage and independent of soil and environmental characteristics. It is very important to note here that the ontogenetic change in the proportion of different organs (leaf vs stem) (Fig 7 D) is strongly related to the size (mass) of the crop. So, the decrease of $N\%_{critical}$ with crop mass (Fig. 8) is very similar for many crops, despite the differences in morphology and crop structure (monocots vs dicots). However, there is substantial difference in $N\%_{critical}$ between C_3 and C_4 crops, explained by the greater efficiency of C_4 photosynthesis with N. From the general relationships determined under field conditions between $N\%_{critical}$ and biomass for a range of crops, fertiliser applications may be quantified. The method accounts for all sources of N available to the plant and for losses due to leaching etc. An index of nitrogen nutrition (INN) of a crop may be defined (Lemaire and Gastal 1997) as the ratio of observed N% ($N\%_{observed}$) to $N\%_{critical}$. This defines the deficiency of N for a particular situation (biomass, N supply etc) relative to the amount

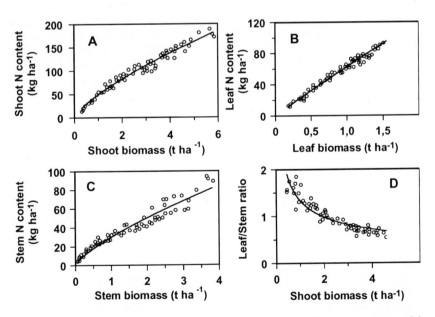

Fig. 7 A-D : Relation between the N accumulated in total biomass and its components, and the leaf to stem biomass ratio, for a lucerne crop. (After Lemaire et al. 1992)

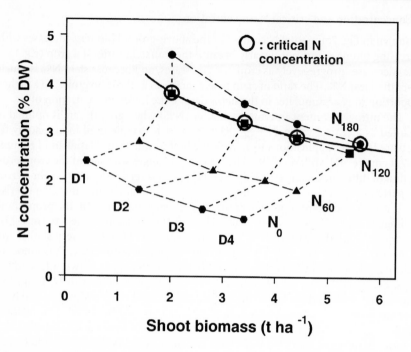

Fig. 8 : Relation between N and shoot biomass with time and N supply (0 to 180 kg N ha[-1]) during growth of a generalized crop.

of N required for maximum production under those conditions. These empirical relationships provide a practical application of value in agriculture, and are based on understanding of plant biochemistry, physiology and crop processes which, as yet are too complex and poorly quantified to be used directly for practical purposes.

Efficiency of N Use

Greatest efficiency of N use in biochemical processes occurs when N is severely limiting and plant growth and production are greatly impaired, because the luxury accumulation of N is stopped (Lawlor et al. 1988). In contrast, if N supply exceeds that required to maintain the potential growth rate, then no further dry matter is produced per unit additional N but more N may be accumulated in unproductive components. Thus, N-use efficiency is small with ample N and decreases progressively and substantially as the N supply increases. Efficiency may depend on the period over which the N is taken up and dry matter produced. If N is stored (see discussion on Rubisco function) this increases N/dry matter ratio and decreases the efficiency. However, later use of the N in grain filling will show an increased efficiency. Therefore, care should be taken not to overemphasise the importance of efficiency without considering the particular processes (see discussion on possible dangers of decreasing Rubisco content). Optimisation of N supply should aim to achieve maximum production per unit of N, linked to the highest N-use efficiency; this is shown in Fig. 6 by points a and b on the vegetative and yield curves, respectively. By combining an understanding of the mechanisms of response to N, with that of the factors determining potential growth rates and critical N contents, it should be possible to describe and quantify the relationship between plant production and N, with agronomic benefits. This will help to achieve the aim of much agronomic research and plant breeding (Khush and Peng 1996), to increase production (specifically economically useful yield) whilst decreasing the use of N.

It is difficult to achieve such optimisation in a complex system, i.e. one with many different inputs and processes affecting the output. The inputs are determined by factors (e.g. temperature, rainfall) beyond agronomic control (although technically and practically achievable in protected cultivation where the crop value is sufficient). Information such as in Fig. 6, and relationships of the N%$_{critical}$ and INN type in Fig. 8, (Lemaire and Gastal 1997) provide valuable *a posteriori*, and more dynamic, information for understanding the responses and modifying them. However, for practical purposes a more dynamic approach is required in which the application of N fertilisers etc. is based on the history of the crop and the desired yield production and quality (Sinclair and Horie 1989; van Keulen et al. 1989). As the future conditions cannot be known exactly, only predicted with a probability based on past experience, forecasting requirements for N and crop production will remain difficult. Indeed, existing fertiliser practice is based on expectation of future conditions and is, therefore, incorrect when those expectations are not met. However, by applying more detailed understanding of mechanisms, the errors in existing practise may be minimised.

Modelling Plant Responses to N

N%$_{critical}$ and INN are top-down approaches to analysing crop responses to N (Monteith 1977; Lemaire and Gastal 1997). A more 'bottom-up' approach is simulation modelling, based on knowledge of mechanisms and short-term responses (van Keulen et al. 1989). Simulation modelling provides a method of describing the dynamics of crop growth and N supply, allowing rapid application of fertilisers to ameliorate N deficiencies etc. Also, it avoids the need, inherent in the top-down approach, to assume fixed correlations between N supply and N in metabolic com-

Table 2. Value of the coefficients a and b of the relationship between critical plant N% (N%$_{critical}$) and crop mass (W in t/ha): N%$_{critical}$ = $a(W)^{-b}$ for different C$_3$ and C$_4$ species. Such a relationship allows the determination of the critical N uptake in kg ha^{-1}: $N = 10a(W)^{1-b}$ where (1-b) is the allometric coefficient between N accumulation rate and crop growth rate

Photosynthetic type	Species of plant (Reference)	Coefficient a	Coefficient b	Reference
C$_3$	Grasses	4.8	0.32	Lemaire and Salette (1984)
C$_3$	Lucerne	4.8	0.34	Lemaire et al. (1985)
C$_3$	Pea	5.1	0.32	Ney et al. (1997)
C$_3$	Wheat	5.3	0.44	Justes et al. (1994)
C$_3$	Oil seed rape	5.1	0.29	Colnenne et al. (1998)
C$_3$	Rice	5.2	0.41	Sheehy et al. (1998)
C$_4$	Maize	3.4	0.37	Plenet and Lemaire (1999)
C$_4$	Sorghum	3.9	0.39	Plenet and Cruz (1997)

ponents and of the leaf. Multiple sources of N exploited by the plant could be accounted for and the impacts of changing environmental conditions determined in real time, based on subprocesses (Van Keulen *et al.* 1989; Grindlay 1997). More efficient use (and decreased losses) of N fertilisers should result when the magnitude and dynamics of alternative sources of N are quantified and related to the dynamics of plant growth. However, mathematical simulation modelling of the mechanisms by which plants respond to N supply from first principals, e.g. from enzyme kinetics and biochemical pathways etc., is impossible with current knowledge. The reductionist approach provides the basis for a conceptual understanding of the system, and should be developed, but the quantitative interactions are still poorly evaluated. Until these are better characterised, assessments of crop-N relations must be empirical, using experimentally determined relationships such as N%$_{critical}$ and INN (Lemaire and Gastal 1997). By combining simulation and empirical models, practical methods for forecasting N needs for yield production may be improved. As a better understanding of quantitative relationships develops, it will become possible to integrate processes, from N supply through biochemistry and cell growth to total dry matter and yield production, using a combination of both basic and applied information.

References

Addiscott TM, Whitmore AP, Powlson DS (1991) Farming, fertilisers and the nitrate problem. CAB International, Wallingford

Bacon PE (1995) Nitrogen fertilization in the environment. Marcel Dekker, New York

Bélanger G, Gastal F, Lemaire G (1992a) Growth analysis of a tall fescue sward fertilized with different rates of nitrogen. Crop Sci 32: 1371-1376

Bélanger G, Gastal F, Warembourg FR (1992b) The effects of nitrogen fertilization and the growing season on carbon partitioning in a sward of tall fescue (Festuca arundinacea Schreb.) Ann Bot 70: 239-244

Bock BR, Hergert GW (1991) Fertilizer nitrogen management. In: Follett RF, Keeney DR, Cruse RM (eds) Managing nitrogen for groundwater quality and farm profitability. Soils Science Society of America, Madison, pp 139-164

Bumb BL (1995) World nitrogen supply and demand: an overview. In: Bacon PE (ed) Nitrogen fertilization in the environment. Marcel Dekker, New York, pp 1-40

Colnenne C, Meynard JM, Reau R, Justes E, Merrien A (1998) Determination of a critical nitrogen dilution curve for winter oilseed rape. Ann Bot 81:311-317

Deregibus V.A., Sanchez R.A., Casal J.J. (1983) Effects of light quality on tiller production in Lolium ssp. Plant Physiol, 72, 900-912

Evans JR (1983) Nitrogen and photosynthesis in the flag leaf of wheat (*Triticum aestivum L.*). Plant Physiol 72: 297-302

Evans JR (1989) Partitioning of nitrogen between and within leaves grown under different irradiances. Aust J Plant Physiol 16: 533-548

Evans JR, Seemann JR (1984) Differences between wheat genotypes in specific activity of ribulose-1,5-bisphosphate carboxylase and the relationship to photosynthesis. Plant Physiol 74: 759-765

Evans JR, Seemann JR (1989)The allocation of protein nitrogen in the photosynthetic apparatus : costs, consequences and control. In: Briggs W (ed) Photosynthesis. Alan R Liss, New York, pp 183-205

Evans LT (1998) Feeding the ten billion. Cambridge University Press, Cambridge

Faure J-D, Meyer C, Caboche M (1997) Assimilation du nitrate: nitrate et nitrite réductases. In: Morot-Gaudry J-F (ed) Assimilation de l'azote chez les plantes. INRA, Paris, pp 45-65

Ferrario S, Foyer, CH, Morot-Gaudry J-F (1997) Coordination entre métabolismes azoté, photosynthétique et respiratoire. In: Morot-Gaudry, J.-F. (ed) Assimilation de l'azote chez les plantes. INRA, Paris, pp 235-248

Follett RF, Keeney DR, Cruse RM (eds) (1991) Managing nitrogen for groundwater quality and farm profitability. Madison, Soils Science Society of America, Inc

Gastal F and Bélanger G 1993. The effects of nitrogen and the growing season on photosynthesis on field grown tall fescue canopies. Ann Bot 72, 401-408

Gastal F, Lemaire G (1997) Nutrition azotée et croissance des peulplements végétaux cultivés In: Morot-Gaudry, J-F (ed) Assimilation de l'azote chez les planes. INRA, Paris, pp 355-367

Gastal F, Nelson CJ (1994) Nitrogen use within the growing leaf blade of tall fescue. Plant Physiology, 105: 191-197

Gastal F, Bélanger G, Lemaire G (1992) A model of leaf extension rate of tall fescue in response to nitrogen and temperature. Ann Bot 70: 437-442

Greenwood DJ, Lemaire G, Gosse G, Cruz P, Draycott A, Neeteson JJ (1990) Decline in percentage N of C_3 and C_4 crops with increasing plant mass. Ann Bot 66: 425-436

Greenwood DJ, Gastal F, Lemaire G, Draycott A, Millard P, Neeteson JJ (1991) Growth rate and N% of field grown crops: theory and experiments. Ann Bot 67: 181-190

Grindlay DJC (1997) Towards an explanation of crop nitrogen demand based on the optimization of leaf nitrogen per unit leaf area. J Agric Sci, Cambridge 128: 377-396

Jeuffroy M-H, Meynard J-M (1997) Azote: production agricole et environnement. In: Morot-Gaudry J-F (ed) Assimilation de l'azote chez les plantes. INRA, Paris, pp 369-380

Justes E, Mary B, Meynard JM, Machet JM, Thelier-Huché L (1994) Determination of critical nitrogen dilution curves for winter wheat crops. Ann Bot 74: 397-407

Khush GS, Peng S (1996) Breaking the yield frontier of rice. In: Reynolds MP, Rajaram S, McNab A (eds) Increasing yield potential in wheat: breaking the barriers. CIMMYT, Mexico, pp 36-51

Kutik J, Natr L, Demmers-Derks HH, Lawlor DW (1995) Chloroplast ultrastructure of sugar beet (Beta vulgaris L.) cultivated in normal and elevated CO_2 concentrations with two contrasted nitrogen supplies. J Exp Bot 46: 1797-1802

Lawlor DW (1991) Concepts of nutrition in relation to cellular processes and environment. In: Porter JR, Lawlor DW (eds) Plant growth: interactions with nutrition and environment. Cambridge University Press, Cambridge, pp 1-32

Lawlor DW (1993) Photosynthesis: molecular, physiological and environmental processes. Longman Scientific and Technical, Harlow

Lawlor DW (1995) Photosynthesis, productivity and environment. J Exp Bot 46: 1449-1461

Lawlor DW, Boyle FA, Keys AJ, Kendall AC, Young AT (1988) Nitrate nutrition and temperature effects on wheat: a synthesis of plant growth and nitrogen uptake in relation to metabolic and physiological processes. J Exp Bot 39: 329-343

Lawlor DW, Kontturi M, Young AT (1989) Photosynthesis by flag leaves of wheat in relation to protein, ribulose bisphosphate carboxylase activity and nitrogen supply. J Exp Bot 40: 43-52

Leigh RA, Johnston AE (1987) The usefulness of expressing nitrogen concentrations in crops on the basis of tissue water. J Sci Food Agric 38: 317-318

Lemaire G, Gastal F (1997) N uptake and distribution in plant canopies. In: Lemaire G (ed) Diagnosis of the nutritional status in crops. Springer Berlin , Heidelberg New York, pp 3-34

Lemaire G, Millard P (1999) An ecophysiological approach to modelling resource fluxes in competing plants. J Exp Bot 50: 15-28.

Lemaire G, Salette J (1984) Relation entre dynamique de croissance et dynamique de prélèvement d'azote pour un peuplement des graminées fourragères. I Étude de l'effet du milieu. Agronomie 4:423-430

Lemaire G, Cruz P, Gosse G, Chartier M (1985) Etude des relations entre la dynamique de prélèvement d'azote et la dynamique de croissance en matière séche d'un peuplement de luzerne. Agronomie 5:685-692

Lemaire G, Onillon B, Gosse G, Chartier M, Allirand JM (1991) Nitrogen distribution within a lucerne canopy during regrowth: relation with light distribution. Ann Bot: 68 483-488

Lemaire G, Khaity M, Onillon B, Allirand JM, Chartier M, Gosse G (1992) Dynamics of accumulation and partitioning of N in leaves, stems and roots of lucerne (*Medicago sativa*) in a dense canopy. Ann Bot: 70 429-435

Lemaire G, Plenet D, Grindlay DJC (1997) Leaf N content as an indicator of crop N nutrition status. In: Lemaire G (ed) Diagnosis of the nutritional status in crops. Springer Berlin, Heidelberg New York, pp 189-199

Makino A, Mae T, Ohira K (1988) Differences between wheat and rice in the enzymic properties of ribulose-1,5-bisphosphate carboxylase/oxygenase and the relationship to photosynthetic gas exchange. Planta 174: 30-38

Millard P (1988) The accumulation and storage of nitrogen by herbaceous plants. Plant, Cell and Environment 11: 1-8

Monteith JL (1977) Climate and the efficiency of crop production in Britain. Philos Trans R Soc Lond B281: 277-294

Nakano H, Makino A, Mae T (1997) The effects of elevated partial pressures of CO_2 on the relationship between photosynthetic capacity and N content in rice leaves. Plant Physiol 115: 191-198

Ney B, Doré T, Sagan M (1997) The nitrogen requirements of major agricultural crops: 6. Grain legumes. In: Lemaire G (ed) Diagnosis of the nitrogen status in crops. Springer Berlin, Heidelberg new York, pp 107-118

Pearman I, Thomas SM, Thorne GN (1977) Effects of nitrogen fertilizer on growth and yield of spring wheat. Ann Bot 41: 93-108

Plenet D, Cruz P (1997) The nitrogen requirement of major agricultural crops: 5. Maize and sorghum. In: Lemaire G (ed) Diagnosis of the nitrogen status in crops. Springer Berlin, Heidelberg New York, pp 93-106

Plenet D, Lemaire G (1999) Relationships between dynamics of nitrogen uptake and dry matter accumulation in maize crops. Determination of critical N concentration. Plant Soil (in press) 216: 65-82

Sage RF, Pearcy RW, Seemann JR (1987) The nitrogen use efficiency of C_3 and C_4 plants. I. Leaf nitrogen, growth, and biomass partitioning in *Chenopodium album* (L.) and *Amaranthus retroflexus* (L.). Plant Physiol 84: 954-958

Sheehy J, Dionara MJA, Mitchell PL, Peng S, Cassman KG, Lemaire G, Williams RL (1998) Critical nitrogen concentrations: implications for high-yielding rice (*Oryza sativa* L) cultivars in the tropics. Field Crops Res 59:31-41

Simon JC, Lemaire G (1987) Tillering and leaf area index in grasses in vegetative phase. Grass Forage Sci 42: 373-380

Sinclair TR, Horie T (1989) Leaf nitrogen, photosynthesis, and crop radiation use efficiency: a review. Crop Sci 29: 90-98

Theobald JC, Mitchell RAC, Parry MAJ, Lawlor DW (1998) Estimating the excess investment in ribulose-1,5-bisphosphate carboxylase/oxygenase in leaves of spring wheat grown under elevated CO_2. Plant Physiol 118: 945-955

Thomas H and Howarth C.J. (2000) Five ways to stay green. Journal of Experimental Botany, Special Issue: Genetic Manipulation of Photosynthesis 51: 329-337

Van Keulen H, Goudriaan J, Seligman NG (1989) Modelling the effects of nitrogen on canopy development and crop growth. In: Russell G, Marshall B, Jarvis PG (eds) Plant canopies, their growth, form and function. Cambridge University Press, Cambridge, pp 83-104

Werger MJA, Hirose T (1991) Leaf nitrogen distribution and whole canopy photosynthetic carbon gain in herbaceous stands. Vegetatio 97: 11-20

Natural Genetic Variability in Nitrogen Metabolism

Anis M. LIMAMI[1] and Dominique de VIENNE[2]

Plant development is the result of various genetically controlled interacting metabolic processes, which are to a large extent modulated by environmental factors. In this context, the key role of absorption, distribution and accumulation of nutrient elements deserves the attention of the geneticist and the plant breeder, because these complex traits display a large variability between and within species, which may be important in terms of adaptation of the genotypes, or species, to various environments (Kjaer and Jensen 1995). Works dedicated to major nutritional variations were carried out to unravel reasons of both inefficiency in nutrient utilisation and efficiency of utilisation under nutrient stress. As early as 1943, Weiss studied the genetic variability of iron absorption efficiency in soybean. A susceptible genotype that developed the typical chlorosis of iron deficiency was crossed with a tolerant genotype. Inheritance studies showed that a single pair of alleles, *Fe/fe*, controlled the efficiency of Fe utilisation. The efficient allele, *Fe*, is dominant over *fe*, the *fefe* plants becoming chlorotic when submitted to Fe-limited conditions. Bell et al. (1958, 1962) demonstrated that the *Zea mays* mutant (*ys*1) discovered by Beadle (1929), exhibited yellow stripes between the main veins due to its inability to use iron (Fe^{3+}). Potassium utilisation under severe K deficiency was investigated by Shea et al. (1967, 1968) in various bean genotypes. Very efficient strains were discovered, the most efficient producing almost 50% more dry weight than the less efficient, when supplied with a 0.13 mM KCl nutrient solution. Efficient and inefficient strains were crossed and their progeny was evaluated for the capacity to grow under low K availability. The results suggested that this trait would be under the control of a single pair of alleles and that "efficiency" was recessive. Similar studies have been expanded to other species and other elements such as P, Ca and Mg. Differences in the use of Ca and Mg were explored in *Zea mays*, showing variations in yield among corn inbreds growing on soils known to induce Ca and Mg deficiencies

1. Unité Nutrition Azotée des Plantes, INRA Versailles, Route de Saint Cyr, 78026 Versailles Cedex, France. *E-mail:* limami@versailles.inra.fr
2. Station de Génétique Végétale, INRA/UPS/INA PG, Ferme du Moulon, 91190 Gif-sur-Yvette, France. *E-mail:* devienne@moulon.inra.fr

(Clark and Brown 1974; Clark 1978). Wide genotypic differences in yield were obtained between bean strains growing on low phosphate (2 mg P), and between tomato strains grown on low nitrogen (35 mg plant^{-1}) (O'Sullivan et al. 1974). Genetic studies coupled to these various investigations showed that the utilisation of major cations and anions could be under polygenic control, except for K in bean, that appeared to be controlled by a single pair of alleles.

In the present chapter we will focus on natural genetic variability in nitrogen use, because this element plays a key role in plant nutrition, and because the excess use of nitrogen fertilisers can cause damage to the environment. We will present examples to illustrate the genetic diversity in the control exerted by nitrogen on several traits, and we will present promising techniques of molecular quantitative genetics for studying these polygenic traits.

Variability in Nitrogen Use

Nitrogen, as an essential element in plant mineral nutrition, is known to affect various traits such as plant growth and architecture. The effect of nitrogen on plant development depends on the species considered, and within species it depends on the genotype under study. Interest in studying N-use efficiency by plants has increased recently as a result of the increasing risks of environment damage by fertilisers: a better understanding of the genetic variation of N uptake and utilisation is required in order to develop genotypes that use N more efficiently. Data on N-use efficiency are becoming available for major economic crops (wheat, maize, soybean, tomato, etc.). Comparison of N economy between wheat, maize and soybean showed that in soybean anthesis occurs after 25% of the total N has accumulated in the crop and half of the vegetative growth has been achieved, whereas in wheat and maize anthesis occurs when about 85 and 70% of the total N has accumulated. It thus appears that competition for N between vegetative and reproductive organs in soybean is higher than in wheat and maize (Cregan and Berkum 1984). Nitrogen harvest index (NHI), which is the proportion of the total N yield contained in the grain at maturity, is higher in soybean (72%) than in wheat (71%) and maize (63%). Nitrogen remobilisation efficiency (NRE), which acts as a measure of the ability of the crop to remove N from vegetative tissue, is higher in wheat (66%) than in soybean (51%) and maize (44%). A comparison of NHI and NRE suggests that in soybean the proportion of the assimilated N used immediately in the developing seed is larger than in wheat, thus avoiding the apparently inefficient process of incorporating N into leaf and stem protein before its subsequent remobilization and translocation.

Rajcan and Tollenaar (1999a, b) reported that differences between two maize hybrids in the efficiency of using nitrogen for filling the grain were entirely attributable to postsilking N uptake. The high-yielding genotype exhibited a larger yield response to N supply than a low-yielding genotype. A higher efficiency of N use for grain filling might be related to a relatively high source/sink ratio, that corresponds to the proportion of source organs sufficient for supplying the root system with assimilates to sustain root growth and N uptake. In favour of this hypothesis, the authors found differences in leaf longevity between the two genotypes; senescence of leaves occurred earlier in the low-yielding maize line, resulting in an earlier decrease in the source/sink ratio during the postsilking period. Furthermore, they have shown that N taken up during postsilking is preferentially allocated to the reproductive organs.

Cox et al. (1986) found that differences in the percentage of grain N between various wheat genotypes (Atlas-derived lines), could be attributed to differences in the efficiency of N translocation from foliage to grain. However, several other workers were unable to relate grain protein content to N translocation, or translocation efficiency. Bertholdsson and Stoy (1995) showed that the exceptionally high grain N content in North American wheat compared to Swedish wheat was not due, as expected, to a more efficient N uptake and distribution. In contrast, they showed that the Swedish genotype was more effective in retrieving N from the soil and utilising it, but the high grain protein content of the American genotype was due to the relatively low dry matter (carbon) accumulation in the grain-filling period.

Several *Arabidopsis thaliana* ecotypes have been tested for their ability to grow on either high or limiting amounts of nitrate, in order to check out for the efficiency of nitrate use. The response of the plants shows that a significant interaction does exist between N availability and genotype. This result indicates that genotypic differences can explain, at least partly, differences in the response of plants to the variation of nitrogen availability (Loudet and Chaillou, unpubl.).

Several lines of chicory were grown on either a high (3 mM) or low (0.6 mM) nitrate medium during the vegetative phase of growth. The tuberised roots were then harvested and placed in a dark, mist-filled chamber at 18 °C and 90% RH for 21 days, where they were fed the same standard nutrient solution (Lesaint and Coïc 1983). This forcing procedure produced an edible, etiolated shoot known as chicon or Belgian endive. This experiment showed that nitrogen nutrition during the vegetative phase can alter the morphology of the chicon and that this effect is genotype-dependent (Richard-Molard et al. 1999). Two groups of lines were distinguished on their response to the changes in nitrate supply. In one group, increasing nitrate fertiliser during the vegetative phase resulted in an increase in the chicon stem length and in a decrease of the tightness of the leaves to the stem, giving the chicon a morphology referred to as opened chicon. In the other group, increasing nitrate fertiliser during the vegetative phase decreased the chicon stem length and increased the tightness of the leaves to the stem, giving the chicon a morphology referred to as a closed chicon.

Considering the large effects of nitrogen nutrition on plant development, the question arises of how the various metabolic events of nitrogen assimilation interact within the plant development programme, to modulate complex traits such as plant growth and architecture. The nitrogen assimilation process, besides being a source of reduced nitrogen, seems to play a signalling role in plant development. NO_3^- and glutamine have been suggested to have signalling properties; however, the mechanisms that would allow these molecules to regulate gene expression is not known (Trewavas 1983; Scheible et al. 1997). The study presented in the next section clearly illustrates the role that nitrogen assimilation could play in plant development.

Genetic Study of the Effect of Glutamine Synthetase (GS) Activity on Plant Growth in the Legume *Lotus* sp.

Root cytosolic glutamine synthetase, a major enzyme of nitrogen metabolism (see Chap. 2.2) has been proposed lately to intervene in plant development in an unpredicted manner (Vincent et al. 1997). Besides its role in ammonium assimilation, root cytosolic GS could be involved in plant growth control. Overexpression of cytosolic GS in *Lotus corniculatus,* that was accompanied by an increase in the enzyme activity in the shoot, led to the activation of a physiological process, resulting in premature flowering under non N-fixing-conditions. When these transgenic plants were grown under N-fixing-conditions, GS activity in the nodules was unexpectedly found to be reduced by 40% on either a fresh weight or protein basis. Nevertheless, plant biomass was twofold higher and nodule dry weight was at least 50% higher in the transgenic plants (Vincent et al. 1997).

Two *Lotus japonicus* ecotypes showing differences in growth rate have been selected, namely Funakura and Gifu, originating respectively from plains and mountains in Japan. When the two ecotypes were grown under the same conditions and fed with a nutrient solution containing nitrate, differences in both growth and root cytosolic GS activity were observed. While Funakura produced 40% more biomass than Gifu, root GS activity was ten times lower than in Gifu, probably due to a decrease in GS transcript in the root. However, differences in growth were not related to any change in shoot GS activity, which remained similar in both ecotypes over the whole period of growth (Limami et al. 1999).

The control of growth by root cytosolic GS activity could be exerted through the modulation of nitrogen assimilation and partitioning between shoot and root, which would finally result in an increase or decrease in nitrogen absorption. This explanation is, however, questionable, because changes in nitrogen metabolism could be a consequence, rather than a cause, of changes in growth.

In order to better understand the involvement of root cytosolic GS activity in plant development, we analysed the distribution of root cytosolic GS activity and plant growth in a population of 50 F_6 RILs (recombinant inbred lines) derived from a cross between Gifu and Funakura (Jiang and Gresshoff 1997). Statistical analyses revealed genetic differences between individuals for plant growth and cytosolic root GS activity. These traits were normally distributed in the progeny, with a significant negative correlation between them ($r = -0.38$), irrespective of the culture conditions (Limami et al. 1999). This result indicates that the GS activity is a polygenic trait, which possibly shares common QTLs (quantitative trait loci – see insert) with growth traits. The negative correlation is consistent with results obtained with transgenic plants, which had a decreased growth due to cytosolic GS overexpression in the root. It is also consistent with the observation that the high growth rate of *L. japonicus* ecotypes is always associated with a low GS activity in the root; but in any case, the physiological explanation for this unexpected negative correlation requires further analysis.

What could be the genetic basis of the continuous variation of GS activity in the progeny? Several isoforms of cytosolic glutamine synthetase have been shown to exist in plants. For example, in *Phaseolus vulgaris* nodules (Cai and Wong 1989; Temple et al. 1993, 1998), nine GS isoforms were found, which differed in the proportion of the two subunits β and γ constituting the octomeric holoenzyme. If this model holds for *L. japonicus,* it may account in part for the variability of the activity in the RILs. Except in the case where there would be only one polymorphic structural gene for the GS (in this case the individuals of the progeny would have either the isoform 8β or the isoform 8γ in the proportion 1:1), the combination of monomers encoded by unlinked polymorphic genes would result in numerous different octo-mers (de Vienne and Rodolphe 1985), their relative abundance differing from one individual to another. In addition the different structural genes may be submitted to various modes of regulation, as diverse as constitutive or substrate inducible expression. Therefore the activity and relative abundance of each glutamine synthetase isoform is the result of the activity, expression and turnover of the subunits, and hence may possibly depend on quite a large number of polymorphic loci. A fraction of this polymorphism could have adaptive significance. For example, the ecotype Funakura, originating from the plains in Japan, may be more adapted to nitrate rich soils when compared to Gifu, originating from mountain soils which are generally characterised by a lower nitrogen content.

Future Prospect: Dissecting Physiological Traits Through the Use of Molecular Quantitative Genetics

The observation of continuous genetic variability for biochemical/physiological traits is now widely documented. For a long time the inheritance of quantitative traits has been supposed to be ascribed to the additive effects of a large number of genes with small and similar actions, modulated by environment. In fact, this assumption relied on practical (to simplify the theoretical models), rather than biological considerations. With the development of molecular markers to construct genetic maps, which allow QTLs to be detected and quantified (see insert), it became clear that usually, even for highly complex traits like tomato fruit size and composition (Paterson et al., 1988) or crop yield (Edwards et al., 1987; 1992), a small number of quantitative trait loci (QTL) could explain a large part of the genetic variability, the rest of the genetic variation being due to a large and variable number of genes of smaller effect.

Although until now QTL analysis has more largely been used for agronomic and morphological traits (see review in Kearsey and Farquhar 1998), the first systematic QTL detection from an RFLP map by Paterson et al. (1988) was applied to physiological traits dealing with tomato fruit growth: mass, soluble solids content (mainly carbohydrate) and pH. Five QTLs controlling fruit mass, four QTLs for the concentration of soluble solids and six QTLs for fruit pH were mapped. These QTLs accounted for 58, 44 and 48% of the phenotypic variance for mass, soluble solids and pH, respectively. On chromosome 6, the same QTL was observed for fruit mass, pH and concentration of solids. Similarly, pH and soluble solids shared the same QTL on chromosomes 3 and 7. This report led to three main conclusions, which are applicable to other QTL analyses: i) a small number of Mendelian factors can explain a large part of the genetic variance, ii) traits for different levels of organisation but for related processes (e.g.. growth and some of its physiological components) frequently seem to share common QTLs, and iii) as a consequence, a complex trait is dissected into discrete factors increasing or decreasing the trait value. Although not discussed in the paper, the apparent co-location of the QTL for tomato fruit pH and sugars may also be given a physiological interpretation: the role of proton-driven translocation of carbohydrate from the cytosol into the vacuole is well established, and a QTL for such a translocator or a proton pump is likely to affect both pH and soluble sugars.

Two main uses of the QTL methodology have emerged. The first deals with the possibility of performing marker-assisted selection aimed at gathering the favourable alleles and breaking their possible linkage with undesirable alleles. The second approach targets the identification of a QTL by determining the contribution of known-function genes to the variation of the traits. This so-called candidate gene approach, applied in particular to physiological traits, has been reviewed in detail by Prioul et al. (1997). A recent example concerning the responses of maize to drought stress was published recently by Pelleschi et al. (1999). Water shortage produces an early and high stimulation of the acid-soluble invertase activity in adult maize leaves, whereas the cell wall invertase activity remains constant. This response was tightly related to the steady-state mRNA level for only one of the invertase genes (*Ivr2*) encoding a vacuolar isoform. In parallel, four QTLs were detected for invertase activity in control plants and nine under stress conditions. One QTL in control plants and one in stressed plants were located within the confidence interval of the *Ivr2* gene position on chromosome 5, suggesting that the polymorphism of *Ivr2* could, in part, account for the variability of invertase activity, and possibly the related traits of the carbohydrate metabolism and stress-responsive traits.

There is no simple and fast method for the validation of the candidate genes. Among the various available strategies, the most straightforward is probably the analysis of sequence polymorphisms of the candidate loci in a range of unrelated genotypes. If the correlation between the trait variation and the gene polymorphism still holds, it may become possible to determine the part of the sequence responsible for the variation. For example, by comparing the various alleles of *Tb1*, a gene controlling the length of the side branches in maize and teosinte, its wild relative, Wang et al. (1999) showed that the polymorphism affecting the trait was limited to the regulatory regions of the gene, the polymorphism in the protein-coding region being apparently neutral.

The QTL approach opens a new and exciting perspective for dissecting and understanding the genetic regulation of complex processes such as nitrogen-use efficiency. Various laboratories are working in this promising direction, and relevant papers are expected to be published soon. Molecular marker technologies bring together the traditional specialisations of physiology, genetics, molecular biology and breeding, and can be used to answer both basic and applied problems. QTL analysis will inevitably lead to better relationships between the physiologists and breeders, who have often in the past been sceptical of claims made by physiologists for the significance of a particular trait for plant improvement. In this connection, it is worth recalling that only the *polymorphic* QTL can be detected in a given genetic background. One can speculate that the physiologically crucial genes will not display much polymorphism, because variations are strongly counter-selected, or are maintained in a narrow range of interallelic differences. Thus, in some instances, it may prove to be necessary to screen not only other varieties or accessions, but also wild relatives which can be crossed with the species of interest, to find polymorphism. In conclusion, QTL analysis should be the physiologist's tool of the future, enabling him/her to understand how to improve plant growth and behaviour in a range of environments.

The Quantitative trait loci (QTL)

A QTL (*quantitative trait locus*) is a locus accounting for a fraction of the variability of a quantitative trait in a population or a progeny. With the advent of the molecular marker technologies, which allow dense genetic maps to be constructed from single progenies, it is now possible to make an inventory of the loci involved in the variation of a given quantitative trait, to determine their mapping position, and to estimate their effects. The effect of a QTL may be quantified as the proportion of variance accounted for by the QTL, or as the additive effect (or substitution effect), which corresponds to the half-difference of homozygote mean values.

The prerequisites for QTL detection are: (1) to have a segregating progeny of plants displaying genetic variability for the trait of interest, (2) to establish genetic linkage groups for that population by analysing the recombination ratios amongst molecular markers, and (3) to score the trait of interest on every individual of the population. The simplest QTL detection method relies on one-way ANOVA (Soller et al. 1976). For every marker, the means of the genotype classes are compared. A significant difference indicates that the marker is likely to be linked to a segregating QTL having an effect on the trait. However, more powerful methods are now commonly used, such as multiple regression or composite interval mapping (Zeng 1994; Jansen and Stam 1994).

F_1-derived populations are the most easily accessible genotypes for QTL analysis. An F_1 generation between two inbred lines may be used to derive (1) backcrosses to one of the parental lines (BC), (2) homozygous doubled haploid (DH) lines, (3) F_2 progenies. Further single seed descent from an F_2, leads to recombinant inbred lines (RILs) which are close to homozygosity after at least five to six generations of selfing. DHs and RILs, which can be identically reproduced by selfing, are "immortalised" populations and hence very useful for accumulating markers and data over time and space. In addition the plants can be replicated across sites to analyse genotype x environment interactions. DHs can rapidly be produced (one generation), when the genotype is capable of regeneration from tissue or anther culture, but segregation distortions may be common in some species. RIL populations are probably the most convenient material, but they take time to obtain. However, neither DH nor RIL populations can be used to estimate dominance effects (*d*), since they do not include heterozygote genotypes. F_2 and BC populations are rapidly obtained (two generations), they allow dominance to be analysed (even though reciprocal progenies are required for BC), but they do not allow replication of the genotypes (except by using vegetative propagation, when applicable). This problem may be partly circumvented by working on the pools of the descendants obtained by selfing from each F_2 or BC individual. Epistatic interactions between QTL may be detected in principle in any progeny, but the power of the statistical tests is quite low, unless the size of the population is very large (several hundreds of individuals).

References

Beadle GW (1929) Yellow-stripe a factor for chlorophyll deficiency in maize located in the Prpr chromosome. Am Nat 63: 189-192

Bell WD, Bogorad L, McIlrath WJ (1958) Response of the yellow-stripe mutant (*ys1*) to ferrous and ferric iron. Bot Gaz 120: 36-39

Bell WD, Bogorad L, McIlrath WJ (1962) Yellow-stripe in maize. I.Effects of *ys1* locus on uptake and utilisation of iron. Bot Gaz 124: 1-8

Bertholdsson NO, Stoy V (1995) Yields of dry matter and nitrogen in highly diverging genotypes of winter wheat in relation to N-uptake and N-utilization. J Agric Crop Sci 175: 285-295

Cai X, Wong PP (1989) Subunit composition of glutamine synthetase isozymes from root nodules of bean (*Phaseolus vulgaris L*). Plant Physiol 91: 1056-1062

Clark RB (1978) Differential response of corn inbreds to calcium. Commun Soil Sci Plant Anal 9: 729-744

Clark RB, Brown JC (1974) Differential mineral uptake by maize inbreds. Commun Soil Sci Plant Anal 5: 213-227

Cox MC, Qualset CO, Rains DW (1985) Genetic variation for nitrogen assimilation and translocation in wheat. II. Nitrogen assimilation in relation to grain yield and protein. Crop Sci 25: 435-440

Cox MC, Qualset CO, Rains DW (1986) Genetic variation for nitrogen assimilation and translocation in wheat. III. Nitrogen translocation in relation to grain yield and protein. Crop Sci 26: 737-740

Cregan PB, van Berkum P (1984) Genetics of nitrogen metabolism and physiological/biochemical selection for increased grain crop productivity. Theor Appl Genet 67: 97-111

de Vienne D, Rodolphe F (1985) Biochemical and genetic properties of oligomeric structures: a general approach. J Theor Biol 116: 527-568

Edwards MD, Stuber CW, Wendel JF (1987) Molecular-marker-facilitated investigations of quantitative-trait loci in maize. 1 Numbers, distribution and types of gene action. Genetics 116: 113-125

Edwards MD, Helentjaris T, Wright S, Stuber CW (1992) Molecular-marker-facilitated investigations of quantitative-trait loci in maize. 4 Analysis based on genome saturation with isozyme and restriction fragment length polyymorphism markers. Theor Appl Genet 83: 765-774

Jansen RC, Stam P (1994) High resolution of quantitative traits into multiple loci via interval mapping. Genetics 136: 1447-1455

Jiang Q, Gresshoff P (1997) Classical and molecular genetics of the model legume *Lotus japonicus*. Mol Plant Micr Interac 10: 59-68

Kearsey MJ, Farquhar GLA (1998) QTL analysis in plants; where are we now? Heredity 80: 137-142

Kjaer B, Jensen J (1995) The inheritance of nitrogen and phosphorus content in barley analysed by genetic markers. Hereditas 123: 109-119

Lesaint C, Coïc Y (1983) Cultures hydroponiques. La Maison Rustique, Flammarion, Paris, 215 pp

Limami MA, Phillipson B, Ameziane R, Pernollet N, Jiang Q, Roy R, Deleens E, Chaumont-Bonnet M, Gresshoff PM, Hirel B (1999) Does root glutamine synthetase control plant biomass production in Lotus japonicus L.? Planta 209: 495-502

O'Sullivan J, Gabelman WH, Gerloff GC (1974) Varitions in efficiency of nitrogen utilization in tomatoes (Lycopersicon esculentum Mill) grown under nitrogen stress. J Am Soc Hortic Sci 99: 543-547

Paterson AH, Lander ES, Hewitt JD, Paterson S, Lincoln SE, Tanksley SD (1988) Resolution of quantitative traits into Mendelian factors by using a complete linkage map of restriction fragment length polymorphisms. Nature 335: 721-726

Pelleschi S, Guy S, Kim JY, Pointe C, Mahé A, Barthes L, Leonardi A, Prioul JL (1999) Ivr2, a candidate gene for a QTL of vacuolar invertase activity in maize leaves. Gene-specific expression under water stress. Plant Mol Biol 39: 373-380

Prioul JL, Quarrie S, Causse M, de Vienne D (1997) Dissecting complex physiological functions through the use of molecular quantitative genetics. J Exp Bot 48: 1151-1163

Rajcan I, Tollenaar M (1999a) Source: sink ratio and leaf senescence in maize: I. Dry matter accumulation and partitioning during grain filling. Field Crops Res. 60: 245-253

Rajcan I, Tollenaar M (1999b) Source: sink ratio and leaf senescence in maize: II. Nitrogen metabolism during grain filling. Field Crops Res 60: 255-265

Richard-Molard C, Wuillème S, Scheel C, Gresshoff PM, Morot-Gaudry JF, Limami MA (1999) Nitrogen-induced changes in morphological development and bacterial susceptibility of Belgian endive (Cichorium intybus L.) are genotype-dependent. Planta 209: 389-398

Scheible WR, Lauerer M, Schulze ED, Caboche M, Stitt M (1997) Accumulation of nitrate in the shoot acts as a signal to regulate shoot-root allocation in tobacco. Plant J 11: 671-691

Shea PF, Gerloff GC, Gabelman WH (1967) The inheritance of efficiency in potassium utilization in snapbeans (Phaseolus vulgaris L.). Proc Am Soc Hortic Sci 91: 286-293

Shea PF, Gerloff GC, Gabelman WH (1968) Differing efficiencies of potassium utilization in strains of snapbeans (Phaseolus vulgaris). L Plant Soil 28: 337-346

Soller M., Brody T., Genizi A. (1976) On the power of experimental designs for the detection of linkage between marker loci and quantitative loci in crosses between inbred lines. Theor Appl Genet, 47: 35-39

Temple SJ, Knight TJ, Unkefer PJ, Sengupta-Gopalan S (1993) Modulation of glutamine synthetase gene expression in tobacco by the introduction of an alfalfa glutamine synthetase gene in sense and antisense orientation: molecular and biochemical analysis. Mol Gen Genet 236: 315-325

Temple SJ, Bagga S, Sengupta-Gopalan S (1998) Down-regulation of specific members of the glutamine syntethase gene family in alfalfa by antisense RNA technology. Plant Mol Biol 37: 535-547

Trewavas AJ (1983) Nitrate as a plant hormone. In: Jackson MB (eds) Interactions between nitrogen and growth regulators in the control of plant development. British Plant Growth Regulator Group, Wantage, Oxfordshire, pp 97-110

Vincent R, Fraisier V, Chaillou S, Limami MA, Deléens E, Phillipson B, Douat C, Boutin JP, Hirel B (1997) Overexpression of a soybean gene encoding cytosolic glutamine synthetase in shoots of transgenic *Lotus corniculatus* L. plants triggers changes in ammonium assimilation and plant development. Planta 201: 424-433

Wang RL, Stec A, Hey J, Lukens L, Doebley J (1999) The limits of selection during maize domestication. Nature 398: 236-239

Zeng Z-B (1994) Precision mapping of quantitative trait loci. Genetics 136: 1457-1468

Zhang H, Forde BG (1998) An *Arabidopsis* MADS box gene that controls nutrient-induced changes in root architecture. Science 279: 407-409

Nitrogen in the Environment

M.W. TER STEEGE[1], Ineke STULEN[2] and Bruno MARY[3]

Nitrate accumulation in vegetables

A Health Hazard

When leafy vegetables such as spinach and lettuce are grown in greenhouses during winter and early spring, i.e. at low light intensity and short day length, they may accumulate a high amount of nitrate in the leaves (Corré and Breimer 1979). A high nitrate content in vegetables is undesirable, because it may be harmful for the consumer. Nitrate itself is not toxic, but it is easily reduced to the toxic compound nitrite. Reduction to nitrite can occur during postharvest storage of vegetables (Aworth et al. 1980), as well as after ingestion as food in saliva and in the gastrointestinal tract (Maynard et al. 1976; Walters and Walker 1979). Acute nitrite toxicity causes a respiratory dysfunction called methaemoglobinaemia. By the oxidation of the ferrous iron of haemoglobin to the ferric form, methaemoglobin is formed which cannot transport oxygen, thereby causing tissue asphyxia. Chronic nitrite poisoning may result in the formation of carcinogenic nitrosamines. These N-nitroso compounds can be formed from nitrite and secondary amine compounds, which often occur in food (Walters and Walker 1979; Vermeer et al. 1998). As yet, the occurrence of (gastric) cancer has not been directly related to the consumption of nitrate, but it is generally accepted that a high nitrate intake should be prevented (Forman et al. 1985; Westgeest 1989).

Consumption of vegetables is the main component of nitrate intake by human beings, representing about 75% (Forman et al. 1985). It is expected that the nitrate burden will seriously increase in the near future, due to increased nitrate levels in drinking water (Westgeest 1989). This is caused by excessive application of manure in agricultural areas. Since nitrate has a high mobility in the soil, the high amounts of manure result in the leakage of nitrate towards surface- and groundwater, which are used for drinking purposes. For health reasons, a limit to the amount of nitrate in drinking water is set (50 mg l^{-1}).

1. Plant Ecophysiology, Utrecht University P.O. Box 800.84, 3508 TB Utrecht The Netherlands.
2. Department of Plant Biology, University of Groningen, Kerklaan 30, 9751 NN Haren, The Netherlands. *E-mail:* g.stulen@biol.rug.nl
3. Unité d'Agronomie, INRA, rue Fernand Christ, 02007 Laon Cedex, France. *E-mail:* mary@laon.inra.fr

In the European Community, regulations for maximum acceptable nitrate concentrations in vegetables have been introduced. The present maximally allowed levels are 2500 to 4500 mg kg^{-1} FW (OJL 108, 26.4.1999, p. 16). In The Netherlands, the nitrate concentration in commercially grown lettuce and spinach often exceeds this permitted level (Van Diest 1986), which puts a high pressure on growers to produce vegetables with acceptable nitrate concentrations.

The problems associated with a high nitrate content in vegetables, as well as the threat of the undesirable nitrate pollution of drinking water, prompted investigations on nitrate utilisation and uptake in plants that aimed at a decrease in the nitrate concentration in plants and environment, without too great sacrifices in terms of yield (Anonymous 1992). Research has focused on (1) elucidation of the causes of nitrate accumulation in plants (Blom-Zandstra and Lampe 1985; Blom-Zandstra et al. 1988; Blom-Zandstra and Eenink 1986; Steingröver et al. 1986a,b; Ter Steege et al. 1998, 1999), (2) improvement of the quality of the vegetables by breeding for low nitrate (Reinink 1988; Reinink and Groenwold 1987), or by overexpression of nitrate reductase, the first enzyme of the nitrate assimilation pathway (Pelsy and Caboche 1992; Quilleré et al. 1994) and (3) management of environmental factors that control the uptake of nitrate from the soil (Smolders, 1993), in order to increase the efficiency of N fertiliser application (Van der Boon et al. 1986; Van Noordwijk and Wadman 1992). The latter aspect will be dealt with in the second part of this chapter.

Physiological Background

Accumulation of nitrate in plants occurs when the rate of nitrate uptake exceeds the rate of reduction and subsequent assimilation into amino acids and proteins for growth. Differences in nitrate accumulation between genotypes and changes in nitrate concentration during the course of the day cannot simply be related to the rate of nitrate reduction (Blom-Zandstra and Eenink 1986; Steingröver et al. 1986 a,b), but are due to interactions between C and N metabolism (Blom-Zandstra et al. 1988; Stulen 1990; Stulen and Ter Steege 1995) and the rate of nitrate uptake (Steingröver et al. 1986 a,b; Ter Steege 1996). In vegetables grown during winter in greenhouses in The Netherlands, high plant nitrate concentrations result mainly from the poor light conditions at that time of the year. The combination of low light intensity and short day length restricts the production of organic osmotic solutes, like soluble sugars and malate via photosynthesis. Nitrate serves as an osmotic alternative (Blom-Zandstra and Lampe 1985; Veen and Kleinendorst 1986) and accumulates in the vacuoles of the leaves during the dark period (Steingröver et al. 1986a,b).

In the spinach cultivar Vroeg Reuzenblad, the accumulation of nitrate in the leaf vacuoles during the dark period was accompanied by an increase in net nitrate uptake rate (NNUR) by the roots (Steingröver et al. 1986 a,b). Since NNUR is the result of two opposite nitrate fluxes, influx and efflux, the observed increase in NNUR in relation to the increased osmotic need for nitrate could have been due to (1) an increase in nitrate influx, (2) a decrease in nitrate efflux, or (3) both.

The uptake of N by the root is a well-regulated process, controlled by shoot-related signals. NNUR is closely linked to the relative growth rate (RGR) of the plant (Rodgers and Barneix 1988), and plants are able to accomplish the same RGR and plant total N concentration (PNC) over a wide range of external nitrate concentrations (Clement et al. 1978). NNUR, therefore, seems to be under the control of an internal regulating mechanism, which adjusts the net N uptake to the N demand of the plant, as determined by RGR and PNC (Touraine et al. 1994; Ter Steege et al. 1998). There has been a long debate on the role of nitrate influx and efflux in the control of NNUR, especially under steady-state conditions (Devienne et al. 1994; Ter Steege 1996). By using a double labelling design, with both ^{13}N and ^{15}N nitrate, it was possible to study the importance of both fluxes in the regulation of NNUR in spinach. Care was taken that physical manipulation of the plants did not affect the results (Ter Steege et al. 1998). These experiments showed that nitrate influx and efflux together regulate NNUR, thereby providing a flexible and sensitive nitrate uptake system (Ter Steege et al. 1998, 1999; Fig. 1). It has been postulated that a growth-related signal down regulates nitrate influx by negative feedback control at the level of the nitrate transporter (Barneix and Causin 1996; Imsande and Touraine 1994; Mueller al. 1995; Ter Steege 1996; Ter Steege et al. 1999). The number of transporters might be influenced by the external nitrate concentration (Ter Steege 1996). Recently plant genes encoding for the transporters for inorganic N have been isolated (Trueman et al. 1996; Von-Wiren et al. 1997; Krapp et al. 1998; see Chap. 1.1). This is of great value for the interpretation of physiological studies, viz. for distinguishing between the effects of internal and external signals on the expression and/or downregulation of the nitrate transporters.

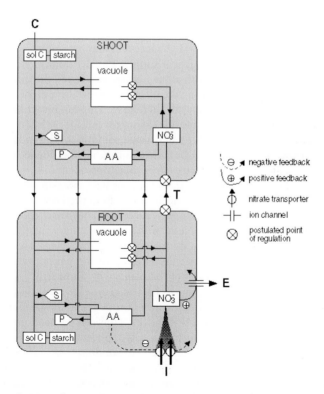

Fig. 1 : Localisation of processes of uptake, translocation, assimilation and storage of nitrate in relation to the availability of C compounds. *I* Nitrate influx; *E* nitrate efflux; *T* nitrate translocation; *AA* amino acids; *P* protein; *S* structural C; *sol C* soluble C

Further experiments with spinach, in which the external nitrate concentration was suddenly increased, showed that a large part of incoming nitrate could be removed from the root by efflux (Ter Steege 1996). The efflux of nitrate might offer a mechanism for rapid reactions to changed internal conditions, and nitrate itself seems a likely candidate for the control (Ter Steege 1996). At present, experimental data on the mechanism of nitrate efflux in higher plant roots is lacking. Theoretical considerations, however, lead to the conclusion that nitrate efflux most likely is a channel-mediated process (Glass and Siddiqi 1995). Nitrate efflux generally increases with increasing internal nitrate concentration (e.g. Clarkson 1986; Zhang and MacKown 1993), which has been interpreted as an indication of nitrate efflux occurring via a channel whose activity is dependent on the nitrate concentration at the cytoplasmic side.

Prevention of Nitrate Accumulation in Commercial Products

The experiments with spinach have shown that growth and nitrate uptake are tightly coupled, resulting in a long-term adjustment of nitrate uptake to the N requirement of the plant (Ter Steege et al. 1998, 1999). Accumulation of nitrate occurs when the osmotic demand for nitrate is high, which is the case under poor light conditions. Stored nitrate in the leaves, therefore, cannot be considered as a surplus or luxury at low light conditions. Because of the tight linkage between the plant's need for nitrate and its uptake (Fig. 1), it will be difficult to manipulate nitrate concentration in plants without affecting growth and yield. A successful manner might be to increase the ammonium : nitrate ratio in the nutrient supply (Van der Boon et al. 1990), but in lettuce, fertilisation with ammonium alone negatively affected the quality of the plants. Another method of manipulating the nitrate concentration in the leaves is to decrease the need for nitrate as an osmotic alternative. This can be achieved by the supply of an osmotic alternative, like chloride (Steingröver et al. 1982; Urrestarazu et al. 1998) or by increasing the production of organic osmotic compounds. Chloride fertilisation should be carefully applied, because chloride might influence the taste of the product. The alternative, stimulation of the production of organic osmotic compounds, can be achieved in two ways: by increasing the light intensity during growth or by increasing the atmospheric CO_2 concentration. Both methods are only applicable in greenhouses or growth cabinets and, if successful, they are effective via stimulation of photosynthesis. The problem with supplementary illumination in commercial spinach production is that it increases production costs. However, it is not necessary to illuminate throughout the entire growing period to reduce the nitrate content in spinach: illumination of the plants with a low light intensity (35 $\mu mol\ m^{-2}\ s^{-1}$) during the last night prior to harvest reduces the nitrate content by 25% (Steingröver et al. 1986). In view of the new European standards on maximum accepted nitrate levels, it might be worthwhile to reconsider the economic costs of this treatment. Increment of the atmospheric CO_2 concentration can result in a lower leaf nitrate concentration only when the light conditions do not limit photosynthesis, which depends on the reduced N concentration in the leaves and thus on N metabolism (Pons et al. 1993). Because the positive effect of elevated atmospheric CO_2 on RGR and on C and N metabolism is transient (Fonseca et al. 1996, 1997), a high CO_2 treatment during the last few days before harvest might have the highest chance of being successful. This suggestion needs further testing under greenhouse conditions. Apart from the osmotic effect, high carbohydrate levels due to a high CO_2 treatment may also positively affect the activity and the expression of nitrate reductase (Fonseca et al. 1997; Vincentz et al. 1993), which may further contribute to a decline in the nitrate concentration. Overexpression of nitrate reductase in *Nicotiana plumbaginifolia* resulted in a 25-150% higher nitrate reductase activity than in the wild type. This was accompanied by a decrease in foliar nitrate content of 32-47% (Quilleré et al. 1994), while dry matter production was not increased. At the same time, there was in increase in glutamine of 74-133% in the transgenic plants. These findings are fully in agreement with the proposed model, in that an increase in one factor, viz. nitrate reductase, does not lead to a higher dry matter production, since the nitrate uptake rate can be expected to be downregulated by the product of nitrate assimilation, viz. glutamine. The use of transgenic plants, which overexpress nitrate reductase, might therefore be a way to improve product quality. In conclusion, nitrate accumulation is the result of complex interactions between N and C metabolism under unfavourable growing conditions. Breeding programs aiming at a selection of low-nitrate cultivars should not only focus on N-metabolism characteristics, but they should also take into account essential characteristics of C metabolism. One-sided efforts to reduce the nitrate content of vegetables, focusing on the N or C metabolism, are not very likely to succeed. There are ways to reduce the nitrate content in the plants, of which additional illumination, thereby increasing photosynthesis, probably will be most successful, but these will increase the price of the product. Maybe the consumers, for the sake of their health, should be prepared to pay a higher price for good-quality food, or only buy the products of the season, which do not need additional lightning.

Nitrogen Losses in Agricultural Systems

Improving N management in agricultural systems is now a general objective in the world, particularly in northern, temperate countries which have developed intensified agriculture with high nitrogen inputs. Improving N management means controlling and reducing nitrogen losses towards surface water, groundwater and the atmosphere. The losses are due to nitrate leaching (NO_3^-), ammonia volatilization (NH_3) and denitrification which results in production in nitrous oxide (N_2O), nitric oxides (NOx) and dinitrogen (N_2).

Nitrate in surface and groundwater must be restricted due to its possible effects on animal and human health (see previous section) and its contribution to eutrophication of rivers, estuaries and coastal seas. Ammonia emissions may have direct toxic effects, for example on coniferous forests, and indirect effects by disturbing N balances and favouring soil acidification (Jarvis and Pain 1990). Nitrous oxide has a double harmful effect: it is a greenhouse gas and it contributes to the destruction of stratospheric ozone layer. Finally, the nitric oxides favour accumulation of ozone near the ground and acid rain.

Sustainable agriculture must be able both to restrict these emissions at an acceptable level and to maintain or increase the quantity and quality of crop production, for example by reducing the nitrate content in the vegetables (see previous section), or improving the protein content of cereal grains. It must also preserve the soil fertility on a long-term basis.

The Global Nitrogen Budget

Nitrogen in soils is mainly present as organic nitrogen, as humified compounds and litter residues. The soil inorganic N is only a minor fraction (0 to 5%) of the total soil N. However, the organic nitrogen stored in agricultural soils represents itself a very small part ($\sim 10^{-5}$) of the total nitrogen present in the biosphere (Table 1). The major N reservoir is composed of the dinitrogen gas in the atmosphere.

Table 1. Nitrogen stocks (in 10^{18} g N) in the biosphere. (McElroy, cited by Bielek et al. 1985)

N in atmosphere	4000
N in minerals and rocks	600
N in oceans	0.8
Organic N in soils	0.06
Plant biomass N	0.01

During the formation of soils, organic nitrogen has accumulated in soils coming from biological fixation, due to either free-living organisms or symbiotic fixation by leguminous plants. These processes are still very active. The present rate of symbiotic fixation for the whole world has been estimated at about 120 Tg per year (Table 2).

During the last 50 years, the application of fertiliser N has increased almost exponentially. It was almost nil until 1940, and reached 10 Tg in 1960, 32 Tg in 1970 and 80 Tg in 1990. The inputs of industrially fixed nitrogen into soils will soon become equivalent to the inputs by natural fixation.

Table 2. Nitrogen inputs and outputs (in Tg N year^{-1})

	Whole world	France[e]	United Kingdom[f]
Inputs			
Symbiotic N_2 fixation	120 [b]	-	0.3
Fertiliser N	80 [c]	2.3	1.5
Atmospheric deposition	-	0.5	0.3
Outputs			
NO_3 leaching	8.0 [i]	0.3	0.3
NH_3 volatilisation[a]	8.4 [d]	0.12	0.04 [g]
N_2O denitrification[a]	1.5 [d]	0.06	0.07 [h]

[a] Originating from the fertilisers (other sources not included).
[b] Ledgard and Giller (1995).
[c] Bumb (1995).
[d] Peoples et al (1995).
[e] IFEN (1998).
[f] Johnston and Jenkinson (1989).
[g] Dragosits et al. (1998).
[h] Skiba et al. (1996).
[i] Estimated.

Such a change necessarily affects the intensity of nitrogen processes in the eco-systems. Table 2 gives estimates of N inputs and outputs in France, in UK and in the world. The ouputs by ammonia volatilisation and nitrous oxide emissions by deni-trification are those associated with fertilisation and do not include other sources (NH_3 emission by animals, N_2O emission by industrial activity, natural sources, etc.). Nitrogen leaching estimates include all sources. These figures are provisional estimates and are continuously updated. However, they show that the losses and particularly the gaseous emissions represent a small proportion of the fertiliser inputs. The challenge is to decrease these agricultural emissions by further improving the efficiency of fertiliser in the soil-plant system.

The Fate of N Fertiliser

Chemical fertilisers used in the world are mainly in the form of ammonium nitrate, urea (solid granules) or a mixture of ammonium nitrate and urea (nitrogen solutions). Urea applied to soil is rapidly hydrolysed into ammonium carbonate due to the action of free urease already present in soil or urease newly synthesized by microorganisms. The transformation is complete in a few hours or a few days. The ammonium then produced, or directly applied by the fertilisers enters in the nitrogen cycle (Fig. 1).

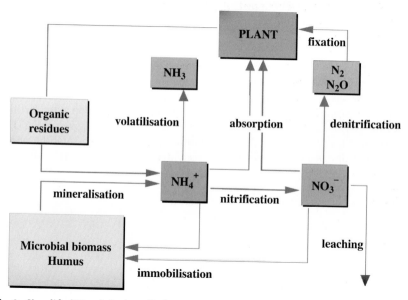

Fig. 2 : Simplified N cycle in the soil-plant-atmosphere. The processes quantitatively dominating are indicated by *bold characters* and *arrows*

The fate of nitrogen fertilisers in the agricultural systems has been studied intensively during the past 20 years by the means of isotopically labelled fertilisers. The use of ^{15}N-labelled fertiliser has allowed the production of a precise budget of the nitrogen applied to the soil. It has revealed four important features:
1. The recovery of the fertiliser-N by the crop is incomplete; the proportion of fertiliser-N found in the crop (aerials and roots) at harvest, usually called nitrogen-use efficiency, most often lies in the range 55-75%.
2. The residual N at harvest, i.e. the mineral nitrogen derived from the fertiliser found in soil at harvest, is usually very small (e.g. McDonald et al. 1989), except for large over fertilisation which increases the residual N.

3. The ^{15}N recovery in the soil-plant system is most frequently incomplete; this indicates that significant gaseous losses occur from the soil or the crop to the atmosphere.
4. The greater is the crop N uptake, the higher is the N-use efficiency; for example the fertiliser applications at booting or heading stages are better utilised by wheat than early applications at the tillering stage (Limaux et al. 1999).

A typical example of ^{15}N budget is shown in Fig. 2. It represents the average recovery of the labelled N measured in three successive maize crops. At harvest, 64% of the applied N was found in the plant, mainly in the aerial parts (59%). A small amount of mineral nitrogen (mainly nitrate) remained in the soil (7%), mainly in the ploughed layer. A significant part of applied N (18%) was immobilised by the soil microorganisms and would contribute to replenish the humified organic matter in soil. The nitrate leached from the fertiliser was negligible during the crop cycle (1%), although the crop was irrigated. The unrecovered [15]N (9%) was attributed mainly to gaseous losses by volatilisation and denitrification.

Similar results have been obtained by other authors with various crops (Pilbeam 1996). Powlson et al. (1992) obtained a mean ^{15}N recovery of 85% in nine experiments with winter wheat. The unrecovered [15]N (15%) was attributed to denitrification and in a smaller proportion to leaching (Addiscott and Powlson 1992). The authors found that a rather constant proportion of the fertiliser (18%) remained in soil at harvest, most of which was immobilized by the soil microbial biomass. Therefore, microbial immobilisation, volatilisation and denitrification appear to be the main processes occurring at the expense of the applied fertiliser, after its absorption by the crop.

Different studies have shown that these processes occur soon after fertiliser application (e.g. Recous et al., 1997). An example of the kinetics is given in Fig. 3. In this experiment, the fate of fertiliser was followed during a crop rotation: winter wheat- intercrop (9 months)-sugarbeet, receiving an optimized fertilisation rate.

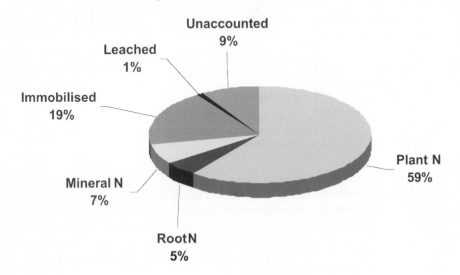

Fig. 3 : Fate of N derived from the fertiliser in irrigated maize crops at harvest. The values are the mean of three successive experiments receiving 260, 160 and 140 kg N ha[-1] labelled with [15]N. Adapted from Normand et al. (1997)

The mineral nitrogen derived from the fertiliser applied to wheat disappeared rapidly following each fertiliser application (50 + 110 kg N ha^{-1}). Nitrification was complete within 3 weeks. Volatilisation, denitrification and microbial immobilisation therefore also occurred during this period.

At wheat flowering and harvest, the mineral nitrogen left in soil was minimal; the nitrogen derived from the fertiliser being almost nil.

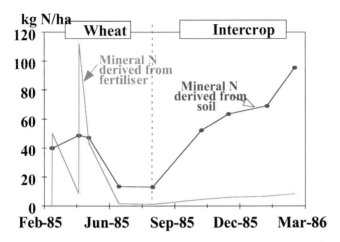

Fig. 4 : Evolution of the mineral N content in soil during a wheat-intercrop-sugarbeet rotation. The fertiliser-N applied to wheat (50 + 110 kg N ha^{-1}) was ^{15}N-labelled, which allowed the fate of soil N and fertilizer N to be distinguished. (Laurent and Mary 1992)

During the intercropping period (August 85 to March 86), the mineral N derived from the soil increased due to mineralization and the absence of crop sink whereas little nitrogen came from the previously applied fertiliser. This example shows that the direct contribution of the N fertiliser to nitrate leaching during the following winter can be a small fraction of the total nitrate leaching.

Nitrate Leaching

Solving the problem of nitrate leaching starts with the optimization of nitrogen fertilisation with respect to the plant demand and the soil supply capacity. The lower the mineral N at harvest, the lower will be the losses of nitrate during the following winter. We have indicated that a well-fertilised crop left very little residual mineral N derived from the fertiliser at harvest. This observation has led some authors to think that "even a drastic reduction in N fertiliser use would have little effect on nitrate leaching" (McDonald et al. 1989). In fact, this conclusion is highly questionable and can be contested, as suggested by Davies and Sylvester-Bradley (1995), for several reasons: *1)* there are interactions between the soil and fertiliser N that complicate the interpretation of ^{15}N experiments, *2)* fertiliser-N may be lost during the crop in very sandy or shallow soils, and *3)* overfertilisation has been a frequent situation in arable farming. The analysis of nitrogen response curves helps to clarify the situation.

The typical response of a crop to fertiliser application rate is shown in Fig. 5. This example is based on a set of 28 field experiments carried out on winter wheat and

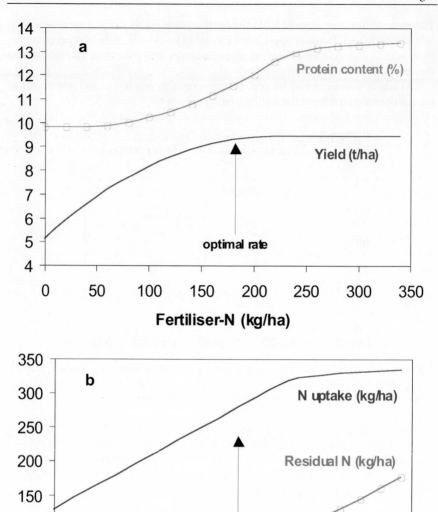

Fig. 5 : Response of wheat crops to N fertiliser application: a) grain yield and grain protein content; b) crop N uptake and mineral N remaining in soil at harvest.
The arrow indicates the optimal fertiliser rate needed to obtain the maximum grain yield. The curves are based on a set of 28 field trials made in France and analysed through a statistical model. Adapted from Makowski et al. (1999)

analysed with a statistical model by Makowski et al. (1999). When fertilisation increases, the grain yield increases up to a maximum (or more exactly an economic optimum), which defines the optimal N rate. The protein content of grains increases with a delay and reaches a maximum for a nitrogen level higher than the optimal rate.

The N uptake by the crop increases almost continuously: this is due to the large capacity of crops to absorb nitrogen, defining a maximal "dilution curve" far beyond the "critical curve" (Justes et al., 1994). The residual mineral N at harvest, which is minimal for the unfertilised crop, is slightly higher at the optimal rate; it increases rapidly when the wheat is over-fertilised. Similar conclusions can be drawn from other studies (Chaney 1990; Guiot and Grevy 1990; Glendining et al. 1992; Recous et al. 1996). These results show that it is difficult to obtain simultaneously the maximum yield, a satisfying grain protein content and a minimal amount of residual inorganic N in soil.

However, the leaching losses are not simply related to the N inputs, and not even simply related to N overfertilisation. This is due to the interaction between nitrogen processes and the water balance in soil. For example, the high concentrations of nitrate which can be found in soil after fertiliser applications (5-50 mol m^{-3}) and which are needed to maximise crop growth rate are much higher than the maximum recommended concentration in percolating water (0.8 mol m^{-3}); but fortunately they occur at a period when water drainage has ceased. In fact, the nitrate losses are mainly determined by the nitrate concentration in the soil during the drainage period which extends in northern Europe from end of autumn to early spring.

The nitrate concentration in the drained water, i.e. the water percolating below the potential rooting depth allowed by the soil and the crop depends on the nature of the crops, the type of soils and the cropping techniques. Crops are harvested at different times in the year and therefore leave variable amounts of nitrate in the soil at the end of autumn when the water drainage starts again.

The effect of crop rotation on nitrate leaching is illustrated in Fig. 6, which shows the mean concentration of the drained water during winter in conventional farmers' fields. Each bar is the mean of about 13 situations over 6 years. The values presented apply only to loamy soils, which yield the lowest nitrate concentrations, but the classification is valid for other soil types with higher concentrations.

A favourable situation is the rotation sugarbeet -winter wheat, producing water at a low concentration (12 mg NO_3 l^{-1}), i.e. much lower than the recommended EC limit (50 mg NO_3 l^{-1}). This is due to the fact that sugarbeet has a deep rooting system and is actively growing and taking up nitrogen until the harvest, which is late. Conversely, the rotation winter peas -winter wheat produces more polluted water (44 mg l^{-1}), although peas – a leguminous crop – receive no N fertiliser. In fact, peas have a rather shallow rooting system and are harvested early; therefore the nitrogen mineralised by the soil during the autumn can accumulate in the soil and part of it is leached. This is also the case for the winter wheat-fallow succession, showing that nitrate leaching is greater in bare soils. In order to reduce leaching, cover crop establishment is now strongly encouraged in most European countries. Indeed, cover crops appeared efficient to reduce nitrate concentration, either when they had a small growth (symbol W-C1) or even better when they were sown early enough to allow sufficient growth, greater than 0.8 t ha^{-1} dry matter (symbol W-C2).

The impact of crop management on nitrate leaching can be evaluated experimentally at the scale of the field and the year, in classical field experiments. This approach is useful but obviously restricted by the number of situations that can be

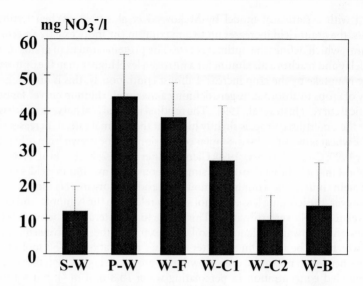

Fig. 6 : Mean nitrate concentration of drained water (water percolating beyond the roots) against crop succession in deep loamy soils. Bars represent the confidence intervals ($p < 0.10$). Crop succession: *S-W* sugarbeet-winter wheat; *P-W* winter peas-winter wheat; *W-F* winter wheat-fallow; *W-C1* winter wheat-cover crop with small growth; *W-C2* winter wheat-cover crop with a higher growth; *W-B* winter wheat-winter barley or oilseed rape. (Mary et al. 1997)

investigated and the difficulty in integrating the experimental results over time and space. A new promising approach consists of simulation studies using crop models at the scale of hydrological catchments and over many years. These models simulate water, carbon and nitrogen dynamics in the soil-plant system. They can predict the impact of agronomic scenarios on crop production and environmental variables. The scenarios may consider changes in N-fertilisation rate and timing, organic wastes applications, crop rotations, set-aside or soil-tillage practices, establishment of cover crops etc. Several studies of this type have been initiated recently (e.g. Beaudoin et al., 1998; Richter et al. 1998; Hoffmann and Johnsson, 1999). The study was conducted by Beaudoin et al. 1998, 1999 to evaluate the interest of the new regulations suggested by the EC community, in order to protect the quality of groundwater, called agri-environmental measures (MAE).

This study compared four basic scenarios: conventional farming without environmental constraints (S1); optimised N fertilisation using the balance-sheet method (Meynard et al. 1997) (S2); optimised N fertilisation and establishment of cover crops before spring crops (S3); reduced N fertilisation (-20%) and establishment of cover crops (S4 = MAE). The four scenarios were simulated in the context of an agricultural catchment during 6 years, using the crop model STICS (Brisson et al. 1998). Fig. 7 gives the results obtained in the four main types of soil and the four scenarios tested.

The mean nitrate concentration in drained water appears to be strongly dependent on the type of soil. The loamy soils, allowing the deepest rooting of crops, are the most favourable situation; the sandy soils which have low water reserves and a compacted structure, and prevent roots than growing below 60 cm, produce very polluted water. The agricultural management also affects nitrate concentration. For the whole catchment, the nitrate concentration was 92, 81, 56 and 53 mg NO_3^- l^{-1} for the scenarios S1, S2, S3 and S4, respectively. The establishment of cover crops would be clearly more efficient in reducing nitrate pollution than optimisation or even reduction in fertiliser rate. Such a result is paradoxical enough to be examined carefully.

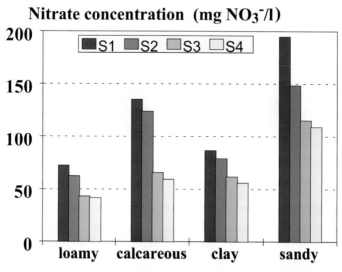

Fig. 7 : Simulated nitrate concentration in percolating water against soil type and for various agronomic scenarios, at the scale of an agricultural catchment during 6 years (1991-1997). The main crops were wheat (38%), sugarbeet (18%), peas (15%), barley (13%) and oilseed rape (9%). The simulated scenarios were: *S1* conventional farming (N fertilization without environmental constraints, no cover crop); *S2* optimised fertilisation using a predictive balance sheet model (no cover crops); *S3* optimised fertilisation and establishment of cover crops before spring crops; *S4* reduced fertilisation (20% less than the optimised rate) and establishment of cover crops before spring crops. (Beaudoin et al. 1998, 1999)

The validity of simulation outputs can be suspect. However, the simulated value for scenario S3, which corresponds to the practices actually carried out by farmers, was similar to the experimentally measured value (52 mg NO_3^- l^{-1}). Furthermore, the values found in the calcareous soils are very close to the values measured in an experiment conducted in the Champagne district on rendzina soils in order to test the four scenarios (Beaudoin et al. 1998).

The small effect of fertiliser reduction is probably underestimated, since only annual effects of fertiliser were accounted for in this simulation. Repeated N fertilisation is likely to exert a long-term effect, for example by enhancing soil mineralisation (Recous et al. 1997) and therefore increasing nitrate leaching. Davies and Sylvester-Bradley (1995) estimated that the increased fertiliser N use on intensive wheat in the UK had resulted in an increase of 36 kg N ha^{-1} $year^{-1}$ during the past 50 years and that more than one third of this increase was due to a long-term effect. The simulation models must be further calibrated or improved in order to account for this long-term effect. The use of finely tuned simulation models is becoming a fruitful approach in order to evaluate the impact of agronomic scenarios on the environment.

References

Addiscott TM, Powlson DS (1992) Partitioning losses of nitrogen fertilizer between leaching and denitrification. J Agric Sci Cambridge 118: 101-107

Anonymous (1992) Nitrogen cycling and leaching in cool and wet regions of Europe. In: François E, Pithan K, Bartiaux-Thill N (eds) for the Management Committee of the COST 814 project. October 22 - 23, Gembloux, Belgium

Aworth OC, Hicks JR, Minotti PL, Lee CY (1980) Effects of plant age and nitrogen fertilization on nitrate accumulation and postharvest nitrite accumulation in fresh spinach. J Am Soc Hortic Sci 105: 18-20

Barneix A, Causin HF (1996) The central role of amino acids on nitrogen utilization and plant growth. J Plant Physiol 149: 358-362

Beaudoin N, Makowski D, Mary B, Wallach D, Parnaudeau V, Parisseaux B, Machet JM, Meynard JM (1998) Evaluation de l'impact économique et environnemental de la mesure agri-environnementale MAE au moyen de modèles agronomiques. Rapport Ministère de l'Agriculture, INRA, Laon, 79pp

Beaudoin N, Mary B, Parnaudeau V (1999) Impact of agricultural scenarios on nitrate pollution at the catchment scale. Communication 10th Nitrogen Workshop, Copenhagen (DK) 1999/08/23-26

Bielek P, Kudeyarov VN, Bashkin VN (1985) Nitrogen balance in the present stage of anthropogenesis. In: Bielek P, Kudeyarov VN (eds) Nitrogen cycles in the present agriculture. Priroda, Bratislava pp 11-38

Blom-Zandstra M, Lampe JEM (1985) The role of nitrate in the osmoregulation of lettuce (*Lactuca sativa* L.) grown at different light intensities. J Exp Bot 36: 1043-1052

Blom-Zandstra M, Eenink AH (1986) Nitrate concentration and reduction in different genotypes of lettuce. J Am Soc Hortic Sci 111: 908-911

Blom-Zandstra M, Lampe JEM, Ammerlaan FHM (1988) C and N utilization of two lettuce genotypes during growth under non-varying light conditions and after changing the light intensity. Physiol Plant 74: 147- 153

Bouwman (1996) Direct emissions of nitrous oxide from agricultural soils. Nutr Cycl Agroecosyst 46: 53-70

Brisson N, Mary B, Ripoche D, Jeuffroy MH, Ruget F, Nicoullaud B, Gate P, Devienne F, Antonioletti R, Dürr C, Richard G, Beaudoin N, Recous S, Tayot X, Plénet D, Cellier P, Machet JM, Meynard JM, Delécolle R (1998) STICS: a generic model for the simulation of crops and their water and nitrogen balance. I. Theory and parameterization applied to wheat and corn. Agronomie 18: 311-346

Bumb BL (1995) World nitrogen supply and demand: an overview. In: Bacon PE (ed) Nitrogen fertilization in the environment. Marcel Dekker, New York, pp 1-40

Chaney K (1990) Effect of nitrogen fertilizer rate on soil nitrogen content after harvesting winter wheat. J Agric Sci Cambridge 114: 171-176

Clarkson DT (1986) Regulation of the absorption and release of nitrate by plant cells: a review of current ideas and methodology. In: Lambers H, Neeteson J, Stulen I (eds) Physiological, ecological and applied aspects of nitrogen metabolism in higher plants. M Nijhoff, Dordrecht, pp 3-28

Clement CR, Hopper MJ, Jones LHP (1978) The uptake of nitrate by *Lolium perenne* from flowing nutrient solution. I. Effect of concentration. J Exp Bot 29: 453-464

Corré WJ, Breimer T (1979) Nitrate and nitrite in vegetables. Pudoc, Wageningen

Davies DB, Sylvester-Bradley R (1995) The contribution of fertilizer nitrogen to leachable nitrogen in the UK: a review. J Sci Food Agric 68: 399-406

Devienne F, Mary B, Lamaze T (1994) Nitrate transport in intact wheat roots. II. Long-term effects of NO_3^- concentration in the nutrient solution on NO_3^- unidirectional fluxes and distribution within the tissues. J Exp Bot 45: 677-684

Dragosits U, Sutton MA, Place CJ, Bayley AA (1998) Modelling the spatial distribution of agricultural ammonia emissions in the UK. Environ Pollut 102: 195-203

Fonseca FG, Den Hertog J, Stulen I (1996) The response of *Plantago major* ssp. *pleiosperma* to elevated CO_2 is modulated by the formation of secondary shoots. New Phytol 133: 627-635

Fonseca FG, Bowsher, CG, Stulen I (1997) Impact of elevated CO_2 on nitrate reductase transcription and activity in leaves and roots of *Plantago major*. Physiol Plant 100: 940-948

Forman D, Al-Dabbagh S, Doll R (1985) Nitrates, nitrites and gastric cancer in Great Britain. Nature 313: 620-625

Glass ADM, Siddiqi MY (1995) Nitrogen absorption by plant roots. In: Srivastava HS, Singh RP (eds) Nitrogen nutrition in higher plants. Associated Publishing, New Delhi, pp 21-56

Glendining MJ, Poulton PR, Powlson DS (1992) The relationship between inorganic N in soil and the rate of fertilizer N applied on the Broadbalk wheat experiment. Aspects Appl Biol 30: 95-102

Granli T, Bøckman OC (1994) Nitrous oxide from agriculture. Norw J Agric Sci 12: 1-129

Guiot J, Grevy L (1990) Evolution des nitrates dans une terre soumise à la rotation betterave-froment-escourgeon. In: Calvet R (ed) Nitrate, agriculture, eau. INRA, Paris, pp 417-423

Hoffmann M, Johnsson H (1999) A method for assessing generalised nitrogen leaching estimates for agricultural land. Environ Mod Assoc 4: 35-44

Imsande J, Touraine B (1994) N demand and the regulation of nitrate uptake. Plant Physiol 105: 3-7

IFEN (1998) Agriculture et Environnement: les indicateurs. Institut Français de l'Environnement, 72 pp

Jarvis SC, Pain BF (1990) Ammonia volatilisation from agricultural land. Fertil Soci Proc 298: 1-35

Johnston AE, Jenkinson DS (1989) The nitrogen cycle in UK arable agriculture. The Fertiliser society Proc. No 286 3-24

Justes E, Jeuffroy MH, Mary B (1997) The nitrogen requirement of major agricultural crops. Wheat, barley and durum wheat. In: Lemaire G (ed) Diagnosis of the nitrogen status in crops vol 4. Springer Berlin Heidelberg New York, 4: 73-92

Krapp A, Fraisler V, Scheible-Wolf R, Quesada A, Gojon A, Stitt M, Caboche M, Daniel-Vedele F (1998) Expression studies of Nrt2: 1Np, a putative high affinity nitrate transporter: evidence for its role in nitrate uptake. Plant J 14: 723-731

Laurent F, Mary B (1992) Management of nitrogen in farming systems and the prevention of nitrate leaching. Aspects Appl Biol 30: 45-61

Ledgard SF, Giller GE (1995) Atmospheric N_2 fixation as an alternative N source. In: Bacon PE (ed) Nitrogen fertilization in the environment. Marcel Dekker, New York, pp 443-486

Limaux F, Recous S, Meynard JM, Guckert A (1999) Relationship between rate of crop growth at date of fertiliser-N application and fate of fertiliser-N applied to winter wheat. Plant Soil 214: 49-59

Makowski D, Wallach D, Meynard JM (1999) Models of yield, grain protein and residual mineral nitrogen responses to applied nitrogen for winter wheat. Agron J 91: 377-385

Mary B, Beaudoin N, Benoît M (1997) Prévention de la pollution nitrique à l'échelle du bassin d'alimentation en eau. In: Lemaire G, Nicolardot B (eds) Maîtrise de l'azote dans les agrosystèmes. Colloques INRA 83: 289-312

Maynard DN, Barker AV, Minotti PL, Peck NH (1976) Nitrate accumulation in vegetables. In: Brady NC (ed) Advances in agronomy. Academic Press, New York, pp 71-118

McDonald AJ, Powlson DS, Poulton PR, Jenkinson DS (1989) Unused fertilizer nitrogen in arable soils – its contribution to nitrate leaching. J Sci Food Agric 46: 407-419

Meynard JM, Justes E, Machet JM, Recous S (1997) Fertilisation azotée des cultures annuelles de plein champ. In: Lemaire G, Nicolardot B (eds) Maîtrise de l'azote dans les agrosystèmes. Colloques INRA 83: 183-200

Müller B, Tillard P, Touraine B (1995) Nitrate fluxes in soybean seedling roots and their response to amino acids: an approach using [15]N. Plant Cell Environ 18:1267-1279

Nicolardot B, Mary B, Houot S, Recous S (1997) La dynamique de l'azote dans les sols cultivés. In: Lemaire G, Nicolardot B (eds) Maîtrise de l'azote dans les agrosystèmes. Colloques INRA 83: 87-104

Normand B, Recous S, Vachaud G, Kengni L, Garino B (1997) ^{15}N tracers combined with tension neutronic method to estimate the nitrogen balance of irrigated maize. Soil Sci Soc Am J 61: 1508-1518

Pelsy F, Caboche M (1992) Molecular genetics of nitrate reductase in higher plants. Adv Genet 30: 1-40

Peoples MB, Mosier AR, Freney JR (1995) Minimizing gaseous losses of nitrogen. In: Bacon PE (ed) Nitrogen fertilization in the environment. Marcel Dekker, New York, pp 565-602

Pilbeam CJ (1996) Effect of climate on the recovery in crop and soil of ^{15}N-labelled fertilizer applied to wheat. Fertil Res 45: 209-215

Pons TL, Van der Werf A, Lambers H (1993) Photosynthetic nitrogen use efficiency of inherently slow- and fast-growing species: possible explanations for observed differences. In: Roy and J, Garnier E (eds) A whole-plant perspective on carbon-nitrogen interactions. SPB Academic Publishing, The Hague, pp 61-78

Powlson DS, Hart PBS, Poulton PR, Johnston AE, Jenkinson DS (1992) Influence of soil type, crop management and weather on the recovery of ^{15}N-labelled fertilizer applied to winter wheat in spring. J Agric Sci Cambridge 118: 83-100

Quilleré I, Dufosse C, Roux Y, Foyer CH, Caboche M, Morot-Gaudry JF (1994) The effects of deregulation of NR gene expression on growth and nitrogen metabolism of *Nicotiana plumbaginifolia* plants. J Exp Bot 45: 1205-1211

Recous S, Jeuffroy MH, Mary B, Meynard JM (1996) Gestion de l'azote en zone d'agriculture intensive. Rapport de contrat INRA-SCGP, Laon, 37 pp

Recous S, Loiseau P, Machet JM, Mary B (1997) Transformations et devenir de l'azote de l'engrais sous cultures annuelles et sous prairies. In: Lemaire G, Nicolardot B (eds) Maîtrise de l'azote dans les agrosystèmes. Colloques INRA 83: 105-120

Reinink K (1988) Improving quality of lettuce by breeding for low nitrate content. Acta Hortic 222: 121-128

Reinink K, Groenwold R (1987) The inheritance of nitrate content in lettuce (*Lactuca sativa* L.). Euphytica 36: 733-744

Richter GM, Beblik AJ, Schmalstieg K, Richter O (1998) N-dynamics and nitrate leaching under rotational and continuous set-aside – a case study at the field and catchment scale. Agric Ecosyst Environ 68: 125-138

Rodgers CO, Barneix AJ (1988) Cultivar differences in the rate of nitrate uptake by intact wheat plants as related to growth rate. Physiol Plant 72: 121-126

Skiba UM, McTaggart IP, Smith KA, Hargreaves KJ, Fowler D (1996) Estimates of nitrous oxide emissions from soil in the UK. Energ Convers Manage 37: 1303-1308

Smolders E (1993) Kinetic aspects of the soil-to-plant transfer of nitrate. PhD Thesis, University of Leuven, Belgium

Steingröver E, Oosterhuis R, Wieringa F (1982) Effect of light treatment and nutrition on nitrate accumulation in spinach (Spinacia oleracea L.). Z Pflanzenphysiol 107: 97-102

Steingröver E, Ratering P, Siesling J (1986a) Daily changes in uptake, reduction and storage of nitrate in spinach at low light intensity. Physiol Plant 65: 50-556

Steingröver E, Siesling J, Ratering P (1986b) Effect of one night with "low light" on uptake, reduction and storage of nitrate in spinach. Physiol Plant 66: 57-562

Stulen I (1990) Interactions between carbon and nitrogen metabolism in relation to plant growth and productivity. In: Abrol YP (ed) Nitrogen in higher plants. Research Studies Press, Taunton, pp 297-312.

Stulen I, Ter Steege MW (1995) Light and nitrogen assimilation. In: Srivastava HS, Singh RP (eds) Nitrogen nutrition of higher plants. Associated Publishing, New Delhi, pp 371-388

Ter Steege MW (1996) Regulation of nitrate uptake in a whole plant perspective. PhD Thesis, University of Groningen, Groningen

Ter Steege MW, Stulen I, Wiersema PK, Paans AJM, Vaalburg W, Kuiper PJC, Clarkson DT (1998) Growth requirement for N as a criterion to assess the effects of physical manipulation on nitrate uptake fluxes in spinach. Physiol Plant 103: 181-192

Ter Steege MW, Stulen I, Wiersema PK, Posthumus F, Vaalburg W (1999) Efficiency of nitrate uptake in spinach impact of external nitrate concentration and relative growth rate on nitrate influx and efflux. Plant Soil 208: 124-134

Touraine B, Clarkson DT, Muller B (1994) Regulation of nitrate uptake at the whole plant level. In: Roy J, Garnier E (eds) A whole-plant perspective on carbon-nitrogen interactions. SPB Academic Publishing, The Hague, pp 11-30

Trueman LJ, Onyeocha I, Forde BG (1996) Recent advances in the molecular biology of a family of eukaryotic high affinity nitrate transporters. Plant Physiol Biochem 34: 621-627

Urrestarazu M, Postigo A, Salas M, Sánchez A, Carrasco G (1998) Nitrate accumulation reduction using chloride in the nutrient solution on lettuce growing by NFT in semiarid climate conditions. J Plant Nutr 21: 1705-1714

Van der Boon J, Pieters JH, Slangen JHG, Titulaer HHH (1986) The effect of nitrogen fertilization on nitrate accumulation and yield of some field vegetables. In: Lambers H, Neeteson J, Stulen I (eds) Physiological, ecological and applied aspects of nitrogen metabolism in higher plants. M Nijhoff, Dordrecht, pp 489-492.

Van der Boon J, Steenhuizen JW, Steingröver E (1990) Growth and nitrate concentration of lettuce as affected by total nitrogen and chloride concentration, NH_4^+/NO_3^- ratio and temperature of the recirculating nutrient solution. J Hortic Sci 65: 309-321

Van Diest A (1986) Means of preventing nitrate accumulation in vegetable and pasture plants. In: Lambers H, Neeteson J, Stulen I (eds) Physiological, ecological and applied aspects of nitrogen metabolism in higher plants. M Nijhoff, Dordrecht, pp 455-471

Van Noordwijk M, Wadman WP (1992) Effects of spatial variability of nitrogen supply on environmentally acceptable nitrogen fertilizer application rates to arable crops. Neth J Agric Sci 40: 51-72

Veen BW, Kleinendorst A (1986) The role of nitrate in the osmoregulation of Italian ryegrass. Plant Soil 91: 433-436

Vermeer ITM, Pachen DMFA, Dallinga JW, Kleinjans JCS, van Maanen JMS (1998) Volatile N-nitrosamine formation after intake of nitrate at the ADI level in combination with an amine-rich diet. Environ Health Perspect 108: 459-463

Vincentz M, Moureaux T, Leydecker MT, Vaucheret H, Caboche M (1993) Regulation of nitrate and nitrite reductase expression in *Nicotiana plumbaginifolia* leaves by nitrogen and carbon metabolites. Plant J 3: 315-324

Von-Wiren N, Gazzarini S, Frommer W (1997) Regulation of mineral nitrogen uptake in plants. Plant Soil 196: 191-199

Walters CL, Walker R (1979) Consequences of accumulation of nitrate in plants. In: Hewitt EJ, Cuttings CV (eds) Nitrogen assimilation of plants. Academic Press, London, pp 637-677

Westgeest P (1989) Scenario-onderzoek nitraatbelasting in Nederland in relatie tot de gezondheid. Katholieke Universiteit Nijmegen, Vakgroep Sociale Geneeskunde en Vakgroep Toxicologie, Nijmegen

Zhang N, MacKown CT (1993) Nitrate fluxes and nitrate reductase activity of suspension-cultured tobacco cells. Effects of internal and external nitrate concentrations. Plant Physiol 102: 851-857

Index

H

I

L

M

O

P

Q

R

S

Printing: Saladruck, Berlin
Binding: H. Stürtz AG, Würzburg